Student's Solutions Manual
to Accompany Gilbert/Gilbert

College Algebra

Richard Semmler
· Northern Virginia Community College

McGraw-Hill, Inc.
New York   St. Louis   San Francisco   Auckland   Bogotá   Caracas
Lisbon   London   Madrid   Mexico City   Milan   Montreal   New Delhi
San Juan   Singapore   Sydney   Tokyo   Toronto

Student's Solutions Manual
to Accompany Gilbert/Gilbert
College Algebra

 This book is printed on recycled paper
containing 10% postconsumer waste.

1234567890   SEM   SEM   90987654

ISBN 0-07-024056-6

The editor was Karen M. Minette;
the production supervisor was Phil Galea.
Semline, Inc., was printer and binder.

TABLE OF CONTENTS

CHAPTER 1 FUNDAMENTALS

Exercises 1.1

1. $A \cup C = \{0, 2, 3, 4, 6, 9\} \cup \{3, 4, 5, 6\} = \{0, 2, 3, 4, 5, 6, 9\}$

3. $C \cap A = \{3, 4, 5, 6\} \cap \{0, 2, 3, 4, 6, 9\} = \{3, 4, 6\}$

5. First, $A \cap B = \{0, 2, 3, 4, 6, 9\} \cap \{0, 2, 4, 6, 8, 10\} = \{0, 2, 4, 6\}$,
   then, $(A \cap B) \cup C = \{0, 2, 4, 6\} \cup \{3, 4, 5, 6\} = \{0, 2, 3, 4, 5, 6\}$

7. First, $B \cup C = \{0, 2, 4, 6, 8, 10\} \cup \{3, 4, 5, 6\} = \{0, 2, 3, 4, 5, 6, 8, 10\}$,
   then, $A \cap (B \cup C) = \{0, 2, 3, 4, 6, 9\} \cap \{0, 2, 3, 4, 5, 6, 8, 10\}$
   $= \{0, 2, 3, 4, 6\}$

9. First, $C \cap A = \{3, 4, 5, 6\} \cap \{0, 2, 3, 4, 6, 9\} = \{3, 4, 6\}$ and
   $C \cap B = \{3, 4, 5, 6\} \cap \{0, 2, 4, 6, 8, 10\} = \{4, 6\}$, then
   $(C \cap A) \cup (C \cap B) = \{3, 4, 6\} \cup \{4, 6\} = \{3, 4, 6\}$

For Exercises 11—41, only answers are required. These are the appropriate answers.

11. N, W, Z, Q, R 13. R, H 15. R, H 17. Q, R 19. Q, R 21. N 23. R

25. Z 27. N 29. Z 31. Q 33. Additive inverse 35. Distributive property

37. Multiplicative identity 39. Additive identity 41. Multiplicative inverse

43. $5 + 7 = 7 + 5$ 45. $x \cdot (y \cdot z) = (x \cdot y) \cdot z$ 47. $a + \underline{0} = a$ 49. $a\left(\dfrac{1}{a}\right) = 1$

51. If $x = y$, then $ax = ay$ 53. $x - \dfrac{3}{x} + 1$ becomes $3 - \dfrac{3}{3} + 1 = 3 - 1 + 1 = 3$

55. $[15(x + 1) - (x + 1)(2 + y)] \div (3 + y)$ becomes
   $[15(3 + 1) - (3 + 1)(2 + 5)] \div (3 + 5) = [15(4) - (4)(7)] \div 8$
   $\qquad\qquad\qquad\qquad\qquad\qquad\qquad = [60 - 28] \div 8$
   $\qquad\qquad\qquad\qquad\qquad\qquad\qquad = 32 \div 8 = 4$

57. $\{8x[3y + 2 - 2(2y - 7)] - 4[12x - 2(2y)]\} \div [y + 2 + x(14 - x)]$ becomes
   $\{8 \cdot 3[3 \cdot 5 + 2 - 2(2 \cdot 5 - 7)] - 4[12 \cdot 3 - 2(2 \cdot 5)]\} \div [5 + 2 + 3(14 - 3)]$
   $= \{24[15 + 2 - 2(10 - 7)] - 4[36 - 2(10)]\} \div [5 + 2 + 3(11)]$
   $= \{24[15 + 2 - 2(3)] - 4[36 - 20]\} \div [5 + 2 + 33]$
   $= \{24[11] - 4[16]\} \div [40]$
   $= \{264 - 64\} \div 40$
   $= 200 \div 40 = 5$

59. $\{z[x - 6(z - x)] + x[z - 2y(3x + y)]\} \div (2xz)$ becomes
$\{-2[3 - 6(-2 - 3)] + 3[-2 - 2 \cdot 5(3 \cdot 3 + 5)]\} \div (2 \cdot 3 \cdot (-2))$

$= \{-2[3 - 6(-5)] + 3[-2 - 10(9 + 5)]\} \div (-12)$

$= \{-2[3 + 30] + 3[-2 - 140]\} \div (-12)$

$= \{-2[33] + 3[-142]\} \div (-12)$

$= \{-66 - 426\} \div (-12) = -492 \div (-12) = 41$

61. $[2y(x - y) + 4xy(x - 2y - 3z)] \div [-yz(2xy + z)]$ becomes
$\{2(4.1)(1.2 - 4.1)] + 4(1.2)(4.1)[1.2 - 2(4.1) - 3(-2.7)\}$

$\div [-(4.1)(-2.7)(2(1.2)(4.1) - 2.7)]$

$= \{8.2(-2.9) + 19.68(1.1)\} \div [11.07(7.14)]$

$= [-23.78 + 21.648] \div 79.0398 = -2.132 \div 79.0398 \approx -0.0270$

63. $y - 1 - x\left(\dfrac{11 - 6x}{7 - 2x}\right)$ becomes $4.1 - 1 - 1.2\left[\dfrac{11 - 6(1.2)}{7 - 2(1.2)}\right]$

$= 4.1 - 1 - 1.2\left(\dfrac{11 - 7.2}{7 - 2.4}\right) = 4.1 - 1 - 1.2\left(\dfrac{3.8}{4.6}\right)$

$= 4.1 - 1 - 1.2(0.826087) \approx 4.1 - 1 - 0.991304 \approx 2.1087$

65. $5(2r - 3) - 7(4 - r) = 10r - 15 - 28 + 7r = 17r - 43$

67. $-2(a - b + 1) - (3a - 4b - 3) = -2a + 2b - 2 - 3a + 4b + 3 = -5a + 6b + 1$

69. $2[7 - 3(q - 2) + 5q] = 2[7 - 3q + 6 + 5q] = 2[2q + 13] = 4q + 26$

71. $3[-(2 - p) - (-p + 3)] = 3[-2 + p + p - 3] = 3[2p - 5] = 6p - 15$

73. $-3[-2(x - 4) - (5x - 1)] = -3[-2x + 8 - 5x + 1] = -3[-7x + 9] = 21x - 27$

75. $2[5 - 3(y + 2)] - 3[2 - 4(y - 2)] = 2[5 - 3y - 6] - 3[2 - 4y + 8]$

$= 2[-3y - 1] - 3[-4y + 10]$

$= -6y - 2 + 12y - 30 = 6y - 32$

77. $3[2 - 4(a - b) + 7a] = 3(2 - 4a + 4b + 7a] = 3[3a + 4b + 2] = 9a + 12b + 6$

79. $2[3(a - b) - 2(a + 2b)] = 2[3a - 3b - 2a - 4b] = 2[a - 7b] = 2a - 14b$

81. $4\{2[3 - 2(x + 2)] - 4\} = 4\{2[3 - 2x - 4] - 4\} = 4\{2[-2x - 1] - 4\}$

$= 4\{-4x - 2 - 4\} = 4(-4x - 6) = -16x - 24$

83. $5\{2 - 3[(2y + 1) - (5y - 2)]\} = 5\{2 - 3[2y + 1 - 5y + 2]\}$

$= 5\{2 - 3[-3y + 3]\} = 5\{2 + 9y - 9\}$

$= 5(9y - 7) = 45y - 35$

85. $6\{5 - 2[(x - 2y) - 3(2x - y)]\} = 6\{5 - 2[x - 2y - 6x + 3y]\}$

$= 6\{5 - 2[-5x + y]\} = 6\{5 + 10x - 2y\}$

$= 60x - 12y + 30$

87. $2[3(4a - b) - 2(5a - 3b)] - \{5[3a - (b - 4a)] + b\}$
$= 2[12a - 3b - 10a + 6b] - \{5[3a - b + 4a] + b\}$
$= 2[2a + 3b] - \{5[7a - b] + b\}$
$= 4a + 6b - \{35a - 5b + b\}$
$= 4a + 6b - (35a - 4b)$
$= 4a + 6b - 35a + 4b$
$= -31a + 10b$

89. $2\{6 - [2 + 3(2p - r)]\} - 3\{4 - 2[3 - 2(4p - 3r)]\}$
$= 2\{6 - [2 + 6p - 3r]\} - 3\{4 - 2[3 - 8p + 6r]\}$
$= 2\{6 - 2 - 6p + 3r\} - 3\{4 - 6 + 16p - 12r\}$
$= 2\{4 - 6p + 3r\} - 3\{-2 + 16p - 12r\}$
$= 8 - 12p + 6r + 6 - 48p + 36r = 14 - 60p + 42r$

91. Subtraction is **not** associative. To see this, the left side of the equation, $a - (b - c)$ simplifies to $a - b + c$ whereas the right side, $(a - b) - c$ simplifies to $a - b - c$. In general $a - b + c \neq a - b - c$.

93. Subtraction is **not** commutative. That is, in general, $a - b \neq b - a$. To illustrate this let $a = 5$ and $b = 9$. Then $a - b = 5 - 9 = -4$ whereas $b - a = 9 - 5 = 4$. Also, you should note that $a - b = -(b - a)$.

95. $\varnothing$ is the empty set whereas $\{\varnothing\}$ is the set containing the empty set.

Exercises 1.2

The answers for Exercises 1—23 are given below:

1.
   $-7 \quad -2$

3.
   $-4 \quad 0$

5.
   $1 \quad 3$

7.
   $-1 \quad 1$

9.
   All real numbers

11.
    $-2 \quad 0 \quad 1$

13.
    $-3 \quad -1 \quad 2$

15.
    $-4 \quad -1 \quad 2$

17. $x < -2$

19. $2 \leq x \leq 5$

21. $x \leq -2$ or $x > 1$

23. $1 < x < 3$ or $x \geq 5$

25. Addition Property where the number 3 is added to each side of the inequality.

27. Multiplication Property where each side of the inequality is multiplied by -2.

29. Addition Property where the number -3 is added to each side of the inequality.

31. Multiplication Property where each side of the inequality is multiplied by $\frac{1}{5}$.

33. Multiplication Property

35. Here, $-|8 - 3| = -|5| = -5$
    $-8 - |3| = -8 - 3 = -11$
    $|8 - 3| = |5| = 5$
    and $|8| + 3 = 8 + 3 = 11$.
    The ordering is $-8 - |3|$, $-|8 - 3|$, 0, $|8 - 3|$, $|8| + 3$

37. Here, $|-2| = 2$, $\sqrt{2} \approx 1.414$, and $-\sqrt{2} \approx -1.414$.
   The ordering is -2, $-\sqrt{2}$, -1, 1, $\sqrt{2}$, $|-2|$.

39. Here, $\sqrt{80} \approx 8.944$, $|-9| = 9$, $-\sqrt{80} \approx -8.944$, $\sqrt{80} - 9 \approx -0.056$, and
   $|\sqrt{80} - 9| \approx |-0.056| = 0.056$.
   The ordering is -9, $-\sqrt{80}$, $\sqrt{80} - 9$, $|\sqrt{80} - 9|$, $\sqrt{80}$, $|-9|$.

41. $|-7| = 7$ 43. $|-7 + 3| = |-4| = 4$ 45. $|-7| + 3 = 7 + 3 = 10$

47. $|9 - 7| = |2| = 2$ 49. $|9| + |-7| = 9 + 7 = 16$ 51. $|7| - |9| = 7 - 9 = -2$

53. $|y - 4| = y - 4$ 55. $|y - 4| = -(y - 4) = -y + 4$ or $4 - y$

57. $|9 - \sqrt{80}| = 9 - \sqrt{80}$ since $9 > \sqrt{80}$

59. $|2x - 10| = -(2x - 10) = -2x + 10$ or $10 - 2x$

61. $|a - b| = -(a - b) = -a + b$ or $b - a$

63. $|2a - b| = -(2a - b) = -2a + b$ or $b - 2a$ 65. $\left|\dfrac{x}{5}\right| = -\left(\dfrac{x}{5}\right)$ or $\dfrac{-x}{5}$

67. $|(-1)^2| = |1| = 1$ 69. $|-a^2| = a^2$ where $a^2 > 0$ 71. $|-5 - 2a| = -5 - 2a$

73. $|3a - 14| = 3a - 14$ 75. Multiplication Property   77. Division Property

79. Triangle Inequality

81. $|x| < 2$ can be written as $-2 < x < 2$ and the graph is:

83. $|x| > 3$ can be written as $x < -3$ or $x > 3$ and the graph is:

85. $|x| \le 4$ can be written as $-4 \le x \le 4$ and the graph is:

87. $|x - 2| < 1$ which can be written as $-1 < x - 2 < 1$ which becomes $1 < x < 3$
   and the graph is:

89. $|x + 1| < 2$ can be written as $-2 < x + 1 < 2$ which becomes $-3 < x < 1$ and
   the graph is:

91. $|x + 6| \le 4$ can be written as $-4 \le x + 6 \le 4$ which becomes $-10 \le x \le -2$ and
   the graph is:

93. $|x - 3| \ge 4$ can be written as $x - 3 \le -4$ or $x - 3 \ge 4$ which becomes $x \le -1$
   or $x \ge 7$ and the graph is:

95. $|x + 4| > 5$ can be written as $x + 4 < -5$ or $x + 4 > 5$ which becomes $x < -9$
   or $x > 1$ and the graph is:

97. To compare two or more numbers, consider their decimal representation. For example, $\pi \approx 3.1416$ and $\frac{22}{7} \approx 3.1429$ which means that $\frac{22}{7} > \pi$.

99. Since $|x|$ is always positive, then there can be no solution to the inequality $|x| < -5$.

Exercises 1.3

1. $(-3)^4 = (-3)(-3)(-3)(-3) = 81$ **3.** $-5^2 = -5 \cdot 5 = -25$ **5.** $-2^{-4} = -\frac{1}{2^4} = -\frac{1}{16}$

7. $\left(\frac{3}{2}\right)^{-2} = \frac{3^{-2}}{2^{-2}} = \frac{2^2}{3^2} = \frac{4}{9}$ **9.** $(2^2)^{-3} = 4^{-3} = \frac{1}{4^3} = \frac{1}{64}$

11. $(4pq^2)^3 = 4^3 p^3 (q^2)^3 = 64p^3 q^6$ **13.** $\left(\frac{2x}{7}\right)^2 = \frac{2^2 x^2}{7^2} = \frac{4x^2}{49}$ **15.** $2y^{-3} = \frac{2}{y^3}$

17. $3^{-2}ab^{-2} = \frac{a}{3^2 b^2} = \frac{a}{9b^2}$ **19.** $\frac{(5r^5)(6r^4)}{2r^3} = \frac{30r^9}{2r^3} = 15r^6$ **21.** $\frac{(6xy^4)^0}{2x^2 y^{-3}} = \frac{1}{2x^2 y^{-3}} = \frac{y^3}{2x^2}$

23. $[(uv^{-1})^3]^{-2} = [u^3 v^{-3}]^{-2} = u^{-6} v^6 = \frac{v^6}{u^6}$

25. $(4xy^2)^{-3}(x^{-2}y^3)^2 = 4^{-3}x^{-3}y^{-6} \cdot x^{-4}y^6 = 4^{-3}x^{-7}y^0 = \frac{1}{64x^7}$ **27.** $\frac{r^{-3}}{s^{-5}} = \frac{s^5}{r^3}$

29. $\frac{2t^{-3}}{(2r)^{-3}} = \frac{2(2r)^3}{t^3} = \frac{2 \cdot 2^3 r^3}{t^3} = \frac{16r^3}{t^3}$ **31.** $\frac{2^{-3}z^{-3}}{5^{-3}z^{-5}} = \frac{5^3 z^5}{2^3 z^3} = \frac{125z^2}{8}$

33. $\frac{4^{-2}a^{-4}}{3^{-2}a^{-4}} = \frac{3^2 a^4}{4^2 a^4} = \frac{9}{16}$ **35.** $\frac{3p^{-3}q^{-1}}{8^{-1}pq^3} = \frac{3 \cdot 8^1}{pp^3 qq^3} = \frac{24}{p^4 q^4}$ **37.** $\left(\frac{x^3}{3y^2}\right)^4 = \frac{(x^3)^4}{3^4(y^2)^4} = \frac{x^{12}}{81y^8}$

39. $\left(\frac{x^2}{y^{-1}z^3}\right)^{-3} = \frac{x^{-6}}{y^3 z^{-9}} = \frac{z^9}{x^6 y^3}$ **41.** $\left(\frac{(-x^2)^3 z}{(2x^4 y)^2}\right)^3 = \left(\frac{-x^6 z}{4x^8 y^2}\right)^3 = \left(\frac{-z}{4x^2 y^2}\right)^3 = \frac{-z^3}{64x^6 y^6}$

43. $\left(\frac{(ab)^{-1}c^2}{(ac^{-2})^{-1}b^2}\right)^{-2} = \left(\frac{a^{-1}b^{-1}c^2}{a^{-1}c^2 b^2}\right)^{-2} = (b^{-3})^{-2} = b^6$

45. $\left(\frac{(x-y)^2(2z^4)}{[z(x-y)]^3}\right)^2 = \left(\frac{(x-y)^2 2z^4}{z^3(x-y)^3}\right)^2 = \left(\frac{2z}{x-y}\right)^2 = \frac{4z^2}{(x-y)^2}$

47. $\left(\frac{3x^m}{y^n}\right)^2 = \frac{3^2(x^m)^2}{(y^n)^2} = \frac{9x^{2m}}{y^{2n}}$ **49.** $\frac{x^{2n-2}}{x^{n-1}} = x^{(2n-2)-(n-1)} = x^{n-1}$

51. $1020 = 1.020 \times 10^3$ or 1.02×10^3 **53.** $41{,}610{,}000 = 4.161 \times 10^7$

55. $0.0035 = 3.5 \times 10^{-3}$ **57.** $(8 \times 10^3)(7 \times 10^4) = 56 \times 10^7 = 5.6 \times 10^8$

59. $(1.2 \times 10^6)(5 \times 10^{-4}) = 6.0 \times 10^2$ **61.** $\frac{7.8 \times 10^{-9}}{1.3 \times 10^{-7}} = 6.0 \times 10^{-2}$

63. $8.22 \times 10^3 = 8,220$ 65. $6.6 \times 10^{-1} = 0.66$ 67. $2.87 \times 10^{-5} = 0.0000287$

69. $\dfrac{4}{5 \times 10^3} = \dfrac{4}{5} \times 10^{-3} = 0.8 \times 10^{-3} = 0.0008$

71. $\dfrac{68,000,000}{1,700,000} = \dfrac{6.8 \times 10^7}{1.7 \times 10^6} = \dfrac{6.8}{1.7} \times 10^1 = 4.0 \times 10^1 = 40$

73. $\dfrac{1,240,000 \times 67,100}{0.0082} = \dfrac{(1.24 \times 10^6)(6.71 \times 10^4)}{8.2 \times 10^{-3}} = \dfrac{1.24 \times 6.71 \times 10^{10}}{8.2 \times 10^{-3}}$
$$\approx 1.015 \times 10^{13}$$

75. $\dfrac{501,484 \times 0.000293}{61,941 \times 0.00275} = \dfrac{(5.01484 \times 10^5)(2.93 \times 10^{-4})}{(6.1941 \times 10^4)(2.75 \times 10^{-3})} = \dfrac{5.01484 \times 2.93 \times 10}{6.1941 \times 2.75 \times 10}$
$$\approx 0.8626 = 8.626 \times 10^{-1}$$

77. $\dfrac{0.004141 \times 73,290 \times 0.0619}{513,500 \times 0.00606} = \dfrac{(4.141 \times 10^{-3})(7.329 \times 10^4)(6.19 \times 10^{-2})}{(5.135 \times 10^5)(6.06 \times 10^{-3})}$
$$= \dfrac{4.141 \times 7.329 \times 6.19 \times 10^{-1}}{5.135 \times 6.06 \times 10^2} = \dfrac{187.8627179 \times 10^{-1}}{31.1181 \times 10^2}$$
$$\approx 6.037 \times \dfrac{10^{-1}}{10^2} = 6.037 \times 10^{-3}$$

79. $\dfrac{4^{-2} \cdot 3^4}{7(4.9)^{-2}} = \dfrac{3^4 (4.9)^2}{7 \cdot 4^2} = \dfrac{(81)(24.01)}{(7)(16)} = \dfrac{1944.81}{112} = 17.364375 \approx 1.736 \times 10^1$

81. $\dfrac{(17^4 \times 8^{-1})^{-2}}{(0.19)^5} = \dfrac{17^{-8} \times 8^2}{(0.19)^5} = \dfrac{8^2}{17^8 (0.19)^5} = \dfrac{64}{(1.7 \times 10)^8 (1.9 \times 10^{-1})^5}$
$$= \dfrac{64}{(69.7575 \times 10^8)(24.7610 \times 10^{-5})}$$
$$= \dfrac{64}{69.7575 \times 24.7610 \times 10^3}$$
$$= \dfrac{64}{1727.276 \times 10^3}$$
$$= 0.03705 \times \dfrac{1}{10^3}$$
$$= 0.03705 \times 10^{-3}$$
$$= 3.705 \times 10^{-5}$$

83. $91,110,000 = 9.111 \times 10^7$ 85. 5.01×10^{-6} gm $= 0.00000501$ gm   87. $10^{-ph} = 10^{-6}$

89. time $= \dfrac{91,110,000}{5.88 \times 10^{12}} = \dfrac{9.111 \times 10^7}{5.88 \times 10^{12}} \approx 1.5493 \times \dfrac{10^7}{10^{12}} = 1.5493 \times 10^{-5}$ yrs.

Now convert to minutes: $(1.5493 \times 10^{-5}$ yr$)\left(\dfrac{365 \text{ days}}{\text{yr}}\right)\left(\dfrac{24 \text{ hr}}{1 \text{ day}}\right)\left(\dfrac{60 \text{ min}}{\text{hr}}\right) \approx 8.14$ min

91. Since $3^2 + 4^2 = 9 + 16 = 25$ whereas $(3 + 4)^2 = 7^2 = 49$, we see that the two expressions are not equal.

Exercises 1.4

1. $(5x^4 - 8x^2 + 4) + (2x^3 + 7x^2 - 9) = 5x^4 + 2x^3 + (-8 + 7)x^2 + (4 - 9)$
    $$= 5x^4 + 2x^3 - x^2 - 5$$

3. $(8r^4 - 3r^5 + 4) - (r^2 + 2r^4 - r^5) = 8r^4 - 3r^5 + 4 - r^2 - 2r^4 + r^5$
    $$= (-3 + 1)r^5 + (8 - 2)r^4 - r^2 + 4$$
    $$= -2r^5 + 6r^4 - r^2 + 4$$

5. $(-3 + 8z + 4z^3) - (-5z^2 + 3z^3 - 12) = -3 + 8z + 4z^3 + 5z^2 - 3z^3 + 12$
    $$= (4 - 3)z^3 + 5z^2 + 8z + (-3 + 12)$$
    $$= z^3 + 5z^2 + 8z + 9$$

7. $5x^2(4x^3 - 3x + 8) = 5x^2(4x^3) + 5x^2(-3x) + 5x^2(8) = 20x^5 - 15x^3 + 40x^2$

9. $(4z + 3)(2z - 7) = 4z(2z) + 4z(-7) + 3(2z) + 3(-7) = 8z^2 - 28z + 6z - 21$
    $$= 8z^2 - 22z - 21$$

11. $(5q + 3)(8q - 1) = 5q(8q) + 5q(-1) + 3(8q) + 3(-1) = 40q^2 - 5q + 24q - 3$
    $$= 40q^2 + 19q - 3$$

13. $(5a - b)(2a + 3b) = 5a(2a) + 5a(3b) + (-b)(2a) + (-b)(3b)$
    $$= 10a^2 + 15ab - 2ab - 3b^2$$
    $$= 10a^2 + 13ab - 3b^2$$

15. $(x - 4)(2x^3 + 3x^2 - 1) = (x - 4)(2x^3) + (x - 4)(3x^2) + (x - 4)(-1)$
    $$= 2x^4 - 8x^3 + 3x^3 - 12x^2 - x + 4$$
    $$= 2x^4 - 5x^3 - 12x^2 - x + 4$$

17. $(2x^2 + 3x - 1)(x^2 - 2x + 4)$
    $$= (2x^2 + 3x - 1)(x^2) + (2x^2 + 3x - 1)(-2x) + (2x^2 + 3x - 1)(4)$$
    $$= 2x^4 + 3x^3 - x^2 - 4x^3 - 6x^2 + 2x + 8x^2 + 12x - 4$$
    $$= 2x^4 - x^3 + x^2 + 14x - 4$$

19. $(3y^2 - 7y - 2)(y^3 - y^2 - 1)$
    $$= (3y^2 - 7y - 2)(y^3) + (3y^2 - 7y - 2)(-y^2) + (3y^2 - 7y - 2)(-1)$$
    $$= 3y^5 - 7y^4 - 2y^3 - 3y^4 + 7y^3 + 2y^2 - 3y^2 + 7y + 2$$
    $$= 3y^5 - 10y^4 + 5y^3 - y^2 + 7y + 2$$

21. $(1 + x)[1 + x(1 - x^2)] = (1 + x)[1 + x - x^3]$
    $$= (1 + x)(1) + (1 + x)(x) + (1 + x)(-x^3)$$
    $$= 1 + x + x + x^2 - x^3 - x^4 = 1 + 2x + x^2 - x^3 - x^4$$

23. $(b - 3)(b + 3) = (b)^2 - (3)^2 = b^2 - 9$

25. $(2r + s)(2r - s) = (2r)^2 - (s)^2 = 4r^2 - s^2$

27. $(y^2 - 2)^2 = (y^2)^2 - 2(y^2)(2) + (2)^2 = y^4 - 4y^2 + 4$

29. $(a + 3bc^2)^2 = (a)^2 + 2(a)(3bc^2) + (3bc^2)^2 = a^2 + 6abc^2 + 9b^2c^4$

31. $[2(3m - 4n)]^2 = [6m - 8n]^2 = (6m)^2 - 2(6m)(8n) + (8n)^2 = 36m^2 - 96mn + 64n^2$

33. $\left(\frac{1}{3}x + \frac{1}{2}y\right)\left(\frac{1}{3}x - \frac{1}{2}y\right) = \left(\frac{1}{3}x\right)^2 - \left(\frac{1}{2}y\right)^2 = \frac{1}{9}x^2 - \frac{1}{4}y^2$ or $\frac{x^2}{9} - \frac{y^2}{4}$

35. $(2x^3 - y)(2x^3 + y) = (2x^3)^2 - (y)^2 = 4x^6 - y^2$

37. $(p - q)(p^2 + pq + q^2) = (p)^3 - (q)^3 = p^3 - q^3$ (Difference of two cubes)

39. $(2x + 1)(4x^2 - 2x + 1) = (2x)^3 + (1)^3 = 8x^3 + 1$ (Sum of two cubes)

41. $(n + k)^3 = (n)^3 + 3(n)^2(k) + 3(n)(k)^2 + (k)^3 = n^3 + 3n^2k + 3nk^2 + k^3$

43. $(x - 2y)^3 = (x)^3 - 3(x)^2(2y) + 3(x)(2y)^2 - (2y)^3 = x^3 - 6x^2y + 12xy^2 - 8y^3$

45. $(3x^2 - 5y^2)(x^2 + 2y^2) = (3x^2 - 5y^2)(x^2) + (3x^2 - 5y^2)(2y^2)$
$$= 3x^4 - 5x^2y^2 + 6x^2y^2 - 10y^4 = 3x^4 + x^2y^2 - 10y^4$$

47. $(2x^3 + 3y)(x^3 - 4y) = (2x^3 + 3y)(x^3) + (2x^3 + 3y)(-4y)$
$$= 2x^6 + 3x^3y - 8x^3y - 12y^2 = 2x^6 - 5x^3y - 12y^2$$

49. $(3a - 2b)(9a^2 + 6ab + 4b^2) = (3a)^3 - (2b)^3 = 27a^3 - 8b^3$

51. $[(x + y) + 1][(x + y) - 1] = (x + y)^2 - (1)^2 = x^2 + 2xy + y^2 - 1$

53. $(a + 2b + c)^2 = [(a + 2b) + c]^2 = (a + 2b)^2 + 2(a + 2b)(c) + (c)^2$
$$= a^2 + 4ab + 4b^2 + 2ac + 4bc + c^2$$

55. $(x + y + z)(x - y - z) = [x + (y + z)][x - (y + z)] = (x)^2 - (y + z)^2$
$$= x^2 - y^2 - 2yz - z^2$$

57. $(2x^n - 1)(3x^n + 4) = (2x^n - 1)(3x^n) + (2x^n - 1)(4)$
$$= 6x^{2n} - 3x^n + 8x^n - 4 = 6x^{2n} + 5x^n - 4$$

59. $(2z^r + 9)(2z^r - 9) = (2z^r)^2 - (9)^2 = 4z^{2r} - 81$

61. $(5y^n + 3)^2 = (5y^n)^2 + 2(5y^n)(3) + (3)^2 = 25y^{2n} + 30y^n + 9$

63. $(2b^k - 5)^2 = (2b^k)^2 - 2(2b^k)(5) + (5)^2 = 4b^{2k} - 20b^k + 25$

65. a) $V = \frac{w(400 - w^2)}{2} = \frac{4(400 - 4^2)}{2} = \frac{4(400 - 16)}{2} = \frac{4(384)}{2} = 768$ cu. in.

    b) $V = \frac{w(400 - w^2)}{2} = \frac{12(400 - 12^2)}{2} = \frac{12(400 - 144)}{2} = \frac{12(256)}{2} = 1536$ cu. in.

67. $R = 8x^2 - 0.02x^3 = 8(100)^2 - 0.02(100)^3 = 8(10,000) - 0.02(1,000,000)$
$$= 80,000 - 20,000 = 60,000 \text{ dollars}$$

69. $x^9 - 7x^6 + 2x^2 - 43 = (2.3)^9 - 7(2.3)^6 + 2(2.3)^2 - 43$
$$\approx 1801.15266 - 1036.25122 + 10.58 - 43 = 732.48144$$
which rounds to 732.4814

71. $x^{11} - 5x^9 + 7x^6 - 345 = (-2.6)^{11} - 5(-2.6)^9 + 7(-2.6)^6 - 345$
$$\approx -36,703.44487 + 27,147.51839 + 2162.41043 - 345$$
$$= -7738.51605 \text{ which rounds to } -7738.5161$$

73. In preparing your writing, be sure to state that the difference of the squares of two quantities is the product of their sum with their difference.

Exercises 1.5

1. $3a - 6b = 3(a - 2b)$ 3. $4x^3 - 16xy = 4x(x^2 - 4y)$

5. $6m^3 - 24m^2 + 6mn = 6m(m^2 - 4m + n)$ 7. $12x^2y - 4xy + 20xy^2 = 4xy(3x - 1 + 5y)$

9. $7(x + 2y) - 3(x + 2y) = (x + 2y)(7 - 3) = (x + 2y)(4)$ or $4(x + 2y)$

11. $x^2 - 25 = x^2 - 5^2 = (x + 5)(x - 5)$ 13. $9y^2 - 1 = (3y)^2 - 1^2 = (3y + 1)(3y - 1)$

15. $16x^2 - 25y^2 = (4x)^2 - (5y)^2 = (4x + 5y)(4x - 5y)$

17. $(x - 1)x^2 - (x - 1)y^2 = (x - 1)(x^2 - y^2) = (x - 1)(x + y)(x - y)$

19. $-y^3x^2 + 9y^3 = -y^3(x^2 - 9) = -y^3(x^2 - 3^2) = -y^3(x + 3)(x - 3)$

21. $u^2 - 8u + 16 = u^2 - 2(u)(4) + 4^2 = (u - 4)(u - 4)$ or $(u - 4)^2$

23. $25y^2 + 10y + 1 = (5y)^2 + 2(5y)(1) + 1^2 = (5y + 1)(5y + 1)$ or $(5y + 1)^2$

25. $4a^2 - 20ab + 25b^2 + (2a)^2 - 2(2a)(5b) + (5b)^2 = (2a - 5b)(2a - 5b)$ or $(2a - 5b)^2$

27. $3a^3 + 6a^2 + 3a = 3a(a^2 + 2a + 1) = 3a(a^2 + 2(a)(1) + 1^2)$
$$= 3a(a + 1)(a + 1) \text{ or } 3a(a + 1)^2$$

29. $1 - 12rs + 36r^2s^2 = 1 - 2(1)(6rs) + (6rs)^2 = (1 - 6rs)(1 - 6rs)$ or $(1 - 6rs)^2$

31. $x^3 - z^3 = (x - z)(x^2 + xz + z^2)$ 33. $8 + u^3 = 2^3 + u^3 = (2 + u)(4 - 2u + u^2)$

35. $8x^3 + 27 = (2x)^3 + 3^3 = (2x + 3)(4x^2 - 6x + 9)$

37. $216x^3 - 8y^3 = 8(27x^3 - y^3) = 8[(3x)^3 - y^3] = 8(3x - y)(9x^2 + 3xy + y^2)$

39. $-4a^4 + 256a = -4a(a^3 - 64) = -4a(a^3 - 4^3) = -4a(a - 4)(a^2 + 4a + 16)$

41. $x^2 - 2x - 24 = (x - 6)(x + 4)$ 43. $x^2 + 10x + 21 = (x + 7)(x + 3)$

45. $4y^2 - 3xy - x^2 = (4y + x)(y - x)$ 47. $3x^2 + 7ax - 6a^2 = (3x - 2a)(x + 3a)$

49. $2x^3 + 4x^2 - 30x = 2x(x^2 + 2x - 15) = 2x(x + 5)(x - 3)$

51. $8y^3 - 20y^2 + 12y = 4y(2y^2 - 5y + 3) = 4y(2y - 3)(y - 1)$

53. $3ac + 3bc + ad + bd = 3c(a + b) + d(a + b) = (a + b)(3c + d)$

55. $ax^2 + bx^2 + ad^2 + bd^2 = x^2(a + b) + d^2(a + b) = (a + b)(x^2 + d^2)$

57. $9x^2 - y^2 + 2yz - z^2 = 9x^2 - (y^2 - 2yz + z^2) = (3x)^2 - (y - z)^2$
$$= [3x + (y - z)][3x - (y - z)] = (3x + y - z)(3x - y + z)$$

59. $2 + 4x - 10x^4 - 5x^3 = 2(1 + 2x) - 5x^3(2x + 1) = (1 + 2x)(2 - 5x^3)$

61. $x^2 - 2x + 1 - y^2 - 2yz - z^2$
$$= (x^2 - 2x + 1) - (y^2 + 2yz + z^2)$$
$$= (x - 1)^2 - (y + z)^2 = [(x - 1) + (y + z)][(x - 1) - (y + z)]$$
$$= (x - 1 + y + z)(x - 1 - y - z)$$

63. $2x^2 - 13x + 21 = (2x - 7)(x - 3)$ 65. $4a^2 + a - 3 = (4a - 3)(a + 1)$

67. $-2x(a + h) - 3y(a + h) = (a + h)(-2x - 3y)$ or $-(a + h)(2x + 3y)$

69. $16a^4 + 64a^3 + 64a^2 = 16a^2(a^2 + 4a + 4) = 16a^2(a + 2)(a + 2)$ or $16a^2(a + 2)^2$

71. $x^3 - x = x(x^2 - 1) = x(x^2 - 1^2) = x(x + 1)(x - 1)$

73. $6y^2 + 11y - 10 = (2y + 5)(3y - 2)$ 75. $6a^2 + 13a - 15 = (6a - 5)(a + 3)$

77. $x^2 + 1$ does **not** factor   79. $3x^2 - 9 = 3(x^2 - 3)$

81. $25x^2 + 10x + 1 - y^2 = (5x + 1)^2 - y^2 = [(5x + 1) + y][(5x + 1) - y]$
$$= (5x + 1 + y)(5x + 1 - y)$$

83. $-x^6 - x^3 = -x^3(x^3 + 1) = -x^3(x^3 + 1^3) = -x^3(x + 1)(x^2 - x + 1)$

85. $ax^2 - 9ay^4) = a(x^2 - 9y^4) = a[x^2 - (3y^2)^2] = a(x + 3y^2)(x - 3y^2)$

87. $(4x - 3y)^2 - 25 = (4x - 3y)^2 - 5^2 = [(4x - 3y) + 5][(4x - 3y) - 5]$
$$= (4x - 3y + 5)(4x - 3y - 5)$$

89. $(4a - b)^2 - (2x - z)^2 = [(4a - b) + (2x - z)][(4a - b) - (2x - z)]$
$$= (4a - b + 2x - z)(4a - b - 2x + z)$$

91. $x^4 - 16x^2 + 64 = (x^2)^2 - 2(x^2)(8) + 8^2 = (x^2 - 8)(x^2 - 8)$ or $(x^2 - 8)^2$

93. $9x^4 - 30x^2 + 25 = (3x^2)^2 - 2(3x^2)(5) + 5^2 = (3x^2 - 5)(3x^2 - 5)$ or $(3x^2 - 5)^2$

95. $4a^4 + 12a^2b^2 + 9b^4 = (2a^2)^2 + 2(2a^2)(3b^2) + (3b^2)^2$
$$= (2a^2 + 3b^2)(2a^2 + 3b^2) \text{ or } (2a^2 + 3b^2)^2$$

97. $u(u^2 - v^2) + v(u^2 - v^2) = (u^2 - v^2)(u + v)$
$$= (u + v)(u - v)(u + v) \text{ or } (u + v)^2(u - v)$$

99. $9w^8 + 27w^5 - 9w^2 = 9w^2(w^6 + 3w^3 - 1)$

101. $x^4 + 4 = (x^4 + 4x^2 + 4) - 4x^2 = (x^2 + 2)^2 - (2x)^2$
$$= [(x^2 + 2) + 2x][(x^2 + 2) - 2x] = (x^2 + 2x + 2)(x^2 - 2x + 2)$$

103. $z^4 + z^2 + 1 = z^4 + z^2 + z^2 + 1 - z^2 = (z^4 + 2z^2 + 1) - z^2 = (z^2 + 1)^2 - z^2$
$$= [(z^2 + 1) + z][(z^2 + 1) - z] = (z^2 + z + 1)(z^2 - z + 1)$$

105. $2x^{2n} + 10x^n + 12 = 2(x^{2n} + 5x^n + 6) = 2(x^n + 2)(x^n + 3)$

107. $9z^{2m} - w^{2n} = (3z^m)^2 - (w^n)^2 = (3z^m + w^n)(3z^m - w^n)$

109. For the large rectangle, the length is $x + y$ and the width is $x + y$ so that the representation of the area is length $\times$ width or $(x + y)(x + y) = (x + y)^2$.

The large diagram is split into four pieces where the bigger square is x by x so that its area is $x \cdot x = x^2$. The smaller square is y by y so that its area is $y \cdot y = y^2$. There are two rectangles given as x by y so that the area of both of them is $2 \cdot x \cdot y = 2xy$.

The sum of the areas of the four regions is
$$x^2 + 2xy + y^2$$
and we now have
$$x^2 + 2xy + y^2 = (x + y)^2$$

Exercises 1.6

1. $x + 2 \neq 0$ or $x \neq -2$ 3. $x(x + 3) \neq 0$ or $x \neq 0$, $x + 3 \neq 0$ or $x \neq 0$, $x \neq -3$

5. $x - y \neq 0$ can be expressed as $x \neq y$

7. $x^2 + 1 \neq 0$ which is always the case. Thus, no restriction is necessary.

9. $\dfrac{x^2 - y^2}{x - y} = \dfrac{(x + y)\cancel{(x - y)}^{1}}{\cancel{x - y}_{1}} = x + y$ 11. $\dfrac{x^2 - x - 12}{x^2 - 2x - 8} = \dfrac{\cancel{(x - 4)}^{1}(x + 3)}{\cancel{(x - 4)}_{1}(x + 2)} = \dfrac{x + 3}{x + 2}$

13. $\dfrac{4\cancel{(a - 1)^2}^{1}\cancel{(a + 2)^3}^{(a + 2)^2}}{\cancel{(a + 2)}_{1}\cancel{(a - 1)^3}_{(a - 1)}} = \dfrac{4(a + 2)^2}{a - 1}$

15. $\dfrac{(x^3 - 27y^3)(x^2 - 4y^2)}{x^2 - 5xy + 6y^2} = \dfrac{\cancel{(x - 3y)}^{1}(x^2 + 3xy + 9y^2)(x + 2y)\cancel{(x - 2y)}^{1}}{\cancel{(x - 3y)}_{1}\cancel{(x - 2y)}_{1}}$

$= (x^2 + 3xy + 9y^2)(x + 2y)$

17. $\dfrac{4pq^3}{p^2 - q^2} \cdot \dfrac{p - q}{12p^2q^2} = \dfrac{\cancel{4pq^3}^{q}\cancel{(p - q)}^{1}}{\cancel{(p - q)}_{1}(p + q)\cancel{12p^2q^2}_{3p}} = \dfrac{q}{3p(p + q)}$

19. $\dfrac{2b - 3y}{3c + 6d} \cdot \dfrac{2c + 4d}{4b - 6y} = \dfrac{\cancel{(2b - 3y)}^{1}\cancel{(2)}^{1}\cancel{(c + 2d)}^{1}}{3\cancel{(c + 2d)}_{1}\cancel{(2)}_{1}\cancel{(2b - 3y)}_{1}} = \dfrac{1}{3}$

21. $(x^2 - 4y^2) \cdot \dfrac{5a}{xy - 2y^2} = \dfrac{(x + 2y)\cancel{(x - 2y)}^{1}(5a)}{y\cancel{(x - 2y)}_{1}} = \dfrac{5a(x + 2y)}{y}$

23. $(cy + dy) \div \dfrac{d^2 - c^2}{3y} = (cy + dy) \cdot \dfrac{3y}{d^2 - c^2} = \dfrac{y\cancel{(c + d)}^{1}(3y)}{\cancel{(d + c)}_{1}(d - c)} = \dfrac{3y^2}{d - c}$

25. $\dfrac{3x - 3y}{4x + 2y} \cdot \dfrac{4x^2 - y^2}{(x - y)^2} = \dfrac{3\cancel{(x - y)}^{1}\cancel{(2x + y)}^{1}(2x - y)}{2\cancel{(2x + y)}_{1}\cancel{(x - y)^2}_{(x - y)}} = \dfrac{3(2x - y)}{2(x - y)}$

27. $\dfrac{y^2 - 2y - 15}{y^2 - 9} \cdot \dfrac{y^2 - 6y + 9}{12 - 4y} = \dfrac{(y - 5)\cancel{(y + 3)}^{1}\cancel{(y - 3)}^{1}\cancel{(y - 3)}^{1}}{\cancel{(y - 3)}_{1}\cancel{(y + 3)}_{1}(-4)\cancel{(y - 3)}_{1}} = \dfrac{y - 5}{-4} = \dfrac{5 - y}{4}$

29. $\dfrac{w + 3}{w^2 + 7w + 10} \div \dfrac{w^2 + 6w + 9}{w^2 + 5w + 6} = \dfrac{w + 3}{w^2 + 7w + 10} \cdot \dfrac{w^2 + 5w + 6}{w^2 + 6w + 9}$

$= \dfrac{\cancel{(w + 3)}^{1}\cancel{(w + 3)}^{1}\cancel{(w + 2)}^{1}}{(w + 5)\cancel{(w + 2)}_{1}\cancel{(w + 3)}_{1}\cancel{(w + 3)}_{1}} = \dfrac{1}{w + 5}$

31. $\dfrac{y^2 + 2y - 15}{y^2 + 11y + 30} \div \dfrac{y^2 - 8y + 15}{y^2 + 2y - 24} = \dfrac{y^2 + 2y - 15}{y^2 + 11y + 30} \cdot \dfrac{y^2 + 2y - 24}{y^2 - 8y + 15}$

$$= \frac{\overset{1}{\cancel{(y + 5)}}\,\overset{1}{\cancel{(y - 3)}}\,\overset{1}{\cancel{(y + 6)}}\,(y - 4)}{\underset{1}{\cancel{(y + 5)}}\,\underset{1}{\cancel{(y + 6)}}\,\underset{1}{\cancel{(y - 3)}}\,(y - 5)} = \frac{y - 4}{y - 5}$$

33. $\dfrac{x^2 - 4y^2}{x^2 + xy - 6y^2} \cdot \dfrac{x^2 + 4xy + 3y^2}{x^2 + xy - 2y^2} = \dfrac{\overset{1}{\cancel{(x + 2y)}}\,\overset{1}{\cancel{(x - 2y)}}\,\overset{1}{\cancel{(x + 3y)}}\,(x + y)}{\underset{1}{\cancel{(x + 3y)}}\,\underset{1}{\cancel{(x - 2y)}}\,\underset{1}{\cancel{(x + 2y)}}\,(x - y)} = \dfrac{x + y}{x - y}$

35. $\dfrac{25 - 4x^2}{x^3 - 1} \cdot \dfrac{x^3 + x^2 + x}{15 - x - 2x^2} = \dfrac{(5 + 2x)\,\overset{1}{\cancel{(5 - 2x)}}\,(x)\,\overset{1}{\cancel{(x^2 + x + 1)}}}{(x - 1)\,\underset{1}{\cancel{(x^2 + x + 1)}}\,\underset{1}{\cancel{(5 - 2x)}}\,(3 + x)} = \dfrac{x(5 + 2x)}{(x - 1)(3 + x)}$

37. $\dfrac{rs^4 + rs^2}{r^2s^2 + s^2} \div \dfrac{r^3s^2 - r^2s^3}{r^2 - rs} = \dfrac{rs^4 + rs^2}{r^2s^2 + s^2} \cdot \dfrac{r^2 - rs}{r^3s^2 - r^2s^3} = \dfrac{\overset{1}{\cancel{rs^2}}\,(s^2 + 1)\,\overset{1}{\cancel{(r)}}\,\overset{1}{\cancel{(r - s)}}}{s^2\,(r^2 + 1)\,\underset{1}{\cancel{(r^2s^2)}}\,\underset{1}{\cancel{(r - s)}}}$

$$= \frac{s^2 + 1}{s^2(r^2 + 1)}$$

39. $\dfrac{x^3 - 27}{2x^2 + 5} \div (3x - 9) = \dfrac{x^3 - 27}{2x^2 + 5} \cdot \dfrac{1}{3x - 9} = \dfrac{\overset{1}{\cancel{(x - 3)}}\,(x^2 + 3x + 9)}{2x^2 + 5} \cdot \dfrac{1}{3\underset{1}{\cancel{(x - 3)}}}$

$$= \frac{x^2 + 3x + 9}{3(2x^2 + 5)}$$

41. $\dfrac{w^3 - 4w^2 + 9w - 36}{w^4 - 81} \cdot \dfrac{2aw^2 - aw - 15a}{w^3 - 64}$

$$= \frac{\overset{1}{\cancel{(w^2 + 9)}}\,\overset{1}{\cancel{(w - 4)}}\,(a)\,(2w + 5)\,\overset{1}{\cancel{(w - 3)}}}{(w + 3)\,\underset{1}{\cancel{(w - 3)}}\,\underset{1}{\cancel{(w^2 + 9)}}\,\underset{1}{\cancel{(w - 4)}}\,(w^2 + 4w + 16)} = \frac{a(2w + 5)}{(w + 3)(w^2 + 4w + 16)}$$

43. $\dfrac{3a + 2b}{2ya - 2xyb} \cdot \dfrac{a^3 - x^3b^3}{xa + 2xb} \cdot \dfrac{a^2 + 4ab + 4b^2}{9a^2 - 4b^2}$

$$= \frac{\overset{1}{\cancel{(3a + 2b)}}\,\overset{1}{\cancel{(a - xb)}}\,(a^2 + axb + x^2b^2)\,\overset{1}{\cancel{(a + 2b)}}\,(a + 2b)}{2y\underset{1}{\cancel{(a - xb)}}\,(x)\,\underset{1}{\cancel{(a + 2b)}}\,\underset{1}{\cancel{(3a + 2b)}}\,(3a - 2b)}$$

$$= \frac{(a + 2b)(a^2 + axb + x^2b^2)}{2xy(3a - 2b)}$$

45. $\dfrac{2x^2 + x - 6}{x^3 + 2x^2 + 4x} \cdot \dfrac{6x^3 + 24x}{3x^4 - 48} \div \dfrac{8x^2 - 18}{x^3 - 8} = \dfrac{2x^2 + x - 6}{x^3 + 2x^2 + 4x} \cdot \dfrac{6x^3 + 24x}{3x^4 - 48} \cdot \dfrac{x^3 - 8}{8x^2 - 18}$

$$= \frac{\overset{1}{\cancel{(2x - 3)}}\,\overset{1}{\cancel{(x + 2)}}\,\overset{1}{\cancel{(6x)}}\,\overset{1}{\cancel{(x^2 + 4)}}\,\overset{1}{\cancel{(x - 2)}}\,\overset{1}{\cancel{(x^2 + 2x + 4)}}}{\underset{1}{\cancel{(x)}}\,\underset{1}{\cancel{(x^2 + 2x + 4)}}\,\underset{1}{\cancel{(3)}}\,\underset{1}{\cancel{(x + 2)}}\,\underset{1}{\cancel{(x - 2)}}\,\underset{1}{\cancel{(x^2 + 4)}}\,\underset{1}{\cancel{(2)}}\,(2x + 3)\,\underset{1}{\cancel{(2x - 3)}}} = \frac{1}{2x + 3}$$

47. $\dfrac{5}{2 + x} - \dfrac{3}{2 + x} = \dfrac{5 - 3}{2 + x} = \dfrac{2}{2 + x}$

49. $\dfrac{x}{x^2 - 1} + \dfrac{1}{x^2 - 1} = \dfrac{x + 1}{x^2 - 1} = \dfrac{\overset{1}{\cancel{x + 1}}}{\cancel{(x + 1)}(x - 1)} = \dfrac{1}{x - 1}$

51. $\dfrac{4}{x - 1} + \dfrac{2}{1 - x} = \dfrac{4}{x - 1} + \dfrac{2(-1)}{(x - 1)(-1)} = \dfrac{4}{x - 1} + \dfrac{-2}{x - 1} = \dfrac{4 - 2}{x - 1} = \dfrac{2}{x - 1}$

53. $\dfrac{a}{a - b} - \dfrac{b}{b - a} = \dfrac{a}{a - b} - \dfrac{b(-1)}{(b - a)(-1)} = \dfrac{a}{a - b} - \dfrac{-b}{a - b} = \dfrac{a - (-b)}{a - b} = \dfrac{a + b}{a - b}$

55. $\dfrac{y}{4x^2} - \dfrac{z}{3x} = \dfrac{(y)(3)}{(4x^2)(3)} - \dfrac{(z)(4x)}{(3x)(4x)} = \dfrac{3y}{12x^2} - \dfrac{4xz}{12x^2} = \dfrac{3y - 4xz}{12x^2}$

57. $\dfrac{7}{2x} - \dfrac{1}{x - 2} = \dfrac{(7)(x - 2)}{(2x)(x - 2)} - \dfrac{(1)(2x)}{(x - 2)(2x)} = \dfrac{7x - 14}{2x(x - 2)} - \dfrac{2x}{2x(x - 2)}$

$= \dfrac{7x - 14 - 2x}{2x(x - 2)} = \dfrac{5x - 14}{2x(x - 2)}$

59. $\dfrac{12}{x^2 - 9} - \dfrac{2}{x - 3} = \dfrac{12}{(x + 3)(x - 3)} - \dfrac{2}{x - 3} = \dfrac{12}{(x + 3)(x - 3)} - \dfrac{2(x + 3)}{(x - 3)(x + 3)}$

$= \dfrac{12 - 2(x + 3)}{(x - 3)(x + 3)} = \dfrac{12 - 2x - 6}{(x - 3)(x + 3)} = \dfrac{6 - 2x}{(x - 3)(x + 3)}$

$= \dfrac{-2\overset{1}{\cancel{(x - 3)}}}{\cancel{(x - 3)}(x + 3)} = \dfrac{-2}{x + 3}$

61. $\dfrac{2y}{2x + 2y} - \dfrac{3}{x^2 - y^2} = \dfrac{\overset{1}{\cancel{2}}y}{\underset{1}{\cancel{2}}(x + y)} - \dfrac{3}{(x + y)(x - y)} = \dfrac{y(x - y)}{(x + y)(x - y)} - \dfrac{3}{(x + y)(x - y)}$

$= \dfrac{y(x - y) - 3}{(x + y)(x - y)} = \dfrac{xy - y^2 - 3}{(x + y)(x - y)}$

63. $\dfrac{x - 28}{x^2 - x - 6} + \dfrac{5}{x - 3} = \dfrac{x - 28}{(x + 2)(x - 3)} + \dfrac{5}{x - 3} = \dfrac{x - 28}{(x + 2)(x - 3)} + \dfrac{5(x + 2)}{(x - 3)(x + 2)}$

$= \dfrac{x - 28 + 5(x + 2)}{(x + 2)(x - 3)} = \dfrac{x - 28 + 5x + 10}{(x + 2)(x - 3)} = \dfrac{6x - 18}{(x + 2)(x - 3)}$

$= \dfrac{6\overset{1}{\cancel{(x - 3)}}}{(x + 2)\cancel{(x - 3)}} = \dfrac{6}{x + 2}$

13

65. $\dfrac{x+5}{x^2+7x+10} - \dfrac{x+5}{2x^2+x-6} = \dfrac{\overset{1}{\cancel{x+5}}}{\cancel{(x+5)}(x+2)} - \dfrac{x+5}{(2x-3)(x+2)}$

$$= \dfrac{1}{x+2} - \dfrac{x+5}{(2x-3)(x+2)} = \dfrac{1(2x-3)}{(x+2)(2x-3)} - \dfrac{x+5}{(2x-3)(x+2)}$$

$$= \dfrac{2x-3-(x+5)}{(2x-3)(x+2)} = \dfrac{2x-3-x-5}{(2x-3)(x+2)} = \dfrac{x-8}{(2x-3)(x+2)}$$

67. $\dfrac{w-1}{2w^2-13w+15} + \dfrac{w+3}{2w^2-15w+18} = \dfrac{w-1}{(2w-3)(w-5)} + \dfrac{w+3}{(2w-3)(w-6)}$

$$= \dfrac{(w-1)(w-6)}{(2w-3)(w-5)(w-6)} + \dfrac{(w+3)(w-5)}{(2w-3)(w-6)(w-5)}$$

$$= \dfrac{(w-1)(w-6)+(w+3)(w-5)}{(2w-3)(w-5)(w-6)} = \dfrac{w^2-7w+6+w^2-2w-15}{(2w-3)(w-5)(w-6)}$$

$$= \dfrac{2w^2-9w-9}{(2w-3)(w-5)(w-6)}$$

69. $\dfrac{z-5}{z^2-9} + \dfrac{z-2}{12-4z} = \dfrac{z-5}{(z-3)(z+3)} + \dfrac{z-2}{4(3-z)}$

$$= \dfrac{(z-5)(4)}{(z-3)(z+3)(4)} + \dfrac{(z-2)(-1)(z+3)}{4(3-z)(-1)(z+3)}$$

$$= \dfrac{(z-5)(4)+(z-2)(-1)(z+3)}{4(z-3)(z+3)} = \dfrac{4z-20-z^2-z+6}{4(z-3)(z+3)}$$

$$= \dfrac{-z^2+3z-14}{4(z-3)(z+3)}$$

71. $y - \dfrac{2y}{y^2-1} + \dfrac{3}{y+1} = \dfrac{y}{1} - \dfrac{2y}{(y-1)(y+1)} + \dfrac{3}{y+1}$

$$= \dfrac{y(y-1)(y+1)}{1(y-1)(y+1)} - \dfrac{2y}{(y-1)(y+1)} + \dfrac{3(y-1)}{(y-1)(y+1)}$$

$$= \dfrac{y(y-1)(y+1)-2y+3(y-1)}{(y-1)(y+1)}$$

$$= \dfrac{y^3-y-2y+3y-3}{(y-1)(y+1)} = \dfrac{y^3-3}{(y-1)(y+1)}$$

73. $(9x^2-25y^2) \div \left(1+\dfrac{5y}{3x}\right) = (9x^2-25y^2) \div \dfrac{3x+5y}{3x} = (9x^2-25y^2) \cdot \dfrac{3x}{3x+5y}$

$$= \dfrac{(3x-5y)\overset{1}{\cancel{(3x+5y)}}(3x)}{\underset{1}{\cancel{3x+5y}}} = 3x(3x-5y)$$

75. $\left(\dfrac{2}{x+2} - \dfrac{1}{x-2}\right)\left(\dfrac{x-2}{x+1}\right) = \left[\dfrac{2(x-2)}{(x+2)(x-2)} - \dfrac{1(x+2)}{(x-2)(x+2)}\right]\left(\dfrac{x-2}{x+1}\right)$

$\qquad\qquad = \dfrac{2(x-2) - (x+2)}{(x-2)(x+2)} \cdot \dfrac{x-2}{x+1} = \dfrac{x-6}{(x-2)(x+2)} \cdot \dfrac{x-2}{x+1}$

$\qquad\qquad = \dfrac{(x-6)\overset{1}{\cancel{(x-2)}}}{\underset{1}{\cancel{(x-2)}}(x+2)(x+1)} = \dfrac{x-6}{(x+2)(x+1)}$

77. $\left(x - 1 - \dfrac{6}{x}\right) \div \left(1 + \dfrac{2}{x} - \dfrac{15}{x^2}\right) = \dfrac{x - 1 - \dfrac{6}{x}}{1 + \dfrac{2}{x} - \dfrac{15}{x^2}} \cdot \dfrac{x^2}{x^2} = \dfrac{x^3 - x^2 - 6x}{x^2 + 2x - 15}$

$\qquad\qquad = \dfrac{x\overset{1}{\cancel{(x-3)}}(x+2)}{(x+5)\underset{1}{\cancel{(x-3)}}} = \dfrac{x(x+2)}{x+5}$

79. $\dfrac{2x+2y}{x^2+2xy-3y^2} - \dfrac{x-2y}{x^2+xy-6y^2} = \dfrac{2(x+y)}{(x-y)(x+3y)} - \dfrac{\overset{1}{\cancel{x-2y}}}{(x+3y)\underset{1}{\cancel{(x-2y)}}}$

$\qquad\qquad = \dfrac{2(x+y)}{(x-y)(x+3y)} - \dfrac{1(x-y)}{(x+3y)(x-y)} = \dfrac{2(x+y) - (x-y)}{(x-y)(x+3y)}$

$\qquad\qquad = \dfrac{\overset{1}{\cancel{x+3y}}}{(x-y)\underset{1}{\cancel{(x+3y)}}} = \dfrac{1}{x-y}$

81. $\dfrac{\dfrac{3w^2+w-2}{4w^2-w-5}}{\dfrac{3w^2-11w+6}{4w^2-7w-15}} = \dfrac{3w^2+w-2}{4w^2-w-5} \div \dfrac{3w^2-11w+6}{4w^2-7w-15} = \dfrac{3w^2+w-2}{4w^2-w-5} \cdot \dfrac{4w^2-7w-15}{3w^2-11w+6}$

$\qquad\qquad = \dfrac{\overset{1}{\cancel{(3w-2)}}\overset{1}{\cancel{(w+1)}}(4w+5)\overset{1}{\cancel{(w-3)}}}{(4w-5)\underset{1}{\cancel{(w+1)}}\underset{1}{\cancel{(3w-2)}}\underset{1}{\cancel{(w-3)}}} = \dfrac{4w+5}{4w-5}$

83. $\dfrac{\dfrac{x^3+y^3}{2x-3y}}{\dfrac{2x+2y}{4x^2-9y^2}} = \dfrac{x^3+y^3}{2x-3y} \div \dfrac{2x+2y}{4x^2-9y^2} = \dfrac{x^3+y^3}{2x-3y} \cdot \dfrac{4x^2-9y^2}{2x+2y}$

$\qquad\qquad = \dfrac{\overset{1}{\cancel{(x+y)}}(x^2-xy+y^2)\overset{1}{\cancel{(2x-3y)}}(2x+3y)}{\underset{1}{\cancel{(2x-3y)}}(2)\underset{1}{\cancel{(x+y)}}} = \dfrac{(2x+3y)(x^2-xy+y^2)}{2}$

85. $\dfrac{1+\dfrac{y}{x}}{1-\dfrac{y}{x}} \cdot \dfrac{x}{x} = \dfrac{x+y}{x-y}$

87. $\dfrac{9x^2 - 4y^2}{\dfrac{x - y}{y - 2x} - 1} = \dfrac{9x^2 - 4y^2}{\dfrac{x - y - (y - 2x)}{y - 2x}} = \dfrac{9x^2 - 4y^2}{\dfrac{3x - 2y}{y - 2x}} = (9x^2 - 4y^2) \div \dfrac{3x - 2y}{y - 2x}$

$\qquad = (9x^2 - 4y^2) \cdot \dfrac{y - 2x}{3x - 2y} = \dfrac{\cancel{(3x - 2y)}(3x + 2y)(y - 2x)}{\cancel{3x - 2y}}$

$\qquad = (3x + 2y)(y - 2x)$

89. $\dfrac{\dfrac{1}{x^2} - \dfrac{1}{y^2}}{\dfrac{1}{x} + \dfrac{1}{y}} \cdot \dfrac{x^2 y^2}{x^2 y^2} = \dfrac{y^2 - x^2}{xy^2 + x^2 y} = \dfrac{(y - x)\cancel{(y + x)}}{xy\cancel{(y + x)}} = \dfrac{y - x}{xy}$

91. $\dfrac{\dfrac{b^2}{a^2} - \dfrac{a}{b}}{\dfrac{1}{2b} - \dfrac{a}{2b^2}} \cdot \dfrac{2a^2 b^2}{2a^2 b^2} = \dfrac{b^2(2b^2) - a(2a^2 b)}{1(a^2 b) - a(a^2)} = \dfrac{2b^4 - 2a^3 b}{a^2 b - a^3} = \dfrac{2b(b^3 - a^3)}{a^2(b - a)}$

$\qquad = \dfrac{2b\cancel{(b - a)}(b^2 + ab + a^2)}{a^2 \cancel{(b - a)}} = \dfrac{2b(b^2 + ab + a^2)}{a^2}$

93. $1 + \dfrac{1}{1 + \dfrac{1}{x}} = 1 + \dfrac{1}{1 + \dfrac{1}{x}} \cdot \dfrac{x}{x} = 1 + \dfrac{x}{x + 1} = \dfrac{x + 1}{x + 1} + \dfrac{x}{x + 1} = \dfrac{2x + 1}{x + 1}$

95. $x - \dfrac{x}{2 - \dfrac{1}{x}} = x - \dfrac{x}{2 - \dfrac{1}{x}} \cdot \dfrac{x}{x} = x - \dfrac{x^2}{2x - 1} = \dfrac{x(2x - 1)}{2x - 1} - \dfrac{x^2}{2x - 1}$

$\qquad = \dfrac{x(2x - 1) - x^2}{2x - 1} = \dfrac{x^2 - x}{2x - 1} \text{ or } \dfrac{x(x - 1)}{2x - 1}$

97. $xy^{-1} = \dfrac{x^{-1}}{y^{-1}} = x\left(\dfrac{1}{y}\right) - \dfrac{\dfrac{1}{x}}{\dfrac{1}{y}} = \dfrac{x}{y} - \dfrac{y}{x} = \dfrac{x(x)}{y(x)} - \dfrac{y(y)}{x(y)} = \dfrac{x^2}{xy} - \dfrac{y^2}{xy} = \dfrac{x^2 - y^2}{xy}$

99. $\dfrac{(ab)^{-2}}{a^{-2} - b^{-2}} = \dfrac{\dfrac{1}{a^2 b^2}}{\dfrac{1}{a^2} - \dfrac{1}{b^2}} = \dfrac{\dfrac{1}{a^2 b^2}}{\dfrac{1}{a^2} - \dfrac{1}{b^2}} \cdot \dfrac{a^2 b^2}{a^2 b^2} = \dfrac{1}{b^2 - a^2}$

101. $\dfrac{x^{-1} - y^{-1}}{x^{-2} - y^{-2}} = \dfrac{\dfrac{1}{x} - \dfrac{1}{y}}{\dfrac{1}{x^2} - \dfrac{1}{y^2}} = \dfrac{\dfrac{1}{x} - \dfrac{1}{y}}{\dfrac{1}{x^2} - \dfrac{1}{y^2}} \cdot \dfrac{x^2 y^2}{x^2 y^2} = \dfrac{xy^2 - x^2 y}{y^2 - x^2} = \dfrac{xy\cancel{(y - x)}}{\cancel{(y - x)}(y + x)} = \dfrac{xy}{y + x}$

103. $\dfrac{x^{-2} - y^{-2}}{(x + y)^2} = \dfrac{\dfrac{1}{x^2} - \dfrac{1}{y^2}}{(x + y)^2} = \dfrac{\dfrac{1}{x^2} - \dfrac{1}{y^2}}{(x + y)^2} \cdot \dfrac{x^2 y^2}{x^2 y^2} = \dfrac{y^2 - x^2}{(x + y)^2 x^2 y^2} = \dfrac{\cancel{(y + x)}(y - x)}{(x + y)\cancel{(x + y)} x^2 y^2}$

$\qquad = \dfrac{y - x}{(x + y)x^2 y^2} \text{ or } \dfrac{y - x}{x^2 y^2(x + y)}$

105. $\dfrac{\dfrac{1}{x} - \dfrac{1}{3}}{\dfrac{1}{x - 3}}$ becomes $\dfrac{\dfrac{1}{2.17} - \dfrac{1}{3}}{\dfrac{1}{2.17 - 3}} \approx \dfrac{0.460829 - 0.333333}{-1.204819} = \dfrac{0.127496}{-1.204819} \approx -0.1058$

107. $\dfrac{x^2 - 9}{x^3 - 5x^2 + 6x} = \dfrac{(x + 3)(x - 3)}{x(x - 2)(x - 3)}$ which can be simplified to $\dfrac{x + 3}{x(x - 2)}$
with the understanding that the original denominator is different from 0.
That is $x(x - 2)(x - 3) \neq 0$
or $x \neq 0$, $x - 2 \neq 0$, $x - 3 \neq 0$
or $x \neq 0$, $x \neq 2$, $x \neq 3$

Exercises 1.7

1. $\sqrt{25} = 5$ since $5^2 = 25$

3. $\sqrt[3]{64} = 4$ since $4^3 = 64$

5. $(\sqrt[5]{73})^5 = 73$ since $73 = 73^{5/5}$

7. $\sqrt[4]{\dfrac{256}{81}} = \dfrac{4}{3}$ since $\left(\dfrac{4}{3}\right)^4 = \dfrac{256}{81}$

9. $\sqrt{49} = 7$ since $7^2 = 49$

11. $\sqrt{\sqrt[5]{1024}} = \sqrt{4} = 2$

13. $\sqrt{25x^2} = \sqrt{(5x)^2} = |5x| = 5|x|$

15. $\sqrt[3]{a^3b^3} = \sqrt[3]{(ab)^3} = ab$

17. $\sqrt[4]{(-2x)^4} = |-2x| = 2|x|$ or $|2x|$

19. $\sqrt{(r + s)^4} = \sqrt{((r + s)^2)^2} = |(r + s)^2| = (r + s)^2$

21. $\sqrt[3]{8b^6} = \sqrt[3]{8}\sqrt[3]{b^6} = 2b^2$

23. $\sqrt{y^4w^6} = \sqrt{y^4}\sqrt{w^6} = y^2w^3$

25. $\sqrt{0.16x^4} = \sqrt{0.16}\sqrt{x^4} = \sqrt{(0.4)^2}\sqrt{x^4} = 0.4x^2$ 27. $\sqrt[7]{-1} = -1$

29. $\sqrt[4]{\dfrac{16x^4}{81y^8}} = \dfrac{\sqrt[4]{16}\sqrt[4]{x^4}}{\sqrt[4]{81}\sqrt[4]{y^8}} = \dfrac{\sqrt[4]{2^4}\sqrt[4]{x^4}}{\sqrt[4]{3^4}\sqrt[4]{y^8}} = \dfrac{2x}{3y^2}$

31. $\sqrt[3]{\dfrac{216x^3}{125}} = \dfrac{\sqrt[3]{216}\sqrt[3]{x^3}}{\sqrt[3]{125}} = \dfrac{\sqrt[3]{6^3}\sqrt[3]{x^3}}{\sqrt[3]{5^3}} = \dfrac{6x}{5}$

33. $\sqrt[4]{x^8y^{-12}} = \sqrt[4]{\dfrac{x^8}{y^{12}}} = \dfrac{\sqrt[4]{x^8}}{\sqrt[4]{y^{12}}} = \dfrac{x^2}{y^3}$

35. $(\sqrt[4]{19y^5})^4 = (19y^5)^{4/4} = 19y^5$

37. $\sqrt{\sqrt[3]{x^6y^{12}}} = \sqrt[6]{x^6y^{12}} = \sqrt[6]{x^6}\sqrt[6]{y^{12}} = xy^2$

39. $\sqrt[3]{\sqrt{\dfrac{x^{18}}{z^6y^0}}} = \sqrt[6]{\dfrac{x^{18}}{z^6}} = \dfrac{\sqrt[6]{x^{18}}}{\sqrt[6]{z^6}} = \dfrac{x^3}{z}$

41. $\sqrt{\left(\dfrac{81x^4}{25z^6}\right)^{-1}} = \sqrt{\dfrac{25z^6}{81x^4}} = \dfrac{\sqrt{25}\sqrt{z^6}}{\sqrt{81}\sqrt{x^4}} = \dfrac{5z^3}{9x^2}$

43. $a^{1/5} = \sqrt[5]{a}$ 45. $3a^{1/4} = 3\sqrt[4]{a}$

47. $\sqrt[4]{a^5} = a^{5/4}$ 49. $\sqrt[9]{x^{10}} = x^{10/9}$ 51. $\sqrt[3]{a^6} = a^{6/3}$ or a^2

53. $(3b)^{1/4} = \sqrt[4]{3b}$ 55. $(2xy^2)^{2/3} = \sqrt[3]{(2xy^2)^2}$ or $(\sqrt[3]{2xy^2})^2$ 57. $\sqrt[7]{bx^2} = (bx^2)^{1/7}$

59. $(16)^{1/4} = \sqrt[4]{16} = 2$

61. $(-64a^6)^{1/6} = \sqrt[6]{-64a^6}$ which is undefined since root is even and radicand is negative

63. $(64)^{-4/3} = (\sqrt[3]{64})^{-4} = 4^{-4} = \dfrac{1}{4^4} = \dfrac{1}{256}$ 65. $(81)^{3/4} = (\sqrt[4]{81})^3 = 3^3 = 27$

67. $\left(\dfrac{1}{32}\right)^{-4/5} = \left(\sqrt[5]{\dfrac{1}{32}}\right)^{-4} = \left(\dfrac{1}{2}\right)^{-4} = \dfrac{1^{-4}}{2^{-4}} = \dfrac{2^4}{1^4} = 16$ 69. $(8p^6)^{2/3} = (\sqrt[3]{8p^6})^2 = (2p^2)^2 = 4p^4$

71. $(a^2b^4)^{3/2} = (\sqrt{a^2b^4})^3 = (ab^2)^3 = a^3b^6$ 73. $\left(-\dfrac{27}{64}\right)^{2/3} = \left(\sqrt[3]{-\dfrac{27}{64}}\right)^2 = \left(-\dfrac{3}{4}\right)^2 = \dfrac{9}{16}$

75. $-\left(\dfrac{4m^2}{n^4}\right)^{3/2} = -\left(\sqrt{\dfrac{4m^2}{n^4}}\right)^3 = -\left(\dfrac{2m}{n^2}\right)^3 = -\dfrac{8m^3}{n^6}$

77. $-\left(\dfrac{x^4}{16y^2}\right)^{-3/2} = -\left(\sqrt{\dfrac{x^4}{16y^2}}\right)^{-3} = -\left(\dfrac{x^2}{4y}\right)^{-3} = -\dfrac{x^{-6}}{4^{-3}y^{-3}} = -\dfrac{64y^3}{x^6}$

79. $x^{2/3}x^{7/3} = x^{2/3+7/3} = x^{9/3} = x^3$ 81. $(x^{1/3}y^{5/6})^{12} = (x^{1/3})^{12}(y^{5/6})^{12} = x^{12/3}y^{60/6} = x^4y^{10}$

83. $(a^{-1/2}b^{3/2})^{-4/3} = (a^{-1/2})^{-4/3}(b^{3/2})^{-4/3} = a^{4/6}b^{-12/6} = a^{2/3}b^{-2} = \dfrac{a^{2/3}}{b^2}$

85. $[(a^{2/3}b^{-1/4})^{-2}]^{-6} = (a^{2/3}b^{-1/4})^{12} = (a^{2/3})^{12}(b^{-1/4})^{12} = a^8b^{-3} = \dfrac{a^8}{b^3}$

87. $[(x^2y^{-2/3})^{-3}]^0 = 1$ since any quantity to the zero power is 1.

89. $x^{1/3}(x^{2/3} - x^{-1/3}) = x^{1/3} \cdot x^{2/3} - x^{1/3}x^{-1/3} = x^{1/3+2/3} - x^{1/3-1/3} = x^{3/3} - x^{0/3} = x - 1$

91. $(a^{1/2} - b^{1/2})(a^{1/2} + b^{1/2}) = (a^{1/2})^2 - (b^{1/2})^2 = a^{2/2} - b^{2/2} = a - b$

93. $\dfrac{ab^0c^{2/5}}{a^{3/4}b^{-1/3}c^{1/5}} = a^{1-3/4}b^{0-(-1/3)}c^{2/5-1/5} = a^{1/4}b^{1/3}c^{1/5}$

95. $(2.5)^{1.4} \approx 3.6067$ which rounds to 3.61

97. $(27.1)^{2/5} = (27.1)^{.4} \approx 3.743$ which rounds to 3.74

99. $\sqrt[5]{731} = (731)^{1/5} = (731)^{0.2} \approx 3.739$ which rounds to 3.74

101. $\left(\dfrac{1}{\sqrt[10]{90.2}}\right)^3 = \dfrac{1}{(\sqrt[10]{90.2})^3} = \dfrac{1}{(90.2)^{3/10}} = \dfrac{1}{(90.2)^{0.3}} \approx 0.259$ which rounds to 0.26

103. $S = \frac{1}{2}(a + b + c) = \frac{1}{2}(114 + 108 + 66) = \frac{1}{2}(288) = 144.$

Then $A = \sqrt{S(S - a)(S - b)(S - c)} = \sqrt{144(144 - 114)(144 - 108)(144 - 66)}$

$= \sqrt{144(30)(36)(78)} = \sqrt{12,130,560} \approx 3482.9$ sq in or $\frac{3482.9}{144} \approx 24$ sq ft.

105. $t = 0.555\sqrt{d} = 0.555\sqrt{2.3} \approx 0.555(1.516575) \approx 0.842$ or 0.84 seconds.

107. $s = 100\sqrt{\frac{P}{A}} = 100\sqrt{\frac{10}{50}} = 100\sqrt{0.2} \approx 100(0.4472136) = 44.72136$ or 45 mph.

109. $r = \sqrt{\frac{3V}{\pi h}} = \sqrt{\frac{3(9.4)}{\pi(3.5)}} \approx \sqrt{\frac{3(9.4)}{(3.14)(3.5)}} = \sqrt{2.56597} \approx 1.6$ inches

111. $v = x^3$ which gives $2.2 = x^3$ or $\sqrt[3]{2.2} = x$ or 1.3 cm $\approx x$

113. $V = \frac{4}{3}\pi r^3$ which gives $65.5 = \frac{4}{3}\pi r^3$ or $\frac{3(65.5)}{4\pi} = r^3$ or $r = \sqrt{\frac{3(65.5)}{4\pi}} \approx 2.5$ ft

115. $\sqrt{a + b} = (a + b)^{1/2}$ which is not equal to $\sqrt{a} + \sqrt{b}$
If it were to be equal, then $(\sqrt{a} + \sqrt{b})^2 = a + b$, by definition.
But $(\sqrt{a} + \sqrt{b})^2 = a + 2\sqrt{a}\sqrt{b} + b$ which is **not** $a + b$. Thus, $\sqrt{a + b} \neq \sqrt{a} + \sqrt{b}$.

117. If $a < 0$, then $a^3 < 0$ and $\sqrt[3]{a^3} = a < 0$ which is not $|a|$ which is $-a$. To see
this, let $a = -2$. Then $a^3 = (-2)^3 = -8$ and $\sqrt[3]{a^3} = \sqrt[3]{-8} = \underline{\underline{-2}}$ which is **not** the
same as $|a| = |-2| = \underline{\underline{2}}$.

119. By definition $a^{m/2} = (\sqrt{a})^m$ which does not exist when a is negative since $\sqrt{a}$
does not exist.

Exercises 1.8

1. $\sqrt{500} = \sqrt{100 \cdot 5} = \sqrt{(10)^2 5} = \sqrt{10^2}\sqrt{5} = 10\sqrt{5}$

3. $\sqrt{32a^3b^4} = \sqrt{(16a^2b^4)(2a)} = \sqrt{(4ab^2)^2(2a)} = \sqrt{(4ab^2)^2}\sqrt{2a} = 4ab^2\sqrt{2a}$

5. $\sqrt[3]{128x^7} = \sqrt[3]{(64x^6)(2x)} = \sqrt[3]{(4x^2)^3(2x)} = \sqrt[3]{(4x^2)^3}\sqrt[3]{2x} = 4x^2\sqrt[3]{2x}$

7. $\sqrt[5]{x^{13}} = \sqrt[5]{(x^{10})(x^3)} = \sqrt[5]{(x^2)^5(x^3)} = \sqrt[5]{(x^2)^5}\sqrt[5]{x^3} = x^2\sqrt[5]{x^3}$

9. $\sqrt[3]{-128a^9} = \sqrt[3]{(-64a^9)(2)} = \sqrt[3]{(-4a^3)^3(2)} = \sqrt[3]{(-4a^3)^3}\sqrt[3]{2} = -4a^3\sqrt[3]{2}$

11. $\sqrt[4]{162(a + b)^5} = \sqrt[4]{[81(a + b)^4][2(a + b)]} = \sqrt[4]{[3(a + b)]^4[2(a + b)]}$
$= \sqrt[4]{[3(a + b)]^4}\sqrt[4]{2(a + b)} = 3(a + b)\sqrt[4]{2(a + b)}$

13. $\sqrt{2} + \sqrt{3}$ remains as $\sqrt{2} + \sqrt{3}$

15. $\sqrt{150} - \sqrt{24} = \sqrt{(25)(6)} - \sqrt{(4)(6)} = 5\sqrt{6} - 2\sqrt{6} = 3\sqrt{6}$

17. $3\sqrt{50} - 2\sqrt{18} = 3\sqrt{(25)(2)} - 2\sqrt{(9)(2)} = 3 \cdot 5\sqrt{2} - 2 \cdot 3\sqrt{2} = 15\sqrt{2} - 6\sqrt{2} = 9\sqrt{2}$

19. $3\sqrt{2} - 2\sqrt[3]{2}$ remains as $3\sqrt{2} - 2\sqrt[3]{2}$

21. $\sqrt[3]{54a^3} + 2b\sqrt[3]{16} = \sqrt[3]{(27a^3)(2)} + 2b\sqrt[3]{(8)(2)} = 3a\sqrt[3]{2} + 4b\sqrt[3]{2} = (3a + 4b)\sqrt[3]{2}$

23. $\sqrt[3]{375x^3y^4} + \sqrt[3]{-3x^3y^4} = \sqrt[3]{(125x^3y^3)(3y)} + \sqrt[3]{(-x^3y^3)(3y)} = 5xy\sqrt[3]{3y} - xy\sqrt[3]{3y} = 4xy\sqrt[3]{3y}$

25. $\sqrt[3]{5} \cdot \sqrt[3]{50} = \sqrt[3]{250} = \sqrt[3]{(125)(2)} = 5\sqrt[3]{2}$ 27. $\dfrac{\sqrt{15}}{\sqrt{3}} = \sqrt{\dfrac{15}{3}} = \sqrt{5}$

29. $\dfrac{\sqrt[3]{20x^4}}{\sqrt[3]{2x}} = \sqrt[3]{\dfrac{20x^4}{2x}} = \sqrt[3]{10x^3} = \sqrt[3]{(x^3)(10)}$ 31. $\sqrt[3]{4x^2} \cdot \sqrt[3]{6x^2y^4} = \sqrt[3]{24x^4y^4}$

$$= x\sqrt[3]{10}$$
$$= \sqrt[3]{(8x^3y^3)(2xy)} = 2xy\sqrt[3]{2xy}$$

33. $(2 - \sqrt{3})(3 + 2\sqrt{3}) = (2 - \sqrt{3})(3) + (2 - \sqrt{3})(2\sqrt{3})$
$$= 6 - 3\sqrt{3} + 4\sqrt{3} - 2\sqrt{9}$$
$$= 6 - 3\sqrt{3} + 4\sqrt{3} - 6$$
$$= 0 + \sqrt{3}$$
$$= \sqrt{3}$$

35. $(\sqrt{5} + 4\sqrt{2})(\sqrt{5} - 4\sqrt{2}) = (\sqrt{5} + 4\sqrt{2})(\sqrt{5}) + (\sqrt{5} + 4\sqrt{2})(-4\sqrt{2})$
$$= \sqrt{25} + 4\sqrt{10} - 4\sqrt{10} - 16\sqrt{4}$$
$$= 5 + 4\sqrt{10} - 4\sqrt{10} - 32$$
$$= -27 + 0\sqrt{10}$$
$$= -27$$

37. $(\sqrt{5} - 3\sqrt{2})(\sqrt{10} - 3) = (\sqrt{5} - 3\sqrt{2})(\sqrt{10}) + (\sqrt{5} - 3\sqrt{2})(-3)$
$$= \sqrt{50} - 3\sqrt{20} - 3\sqrt{5} + 9\sqrt{2}$$
$$= \sqrt{(25)(2)} - 3\sqrt{(4)(5)} - 3\sqrt{5} + 9\sqrt{2}$$
$$= 5\sqrt{2} - 6\sqrt{5} - 3\sqrt{5} + 9\sqrt{2}$$
$$= 14\sqrt{2} - 9\sqrt{5}$$

39. $\sqrt[9]{125} = \sqrt[9]{5^3} = 5^{3/9} = 5^{1/3} = \sqrt[3]{5}$ 41. $\sqrt[6]{u^4} = u^{4/6} = u^{2/3} = \sqrt[3]{u^2}$

43. $\sqrt[12]{y^8} = y^{8/12} = y^{2/3} = \sqrt[3]{y^2}$

45. $\sqrt[10]{25a^6b^4} = \sqrt[10]{5^2a^6b^4} = 5^{2/10}a^{6/10}b^{4/10} = 5^{1/5}a^{3/5}b^{2/5} = \sqrt[5]{5a^3b^2}$

47. $\sqrt{\dfrac{1}{5}} = \sqrt{\dfrac{1 \cdot 5}{5 \cdot 5}} = \sqrt{\dfrac{5}{25}} = \dfrac{\sqrt{5}}{\sqrt{25}} = \dfrac{\sqrt{5}}{5}$ 49. $\sqrt{\dfrac{5}{8}} = \sqrt{\dfrac{5 \cdot 2}{8 \cdot 2}} = \sqrt{\dfrac{10}{16}} = \dfrac{\sqrt{10}}{\sqrt{16}} = \dfrac{\sqrt{10}}{4}$

51. $\dfrac{5}{\sqrt{7}} = \dfrac{5\sqrt{7}}{\sqrt{7}\sqrt{7}} = \dfrac{5\sqrt{7}}{7}$

53. $\sqrt[3]{\dfrac{1}{4}} = \sqrt[3]{\dfrac{1\cdot 2}{4\cdot 2}} = \sqrt[3]{\dfrac{2}{8}} = \dfrac{\sqrt[3]{2}}{\sqrt[3]{8}} = \dfrac{\sqrt[3]{2}}{2}$

55. $\dfrac{3\sqrt{5}}{2\sqrt{7}} = \dfrac{3\sqrt{5}\sqrt{7}}{2\sqrt{7}\sqrt{7}} = \dfrac{3\sqrt{35}}{2\cdot 7} = \dfrac{3\sqrt{35}}{14}$

57. $\dfrac{\sqrt[3]{12}}{\sqrt[3]{4}} = \sqrt[3]{\dfrac{12}{4}} = \sqrt[3]{3}$

59. $\sqrt{\dfrac{a}{3}} = \sqrt{\dfrac{a\cdot 3}{3\cdot 3}} = \sqrt{\dfrac{3a}{9}} = \dfrac{\sqrt{3a}}{\sqrt{9}} = \dfrac{\sqrt{3a}}{3}$

61. $\sqrt{\dfrac{5a}{2b}} = \sqrt{\dfrac{(5a)(2b)}{(2b)(2b)}} = \sqrt{\dfrac{10ab}{4b^2}} = \dfrac{\sqrt{10ab}}{\sqrt{4b^2}} = \dfrac{\sqrt{10ab}}{2b}$

63. $\sqrt[3]{\dfrac{2cd}{9x^2}} = \sqrt[3]{\dfrac{(2cd)(3x)}{(9x^2)(3x)}} = \sqrt[3]{\dfrac{6cdx}{27x^3}} = \dfrac{\sqrt[3]{6cdx}}{\sqrt[3]{27x^3}} = \dfrac{\sqrt[3]{6cdx}}{3x}$

65. $\sqrt{\dfrac{1}{5u^5}} = \sqrt{\dfrac{1(5u)}{(5u^5)(5u)}} = \sqrt{\dfrac{5u}{25u^6}} = \dfrac{\sqrt{5u}}{\sqrt{25u^6}} = \dfrac{\sqrt{5u}}{5u^3}$

67. $\sqrt[4]{\dfrac{2a^2y^6}{x^3z^5}} = \sqrt[4]{\dfrac{(2a^2y^6)(xz^3)}{(x^3z^5)(xz^3)}} = \sqrt[4]{\dfrac{2a^2xy^6z^3}{x^4z^8}} = \dfrac{\sqrt[4]{2a^2xy^6z^3}}{\sqrt[4]{x^4z^8}}$

$= \dfrac{\sqrt[4]{2a^2xy^6z^3}}{xz^2} = \dfrac{\sqrt[4]{y^4(2a^2xy^2z^3)}}{xz^2} = \dfrac{y\sqrt[4]{2a^2xy^2z^3}}{xz^2}$

69. $\dfrac{2\sqrt{7}-3}{\sqrt{7}-2} = \dfrac{2\sqrt{7}-3}{\sqrt{7}-2}\cdot\dfrac{\sqrt{7}+2}{\sqrt{7}+2} = \dfrac{(2\sqrt{7}-3)(\sqrt{7}+2)}{(\sqrt{7}-2)(\sqrt{7}+2)} = \dfrac{2\sqrt{49}-3\sqrt{7}+4\sqrt{7}-6}{7-4}$

$= \dfrac{14-3\sqrt{7}+4\sqrt{7}-6}{3} = \dfrac{8+\sqrt{7}}{3}$

71. $\dfrac{3}{5\sqrt{2}-\sqrt{7}} = \dfrac{3}{5\sqrt{2}-\sqrt{7}}\cdot\dfrac{5\sqrt{2}+\sqrt{7}}{5\sqrt{2}+\sqrt{7}} = \dfrac{3(5\sqrt{2}+\sqrt{7})}{25\cdot 2-7} = \dfrac{3(5\sqrt{2}+\sqrt{7})}{50-3} = \dfrac{3(5\sqrt{2}+\sqrt{7})}{43}$

73. $\dfrac{\sqrt{5}-\sqrt{3}}{3\sqrt{3}+\sqrt{2}} = \dfrac{\sqrt{5}-\sqrt{3}}{3\sqrt{3}+\sqrt{2}}\cdot\dfrac{3\sqrt{3}-\sqrt{2}}{3\sqrt{3}-\sqrt{2}} = \dfrac{(\sqrt{5}-\sqrt{3})(3\sqrt{3}-\sqrt{2})}{(3\sqrt{3}+2)(3\sqrt{3}-2)}$

$= \dfrac{3\sqrt{15}-\sqrt{10}-3\sqrt{9}+\sqrt{6}}{9\cdot 3-2} = \dfrac{3\sqrt{15}-\sqrt{10}-9+\sqrt{6}}{27-2}$

$= \dfrac{3\sqrt{15}-\sqrt{10}-9+\sqrt{6}}{25}$

75. $\sqrt[5]{32uv^7} = \sqrt[5]{(32v^5)(uv^2)} = \sqrt[5]{32v^5}\sqrt[5]{uv^2} = 2v\sqrt[5]{uv^2}$

77. $\sqrt[4]{3y^{-8}} = \sqrt[4]{\dfrac{3}{y^8}} = \dfrac{\sqrt[4]{3}}{\sqrt[4]{y^8}} = \dfrac{\sqrt[4]{3}}{y^2}$

79. $\sqrt{x^9 y^{-2} z^0} = \sqrt{\dfrac{x^9}{y^2}} = \dfrac{\sqrt{x^9}}{\sqrt{y^2}} = \dfrac{\sqrt{(x^8)(x)}}{\sqrt{y^2}} = \dfrac{\sqrt{x^8}\sqrt{x}}{\sqrt{y^2}} = \dfrac{x^4\sqrt{x}}{y}$

81. $\sqrt{(a+2b)^6} = (a+2b)^{6/2} = (a+2b)^3$

83. $\sqrt[4]{\dfrac{5}{8}} = \sqrt[4]{\dfrac{5\cdot 2}{8\cdot 2}} = \sqrt[4]{\dfrac{10}{16}} = \dfrac{\sqrt[4]{10}}{\sqrt[4]{16}} = \dfrac{\sqrt[4]{10}}{2}$

85. $\dfrac{\sqrt[3]{3c^2 v^3}}{\sqrt[3]{15c^2 v^2}} = \sqrt[3]{\dfrac{3c^2 v^3}{15c^2 v^2}} = \sqrt[3]{\dfrac{v}{5}} = \sqrt[3]{\dfrac{v\cdot 25}{5\cdot 25}} = \sqrt[3]{\dfrac{25v}{125}} = \dfrac{\sqrt[3]{25v}}{\sqrt[3]{125}} = \dfrac{\sqrt[3]{25v}}{5}$

87. $\dfrac{x}{\sqrt{4x^{-2}}} = \dfrac{x}{2x^{-1}} = \dfrac{x^{1+1}}{2} = \dfrac{x^2}{2}$

89. $\sqrt{\dfrac{27(a+b)^2}{(a+b)^{-3}}} = \sqrt{27(a+b)^5} = \sqrt{[9(a+b)^4][3(a+b)]}$

$= \sqrt{9(a+b)^4}\sqrt{3(a+b)} = 3(a+b)^2\sqrt{3(a+b)}$

91. $(\sqrt{y} - 3\sqrt{x})^2 = (\sqrt{y})^2 - 2(\sqrt{y})(3\sqrt{x}) + (3\sqrt{x})^2 = y - 6\sqrt{xy} + 9x$

93. $(1+\sqrt{2})^{-2} = \dfrac{1}{(1+\sqrt{2})^2} = \dfrac{1}{1^2 + 2(1)(\sqrt{2}) + (\sqrt{2})^2} = \dfrac{1}{1 + 2\sqrt{2} + 2} = \dfrac{1}{3 + 2\sqrt{2}}$

$= \dfrac{1}{3+2\sqrt{2}} \cdot \dfrac{3-2\sqrt{2}}{3-2\sqrt{2}} = \dfrac{3-2\sqrt{2}}{9 - 4\cdot 2} = \dfrac{3-2\sqrt{2}}{9-8} = 3 - 2\sqrt{2}$

95. $\sqrt[3]{32y^{-4}} - \sqrt[3]{-4y^2} = \sqrt[3]{\dfrac{32}{y^4}} - \sqrt[3]{-4y^2} = \sqrt[3]{\dfrac{32\cdot y^2}{y^4 y^2}} - \sqrt[3]{-4y^2}$

$= \dfrac{\sqrt[3]{32y^2}}{\sqrt[3]{y^6}} - \sqrt[3]{-4y^2} = \dfrac{\sqrt[3]{32y^2}}{y^2} - \sqrt[3]{-4y^2} = \dfrac{\sqrt[3]{(8)(4y^2)}}{y^2} - \sqrt[3]{(-1)(4y^2)}$

$= \dfrac{2\sqrt[3]{4y^2}}{y^2} + \sqrt[3]{4y^2} = \dfrac{2\sqrt[3]{4y^2}}{y^2} + \dfrac{y^2\sqrt[3]{4y^2}}{y^2} = \dfrac{(2+y^2)\sqrt[3]{4y^2}}{y^2}$

97. $\sqrt[3]{-2\sqrt{16x^8}} = \sqrt[3]{-2(4x^4)} = \sqrt[3]{-8x^4} = \sqrt[3]{(-8x^3)(x)} = \sqrt[3]{-8x^3}\sqrt[3]{x} = -2x\sqrt[3]{x}$

99. $\sqrt{x^3\sqrt[3]{x^{-3}}} = \sqrt{x^3(x^{-1})} = \sqrt{x^2} = x$

101. $\dfrac{\sqrt{3}}{6} = \dfrac{\sqrt{3}\cdot\sqrt{3}}{6\cdot\sqrt{3}} = \dfrac{\cancel{3}^{\,1}}{\cancel{6}_{\,2}\sqrt{3}} = \dfrac{1}{2\sqrt{3}}$

103. $\dfrac{1-\sqrt{3}}{2} \cdot \dfrac{1+\sqrt{3}}{1+\sqrt{3}} = \dfrac{1-3}{2(1+\sqrt{3})} = \dfrac{\cancel{-2}^{\,-1}}{\cancel{2}_{\,1}(1+\sqrt{3})} = \dfrac{-1}{1+\sqrt{3}}$

105. $\dfrac{\sqrt{x}-2}{x} \cdot \dfrac{\sqrt{x}+2}{\sqrt{x}+2} = \dfrac{x-4}{x(\sqrt{x}+2)}$

...

107. $\dfrac{\sqrt{a}+\sqrt{b}}{a-b}\cdot\dfrac{\sqrt{a}-\sqrt{b}}{\sqrt{a}-\sqrt{b}}=\dfrac{\overset{1}{\cancel{a-b}}}{\underset{1}{\cancel{(a-b)}}(\sqrt{a}-\sqrt{b})}=\dfrac{1}{\sqrt{a}-\sqrt{b}}$

109. $\sqrt[3]{x^2}\sqrt[4]{x^3}=x^{2/3}x^{3/4}=x^{2/3+3/4}=x^{8/12+9/12}=x^{17/12}=\sqrt[12]{x^{17}}$

$=\sqrt[12]{(x^{12})(x^5)}=\sqrt[12]{x^{12}}\sqrt[12]{x^5}=x\sqrt[12]{x^5}$

111. $\sqrt[4]{293}=\sqrt{\sqrt{293}}\approx\sqrt{17.11724}\approx 4.137$ which rounds to 4.14

113. $\sqrt[8]{5691}=\sqrt{\sqrt{\sqrt{5691}}}\approx\sqrt{\sqrt{75.438717}}\approx\sqrt{8.68555}\approx 2.947$ which rounds to 2.95

115. $\sqrt[16]{5772}=\sqrt{\sqrt{\sqrt{\sqrt{5772}}}}\approx\sqrt{\sqrt{\sqrt{75.973680}}}\approx\sqrt{\sqrt{8.716288}}\approx\sqrt{2.952336}\approx 1.718$
which rounds to 1.72

117. $\dfrac{1}{\sqrt{2}}\approx\dfrac{1}{1.414}\approx 0.707$

$\dfrac{\sqrt{2}}{4}\approx\dfrac{1.414}{2}=0.707$

Using computations by hand, the second division is easier to perform.

Exercises 1.9

1. $2-yi=x+6i$ produces $x=2$ and $y=-6$

3. $x-7i=yi$ produces $x=0$ and $y=-7$

5. $x-yi=4+i$ produces $x=4$ and $y=-1$

7. $\sqrt{-16}=4i$ 9. $-\sqrt{-49}=-7i$ 11. $\sqrt{-25}\sqrt{-9}=(5i)(3i)=15i^2=15(-1)=-15$

13. $\dfrac{\sqrt{-45}}{\sqrt{-5}}=\dfrac{3i\sqrt{5}}{i\sqrt{5}}=3$ 15. $(3+2i)+(6-3i)=(3+6)+(2-3)i=9-i$

17. $(65-3i)-(50-70i)=65-3i-50+70i=(65-50)+(-3+70)i=15+67i$

19. $(2-5i)-(6-3i)=2-5i-6+3i=(2-6)+(-5+3)i=-4-2i$

21. $(2-3i)-(7-6i)=2-3i-7+6i=(2-7)+(-3+6)i=-5+3i$

23. $(3-7i)(2-i)=(3)(2)+(3)(-i)+(-7i)(2)+(-7i)(-i)$
$=6-3i-14i+7i^2=6-17i+7(-1)=-1-17i$

25. $(2+3i)(3-i)=(2)(3)+(2)(-i)+(3i)(3)+(3i)(-i)$
$=6-2i+9i-3i^2=6+7i-3(-1)=9+7i$

27. $(6+4i)(-7+3i)=(6)(-7)+(6)(3i)+(4i)(-7)+(4i)(3i)$
$=-42+18i-28i+12i^2=-42-10i+12(-1)=-54-10i$

29. $(6 - 3i)^2 = 6^2 - 2(6)(3i) + (3i)^2 = 36 - 36i + 9i^2 = 36 - 36i + 9(-1) = 27 - 36i$

31. $i(12 - 4i)(3 + i) = i[(12)(3) + (12)(i) + (-4i)(3) + (-4i)(i)]$
$$= i[36 + 12i - 12i - 4i^2]$$
$$= i[36 + 0i - 4(-1)]$$
$$= i(40)$$
$$= 40i$$

33. $-i(-2 + 3i)(4 - i) = -i[(-2)(4) + (-2)(-i) + (3i)(4) + (3i)(-i)]$
$$= -i[-8 + 2i + 12i - 3i^2] = -i[-8 + 14i - 3(-1)]$$
$$= -i[-5 + 14i] = 5i - 14i^2 = 5i - 14(-1) = 14 + 5i$$

35. $(6 - 7i) \div 3 = \dfrac{6 - 7i}{3} = \dfrac{6}{3} - \dfrac{7i}{3} = 2 - \dfrac{7}{3}i$

37. $\dfrac{2}{i} = \dfrac{(2)(i)}{(i)(i)} = \dfrac{2i}{i^2} = \dfrac{2i}{-1} = -2i$ 39. $\dfrac{4}{3i} = \dfrac{(4)(i)}{(3i)(i)} = \dfrac{4i}{3i^2} = \dfrac{4i}{3(-1)} = -\dfrac{4}{3}i$

41. $\dfrac{6 - i}{-3i} = \dfrac{(6 - i)(i)}{(-3i)(i)} = \dfrac{6i - i^2}{-3i^2} = \dfrac{6i - (-1)}{-3(-1)} = \dfrac{6i + 1}{3} = \dfrac{1}{3} + 2i$

43. $\dfrac{1}{3 + 4i} = \dfrac{1(3 - 4i)}{(3 + 4i)(3 - 4i)} = \dfrac{3 - 4i}{(3)^2 - (4i)^2} = \dfrac{3 - 4i}{9 - 16i^2}$
$$= \dfrac{3 - 4i}{9 - 16(-1)} = \dfrac{3 - 4i}{25} = \dfrac{3}{25} - \dfrac{4}{25}i$$

45. $\dfrac{1}{5 - 12i} = \dfrac{1(5 + 12i)}{(5 - 12i)(5 + 12i)} = \dfrac{5 + 12i}{(5)^2 - (12i)^2} = \dfrac{5 + 12i}{25 - 144i^2} = \dfrac{5 + 12i}{25 - 144(-1)}$
$$= \dfrac{5 + 12i}{169} = \dfrac{5}{169} + \dfrac{12}{169}i$$

47. $\dfrac{6}{3 - i} = \dfrac{6(3 + i)}{(3 - i)(3 + i)} = \dfrac{18 + 6i}{(3)^2 - (i)^2} = \dfrac{18 + 6i}{9 - i^2} = \dfrac{18 + 6i}{9 - (-1)} = \dfrac{18 + 6i}{10}$
$$= \dfrac{18}{10} + \dfrac{6}{10}i = \dfrac{9}{5} + \dfrac{3}{5}i$$

49. $\dfrac{5 - 3i}{6 + i} = \dfrac{(5 - 3i)(6 - i)}{(6 + i)(6 - i)} = \dfrac{30 - 5i - 18i + 3i^2}{(6)^2 - (i)^2} = \dfrac{30 - 23i + 3(-1)}{36 - (-1)}$
$$= \dfrac{27 - 23i}{37} = \dfrac{27}{37} - \dfrac{23}{37}i$$

51. $\dfrac{4 - 7i}{1 - 2i} = \dfrac{(4 - 7i)(1 + 2i)}{(1 - 2i)(1 + 2i)} = \dfrac{4 + 8i - 7i - 14i^2}{(1)^2 - (2i)^2} = \dfrac{4 + i - 14i^2}{1 - 4i^2}$
$$= \dfrac{4 + i - 14(-1)}{1 - 4(-1)} = \dfrac{18 + i}{5} = \dfrac{18}{5} + \dfrac{1}{5}i$$

53. $\dfrac{4 - 3i}{3 + 4i} = \dfrac{(4 - 3i)(3 - 4i)}{(3 + 4i)(3 - 4i)} = \dfrac{12 - 16i - 9i + 12i^2}{(3)^2 - (4i)^2} = \dfrac{12 - 25i + 12i^2}{9 - 16i^2}$
$$= \dfrac{12 - 25i + 12(-1)}{9 - 16(-1)} = \dfrac{-25i}{25} = -i$$

55. $\dfrac{-1 + 4i}{1 - 3i} = \dfrac{(-1 + 4i)(1 + 3i)}{(1 - 3i)(1 + 3i)} = \dfrac{-1 - 3i + 4i + 12i^2}{(1)^2 - (3i)^2} = \dfrac{-1 + i + 12i^2}{1 - 9i^2}$

$= \dfrac{-1 + i + 12(-1)}{1 - 9(-1)} = \dfrac{-1 + i - 12}{10} = \dfrac{-13 + i}{10} = -\dfrac{13}{10} + \dfrac{1}{10}i$

57. $(1 - i)^3 = (1 - i)(1 - i)(1 - i)$

$= (1 - i)[1 - i - i + i^2] = (1 - i)[1 - 2i + (-1)]$

$= (1 - i)[-2i]$

$= -2i + 2i^2 = -2i + 2(-1) = -2 - 2i$

59. $\left[\dfrac{i}{1 - i}\right]^2 = \dfrac{i^2}{(1 - i)^2} = \dfrac{-1}{-2i} = \dfrac{1}{2i} = \dfrac{(1)(i)}{(2i)(i)} = \dfrac{i}{2i^2} = \dfrac{i}{2(-1)} = -\dfrac{1}{2}i$

       ↑

     solution to #57

61. $i^{72} = (i^4)^{18} = 1^{18} = 1$ 63. $i^{29} = i^{28} \cdot i = (i^4)^7 \cdot i = (1)^7 \cdot i = 1 \cdot i = i$

65. $i^{91} = i^{88} \cdot i^3 = (i^4)^{22} \cdot i^3 = (1)^{22} \cdot i^3 = 1 \cdot i^3 = i^3 = -i$

67. $\dfrac{1}{i^{15}} = \dfrac{(1)(i)}{(i^{15})(i)} = \dfrac{i}{i^{16}} = \dfrac{i}{(i^4)^4} = \dfrac{i}{(1)^4} = \dfrac{i}{1} = i$

69. $i^{-62} = \dfrac{1}{i^{62}} = \dfrac{(1)(i^2)}{(i^{62})(i^2)} = \dfrac{i^2}{i^{64}} = \dfrac{i^2}{(i^4)^{16}} = \dfrac{i^2}{(1)^{16}} = \dfrac{i^2}{1} = i^2 = -1$

71. $i^{915} = i^{912} \cdot i^3 = (i^4)^{228} \cdot i^3 = (1)^{228} \cdot i^3 = 1 \cdot i^3 = i^3 = -i$

73. $P = z^2 + 9 = (3i)^2 + 9 = 9i^2 + 9 = 9(-1) + 9 = -9 + 9 = 0$.
Thus, $z = 3i$ is a zero of $P = z^2 + 9$.

75. $P = z^2 - 4z + 5 = (2 - i)^2 - 4(2 - i) + 5$

$= 3 - 4i - 8 + 4i + 5$

$= 0 + 0i$

$= 0$

Thus, $z = 2 - i$ is a zero of $P = z^2 - 4z + 5$.

77. $P = z^3 - 8 = (-1 + \sqrt{3}\,i)^3 - 8$

$= (-1 + \sqrt{3}\,i)(-1 + \sqrt{3}\,i)^2 - 8$

$= (-1 + \sqrt{3}\,i)(-2 - 2\sqrt{3}\,i) - 8$

$= (8 + 0i) - 8$

$= 8 - 8 = 0$

Thus, $z = -1 + \sqrt{3}\,i$ is a zero of $P = z^3 - 8$.

79. Observe that $|a + bi| = \sqrt{a^2 + b^2}$. Now $\sqrt{a^2 + b^2} = 0$ only when both a and b are 0. Thus, $a = 0$ and $b = 0$ so that the complex number $z = a + bi$ becomes

$z = 0 + 0i$

$= 0$

CHAPTER REVIEW

1. Z, Q, R 2. W, Z, Q, R 3. Q, R 4. R, H 5. Q, R 6. N, W, Z, Q, R

7. $c(p + q) = cp + cq$ 8. $ab = ba$ 9. $a(bc) = (ab)c$ 10. $x\left(\dfrac{1}{x}\right) = 1$

11. $ax + ay = ay + ax$ 12. $1 \cdot x + 0 = 1 \cdot x$ 13. Additive identity

14. Multiplicative identity   15. Distributive property

16. Commutative property   17. Associative property

18. Associative property for addition   19. Additive inverse

20. Commutative property for addition

21. $[2x(3 - y)] - (7 - y)(2x - 3y) = [2 \cdot 2(3 - 3)] - (7 - 3)(2 \cdot 2 - 3 \cdot 3)$

$$= [4(0)] - (4)(4 - 9)$$
$$= 0 - 4(-5)$$
$$= 20$$

22. $\dfrac{x + 10}{y} + x^2 - 2y^2 = \dfrac{2 + 10}{3} + 2^2 - 2 \cdot 3^2 = \dfrac{2 + 10}{3} + 4 - 2 \cdot 9$

$$= \dfrac{12}{3} + 4 - 2 \cdot 9$$
$$= 4 + 4 - 18$$
$$= -10$$

23. $\dfrac{[(x - 4)(y + 2) - (3x - y)(7 + y)]}{x + y} = \dfrac{(2 - 4)(3 + 2) - (3 \cdot 2 - 3)(7 + 3)}{2 + 3}$

$$= \dfrac{(2 - 4)(3 + 2) - (6 - 3)(7 + 3)}{2 + 3}$$

$$= \dfrac{(-2)(5) - (3)(10)}{5} = \dfrac{-10 - 30}{5} = -\dfrac{40}{5} = -8$$

24. $\dfrac{(-x)(-y) - y\left[\dfrac{2x + y}{2x - y}\right]}{x - x\left[\dfrac{x + 2y}{y - 2x}\right]} = \dfrac{(-2)(-3) - 3\left[\dfrac{2(2) + 3}{2(2) - 3}\right]}{2 - 2\left[\dfrac{2 + 2(3)}{3 - 2(2)}\right]}$

$$= \dfrac{(-2)(-3) - 3\left[\dfrac{4 + 3}{4 - 3}\right]}{2 - 2\left[\dfrac{2 + 6}{3 - 4}\right]} = \dfrac{(-2)(-3) - 3[7]}{2 - 2[-8]}$$

$$= \dfrac{6 - 21}{2 + 16} = -\dfrac{15}{18} = -\dfrac{5}{6}$$

25. $2[x - 2(x - 3) - 4] = 2[x - 2x + 6 - 4] = 2(-x + 2) = -2x + 4$

26. $3\{2 - [3z - 4(z + 2)] + z\} = 3\{2 - [3z - 4z - 8] + z\}$

$$= 3\{2 - [-z - 8] + z\}$$
$$= 3\{2 + z + 8 + z\} = 3\{2z + 10\} = 6z + 30$$

27. -2 28. $0\ 4$ 29. $-4\ -2\ \ 1$

30. $-2\ \ 2\ \ 5$ 31. $-1 \le x < 4$ 32. $x < 1$ or $x > 5$

33. $x < -1$ or $2 < x < 5$ 34. $-2 \le x < 0$ or $x \ge 3$

35. $|x| \leq 1$ becomes $-1 \leq x \leq 1$ and the graph is: $\longleftarrow\!\!\underset{-1\quad\;1}{\rule{0pt}{0pt}\bullet\!\!-\!\!\bullet}\!\!\longrightarrow$

36. $|x| > 2$ becomes $x < -2$ or $x > 2$ and the graph is: $\longleftarrow\!\!\blacktriangleleft\underset{-2\qquad\;2}{\rule{0pt}{0pt}\circ\qquad\circ}\blacktriangleright\!\!\longrightarrow$

37. $|x - 3| < 4$ becomes $-4 < x - 3 < 4$ or $-1 < x < 7$ and the graph
    is: $\longleftarrow\!\!\underset{-1\qquad\;7}{\rule{0pt}{0pt}\circ\!-\!\!-\!\!\circ}\!\!\longrightarrow$

38. $|x + 2| \geq 3$ becomes $x + 2 \leq -3$ or $x + 2 \geq 3$
    $\qquad\qquad\qquad\qquad x \leq -5$ or $x \geq 1$
    $\qquad\qquad$ and the graph is: $\longleftarrow\!\!\blacktriangleleft\underset{-5\qquad\;1}{\rule{0pt}{0pt}\bullet\qquad\bullet}\blacktriangleright\!\!\longrightarrow$

39. $|5 - 7| = |-2| = -(-2) = 2$ 40. $|\sqrt{35} - 6| = -(\sqrt{35} - 6) = 6 - \sqrt{35}$ since $6 > \sqrt{35}$

41. $|6 - \sqrt{37}| = -(6 - \sqrt{37}) = -6 + \sqrt{37}$ since $\sqrt{37} > 6$ 42. $\left|\dfrac{x}{3}\right| = -\dfrac{x}{3}$

43. $|y + 3| = y + 3$ 44. $|14 - 3x| = 14 - 3x$ when $x < 4$

45. False since $|x| < 3$ means $-3 < x < 3$ which is **not** the same as $x < 3$.

46. True

47. False since if $x = 0$ then $|x - 3|$ becomes $|0 - 3| = |-3| = 3$ whereas
    $|x| - |3|$ becomes $|0| - |3| = 0 - 3 = -3$ and we see that
    $|x - 3| \neq |x| - |3|$. This is the case for $x < 3$.

48. True since $|-3x| > 0$ whereas $-3|x| < 0$ 49. $4^{-2} = \dfrac{1}{4^2} = \dfrac{1}{16}$

50. $(-5x^3)^{-2} = \dfrac{1}{(-5x^3)^2} = \dfrac{1}{25x^6}$ 51. $x^2 y^{-2} = \dfrac{x^2}{y^2}$ 52. $\dfrac{(3x^4)(5x^2)}{6x^5} = \dfrac{15x^6}{6x^5} = \dfrac{5}{2}x$ or $\dfrac{5x}{2}$

53. $(x^2 y)^2 (2xy^3)^2 = (x^2)^2 (y)^2 (2)^2 (x)^2 (y^3)^2 = x^4 y^2 \cdot 4 \cdot x^2 y^6 = 4x^6 y^8$

54. $(a + b)^0 = \begin{cases} 1 \text{ if } a \neq -b \\ \text{undefined if } a = -b \end{cases}$

55. $\left[\dfrac{3a^{-1}b^2}{2ab^{-3}}\right]^{-1} = \dfrac{3^{-1}(a^{-1})^{-1}(b^2)^{-1}}{2^{-1}(a)^{-1}(b^{-3})^{-1}} = \dfrac{3^{-1}(a)(b^{-2})}{2^{-1}(a^{-1})(b^3)} = \dfrac{\dfrac{a}{3b^2}}{\dfrac{b^3}{2a}} = \dfrac{a}{3b^2} \cdot \dfrac{2a}{b^3} = \dfrac{2a^2}{3b^5}$

56. $29{,}300 = 2.93 \times 10^4$ 57. $5{,}790{,}000 = 5.79 \times 10^6$ 58. $0.00016 = 1.6 \times 10^{-4}$

59. $0.0706 = 7.06 \times 10^{-2}$ 60. $4.93 \times 10^5 = 493{,}000$ 61. $8.07 \times 10^{-3} = 0.00807$

62. $7.68 \times 10^{-1} = 0.768$ 63. $1.19 \times 10 = 11.9$

64. $(5x^2 - 3x + 2) + (-7x^3 - 3x^2 + x - 1) = -7x^3 + (5 - 3)x^2 + (-3 + 1)x + (2 - 1)$
    $\qquad\qquad\qquad\qquad\qquad\qquad\qquad\qquad = -7x^3 + 2x^2 - 2x + 1$

65. $(9x^4 - 3x^2 + 7x) - (5x^3 + 4x^2 - 3x - 2) = 9x^4 - 3x^2 + 7x - 5x^3 - 4x^2 + 3x + 2$
    $\qquad\qquad\qquad\qquad\qquad\qquad\qquad\qquad\; = 9x^4 - 5x^3 - 7x^2 + 10x + 2$

66. $(5r - 3)(7r^2 - 3) = (5r - 3)(7r^2) + (5r - 3)(-3)$
$$= (5r)(7r^2) + (-3)(7r^2) + (5r)(-3) + (-3)(-3)$$
$$= 35r^3 - 21r^2 - 15r + 9$$

67. $(6x^2 - 3x + 1)(5x^3 - 3x) = (6x^2 - 3x + 1)(5x^3) + (6x^2 - 3x + 1)(-3x)$
$$= 30x^5 - 15x^4 + 5x^3 - 18x^3 + 9x^2 - 3x$$
$$= 30x^5 - 15x^4 - 13x^3 + 9x^2 - 3x$$

68. $(2p - 3q)(2p + 3q) = (2p)^2 - (3q)^2 = 4p^2 - 9q^2$

69. $(3x - y)^2 = (3x)^2 - 2(3x)(y) + (y)^2 = 9x^2 - 6xy + y^2$

70. $(u - v)(u^2 + uv + v^2) = u^3 - v^3$

71. $(4x - 2y)^3 = (4x)^3 - 3(4x)^2(2y) + 3(4x)(2y)^2 - (2y)^3 = 64x^3 - 96x^2y + 48xy^2 - 8y^3$

72. $(2x - y)(2x + y)^2 = (2x - y)(2x + y)(2x + y)$
$$= [(2x)^2 - (y)^2](2x + y)$$
$$= (4x^2 - y^2)(2x + y)$$
$$= (4x^2 - y^2)(2x) + (4x^2 - y^2)(y)$$
$$= 8x^3 - 2xy^2 + 4x^2y - y^3$$

73. $(3a + b)(9a^2 - 3ab + b^2) = (3a)^3 + (b)^3 = 27a^3 + b^3$

74. $a^2 - 16b^2 = a^2 - (4b)^2 = (a + 4b)(a - 4b)$

75. $49x^2 - 14x + 1 = (7x)^2 - 2(7x)(1) + 1^2 = (7x - 1)^2$

76. $4x^2 - 8xy + 4y^2 = 4(x^2 - 2xy + y^2) = 4(x - y)^2$

77. $8pq^2 + 2p^2q - 4pq = 2pq(4q + p - 2)$ 78. $m^2 - 3mn + 2n^2 = (m - n)(m - 2n)$

79. $8x^3 - 27 = (2x)^3 - (3)^3 = (2x - 3)[(2x)^2 + (2x)(3) + (3)^2]$
$$= (2x - 3)(4x^2 + 6x + 9)$$

80. $a^4 + 64ab^3 = a(a^3 + 64b^3) = a[a^3 + (4b)^3] = a(a + 4b)(a^2 - 4ab + 16b^2)$

81. $6t^3 - 14t^2 - 12t = 2t(3t^2 - 7t - 6) = 2t(3t + 2)(t - 3)$

82. $z^3 - 4z^2 - 9z + 36 = z^2(z - 4) - 9(z - 4)$
$$= (z - 4)(z^2 - 9)$$
$$= (z - 4)(z + 3)(z - 3)$$

83. $4x^2 - y^2 + 4yz - 4z^2 = 4x^2 - (y^2 - 4yz + 4z^2)$
$$= 4x^2 - (y - 2z)^2$$
$$= (2x)^2 - (y - 2z)^2$$
$$= [2x + (y - 2z)][2x - (y - 2z)]$$
$$= (2x + y - 2z)(2x - y + 2z)$$

84. $(9x^2 - 16y^2) \cdot \left[\dfrac{4x}{3x - 4y} \right] = \dfrac{(3x + 4y)\overset{1}{\cancel{(3x - 4y)}}(4x)}{\underset{1}{\cancel{3x - 4y}}} = 4x(3x + 4y)$

85. $\dfrac{x^2 - y^2}{4x^2y} \cdot \dfrac{20xy^3}{(x - y)^2} = \dfrac{\overset{1}{\cancel{(x - y)}}(x + y)(\overset{5y^2}{\cancel{20xy^3}})}{\underset{x}{4x^2y}\,\underset{1}{\cancel{(x - y)}}(x - y)} = \dfrac{5y^2(x + y)}{x(x - y)}$

86. $\dfrac{2x + 4}{x^2 - 4x + 4} \cdot \dfrac{x^2 - 4}{x + 2} = \dfrac{2(x + 2)\overset{1}{\cancel{(x + 2)}}\,\overset{1}{\cancel{(x - 2)}}}{(x - 2)\underset{1}{\cancel{(x - 2)}}\,\underset{1}{\cancel{(x + 2)}}} = \dfrac{2(x + 2)}{x - 2}$

87. $\dfrac{16 - 9x^2}{x^3 - 8} \cdot \dfrac{x^3 + 2x^2 + 4x}{4 + x - 3x^2} = \dfrac{\overset{1}{\cancel{(4 - 3x)}}(4 + 3x)(x)(\overset{1}{\cancel{x^2 + 2x + 4}})}{(x - 2)\underset{1}{\cancel{(x^2 + 2x + 4)}}\,\underset{1}{\cancel{(4 - 3x)}}(1 + x)}$

$= \dfrac{x(3x + 4)}{(x - 2)(x + 1)}$

88. $\dfrac{w^2 - 5w + 6}{w^2 - 7w + 10} \cdot \dfrac{w^2 + 6w + 9}{w^2 - 9} = \dfrac{\overset{1}{\cancel{(w - 3)}}\,\overset{1}{\cancel{(w - 2)}}\,\overset{1}{\cancel{(w + 3)}}(w + 3)}{(w - 5)\underset{1}{\cancel{(w - 2)}}\,\underset{1}{\cancel{(w - 3)}}\,\underset{1}{\cancel{(w + 3)}}} = \dfrac{w + 3}{w - 5}$

89. $(5x^2 - 20) \div \left(1 + \dfrac{x}{2}\right)$

$(5x^2 - 20) \div \left(\dfrac{2 + x}{2}\right) = (5x^2 - 20) \cdot \dfrac{2}{2 + x} = \dfrac{5\overset{1}{\cancel{(x + 2)}}(x - 2)(2)}{\underset{1}{\cancel{2 + x}}} = 10(x - 2)$

90. $\dfrac{(x - y)^2}{2x^2 + 3xy + y^2} \div \dfrac{x^2 - y^2}{x^2 + 2xy + y^2} = \dfrac{(x - y)^2}{2x^2 + 3xy + y^2} \cdot \dfrac{x^2 + 2xy + y^2}{x^2 - y^2}$

$= \dfrac{(x - y)\overset{1}{\cancel{(x - y)}}\,\overset{1}{\cancel{(x + y)}}\,\overset{1}{\cancel{(x + y)}}}{(2x + y)\underset{1}{\cancel{(x + y)}}\,\underset{1}{\cancel{(x + y)}}\,\underset{1}{\cancel{(x - y)}}} = \dfrac{x - y}{2x + y}$

91. $\dfrac{2a - 3}{a - 3} - \dfrac{2a^2}{a^2 - 9} = \dfrac{(2a - 3)(a + 3)}{(a - 3)(a + 3)} - \dfrac{2a^2}{(a - 3)(a + 3)} = \dfrac{(2a - 3)(a + 3) - 2a^2}{(a - 3)(a + 3)}$

$= \dfrac{2a^2 + 3a - 9 - 2a^2}{(a - 3)(a + 3)} = \dfrac{3a - 9}{(a - 3)(a + 3)}$

$= \dfrac{3\overset{1}{\cancel{(a - 3)}}}{\underset{1}{\cancel{(a - 3)}}(a + 3)} = \dfrac{3}{a + 3}$

92. $\dfrac{y}{6x^2} - \dfrac{x}{2xy} = \dfrac{y(y)}{6x^2(y)} - \dfrac{x(3x)}{2xy(3x)} = \dfrac{y^2}{6x^2y} - \dfrac{3x^2}{6x^2y} = \dfrac{y^2 - 3x^2}{6x^2y}$

93. $\dfrac{3}{2x - 2y} - \dfrac{4y}{x^2 - y^2} = \dfrac{3}{2(x - y)} - \dfrac{4y}{(x - y)(x + y)}$

$= \dfrac{3(x + y)}{2(x - y)(x + y)} - \dfrac{4y(2)}{(x - y)(x + y)(2)} = \dfrac{3(x + y) - 8y}{2(x - y)(x + y)}$

$= \dfrac{3x + 3y - 8y}{2(x - y)(x + y)} = \dfrac{3x - 5y}{2(x - y)(x + y)}$

94. $\dfrac{1}{2x - 2} - \dfrac{x}{x^2 - 4x + 3} = \dfrac{1}{2(x - 1)} - \dfrac{x}{(x - 1)(x - 3)}$

$$= \dfrac{1(x - 3)}{2(x - 1)(x - 3)} - \dfrac{x(2)}{(x - 1)(x - 3)2}$$

$$= \dfrac{x - 3 - 2x}{2(x - 1)(x - 3)} = \dfrac{-x - 3}{2(x - 1)(x - 3)} \text{ or } -\dfrac{x + 3}{2(x - 1)(x - 3)}$$

95. $\dfrac{2}{x - 2} + \dfrac{3}{x + 4} + \dfrac{18}{x^2 + 2x - 8} = \dfrac{2}{x - 2} + \dfrac{3}{x + 4} + \dfrac{18}{(x - 2)(x + 4)}$

$$= \dfrac{2(x + 4)}{(x - 2)(x + 4)} + \dfrac{3(x - 2)}{(x + 4)(x - 2)} + \dfrac{18}{(x - 2)(x + 4)}$$

$$= \dfrac{2(x + 4) + 3(x - 2) + 18}{(x - 2)(x + 4)} = \dfrac{2x + 8 + 3x - 6 + 18}{(x - 2)(x + 4)}$$

$$= \dfrac{5x + 20}{(x - 2)(x + 4)} = \dfrac{5\cancel{(x + 4)}^{1}}{(x - 2)\cancel{(x + 4)}_{1}} = \dfrac{5}{x - 2}$$

96. $\dfrac{3x + 2}{2x^2 - x - 3} + \dfrac{4 + 3x}{5x^2 + 3x - 2} = \dfrac{3x + 2}{(2x - 3)(x + 1)} + \dfrac{4 + 3x}{(5x - 2)(x + 1)}$

$$= \dfrac{(3x + 2)(5x - 2)}{(2x - 3)(x + 1)(5x - 2)} + \dfrac{(4 + 3x)(2x - 3)}{(5x - 2)(x + 1)(2x - 3)}$$

$$= \dfrac{(3x + 2)(5x - 2) + (4 + 3x)(2x - 3)}{(5x - 2)(x + 1)(2x - 3)}$$

$$= \dfrac{15x^2 + 10x - 6x - 4 + 8x + 6x^2 - 12 - 9x}{(5x - 2)(x + 1)(2x - 3)} = \dfrac{21x^2 + 3x - 16}{(5x - 2)(x + 1)(2x - 3)}$$

97. $(25x^2 - 9y^2) \div \left[1 + \dfrac{3y}{5x}\right] = (25x^2 - 9y^2) \div \left[\dfrac{5x + 3y}{5x}\right]$

$$= (25x^2 - 9y^2) \cdot \dfrac{5x}{5x + 3y} = \dfrac{(5x - 3y)\cancel{(5x + 3y)}^{1}(5x)}{\cancel{5x + 3y}_{1}}$$

$$= 5x(5x - 3y) \text{ or } 25x^2 - 15xy$$

98. $\dfrac{1 - \dfrac{2}{x}}{x - \dfrac{4}{x}} = \dfrac{\left(1 - \dfrac{2}{x}\right)x}{\left(x - \dfrac{4}{x}\right)x} = \dfrac{x - 2}{x^2 - 4} = \dfrac{\cancel{x - 2}^{1}}{(x + 2)\cancel{(x - 2)}_{1}} = \dfrac{1}{x + 2}$

99. $\dfrac{4x^2y}{\dfrac{x - y}{x - 2y} - 1} = \dfrac{4x^2y}{\dfrac{x - y}{x - 2y} - 1} \cdot \dfrac{x - 2y}{x - 2y} = \dfrac{4x^2y(x - 2y)}{x - y - (x - 2y)}$

$$= \dfrac{4x^2\cancel{y}^{1}(x - 2y)}{\cancel{y}_{1}} = 4x^2(x - 2y) \text{ or } 4x^3 - 8x^2y$$

100.
$$\frac{9x^2 - 16y^2}{\frac{x - 2y}{2y - 2x} - 1} = (9x^2 - 16y^2) \div \left[\frac{x - 2y}{2y - 2x} - 1\right]$$

$$= (9x^2 - 16y^2) \div \left[\frac{x - 2y - 1(2y - 2x)}{2y - 2x}\right]$$

$$= (9x^2 - 16y^2) \div \frac{3x - 4y}{2y - 2x} = (9x^2 - 16y^2) \cdot \frac{2y - 2x}{3x - 4y}$$

$$= \frac{(3x + 4y)(\cancel{3x - 4y})(2)(y - x)}{\cancel{3x - 4y}} = 2(y - x)(3x + 4y)$$

101.
$$\left[\frac{x^{-1} - y^{-1}}{x^{-2} - y^{-2}}\right]^{-1} = \left[\frac{\frac{1}{x} - \frac{1}{y}}{\frac{1}{x^2} - \frac{1}{y^2}}\right]^{-1} = \left[\frac{\left(\frac{1}{x} - \frac{1}{y}\right)}{\left(\frac{1}{x^2} - \frac{1}{y^2}\right)} \cdot \frac{x^2y^2}{x^2y^2}\right]^{-1} = \left[\frac{xy^2 - x^2y}{y^2 - x^2}\right]^{-1}$$

$$= \left[\frac{xy\cancel{(y - x)}}{\cancel{(y - x)}(y + x)}\right]^{-1} = \left[\frac{xy}{y + x}\right]^{-1} = \frac{1}{\frac{xy}{y + x}} = \frac{y + x}{xy} \text{ or } \frac{x + y}{xy}$$

102. $\sqrt{\frac{121}{36}} = \sqrt{\left(\frac{11}{6}\right)^2} = \frac{11}{6}$

103. $(\sqrt[4]{79})^4 = 79^{4/4} = 79^1 = 79$

104. $\sqrt[3]{\frac{-64}{125}} = \sqrt[3]{\left(-\frac{4}{5}\right)^3} = -\frac{4}{5}$

105. $\sqrt[3]{-27x^3y^6} = \sqrt[3]{(-3xy^2)^3} = -3xy^2$

106. $\sqrt{\sqrt[3]{64}} = \sqrt{4} = 2$

107. $\sqrt{\sqrt[3]{729}} = \sqrt{9} = 3$

108. $\sqrt[3]{-27x^{-3}y^6} = \sqrt[3]{\frac{-27y^6}{x^3}} = -\frac{3y^2}{x}$

109. $\sqrt{\sqrt[3]{64x^6y^{12}}} = \sqrt{4x^2y^4} = 2xy^2$

110. $(\sqrt[3]{5ab^3})^6 = (5ab^3)^{6/3} = (5ab^3)^2 = 25a^2b^6$

111. $\sqrt[5]{\left[-\frac{32x^5}{y^{10}}\right]^{-2}} = \sqrt[5]{\frac{y^{20}}{1024x^{10}}} = \frac{y^4}{4x^2}$ or $\sqrt[5]{\left[-\frac{32x^5}{y^{10}}\right]^{-2}} = \left(\sqrt[5]{-\frac{32x^5}{y^{10}}}\right)^{-2} = \left(-\frac{2x}{y^2}\right)^{-2} = \frac{y^4}{4x^2}$

112. $\sqrt[7]{x^3} = x^{3/7}$

113. $(5ab^2)^{2/5} = \sqrt[5]{(5ab^2)^2}$ or $(\sqrt[5]{5ab^2})^2$

114. $(-32)^{3/5} = (\sqrt[5]{-32})^3 = (-2)^3 = -8$

115. $\left(\frac{1}{125}\right)^{-2/3} = \frac{1}{\left(\frac{1}{125}\right)^{2/3}} = (125)^{2/3} = (\sqrt[3]{125})^2 = 5^2 = 25$

116. $(36)^{-3/2} = \dfrac{1}{(36)^{3/2}} = \dfrac{1}{(\sqrt{36})^3} = \dfrac{1}{6^3} = \dfrac{1}{216}$ 117. $(-32a^5)^{1/5} = \sqrt[5]{-32a^5} = -2a$

118. $(0.04a^4)^{1/2} = \sqrt{0.04a^4} = 0.2a^2$

119. $(-125a^6)^{-2/3} = \dfrac{1}{(-125a^6)^{2/3}} = \dfrac{1}{(\sqrt[3]{-125a^6})^2} = \dfrac{1}{(-5a^2)^2} = \dfrac{1}{25a^4}$

120. $\sqrt{27a^3} = \sqrt{9a^2 \cdot 3a} = \sqrt{9a^2}\sqrt{3a} = 3a\sqrt{3a}$

121. $\sqrt[6]{64x^6y^5} = \sqrt[6]{64x^6 \cdot y^5} = \sqrt[6]{64x^6}\sqrt[6]{y^5} = 2x\sqrt[6]{y^5}$

122. $\sqrt[4]{128a^5b^4} = \sqrt[4]{16a^4b^4 \cdot 8a} = \sqrt[4]{16a^4b^4}\sqrt[4]{8a} = 2ab\sqrt[4]{8a}$

123. $\sqrt[5]{-x^7y^4} = \sqrt[5]{(-x^5)(x^2y^4)} = \sqrt[5]{-x^5}\sqrt[5]{x^2y^4} = -x\sqrt[5]{x^2y^4}$

124. $\sqrt[3]{81a^4b^5} = \sqrt[3]{27a^3b^3 \cdot 3ab^2} = \sqrt[3]{27a^3b^3}\sqrt[3]{3ab^2} = 3ab\sqrt[3]{3ab^2}$

125. $\sqrt[3]{\dfrac{216x^6}{64y^3}} = \sqrt[3]{\dfrac{27x^6}{8y^3}} = \dfrac{\sqrt[3]{27x^6}}{\sqrt[3]{8y^3}} = \dfrac{3x^2}{2y}$

126. $\sqrt{8} + \sqrt{32} = \sqrt{4 \cdot 2} + \sqrt{16 \cdot 2}$

$= \sqrt{4}\sqrt{2} + \sqrt{16}\sqrt{2}$

$= 2\sqrt{2} + 4\sqrt{2}$

$= 6\sqrt{2}$

127. $3\sqrt{18} - 5\sqrt{50} = 3\sqrt{9 \cdot 2} - 5\sqrt{25 \cdot 2} = 3 \cdot 3\sqrt{2} - 5 \cdot 5\sqrt{2}$

$= 9\sqrt{2} - 25\sqrt{2}$

$= -16\sqrt{2}$

128. $\sqrt{8a^2b} - \sqrt{32a^4b^3} = \sqrt{4a^2 \cdot 2b} - \sqrt{16a^4b^2 \cdot 2b}$

$= \sqrt{4a^2}\sqrt{2b} - \sqrt{16a^4b^2}\sqrt{2b}$

$= 2a\sqrt{2b} - 4a^2b\sqrt{2b}$

$= (2a - 4a^2b)\sqrt{2b}$

$= 2a(1 - 2ab)\sqrt{2b}$

129. $\sqrt[3]{8ab^4} + \sqrt[3]{64a^4b} = \sqrt[3]{8b^3 \cdot ab} + \sqrt[3]{64a^3 \cdot ab} = \sqrt[3]{8b^3}\sqrt[3]{ab} + \sqrt[3]{64a^3}\sqrt[3]{ab} = 2b\sqrt[3]{ab} + 4a\sqrt[3]{ab}$

$= (2b + 4a)\sqrt[3]{ab}$ or $2(b + 2a)\sqrt[3]{ab}$

130. $\sqrt[3]{4}\sqrt[3]{16a^4} = \sqrt[3]{64a^4} = \sqrt[3]{64a^3 \cdot a} = \sqrt[3]{64a^3}\sqrt[3]{a} = 4a\sqrt[3]{a}$

131. $\sqrt[3]{9} \cdot \sqrt[3]{9} = \sqrt[3]{9 \cdot 9} = \sqrt[3]{81} = \sqrt[3]{27 \cdot 3} = \sqrt[3]{27}\sqrt[3]{3} = 3\sqrt[3]{3}$

132. $\sqrt[4]{16a^2}\sqrt[4]{4a^2b} = \sqrt[4]{64a^4b} = \sqrt[4]{16a^4 \cdot 4b} = \sqrt[4]{16a^4}\sqrt[4]{4b} = 2a\sqrt[4]{4b}$

133. $\dfrac{\sqrt[3]{40}}{\sqrt[3]{5a^3}} = \sqrt[3]{\dfrac{40}{5a^3}} = \sqrt[3]{\dfrac{8}{a^3}} = \dfrac{\sqrt[3]{8}}{\sqrt[3]{a^3}} = \dfrac{2}{a}$

134. $(3 - 2\sqrt{5})(2 + 4\sqrt{5}) = (3 - 2\sqrt{5})(2) + (3 - 2\sqrt{5})(4\sqrt{5})$

$\qquad\qquad\qquad\qquad\quad = 6 - 4\sqrt{5} + 12\sqrt{5} - 40$

$\qquad\qquad\qquad\qquad\quad = -34 + 8\sqrt{5}$

135. $(\sqrt{5} - \sqrt{3})(\sqrt{5} - 2\sqrt{3}) = (\sqrt{5} - \sqrt{3})(\sqrt{5}) + (\sqrt{5} - \sqrt{3})(-2\sqrt{3})$

$\qquad\qquad\qquad\qquad\qquad = \sqrt{25} - \sqrt{15} - 2\sqrt{15} + 2\sqrt{9}$

$\qquad\qquad\qquad\qquad\qquad = 5 - \sqrt{15} - 2\sqrt{15} + 6$

$\qquad\qquad\qquad\qquad\qquad = 11 - 3\sqrt{15}$

136. $\sqrt[12]{125} = \sqrt[12]{5^3} = 5^{3/12} = 5^{1/4}$ or $\sqrt[4]{5}$

137. $\sqrt[9]{125x^{12}} = \sqrt[9]{x^9(125x^3)} = x\sqrt[9]{125x^3} = x\sqrt[9]{(5x)^3} = x(5x)^{3/9} = x(5x)^{1/3}$ or $x\sqrt[3]{5x}$

138. $\sqrt[12]{4x^4y^6} = \sqrt[12]{(2x^2y^3)^2} = (2x^2y^3)^{2/12} = (2x^2y^3)^{1/6}$ or $\sqrt[6]{2x^2y^3}$

139. $\sqrt[15]{27x^6y^{12}} = \sqrt[15]{(3x^2y^4)^3} = (3x^2y^4)^{3/15} = (3x^2y^4)^{1/5}$ or $\sqrt[5]{3x^2y^4}$

140. $\sqrt{\dfrac{3}{5}} = \sqrt{\dfrac{3 \times 5}{5 \times 5}} = \sqrt{\dfrac{15}{25}} = \dfrac{\sqrt{15}}{\sqrt{25}} = \dfrac{\sqrt{15}}{5}$ 141. $\sqrt[3]{\dfrac{3}{4}} = \sqrt[3]{\dfrac{3 \cdot 2}{4 \cdot 2}} = \sqrt[3]{\dfrac{6}{8}} = \dfrac{\sqrt[3]{6}}{\sqrt[3]{8}} = \dfrac{\sqrt[3]{6}}{2}$

142. $\dfrac{\sqrt{3} - 2}{2\sqrt{3} + 1} \cdot \dfrac{2\sqrt{3} - 1}{2\sqrt{3} - 1} = \dfrac{(\sqrt{3} - 2)(2\sqrt{3} - 1)}{(2\sqrt{3})^2 - 1^2} = \dfrac{6 - 4\sqrt{3} - \sqrt{3} + 2}{12 - 1} = \dfrac{8 - 5\sqrt{3}}{11}$

143. $\sqrt{27x^4y^{-2}} = \sqrt{\dfrac{27x^4}{y^2}} = \dfrac{\sqrt{27x^4}}{\sqrt{y^2}} = \dfrac{\sqrt{9x^4 \cdot 3}}{\sqrt{y^2}} = \dfrac{3x^2\sqrt{3}}{y}$

144. $\dfrac{\sqrt[3]{2a^2b^4}}{\sqrt[3]{16ab^5}} = \sqrt[3]{\dfrac{2a^2b^4}{16ab^5}} = \sqrt[3]{\dfrac{a}{8b}} = \sqrt[3]{\dfrac{a \times b^2}{8b \times b^2}} = \sqrt[3]{\dfrac{ab^2}{8b^3}} = \dfrac{\sqrt[3]{ab^2}}{\sqrt[3]{8b^3}} = \dfrac{\sqrt[3]{ab^2}}{2b}$

145. $\sqrt[3]{40x^2y^{-1}} = \sqrt[3]{\dfrac{40x^2}{y}} = \sqrt[3]{\dfrac{40x^2 \cdot y^2}{y \cdot y^2}} = \sqrt[3]{\dfrac{40x^2y^2}{y^3}} = \dfrac{\sqrt[3]{40x^2y^2}}{\sqrt[3]{y^3}} = \dfrac{\sqrt[3]{8(5x^2y^2)}}{\sqrt[3]{y^3}} = \dfrac{2\sqrt[3]{5x^2y^2}}{y}$

146. $\dfrac{3 + \sqrt{y}}{2 - \sqrt{y}} \cdot \dfrac{2 + \sqrt{y}}{2 + \sqrt{y}} = \dfrac{(3 + \sqrt{y})(2 + \sqrt{y})}{2^2 - (\sqrt{y})^2} = \dfrac{6 + 2\sqrt{y} + 3\sqrt{y} + y}{4 - y} = \dfrac{6 + 5\sqrt{y} + y}{4 - y}$

147. $\sqrt{\dfrac{2}{x - 2}} = \sqrt{\dfrac{2(x - 2)}{(x - 2)(x - 2)}} = \dfrac{\sqrt{2(x - 2)}}{\sqrt{(x - 2)^2}} = \dfrac{\sqrt{2(x - 2)}}{x - 2}$

148. $\sqrt[3]{9\sqrt{36}} = \sqrt[3]{9 \cdot 6} = \sqrt[3]{54} = \sqrt[3]{27 \cdot 2} = \sqrt[3]{27}\sqrt[3]{2} = 3\sqrt[3]{2}$

149. $\sqrt[5]{x^2 y^3}\sqrt{xy} = x^{2/5}y^{3/5} \cdot x^{1/2}y^{1/2} = x^{2/5+1/2}y^{3/5+1/2} = x^{9/10}y^{11/10} = \sqrt[10]{x^9 y^{11}}$

$= \sqrt[10]{y^{10}(x^9 y)} = \sqrt[10]{y^{10}}\sqrt[10]{x^9 y} = y\sqrt[10]{x^9 y}$

150. $(8 + 3i) - (5 - 7i) = 8 + 3i - 5 + 7i = 3 + 10i$

151. $\sqrt{-36}\sqrt{-25} = (6i)(5i) = 30i^2 = 30(-1) = -30$

152. $(2 - i)(3 + 4i) = (2 - i)(3) + (2 - i)(4i)$

$= 6 - 3i + 8i - 4i^2$

$= 6 + 5i - 4i^2$

$= 6 + 5i - 4(-1)$

$= 10 + 5i$

153. $\dfrac{3 - i}{2 + 5i} = \dfrac{(3 - i)(2 - 5i)}{(2 + 5i)(2 - 5i)} = \dfrac{6 - 2i - 15i + 5i^2}{2^2 + 5^2} = \dfrac{6 - 17i + 5i^2}{29}$

$= \dfrac{6 - 17i + 5(-1)}{29} = \dfrac{1 - 17i}{29} = \dfrac{1}{29} - \dfrac{17}{29}i$

154. $i^{-5} = \dfrac{1}{i^5} = \dfrac{1}{i^5} \cdot \dfrac{i}{i} = \dfrac{i}{i^6} = \dfrac{i}{i^4 \cdot i^2} = \dfrac{i}{(1)(-1)} = \dfrac{i}{-1} = -i$

CHAPTER 2 EQUATIONS AND INEQUALITIES IN ONE VARIABLE

Exercises 2.1

1. $9x + 5 = 32$
$$9x = 27$$
$$x = \frac{27}{9} = 3$$

3. $5y + 3(1 - y) = 3y - 3$
$$5y + 3 - 3y = 3y - 3$$
$$2y + 3 = 3y - 3$$
$$-y + 3 = -3$$
$$-y = -6$$
$$y = \frac{-6}{-1} = 6$$

5. $\frac{10}{3} - \frac{1}{2}z = \frac{3}{4}z$
$$\frac{10}{3}(12) - \frac{1}{2}z(12) = \frac{3}{4}z(12)$$
$$40 - 6z = 9z$$
$$40 = 15z$$
$$\frac{40}{15} = z$$
$$\text{or } \frac{8}{3} = z$$

7. $3.6 - 0.2x = -1.4x$
$$3.6 = -1.2x$$
$$\frac{3.6}{-1.2} = x$$
$$\text{or } -3 = x$$

9. $-1.09 + 1.21x = 0.13x + 1.07$
$$-1.09 + 1.08x = 1.07$$
$$1.08x = 2.16$$
$$x = \frac{2.16}{1.08} = 2$$

11. $6w - [5w - 3(w - 4)] + 1 = 2(1 - w) + 5$
$$6w - [5w - 3w + 12] + 1 = 2 - 2w + 5$$
$$6w - [2w + 12] + 1 = -2w + 7$$
$$6w - 2w - 12 + 1 = -2w + 7$$
$$4w - 11 = -2w + 7$$
$$6w - 11 = 7$$
$$6w = 18$$
$$w = \frac{18}{6} = 3$$

13. $(5w - 1)(4w + 2) = (10w - 1)(2w + 3) - 2(w + 13)$
$$20w^2 + 6w - 2 = 20w^2 + 28w - 3 - 2w - 26$$
$$6w - 2 = 26w - 29$$
$$-20w - 2 = -29$$
$$-20w = -27$$
$$w = \frac{-27}{-20} = \frac{27}{20}$$

15. $\frac{t - 2}{2} = \frac{3t - 1}{4}$ Cross multiply to get $4(t - 2) = 2(3t - 1)$
$$4t - 8 = 6t - 2$$
$$-2t - 8 = -2$$
$$-2t = 6$$
$$t = \frac{6}{-2} = -3$$

17. $\frac{7x - 4}{3} = x - \frac{2}{5}$ Multiply each term by 15 to get $\frac{7x - 4}{3}(15) = x(15) - \frac{2}{5}(15)$
$$5(7x - 4) = 15x - 6$$
$$35x - 20 = 15x - 6$$
$$20x - 20 = -6$$
$$20x = 14$$
$$x = \frac{14}{20} = \frac{7}{10}$$

19. $\dfrac{1}{2 - t} = \dfrac{-3}{2 + t}$ Cross multiply: $1(2 + t) = -3(2 - t)$

$$2 + t = -6 + 3t$$
$$2 - 2t = -6$$
$$-2t = -8$$
$$t = \dfrac{-8}{-2} = 4$$

21. $\dfrac{1}{r + 1} = \dfrac{-r}{r + 1}$ Since the denominators are the same, multiply each side by $r + 1$.

$$\dfrac{1}{\cancel{r + 1}}\overset{1}{\cancel{(r + 1)}} = \dfrac{-r}{\cancel{r + 1}}\overset{1}{\cancel{(r + 1)}}$$

$1 = -r$ and this value does not check due to division by 0.

23. $\dfrac{-3}{p - 2} + 1 = \dfrac{2p}{3(p - 2)}$ Multiply each term by $3(p - 2)$ to obtain

$$\dfrac{-3}{\cancel{p - 2}}[3\overset{1}{\cancel{(p - 2)}}] + 1[3(p - 2)] = \dfrac{2p}{3\cancel{(p - 2)}}[3\overset{1}{\cancel{(p - 2)}}]$$

$$-9 + 3(p - 2) = 2p$$
$$-9 + 3p - 6 = 2p$$
$$-15 + 3p = 2p$$
$$-15 = -p$$
$$\dfrac{-15}{-1} = p$$
$$\text{or } 15 = p$$

25. $\dfrac{y + 2}{y + 3} - 1 = \dfrac{3y + 8}{y + 3}$ Multiply each term by $y + 3$ to get

$$\dfrac{y + 2}{\cancel{y + 3}}\overset{1}{\cancel{(y + 3)}} - 1(y + 3) = \dfrac{3y + 8}{\cancel{y + 3}}\overset{1}{\cancel{(y + 3)}}$$

$$y + 2 - 1(y + 3) = 3y + 8$$
$$y + 2 - y - 3 = 3y + 8$$
$$-1 = 3y + 8$$
$$-9 = 3y$$

$$-\dfrac{9}{3} = y \text{ or } y = -3 \text{ and this value does not check}$$
$$\text{due to division by 0.}$$

27. $\dfrac{1}{1 - z} + \dfrac{3}{z} = \dfrac{-1}{z}$

$$\dfrac{1}{1 - z} = \dfrac{-3}{z} - \dfrac{1}{z}$$

$$\dfrac{1}{1 - z} = \dfrac{-4}{z}$$ Now cross multiply: $1(z) = -4(1 - z)$

$$z = -4 + 4z$$
$$-3z = -4$$
$$z = \dfrac{-4}{-3} = \dfrac{4}{3}$$

29. $\dfrac{-1}{2x + 3} - \dfrac{2}{x - 2} = \dfrac{3x + 4}{(2x + 3)(x - 2)}$ Multiply each term by $(2x + 3)(x - 2)$ to get

$\dfrac{-1}{2x + 3} [(2x + 3)(x - 2)] - \dfrac{2}{x - 2} [(2x + 3)(x - 2)]$

$= \dfrac{3x + 4}{(2x + 3)(x - 2)} [(2x + 3)(x - 2)]$

$-1(x - 2) - 2(2x + 3) = 3x + 4$
$-x + 2 - 4x - 6 = 3x + 4$
$-5x - 4 = 3x + 4$
$-8x - 4 = 4$
$-8x = 8$
$x = \dfrac{8}{-8} = -1$

31. $\dfrac{3}{p + 4} + \dfrac{4}{p + 3} = \dfrac{4}{(p + 4)(p + 3)}$ Multiply each term by $(p + 4)(p + 3)$ to get

$\dfrac{3}{p + 4} [(p + 4)(p + 3)] + \dfrac{4}{p + 3} [(p + 4)(p + 3)]$

$= \dfrac{4}{(p + 4)(p + 3)} [(p + 4)(p + 3)]$

$3(p + 3) + 4(p + 4) = 4$
$3p + 9 + 4p + 16 = 4$
$7p + 25 = 4$
$7p = -21$

$p = \dfrac{-21}{7} = -3$ and this value
does not check

33. $ax + by = c$
$ax = c - by$
$x = \dfrac{c - by}{a}$

35. $A = P + Prt$
$A - P = Prt$
$\dfrac{A - P}{Pt} = r$

37. $A = \dfrac{h}{2}(a + b)$
$2A = h(a + b)$
$2A = ha + hb$
$2A - ha = hb$
$\dfrac{2A - ha}{h} = b$

39. $V = \dfrac{1}{6}h(b + 4M + B)$
$6V = h(b + 4M + B)$
$6V = hb + 4hM + hB$
$6V - hb - 4hM = hB$
$\dfrac{6V - hb - 4hM}{h} = B$

41. $p = \dfrac{S}{S + F}$
$p(S + F) = S$
$pS + pF = S$
$pF = S - pS$
$pF = S(1 - p)$
$\dfrac{pF}{1 - p} = S$

43. $S = \dfrac{a - rL}{1 - r}$
$S(1 - r) = a - rL$
$S - rS = a - rL$
$-rS = a - S - rL$
$rL - rS = a - S$
$r(L - S) = a - S$
$r = \dfrac{a - S}{L - S}$

45. $|x - 3| = 6$
$x - 3 = 6$ or $x - 3 = -6$
$x = 9$ or $\quad x = -3$

47. $|2r + 7| = 11$
$2r + 7 = 11$ or $2r + 7 = -11$
$2r = 4 \qquad\qquad 2r = -18$
$r = 2$ or $\qquad r = -9$

49. $|2z - 7| = 0$ gives $2z - 7 = 0$
$$2z = 7$$
$$z = \frac{7}{2}$$

51. $|4w + 13| = -3$ has no solution since the left side, $|4w + 13|$, is always positive whereas the right side, -3, is negative.

53. $|3y + 8| = |2 - y|$

$3y + 8 = 2 - y$ or $\quad 3y + 8 = -(2 - y)$
$4y + 8 = 2 \qquad\qquad 3y + 8 = -2 + y$
$\quad 4y = -6 \qquad\qquad 2y + 8 = -2$
$\quad\quad y = -\frac{6}{4} = -\frac{3}{2} \qquad\quad 2y = -10$
$$\text{or} \qquad y = -\frac{10}{2} = -5$$

55. $\dfrac{|2x - 5|}{|x + 2|} = 1$ can be expressed as $|2x - 5| = |x + 2|$

$2x - 5 = x + 2$ or $2x - 5 = -(x + 2)$
$\quad x - 5 = 2 \qquad\qquad 2x - 5 = -x - 2$
$\quad\quad x = 7 \qquad\qquad\quad 3x = 3$
$$\text{or} \qquad x = 1$$

57. $\dfrac{|9 - 4t|}{|t + 2|} = 0$ which is possible when the numerator is 0: $|9 - 4t| = 0$
$$9 - 4t = 0$$
$$9 = 4t$$
$$\frac{9}{4} = t$$

59. $|x| + 3x - 9 = 0$

If $x > 0$, then $|x| = x$ and the equation is
$x + 3x - 9 = 0$
$4x - 9 = 0$
$4x = 9$
$x = \frac{9}{4}$

If $x < 0$, then $|x| = -x$ and the equation is
$-x + 3x - 9 = 0$
$2x - 9 = 0$
$2x = 9$
$x = \frac{9}{2}$

which does **not** check since for this case $x = \frac{9}{2}$ is **not** a negative number

61. $2.79 - 4.20x = 6.44x + 9.56$
$2.79 - 10.64x = 9.56$
$-10.64x = 6.77$
$$x = \frac{6.77}{-10.64} \approx -0.64$$

63. $\sqrt{2.41}\,x - \sqrt{7.63} = \sqrt{0.42} + 3\sqrt{5.87}\,x$
$$\sqrt{2.41}\,x = \sqrt{0.42} + \sqrt{7.63} + 3\sqrt{5.87}\,x$$
$$\sqrt{2.41}\,x - 3\sqrt{5.87}\,x = \sqrt{0.42} + \sqrt{7.63}$$
$$(\sqrt{2.41} - 3\sqrt{5.87})x = \sqrt{0.42} + \sqrt{7.63}$$
$$x = \frac{\sqrt{0.42} + \sqrt{7.63}}{\sqrt{2.41} - 3\sqrt{5.87}} \approx -0.60$$

65. Use the formula $F = \frac{9}{5}C + 32$ with $F = 74$

$$74 = \frac{9}{5}C + 32$$

$$42 = \frac{9}{5}C$$

$$\frac{5}{9}(42) = \frac{5}{9}\left[\frac{9}{5}C\right]$$

$$\frac{210}{9} = C$$

$$\frac{70}{3} = C$$

$$23\frac{1}{3}° = C$$

67. Use $A = P(1 + rt)$ with
$P = 2800$, $t = 1$, and
$r = 1.5\%(12) = 0.015(12) = 0.18$
$$A = 2800[1 + (0.18)(1)]$$
$$= 2800(1.18)$$
$$= 3304 \text{ dollars}$$

69. Use $P = A(1 - rn)$ with $n = 3$,
$A = 3000$, and $P = 1470$
$$1470 = 3000(1 - r \cdot 3)$$
$$1470 = 3000(1 - 3r)$$
$$\frac{1470}{3000} = 1 - 3r$$
$$0.49 = 1 - 3r$$
$$-0.51 = -3r$$
$$\frac{-0.51}{-3} = r$$
$$0.17 = r \text{ or } r = 17\%$$

71. Use $R = C + rC$ with $r = 25\% = 0.25$
$$245 = C + 0.25C$$
$$245 = (1 + 0.25)C$$
$$245 = (1.25)C$$
$$\frac{245}{1.25} = C$$
$$\$196 = C$$

73. If $|R| = |S|$ then $|R|^2 = |S|^2$
or $R^2 = S^2$.
Also, if $R^2 = S^2$, then $\sqrt{R^2} = \sqrt{S^2}$
or $|R| = |S|$

Exercises 2.2

1. Let x, $x + 1$, and $x + 2$ represent the three consecutive integers. Since the sum must be 147, the equation is
$$x + (x + 1) + (x + 2) = 147$$
$$3x + 3 = 147$$
$$3x = 144$$
$$x = \frac{144}{3} = 48$$
The numbers are 48, 49, and 50.

3. Let x represent the units digit. Then the tens digit is $4x + 1$. Since the sum of the digits is 11, the equation is
$$x + (4x + 1) = 11$$
$$5x + 1 = 11$$
$$5x = 10$$
$$x = \frac{10}{5} = 2$$
The units digit is $x = 2$ and the tens digit is $4x + 1 = 4 \cdot 2 + 1 = 8 + 1 = 9$ and the desired number is 92.

5. Let x be the tens digit. Then the units digit for the original number is $2x - 1$ and the original number expressed in terms of x has a value of $10x + (2x - 1)$. For the new number, x is the units digit and $2x - 1$ is the tens digit, so that its value is $10(2x - 1) + x$. The equation is

$$10(2x - 1) + x = 2[10x + (2x - 1)] - 20$$
$$10(2x - 1) + x = 2(12x - 1) - 20$$
$$20x - 10 + x = 24x - 2 - 20$$
$$21x - 10 = 24x - 22$$
$$-3x - 10 = -22$$
$$-3x = -12$$
$$x = \frac{-12}{-3} = 4$$

The tens digit is $x = 4$ and the units digit is $2x - 1 = 2 \cdot 4 - 1 = 8 - 1 = 7$, and the desired number is 47.

7. For example 3, the equation is
$$S = |80 - 32t|.$$
Let S be 16 to obtain
$$16 = |80 - 32t|.$$
Solve to obtain

$$80 - 32t = 16 \quad \text{or} \quad 80 - 32t = -16$$
$$-32t = -64 \qquad\qquad -32t = -96$$
$$t = \frac{-64}{-32} \qquad\qquad -32t = -96$$
$$t = 2 \text{ sec} \qquad\qquad t = \frac{-96}{-32}$$
$$\text{or} \qquad\qquad t = 3 \text{ sec}$$

9. Let y be the additional amount of money to be invested at 12%. We can describe the situation in the following table:

Principal	Rate	Time	Interest
4,000	0.08	1	320
y	0.12	1	$0.12y$

Equation is
$$320 + 0.12y = 968$$
$$0.12y = 648$$
$$y = \frac{648}{0.12} = \$5400$$

11. Let x be the amount invested at 9%. The remaining amount, $14,000 - x$ must be invested at 12%. The table is

Principal	Rate	Time	Interest
x	0.09	1	$0.09x$
$14,000 - x$	0.12	1	$0.12(14,000 - x)$

The equation is
$$0.09x + 0.12(14,000 - x) = 1500$$
$$0.09x + 1680 - 0.12x = 1500$$
$$-0.03x + 1680 = 1500$$
$$-0.03x = -180$$
$$x = \frac{-180}{-0.03} = \$6000$$

Thus, $6,000 was invested at 9% and $14,000 - x = 14,000 - 6,000 = \$8,000$ was invested at 12%.

13. Let x be the amount invested at 12%. Then $2x$ must be the amount invested at 10% and the table is

Principal	Rate	Time	Interest
x	0.12	1	$0.12x$
$2x$	0.10	1	$0.10(2x) = 0.20x$

Equation is
$$0.12x + 0.20x = 8640$$
$$0.32x = 8640$$
$$x = \frac{8640}{0.32} = \$27,000$$

Thus, \$27,000 was invested at 12% and $2(\$27,000) = \$54,000$ was invested at 10%. The total investment was $\$27,000 + \$54,000 = \$81,000$.

15. Let x be the number of dimes. Then $12 - x$ must be the number of nickels and the table is

	Number	Value	
dimes:	x	$10x$¢	← each dime is 10¢
nickels:	$12 - x$	$5(12 - x)$¢	← each nickel is 5¢

The equation is $10x$¢ $+ 5(12 - x)$¢ $= 95$¢
or simply
$$10x + 5(12 - x) = 95$$
$$10x + 60 - 5x = 95$$
$$5x + 60 = 95$$
$$5x = 35$$
$$x = 7$$

There are 7 dimes and $12 - 7 = 5$ nickels.

17. Let x be the number of cartons of milk. Then $8 - x$ is the number of cartons of eggs and the table is

	Number	Value
milk:	x	$93x$¢
eggs:	$8 - x$	$99(8 - x)$¢

The equation is
$$93x + 99(8 - x) = 774$$
$$93x + 792 - 99x = 774$$
$$-6x + 792 = 774$$
$$-6x = -18$$
$$x = \frac{-18}{-6} = 3$$

There are 3 cartons of milk and $8 - 3 = 5$ cartons of eggs.

19. Let x be the width. Then $2x - 4$ is the length and the equation $2W + 2L = P$ becomes
$$2(x) + 2(2x - 4) = 112$$
$$2x + 4x - 8 = 112$$
$$6x - 8 = 112$$
$$6x = 120$$
$$x = \frac{120}{6} = 20 \text{ meters}$$

Thus, the width is 20 meters and the length is $2(20) - 4 = 40 - 4 = 36$ meters.

21. Let x be the number of pounds of peanuts. Then $12 - x$ is the number of pounds of cashews and the table is

	Amount	Rate	Value
peanuts:	x	\$2.10/lb	$\$2.10x$
cashews:	$12 - x$	\$2.40/lb	$\$2.40(12 - x)$
mixture:	12	\$2.30/lb	$\$2.30(12)$

The equation (without $) is $2.10x + 2.40(12 - x) = 2.30(12)$
$$2.10x + 28.80 - 2.40x = 27.60$$
$$-0.30x + 28.80 = 27.60$$
$$-0.30x = -1.20$$
$$x = \frac{-1.20}{-0.30} = 4$$

Thus, we need 4 pounds of peanuts and 12 - 4 = 8 pounds of cashews.

23. Let x be the amount of 20% silver. Then 40 - x is the amount of 12% silver and the table is

	Amount	Rate	Value
20% silver:	x	20% = 0.20	0.20x
12% silver:	40 - x	12% = 0.12	0.12(40 - x)
Mixture:	40	15% = 0.15	0.15(40)

The equation is $0.20x + 0.12(40 - x) = 0.15(40)$
$$0.20x + 4.80 - 0.12x = 6.00$$
$$0.08x + 4.80 = 6.00$$
$$0.08x = 1.20$$
$$x = \frac{1.20}{0.08} = 15 \text{ grams}$$

Thus, we need 15 grams of 20% silver and 40 - 15 = 25 grams of the 12% silver.

25. Let x be the amount of water added. The table is

Amount	Rate	Acid(Value)
40	80% = 0.80	0.80(40)
x	0% = 0	0(x)
40 + x	50% = 0.50	0.50(40 + x)

The equation is $0.80(40) + 0x = 0.50(40 + x)$
$$32 = 20 + 0.50x$$
$$12 = 0.50x$$
$$\frac{12}{0.05} = x$$
$$24 \text{ liters} = x$$

Thus, 24 liters of water must be added.

27. Let t be the time for them to meet. The table is

	Rate(km/hr)	Time(hr)	Distance(km)
Dan:	50	t	50t
Fran:	55	t	55t

Since the total distance is 450 kilometers, the equation is
$$50t + 55t = 450$$
$$105t = 450$$
$$t = \frac{450}{105} = \frac{30}{7} = 4\frac{2}{7} \text{ hours}$$

29. Let x be the speed of the boat in still water. Now 1 hr 20 minutes = $1\frac{20}{60}$ = $1\frac{1}{3}$ hr and the table is

	Rate	Time	Distance
upstream:	x - 3	$1\frac{1}{3}$	$(x - 3)1\frac{1}{3}$
downstream:	x + 3	1	$(x + 3)(1)$

Since the two distances are the same, the equation is

$$(x - 3)1\tfrac{1}{3} = (x + 3)(1)$$

$$(x - 3)\tfrac{4}{3} = x + 3$$

$$\tfrac{4}{3}x - 4 = x + 3$$

$$\tfrac{4}{3}x = x + 7$$

$$\tfrac{1}{3}x = 7$$

$$x = 7 \cdot 3 = 21 \text{ km/hr}$$

The speed of the boat was 21 km/hr and the distance is 21 + 3 = 24 km.

31. Let x be the speed of the slower car. Then $x + 5$ is the speed of the faster car and the table is

	Rate	Time	Distance
slower:	x	5	$5x$
faster:	$x + 5$	5	$5(x + 5)$

Since the total distance is 725 km, the equation is

$$5x + 5(x + 5) = 725$$
$$5x + 5x + 25 = 725$$
$$10x + 25 = 725$$
$$10x = 700$$
$$x = 70 \text{ km/hr}$$

The slower car travels at 70 km/hr and the faster car at 70 + 5 = 75 km/hr.

33. Let x be the time it takes both to mow the grass. Then $\tfrac{1}{x}$ is the portion of the lawn mowed in one hour working together. Also, Jill mows $\tfrac{1}{8}$ of the grass each hour and Martin mows $\tfrac{1}{5}$ of the grass each hour. Working together, they would mow $\tfrac{1}{8} + \tfrac{1}{5}$ of the grass each hour. The equation is

$$\tfrac{1}{8} + \tfrac{1}{5} = \tfrac{1}{x} \quad \text{Multiply by } 40x \text{ to get} \quad \tfrac{1}{8} \cdot 40x + \tfrac{1}{5} \cdot 40x = \tfrac{1}{x} 40x$$

$$5x + 8x = 40$$
$$13x = 40$$
$$x = \tfrac{40}{13} = 3\tfrac{1}{13} \text{ hours}$$

35. Let x be the time it takes Ruth to do the job. Then $\tfrac{1}{x}$ is the portion of the roof done (in one hour) by Ruth, $\tfrac{1}{36}$ is the portion of the roof done by Jim and $\tfrac{1}{20}$ is the portion done when they work together. The equation is

$$\tfrac{1}{x} + \tfrac{1}{36} = \tfrac{1}{20} \quad \text{Multiply by } 180x \text{ to get}$$

$$\tfrac{1}{x} \cdot 180x + \tfrac{1}{36} \cdot 180x = \tfrac{1}{20} \cdot 180x$$

$$180 + 5x = 9x$$
$$180 = 4x$$
$$\tfrac{180}{4} = x$$

$$\text{or} \quad x = 45 \text{ hours}$$

Working alone, Ruth can roof a house in 45 hours.

37. Let x be the time for the leak to empty the tank. The equation is
$$\frac{1}{10} - \frac{1}{12} = \frac{1}{x} \quad \text{Multiply by } 120x \text{ to get}$$
$$\frac{1}{10} \cdot 120x - \frac{1}{12} \cdot 120x = \frac{1}{x} \cdot 120x$$
$$12x - 10x = 120$$
$$2x = 120$$
$$x = \frac{120}{2} = 60 \text{ hours}$$

39. Measurements such as height and weight cannot be negative. Perimeter, area, and volume are also examples of quantities which cannot be negative.

Exercises 2.3

1. $2x < 8$
$x < 4$ which is $(-\infty, 4)$

3. $25 - 5x < 0$
$25 < 5x$
$5 < x$ which is $(5, \infty)$

5. $2y > \frac{8}{3}y - 4$
Multiply by 3 to obtain
$6y > 8y - 12$
$-2y > -12$
$y < 6$ which is $(-\infty, 6)$

7. $17 - 4x > 2x + 5$
$12 - 4x > 2x$
$12 > 6x$
$2 > x$ which is $(-\infty, 2)$

9. $-3x + 2\sqrt{2} < x - 2\sqrt{2}$
$-4x + 2\sqrt{2} < -2\sqrt{2}$
$-4x < -4\sqrt{2}$
$x > \sqrt{2}$ which is $(\sqrt{2}, \infty)$

11. $\frac{1}{2}x \leq \frac{3}{4}x + 9$
Multiply by 4 to obtain
$2x \leq 3x + 36$
$-x \leq 36$
$x \geq -36$ which is $[-36, \infty)$

13. $17 + 5(x + 2) \leq x - 3(x - 2)$
$17 + 5x + 10 \leq x - 3x + 6$
$27 + 5x \leq -2x + 6$
$27 + 7x \leq 6$
$7x \leq -21$
$x \leq -3$ which is $(-\infty, -3]$

15. $\frac{x}{3} + \frac{x}{2} - \frac{x}{4} \geq \frac{x}{5}$
Multiply by 60 to obtain
$20x + 30x - 15x \geq 12x$
$35x \geq 12x$
$23x \geq 0$
$x \geq 0$ which is $[0, \infty)$

17. $\frac{7 - 2x}{3} \geq 11$
Multiply by 3 to obtain
$7 - 2x \geq 33$
$-2x \geq 26$
$x \leq -13$ which is $(-\infty, -13]$

19. $\frac{2 - x}{3} > \frac{1 - 2x}{5}$
Multiply by 15 to obtain
$5(2 - x) > 3(1 - 2x)$
$10 - 5x > 3 - 6x$
$10 + x > 3$
$x > -7$ which is $(-7, \infty)$

21. $-17 < 5x - 7 \leq 13$
$-10 < 5x \leq 20$
$-2 < x \leq 4$
or $(-2, 4]$

23. $-3 < 5 - 2t \leq 3$
$-8 < -2t \leq -2$
$4 > t \geq 1$
or $[1, 4)$

25. $x - 1 \geq 0$ and $3 \geq x - 1$
     $x \geq 1$ and $4 \geq x$
which produces $1 \leq x \leq 4$
$[1, 4]$

27. $1 - 2x < 3$ and $2(x - 2) < x - 1$
    $-2x < 2$ and $2x - 4 < x - 1$
    $x > -1$ and $x - 4 < -1$
    $x > -1$ and $x < 3$
which produces $-1 < x < 3$
$(-1, 3)$

29. $7 - 2x \leq 5$ and $1 - 3x > 2(2 - x)$
   $-2x \leq -2$ and $1 - 3x > 4 - 2x$
    $x \geq 1$ and $1 - x > 4$
    $x \geq 1$ and $-x > 3$
    $x \geq 1$ and $x < -3$
which has no region of
intersection so that the solution
is $\varnothing$.

31. $3x + 5 > 2$ and $9x + 2 \geq 4(x + 3)$
   $3x > -3$ and $9x + 2 \geq 4x + 12$
    $x > -1$ and $5x \geq 10$
    $x > -1$ and $x \geq 2$
which produces $x \geq 2$ for the region
of intersection. The solution is
written as $[2, \infty)$

33. $3x + 11 \leq 5$ or $4x - 3 > x + 6$
    $3x \leq -6$ or $4x > x + 9$
    $x \leq -2$ or $3x > 9$
    $x \leq -2$ or $x > 3$
which produces $(-\infty, -2] \cup (3, \infty)$

35. $-2x + 7 \geq 3$ or $5x - 2 > 2(x + 5)$
   $-2x \geq -4$ or $5x - 2 > 2x + 10$
    $x \leq 2$ or $3x - 2 > 10$
    $x \leq 2$ or $3x > 12$
    $x \leq 2$ or $x > 4$
which produces $(-\infty, 2] \cup (4, \infty)$

37. $\dfrac{7x + 6}{6} > \dfrac{x + 2}{2}$ or $4(x + 4) > 2(2 - x)$
$7x + 6 > 3(x + 2)$ or $4x + 16 > 4 - 2x$
$7x + 6 > 3x + 6$ or $6x + 16 > 4$
   $4x > 0$ or $6x > -12$
    $x > 0$ or $x > -2$
which produces $x > -2$ which is $(-2, \infty)$

39. $7x - 8 \geq -43$ or $-x > 8 + x$
   $7x \geq -35$ or $-2x > 8$
    $x \geq -5$ or $x < -4$
which is the entire real number
line: R

41. $|2x - 5| < 3$
$-3 < 2x - 5 < 3$
  $2 < 2x < 8$
  $1 < x < 4$
and the graph is
$(1, 4)$

    ⟵———(━━━)———⟶
       1   4

43. $|7 - x| \leq 2$
$-2 \leq 7 - x \leq 2$
$-9 \leq -x \leq -5$
 $9 \geq x \geq 5$
and the graph is
$[5, 9]$

    ⟵———[━━━]———⟶
       5   9

45. $|8x + 5| < 25$
$-25 < 8x + 5 < 25$
$-30 < 8x < 20$
$\dfrac{-30}{8} < x < \dfrac{20}{8}$
$\dfrac{-15}{4} < x < \dfrac{5}{2}$
and the graph is
$\left(\dfrac{-15}{4}, \dfrac{5}{2}\right)$

    ⟵———(━━)———⟶
      $-\frac{15}{4}$ $\frac{5}{2}$

47. $|2x - 4| \geq 3$
$2x - 4 \geq 3$ or $2x - 4 \leq -3$
  $2x \geq 7$ or $2x \leq 1$
   $x \geq \dfrac{7}{2}$ or $x \leq \dfrac{1}{2}$
and the graph is
$\left(-\infty, \dfrac{1}{2}\right] \cup \left[\dfrac{7}{2}, \infty\right)$

    ⟵━━]———[━━⟶
      $\frac{1}{2}$ $\frac{7}{2}$

49. $|3 - 4x| > 2$
$3 - 4x > 2$ or $3 - 4x < -2$
 $-4x > -1$ or $-4x < -5$
   $x < \dfrac{1}{4}$ or $x > \dfrac{5}{4}$
and the graph is
$\left(-\infty, \dfrac{1}{4}\right) \cup \left(\dfrac{5}{4}, \infty\right)$

    ⟵━━)———(━━⟶
      $\frac{1}{4}$ $\frac{5}{4}$

51. $|5x + 3| > 7$
$5x + 3 > 7$ or $5x + 3 < -7$
$5x > 4$ or $5x < -10$
$x > \dfrac{4}{5}$ or $x < -2$
and the graph is

$(-\infty, -2) \cup \left(\dfrac{4}{5}, \infty\right)$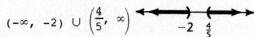
$-2 \quad \frac{4}{5}$

53. $|4 - 5x| < 0$
Since $|4 - 5x|$ is always ≥ 0 it is **not** possible for $|4 - 5x| < 0$. The solution is the empty set and is represented as $\varnothing$.

55. $|3x - 6| \leq 0$
Since $|3x - 6|$ is always ≥ 0 the only solution occurs when
$|3x - 6| = 0$
$3x - 6 = 0$
$3x = 6$
$x = 2$
and the graph is simply a point
$\{2\}$
    2

57. $|1 - x| > 0$
$1 - x > 0$ or $1 - x < 0$
$-x > -1$ or $-x < -1$
$x < 1$ or $x > 1$
and the graph is

$(-\infty, 1) \cup (1, \infty)$
        1

59. $|2x + 5| > -5$
Since $|2x - 5|$ is always ≥ 0 then $|2x - 5| > -5$ produces all values for x and the graph is the entire real line.

R

61. $4.38x - 6.15 < 2.17 - 1.95x$
$6.33x - 6.15 < 2.17$
$6.33x < 8.32$
$x < \dfrac{8.32}{6.33}$
$x < 1.31$
which is written as $(-\infty, 1.31)$

63. $-7.21 < \dfrac{1.55x - 2.74}{6.01} < 10.2$
$-43.3321 < 1.55x - 2.74 < 61.302$
$-40.5921 < 1.55x < 64.042$
$\dfrac{-40.5921}{1.55} < x < \dfrac{64.042}{1.55}$
$-26.19 < x < 41.32$
which is written as $(-26.19, 41.32)$

65. $|x - \sqrt{2}| < \sqrt{3}$
$-\sqrt{3} < x - \sqrt{2} < \sqrt{3}$
$-\sqrt{3} + \sqrt{2} < x < \sqrt{3} + \sqrt{2}$
$-1.732 + 1.414 < x < 1.732 + 1.414$
$-0.32 < x < 3.15$
which is written as $(-0.32, 3.15)$

67. $|10.2^{0.56}x - \sqrt{2}| \geq 17.3$
$10.2^{0.56}x - \sqrt{2} \geq 17.3$ or $10.2^{0.56}x - \sqrt{2} \leq -17.3$
$3.671x - 1.414 \geq 17.3$ or $3.671x - 1.414 \leq -17.3$
$3.671x \geq 18.714$ $3.671x \leq -15.886$
$x \geq \dfrac{18.714}{3.671}$ or $x \leq \dfrac{-15.886}{3.671}$
$x \geq 5.10$ or $x \leq -4.33$
which is written as $(-\infty, -4.33] \cup [5.10, \infty)$

69. Let x be the number of hours working at the house.
The inequality is
$40x + 28 \leq 200$
$40x \leq 172$
$x \leq \dfrac{172}{40}$
$x \leq 4.3$ hours
The maximum time is 4.3 hours.

71. Let t be the time in years. The inequality is
$16,000 - 3,000t \geq 5,500$
$-3,000t \geq -10,500$
$t \leq \dfrac{-10,500}{-3,000}$
$t \leq 3.5$ years,
or 3 years and 6 months

73. Since $p = 62.5(d - 2)$ and $p > 250$, the inequality becomes

$$62.5(d - 2) > 250$$
$$d - 2 > \frac{250}{62.5}$$
$$d - 2 > 4$$
$$d > 6$$

Also, $d \leq 12$ since the distance cannot go below the bottom of the cylinder. The final answer is $6 < d \leq 12$.

75. Since $F = 94 - 3|t - 15|$ and $F > 82°$, the inequality becomes

$$94 - 3|t - 15| > 82$$
$$-3|t - 15| > -12$$
$$|t - 15| < 4$$
$$-4 < t - 15 < 4$$
$$11 < t < 19$$

and the interval is $(11, 19)$

77. The same number may be added to (or subtracted from) both sides of an inequality. The sense (or direction) of the inequality is maintained.

Exercises 2.4

1. $4x^2 - 25 = 0$
$(2x + 5)(2x - 5) = 0$
$2x + 5 = 0, \; 2x - 5 = 0$
$2x = -5, \quad 2x = 5$
$x = -\dfrac{5}{2}, \quad x = \dfrac{5}{2}$

3. $r^2 + 3r = 0$
$r(r + 3) = 0$
$r = 0, \; r + 3 = 0$
$r = 0, \quad r = -3$

5. $y^2 + 3y - 10 = 0$
$(y + 5)(y - 2) = 0$
$y + 5 = 0, \; y - 2 = 0$
$y = -5, \quad y = 2$

7. $x^2 + 6x + 9 = 0$
$(x + 3)^2 = 0$
$x + 3 = 0$
$x = -3$

9. $3t^2 - 10t + 3 = 0$
$(3t - 1)(t - 3) = 0$
$3t - 1 = 0, \; t - 3 = 0$
$3t = 1, \quad t = 3$
$t = \dfrac{1}{3}, \quad t = 3$

11. $49x^3 + 7x^2 - 2x = 0$
$x(49x^2 + 7x - 2) = 0$
$x(7x + 2)(7x - 1) = 0$
$x = 0, \; 7x + 2 = 0, \; 7x - 1 = 0$
$x = 0, \quad 7x = -2, \quad 7x = 1$
$x = 0, \quad x = -\dfrac{2}{7}, \quad x = \dfrac{1}{7}$

13. $p^2 + 4p + 3 = 0$
$p^2 + 4p = -3$
$p^2 + 4p + 4 = -3 + 4$
$(p + 2)^2 = 1$
$p + 2 = \pm 1$
$p = -2 \pm 1$ which gives
$p = -2 + 1 = -1,$
$p = -2 - 1 = -3$

15. $x^2 + x - 1 = 0$
$x^2 + x = 1$
$x^2 + x + \dfrac{1}{4} = 1 + \dfrac{1}{4}$
$\left(x + \dfrac{1}{2}\right)^2 = \dfrac{5}{4}$
$x + \dfrac{1}{2} = \pm\sqrt{\dfrac{5}{4}}$
$x + \dfrac{1}{2} = \pm\dfrac{\sqrt{5}}{2}$
$x = -\dfrac{1}{2} \pm \dfrac{\sqrt{5}}{2} = \dfrac{-1 \pm \sqrt{5}}{2}$

17. $w^2 - 2w + 2 = 0$
$w^2 - 2w = -2$
$w^2 - 2w + 1 = -2 + 1$
$(w - 1)^2 = -1$
$w - 1 = \pm\sqrt{-1}$
$w - 1 = \pm i$
$w = 1 \pm i$

19. $2x^2 - 8x = -16$
$x^2 - 4x = -8$
$x^2 - 4x + 4 = -8 + 4$
$(x - 2)^2 = -4$
$x - 2 = \pm\sqrt{-4}$
$x - 2 = \pm 2i$
$x = 2 \pm 2i$

21. $2y^2 + 5y + 1 = 0$

$$2y^2 + 5y = -1$$

$$y^2 + \frac{5}{2}y = -\frac{1}{2}$$

$$y^2 + \frac{5}{2}y + \frac{25}{16} = -\frac{1}{2} + \frac{25}{16}$$

$$\left(y + \frac{5}{4}\right)^2 = \frac{17}{16}$$

$$y + \frac{5}{4} = \pm\sqrt{\frac{17}{16}}$$

$$y + \frac{5}{4} = \pm\frac{\sqrt{17}}{4}$$

$$y = -\frac{5}{4} \pm \frac{\sqrt{17}}{4} = \frac{-5 \pm \sqrt{17}}{4}$$

23. $6x^2 + 13x = -6$

$$x^2 + \frac{13}{6}x = -1$$

$$x^2 + \frac{13}{6}x + \frac{169}{144} = -1 + \frac{169}{144}$$

$$\left(x + \frac{13}{12}\right)^2 = \frac{25}{144}$$

$$x + \frac{13}{12} = \pm\sqrt{\frac{25}{144}}$$

$$x + \frac{13}{12} = \pm\frac{5}{12}$$

$$x = -\frac{13}{12} \pm \frac{5}{12} \text{ which gives}$$

$$x = \frac{-13 + 5}{12} = \frac{-8}{12} = -\frac{2}{3},$$

$$x = \frac{-13 - 5}{12} = \frac{-18}{12} = -\frac{3}{2}$$

25. $7 - 15x + 2x^2 = 0$

can be written as

$$2x^2 - 15x + 7 = 0$$

$$(2x - 1)(x - 7) = 0$$

$$2x - 1 = 0, \quad x - 7 = 0$$

$$2x = 1, \qquad x = 7$$

$$x = \frac{1}{2}, \qquad x = 7$$

27. $x^2 + 2x - 1 = 0$

$$x^2 + 2x = 1$$

$$x^2 + 2x + 1 = 1 + 1$$

$$(x + 1)^2 = 2$$

$$x + 1 = \pm\sqrt{2}$$

$$x = -1 \pm \sqrt{2}$$

29. $4t^2 - 5t - 6 = 0$

$$(4t + 3)(t - 2) = 0$$

$$4t + 3 = 0, \quad t - 2 = 0$$

$$4t = -3, \qquad t = 2$$

$$t = -\frac{3}{4}, \qquad t = 2$$

31. $-27y^2 + 3y + 2 = 0$ can be written as

$$27y^2 - 3y - 2 = 0$$

$$(9y + 2)(3y - 1) = 0$$

$$9y + 2 = 0, \quad 3y - 1 = 0$$

$$9y = -2, \qquad 3y = 1$$

$$y = -\frac{2}{9}, \qquad y = \frac{1}{3}$$

33. $2z^2 + 2z - 5 = 0$

$$2z^2 + 2z = 5$$

$$z^2 + z = \frac{5}{2}$$

$$z^2 + z + \frac{1}{4} = \frac{5}{2} + \frac{1}{4}$$

$$\left(z + \frac{1}{2}\right)^2 = \frac{11}{4}$$

$$z + \frac{1}{2} = \pm\sqrt{\frac{11}{4}}$$

$$z + \frac{1}{2} = \pm\frac{\sqrt{11}}{2}$$

$$z = -\frac{1}{2} \pm \frac{\sqrt{11}}{2} = \frac{-1 \pm \sqrt{11}}{2}$$

35. $4x^2 - 8x + 5 = 0$

$$4x^2 - 8x = -5$$

$$x^2 - 2x = -\frac{5}{4}$$

$$x^2 - 2x + 1 = -\frac{5}{4} + 1$$

$$(x - 1)^2 = -\frac{1}{4}$$

$$x - 1 = \pm\sqrt{-\frac{1}{4}}$$

$$x - 1 = \pm\frac{i}{2}$$

$$x = 1 \pm \frac{i}{2} = \frac{2 \pm i}{2}$$

37. Let $z = y^2$ so that $z^2 = y^4$ and the equation becomes

$$z^2 - 4z + 4 = 0$$
$$(z - 2)^2 = 0$$
$$z - 2 = 0$$
$$z = 2$$

or, since $z = y^2$, we have $y^2 = 2$

$$y = \pm\sqrt{2}$$

39. Let $x = z^3$ so that $x^2 = z^6$ and the equation becomes

$$27x^2 + 215x - 8 = 0$$
$$(x + 8)(27x - 1) = 0$$
$$x + 8 = 0, \ 27x - 1 = 0$$
$$x = -8, \qquad 27x = 1$$
$$x = -8, \qquad x = \frac{1}{27}$$

Since $x = z^3$, we have

$z^3 = -8$ \qquad or \qquad $z^3 = \dfrac{1}{27}$

$z^3 + 8 = 0$ \qquad\qquad\qquad $27z^3 = 1$

$(z + 2)(z^2 - 2z + 4) = 0$ \qquad\qquad $27z^3 - 1 = 0$

$z = 2$ or $z^2 - 2z + 4 = 0$ \qquad\quad $(3z - 1)(9z^2 + 3z + 1) = 0$

$\qquad\qquad z^2 - 2z + 1 = -4 + 1$ \qquad $3z - 1 = 0$ or $9z^2 + 3z + 1 = 0$

$\qquad\qquad\qquad (z - 1)^2 = -3$ \qquad\qquad $z = \dfrac{1}{3}$ or $z^2 + \dfrac{1}{3}z + \dfrac{1}{9} = 0$

$\qquad\qquad z - 1 = \pm\sqrt{-3} = \pm\sqrt{3}\,i$ \qquad $z^2 + \dfrac{1}{3}z + \dfrac{1}{36} = -\dfrac{1}{9} + \dfrac{1}{36}$

$\qquad\qquad\qquad z = 1 \pm \sqrt{3}\,i$ \qquad\qquad $\left(z + \dfrac{1}{6}\right)^2 = -\dfrac{3}{36}$

$\qquad\qquad\qquad\qquad\qquad\qquad\qquad z + \dfrac{1}{6} = \pm\sqrt{\dfrac{-3}{36}} = \pm\dfrac{\sqrt{3}}{6}\,i$

$\qquad\qquad\qquad\qquad\qquad\qquad\qquad z = -\dfrac{1}{6} \pm \dfrac{\sqrt{3}}{6}\,i$

$\qquad\qquad\qquad\qquad\qquad\qquad\qquad = \dfrac{-1 \pm \sqrt{3}\,i}{6}$

The values are 2, $1 \pm \sqrt{3}\,i$, $\dfrac{1}{3}$, $\dfrac{-1 \pm \sqrt{3}\,i}{6}$.

41. Let $p = y + 2$. The equation becomes

$$p^2 - 5p = 14$$
$$p^2 - 5p - 14 = 0$$
$$(p - 7)(p + 2) = 0$$
$$p - 7 = 0, \ p + 2 = 0$$
$$p = 7, \qquad p = -2$$

Since $p = y + 2$, we have $y + 2 = 7$, $y + 2 = -2$

$$y = 5, \qquad y = -4$$

43. Let $y = \dfrac{1}{x}$. The equation becomes

$$y^2 + 6y + 5 = 0$$
$$(y + 5)(y + 1) = 0$$
$$y + 5 = 0, \ y + 1 = 0$$
$$y = -5, \qquad y = -1$$

Since $y = \dfrac{1}{x}$, we have $\dfrac{1}{x} = -5$, $\dfrac{1}{x} = -1$

$$x = -\dfrac{1}{5}, \ x = -1$$

45. Let $x = y^{1/3}$ so that $x^2 = y^{2/3}$ and the equation becomes
$$2x^2 + x - 1 = 0$$
$$(2x - 1)(x + 1) = 0$$
$$2x - 1 = 0, \quad x + 1 = 0$$
$$2x = 1, \quad x = -1$$
$$x = \frac{1}{2}, \quad x = -1$$

Since $x = y^{1/3}$, we have $y^{1/3} = \frac{1}{2}$, $y^{1/3} = -1$
$$y = \left(\frac{1}{2}\right)^3, \quad y = (-1)^3$$
$$y = \frac{1}{8}, \quad y = -1$$

47. Let $p = \sqrt{x}$ so that $p^2 = x$ and the equation becomes
$$p^2 = 8p - 15$$
$$p^2 - 8p + 15 = 0$$
$$(p - 5)(p - 3) = 0$$
$$p - 5 = 0, \quad p - 3 = 0$$
$$p = 5, \quad p = 3$$

Since $p = \sqrt{x}$, we have $\sqrt{x} = 5$, $\sqrt{x} = 3$
$$x = 5^2, \quad x = 3^2$$
$$x = 25, \quad x = 9$$

49. Let x represent the distance in yards across the sinkhole. By the Pythagorean Theorem, the equation is
$$(60)^2 + x^2 = (90)^2$$
$$3600 + x^2 = 8100$$
$$x^2 = 4500$$
$$x = \sqrt{4500} \approx 67 \text{ yds}$$

51. Let x represent the width in meters and $2x$ represent the length in meters of the field.
Since (Length)(Width) = Area, the equation is
$$(2x) \quad (x) = 1800$$
$$2x^2 = 1800$$
$$x^2 = 900$$
$$x = \sqrt{900} = 30 \text{ m} \ (\text{Discard } -\sqrt{900} \text{ as meaningless})$$
Thus, the width is 30 m and the length is $2x = 2 \cdot 30 = 60$ m.

53. Yes, since $3^2 + 4^2 \overset{?}{=} 5^2$
$$9 + 16 \overset{?}{=} 25$$
$$25 \overset{?}{=} 25$$
No other list of three consecutive positive integers has this condition.

55. Let x be the width of the strip. The area of the original field is $(100)(200)m^2 = 20,000$ m^2. Now, 52% of this amount is $(0.52)(20,000$ m$^2) = 10,400$ m^2. The unplowed part (center rectangle) is $(20,000 - 10,400)m^2 = 9,600$ m^2. The diagram is

The equation is

$$(100 - 2x)(200 - 2x) = 9,600$$
$$20,000 - 600x + 4x^2 = 9,600$$
$$10,400 - 600x + 4x^2 = 0$$
$$\text{or} \quad 4x^2 - 600x + 10,400 = 0$$

Now divide by 4:

$$x^2 - 150x + 2600 = 0$$
$$(x - 130)(x - 20) = 0$$
$$x - 130 = 0, \quad x - 20 = 0$$
$$x = 130, \qquad x = 20$$

The only meaningful value is $x = 20$ m.

57. a) $x = 3, \; x = 7$
$x - 3 = 0, \; x - 7 = 0$
$(x - 3)(x - 7) = 0$
$x^2 - 10x + 21 = 0$

  b) $x = r, \; x = s$
$x - r = 0, \; x - s = 0$
$(x - r)(x - s) = 0$
$x^2 - (r + s)x + rs = 0$

Exercises 2.5

1. $x^2 + 3x - 28 = 0$. Let $a = 1$, $b = 3$, and $c = -28$.

$$x = \frac{-3 \pm \sqrt{3^2 - 4(1)(-28)}}{2 \cdot 1} = \frac{-3 \pm \sqrt{9 + 112}}{2} = \frac{-3 \pm \sqrt{121}}{2} = \frac{-3 \pm 11}{2}$$

This gives $x = \dfrac{-3 + 11}{2} = \dfrac{8}{2} = 4$ as well as $x = \dfrac{-3 - 11}{2} = \dfrac{-14}{2} = -7$

3. $4x^2 - 8x + 3$. Let $a = 4$, $b = -8$, and $c = 3$.

$$x = \frac{-(-8) \pm \sqrt{(-8)^2 - 4(4)(3)}}{2 \cdot 4} = \frac{8 \pm \sqrt{64 - 48}}{8} = \frac{8 \pm \sqrt{16}}{8} = \frac{8 \pm 4}{8}$$

This gives $x = \dfrac{8 + 4}{8} = \dfrac{12}{8} = \dfrac{3}{2}$ as well as $x = \dfrac{8 - 4}{8} = \dfrac{4}{8} = \dfrac{1}{2}$

5. $x^2 + x = 0$. Let $a = 1$, $b = 1$, $c = 0$.

$$x = \frac{-1 \pm \sqrt{1^2 - 4(1)(0)}}{2 \cdot 1} = \frac{-1 \pm \sqrt{1 - 0}}{2} = \frac{-1 \pm 1}{2}$$

This gives $x = \dfrac{-1 + 1}{2} = \dfrac{0}{2} = 0$ as well as $x = \dfrac{-1 - 1}{2} = \dfrac{-2}{2} = -1$

7. $-x^2 - 4x + 12 = 0$. Let $a = -1$, $b = -4$, and $c = 12$.

$$x = \frac{-(-4) \pm \sqrt{(-4)^2 - 4(-1)(12)}}{2(-1)} = \frac{4 \pm \sqrt{16 + 48}}{-2} = \frac{4 \pm \sqrt{64}}{-2} = \frac{4 \pm 8}{-2}$$

This gives $x = \dfrac{4 + 8}{-2} = \dfrac{12}{-2} = -6$ as well as $x = \dfrac{4 - 8}{-2} = \dfrac{-4}{-2} = 2.$

9. $16x^2 - 25 = 0$. Let $a = 16$, $b = 0$, and $c = -25$.

$$x = \frac{-0 \pm \sqrt{0^2 - 4(16)(-25)}}{2 \cdot 16} = \frac{0 \pm \sqrt{1600}}{32} = \pm \frac{40}{32} = \pm \frac{5}{4}$$

11. $4x^2 = -7x - 2$

$4x^2 + 7x + 2 = 0$. Let $a = 4$, $b = 7$, and $c = 2$.

$$x = \frac{-7 \pm \sqrt{7^2 - 4(4)(2)}}{2 \cdot 4} = \frac{-7 \pm \sqrt{49 - 32}}{8} = \frac{-7 \pm \sqrt{17}}{8}$$

13. $x^2 + 5x + 5 = 0$. Let $a = 1$, $b = 5$, and $c = 5$.

$$x = \frac{-5 \pm \sqrt{5^2 - 4(1)(5)}}{2 \cdot 1} = \frac{-5 \pm \sqrt{25 - 20}}{2} = \frac{-5 \pm \sqrt{5}}{2}$$

15. $3x^2 + 8x + 3 = 0$. Let $a = 3$, $b = 8$, and $c = 3$.

$$x = \frac{-8 \pm \sqrt{8^2 - 4(3)(3)}}{2 \cdot 3} = \frac{-8 \pm \sqrt{64 - 36}}{6} = \frac{-8 \pm \sqrt{28}}{6} = \frac{-8 \pm 2\sqrt{7}}{6}$$

$$= \frac{2(-4 \pm \sqrt{7})}{6} = \frac{-4 \pm \sqrt{7}}{3}$$

17. $x^2 - 2x + 4 = 0$. Let $a = 1$, $b = -2$, and $c = 4$.

$$x = \frac{-(-2) \pm \sqrt{(-2)^2 - 4(1)(4)}}{2 \cdot 1} = \frac{2 \pm \sqrt{4 - 16}}{2} = \frac{2 \pm \sqrt{-12}}{2} = \frac{2 \pm 2i\sqrt{3}}{2}$$

$$= \frac{2(1 \pm i\sqrt{3})}{2} = 1 \pm i\sqrt{3}$$

19. $28x^2 = 45 - x$

$28x^2 + x - 45 = 0$. Let $a = 28$, $b = 1$, and $c = -45$.

$$x = \frac{-1 \pm \sqrt{1^2 - 4(28)(-45)}}{2 \cdot 28} = \frac{-1 \pm \sqrt{1 + 5040}}{56} = \frac{-1 \pm \sqrt{5041}}{56} = \frac{-1 \pm 71}{56}$$

This gives $x = \frac{-1 + 71}{56} = \frac{70}{56} = \frac{5}{4}$ as well as $x = \frac{-1 - 71}{56} = \frac{-72}{56} = -\frac{9}{7}$

21. $0.26x^2 + 1.02x + 0.82 = 0$. Let $a = 0.26$, $b = 1.02$, and $c = 0.82$.

$$x = \frac{-1.02 \pm \sqrt{(1.02)^2 - 4(0.26)(0.82)}}{2(0.26)} = \frac{-1.02 \pm \sqrt{1.0404 - .8528}}{0.52}$$

$$= \frac{-1.02 \pm \sqrt{0.1876}}{0.52} = \frac{-1.02 \pm 0.43313}{0.52}$$

This gives $x = \frac{-1.02 + 0.43313}{0.52} = \frac{-0.58687}{0.52} \approx -1.128 \approx -1.13$ as well as

$$x = \frac{-1.02 - 0.43313}{0.52} = \frac{-1.45313}{0.52} \approx -2.794 \approx -2.79$$

23. $1.01x^2 + 3.72x + 1.82 = 0$. Let $a = 1.01$, $b = 3.72$, and $c = 1.82$.

$$x = \frac{-3.72 \pm \sqrt{(3.72)^2 - 4(1.01)(1.82)}}{2(1.01)} = \frac{-3.72 \pm \sqrt{13.8384 - 7.3528}}{2.02}$$

$$= \frac{-3.72 \pm \sqrt{6.4856}}{2.02} \approx \frac{-3.72 \pm 2.5467}{2.02}$$

This gives $x = \frac{-3.72 + 2.5467}{2.02} = \frac{-1.1733}{2.02} \approx -0.581 \approx -0.58$ as well as

$$x = \frac{-3.72 - 2.5467}{2.02} = \frac{-6.2667}{2.02} \approx -3.102 \approx -3.10$$

25. $b^2 - 4ac = 0^2 - 4(1)(-4) = 0 + 16 = 16$; two distinct real roots

27. $b^2 - 4ac = (-5)^2 - 4(2)(7) = 25 - 56 = -31$; two distinct complex roots

29. $b^2 - 4ac = 5^2 - 4(3)(-1) = 25 + 12 = 37$; two distinct real roots

31. $b^2 - 4ac = (14)^2 - 4(49)(1) = 196 - 196 = 0$; two equal roots

33. $b^2 - 4ac = (9.21)^2 - 4(5.09)(1.82) = 84.8241 - 37.0552 = 47.7689$; two distinct real roots

35. $b^2 - 4ac = (-15.25)^2 - 4(27.81)(11.84) = 232.5625 - 1317.0816 = -1084.5191$; two distinct complex roots

37.
$$b^2 - 4ac = 0$$
$$6^2 - 4(k)(-2) = 0$$
$$36 + 8k = 0$$
$$8k = -36$$
$$k = \frac{-36}{8} = -\frac{9}{2}$$

39.
$$b^2 - 4ac = 0$$
$$k^2 - 4(4)(3) = 0$$
$$k^2 - 48 = 0$$
$$k^2 = 48$$
so that $k = \pm\sqrt{48} = \pm\sqrt{16\cdot 3} = \pm4\sqrt{3}$

41.
$$(x - 1)[x - (-2)] = 0$$
$$(x - 1)(x + 2) = 0$$
$$x^2 + x - 2 = 0$$

43.
$$\left(x - \frac{3}{2}\right)(x - 4) = 0$$
$$x^2 - \frac{11}{2}x + 6 = 0$$
If you multiply by 2, the resulting equation is
$$2x^2 - 11x + 12 = 0$$

45.
$$[x - (-2)][x - (-2)] = 0$$
$$(x + 2)(x + 2) = 0$$
$$x^2 + 4x + 4 = 0$$

47. $x^2 - 2x + 1 = 0$ 49. $x^2 + 3x - 4 = 0$ 51. $x^2 - 2x = 0$

53. Let x and $x + 1$ represent the two numbers. The equation is
$$x(x + 1) = 462$$
$$x^2 + x = 462$$
$$x^2 + x - 462 = 0$$
$$(x + 22)(x - 21) = 0$$
$$x + 22 = 0, \quad x - 21 = 0$$
$$x = -22, \quad x = 21$$
When $x = -22$, the two numbers are -22 and -21.
When $x = 21$, the two numbers are 21 and 22.

55. Let x, $x + 1$, $x + 2$, and $x + 3$ represent the four numbers. The equation is
$$x^2 + (x + 1)^2 + (x + 2)^2 + (x + 3)^2 = 174$$
$$x^2 + x^2 + 2x + 1 + x^2 + 4x + 4 + x^2 + 6x + 9 = 174$$
$$4x^2 + 12x + 14 = 174$$
$$4x^2 + 12x - 160 = 0$$
$$x^2 + 3x - 40 = 0$$
$$(x + 8)(x - 5) = 0$$
$$x + 8 = 0, \quad x - 5 = 0$$
$$x = -8, \quad x = 5$$
When $x = -8$, the four numbers are -8, -7, -6, -5.
When $x = 5$, the four numbers are 5, 6, 7, 8.

57. $h = 48t - 16t^2$. Let $h = 32$ to obtain
$$32 = 48t - 16t^2$$
$$16t^2 - 48t + 32 = 0$$

Now $\div$ by 16 to get:
$$t^2 - 3t + 2 = 0$$
$$(t - 2)(t - 1) = 0$$
$$t - 2 = 0, \quad t - 1 = 0$$
$$t = 2, \qquad t = 1$$
It takes 1 second on the way up and 2 seconds on the way down.

59. $C = 1500 - 60x + 3x^2$. Let $C = 1200$ to obtain
$$1200 = 1500 - 60x + 3x^2$$
$$0 = 300 - 60x + 3x^2$$

Now $\div$ by 3:
$$x^2 - 20x + 100 = 0$$
$$(x - 10)(x - 10) = 0$$
$$x - 10 = 0$$
$$x = 10 \text{ Lawnmowers}$$

61. $R = 15,000n - 150n^2$.
Let $R = 240,000$ to obtain
$$240,000 = 15,000n - 150n^2$$
$$150n^2 - 15,000n - 240,000 = 0$$

Now $\div$ by 150:
$$n^2 - 100 - 1600 = 0$$
$$(n - 20)(n - 80) = 0$$
$$n - 20 = 0, \quad n - 80 = 0$$
$$n = 20, \qquad n = 80$$

63. $P = 4t - 150 - 0.02t^2$. Let $P = 50$ to obtain
$$50 = 4t - 150 - 0.02t^2$$
$$0.02t^2 - 4t + 200 = 0$$

Multiply by 50 to get:
$$t^2 - 200t + 10,000 = 0$$
$$(t - 100)(t - 100) = 0$$
$$t - 100 = 0$$
$$t = 100 \text{ tons}$$

65. $C = 5 + 10p - p^2$. Let $C = 29$ to obtain
$$29 = 5 + 10p - p^2$$
$$p^2 - 10p + 24 = 0$$
$$(p - 6)(p - 4) = 0$$
$$p - 6 = 0, \quad p - 4 = 0$$
$$p = 6, \qquad p = 4$$

67. $S = p^2 + 4p - 31$. Let $S = 29$
$$29 = p^2 + 4p - 31$$
$$0 = p^2 + 4p - 60$$
$$0 = (p - 6)(p + 10)$$
$$p - 6 = 0, \quad p + 10 = 0$$
$$p = 6, \qquad p = -10$$
The only meaningful value is $p = 6$ which translates into \$6,000. (Remember that p is measured in thousands of dollars.)

69. Let x be additional people beyond the 20 children. Then $20 + x$ is the new size of the class and the cost per child is $3.00 - 0.10x$. The cost for the class is $(20 + x)(3.00 - 0.10x)$. Set this equal to 62.50 and solve
$$(20 + x)(3.00 - 0.10x) = 62.50$$
$$60 + x - 0.10x^2 = 62.50$$
$$-0.10x^2 + x - 2.50 = 0$$
Multiply by -10:
$$x^2 - 10x + 25 = 0$$
$$(x - 5)^2 = 0$$
$$x - 5 = 0$$
$$x = 5$$
Thus, 5 additional children should join the class making the total size $20 + 5 = 25$ children.

71. Let x be the side of one corner cut out. Then the length is $8 - 2x$ in. and the width is $8 - 2x$ in. Since the area must be 49 sq. in., the equation is
$$(8 - 2x)(8 - 2x) = 49$$
$$64 - 32x + 4x^2 = 49$$
$$4x^2 - 32x + 15 = 0$$
$$(2x - 1)(2x - 15) = 0$$
$$2x - 1 = 0, \quad 2x - 15 = 0$$
$$2x = 1, \qquad 2x = 15$$
$$x = \frac{1}{2} \text{ in}, \qquad x = \frac{15}{2} \text{ in} \rightarrow \text{Discard as a meaningless value}$$

Thus, we need to cut out a square of $\frac{1}{2}$ inch by $\frac{1}{2}$ inch from each corner.

73. Let r and s represent the two numbers. Then
$$r + s = c_1$$
$$r \cdot s = c_2$$
Solve the first equation for r to get $r = c_1 - s$. Then, substitute this expression for r into the second equation:
$$r \cdot s = c_2$$
$$(c_1 - s)s = c_2$$
$$c_1 s - s_2 = c_2$$
This results in a quadratic equation which can be solved by factoring, completing the square, or by using the quadratic formula.

Exercises 2.6

1. $x = \dfrac{6}{x + 1}$

   Multiply each side by $x + 1$:

   $x(x + 1) = \dfrac{6}{x + 1} (x + 1)$

   $x^2 + x = 6$
   $x^2 + x - 6 = 0$
   $(x + 3)(x - 2) = 0$
   $x + 3 = 0, \ x - 2 = 0$
   $x = -3, \quad x = 2$ and both values check

3. $2x = \dfrac{4}{x} - 3 + x$

   Multiply each side by x:

   $2x(x) = \dfrac{4}{x} (x) - 3(x) + x(x)$

   $2x^2 = 4 - 3x + x^2$
   $x^2 + 3x - 4 = 0$
   $(x + 4)(x - 1) = 0$
   $x + 4 = 0, \ x - 1 = 0$
   $x = -4, \quad x = 1$ and both values check

5. $2x - 1 = \dfrac{-5(3x + 2)}{2x + 1}$ Multiply each side by $2x + 1$

   $2x(2x + 1) - 1(2x + 1) = \dfrac{-5(3x + 2)}{2x + 1} (2x + 1)$

   $4x^2 + 2x - 2x - 1 = -15x - 10$
   $4x^2 + 15x + 9 = 0$
   $(4x + 3)(x + 3) = 0$
   $4x + 3 = 0, \ x + 3 = 0$
   $4x = -3, \quad x = -3$
   $x = -\dfrac{3}{4}, \quad x = -3$ and both values check.

7. $\dfrac{1}{x} = \dfrac{x}{x + 1}$ Here you can cross multiply to obtain

   $1(x + 1) = x(x)$
   $x + 1 = x^2$
   $0 = x^2 - x - 1$

   Use the quadratic formula to solve for x:

   $$x = \frac{-(-1) \pm \sqrt{(-1)^2 - 4(1)(-1)}}{2 \cdot 1} = \frac{1 \pm \sqrt{1 + 4}}{2} = \frac{1 \pm \sqrt{5}}{2}$$

9. $\dfrac{6}{(y+2)^2} + \dfrac{1}{y+2} = 1$ Multiply each side by $(y+2)^2$

$$\dfrac{6}{\cancel{(y+2)^2}}\,\cancel{(y+2)^2} + \dfrac{1}{\cancel{y+2}}\,\overset{y+2}{\cancel{(y+2)^2}} = 1(y+2)^2$$

$$6 + y + 2 = (y+2)^2$$
$$y + 8 = y^2 + 4y + 4$$
$$0 = y^2 + 3y - 4$$
$$0 = (y+4)(y-1)$$
$$y + 4 = 0, \quad y - 1 = 0$$
$$y = -4, \qquad y = 1 \quad \text{and both values check}$$

11. $\dfrac{4}{(t-2)^2} + \dfrac{7}{t-2} + 3 = 0$ Multiply each side by $(t-2)^2$

$$\dfrac{4}{\cancel{(t-2)^2}}\,\cancel{(t-2)^2} + \dfrac{7}{\cancel{t-2}}\,\overset{t-2}{\cancel{(t-2)^2}} + 3(t-2)^2 = 0(t-2)^2$$

$$4 + 7(t-2) + 3(t-2)^2 = 0$$
$$4 + 7t - 14 + 3t^2 - 12t + 12 = 0$$
$$3t^2 - 5t + 2 = 0$$
$$(3t-2)(t-1) = 0$$
$$3t - 2 = 0, \quad t - 1 = 0$$
$$3t = 2, \qquad t = 1$$
$$t = \dfrac{2}{3}, \qquad t = 1 \quad \text{and both values check}$$

13. $x - \dfrac{2x}{x+1} = \dfrac{2}{x+1}$ Multiply each side by $x+1$

$$x(x+1) - \dfrac{2x}{\cancel{x+1}}\,\overset{1}{\cancel{(x+1)}} = \dfrac{2}{\cancel{x+1}}\,\overset{1}{\cancel{(x+1)}}$$

$$x^2 + x - 2x = 2$$
$$x^2 - x = 2$$
$$x^2 - x - 2 = 0$$
$$(x-2)(x+1) = 0$$
$$x - 2 = 0, \quad x + 1 = 0$$
$$x = 2, \qquad x = -1$$

Only $x = 2$ checks since $x = -1$ results in division by 0.

15. $\dfrac{2z}{3} - \dfrac{4-z}{z+4} = \dfrac{2z}{z+4}$ Multiply each side by $3(z+4)$

$$\dfrac{2z}{\cancel{3}}\,[\overset{1}{\cancel{3}}(z+4)] - \dfrac{4-z}{\cancel{z+4}}\,[3\,\overset{1}{\cancel{(z+4)}}] = \dfrac{2z}{\cancel{z+4}}\,[3\,\overset{1}{\cancel{(z+4)}}]$$

$$2z(z+4) - (4-z)3 = 2z(3)$$
$$2z^2 + 8z - 12 + 3z = 6z$$
$$2z^2 + 11z - 12 = 6z$$
$$2z^2 + 5z - 12 = 0$$
$$(2z-3)(z+4) = 0$$
$$2z - 3 = 0, \quad z + 4 = 0$$
$$2z = 3, \qquad z = -4$$
$$z = \dfrac{3}{2}, \qquad z = -4$$

Only $z = \dfrac{3}{2}$ checks since $z = -4$ results in division by 0.

17. $\sqrt{x + 7} = \sqrt{2x + 1}$
Square each side:

$$(\sqrt{x + 7})^2 = (\sqrt{2x + 1})^2$$
$$x + 7 = 2x + 1$$
$$7 = x + 1$$
$$6 = x$$

and it checks since the left side
is $\sqrt{6 + 7} = \sqrt{13}$
and the right side is
$\sqrt{2 \cdot 6 + 1} = \sqrt{12 + 1} = \sqrt{13}$

19. $x = \sqrt{8 - 2x}$ Square each side:

$$x^2 = (\sqrt{8 - 2x})^2$$
$$x^2 = 8 - 2x$$
$$x^2 + 2x - 8 = 0$$
$$(x + 4)(x - 2) = 0$$
$$x + 4 = 0, \quad x - 2 = 0$$
$$x = -4, \qquad x = 2$$

Here $x = 2$ checks since the left side
is 2 and the right side is $\sqrt{8 - 2 \cdot 2} =$
$\sqrt{4} = 2$, but $x = -4$ does not check
since the left side is -4 while the
right side is $\sqrt{8 - 2(-4)} = \sqrt{8 + 8} =$
$\sqrt{16} = 4$.

21. $\sqrt{x + 10} = x - 2$ Square each side:
$$(\sqrt{x + 10})^2 = (x - 2)^2$$
$$x + 10 = x^2 - 4x + 4$$
$$0 = x^2 - 5x - 6$$
$$0 = (x - 6)(x + 1)$$
$$x - 6 = 0, \quad x + 1 = 0$$
$$x = 6, \qquad x = -1$$

Here $x = 6$ checks since the left
side is $\sqrt{6 + 10} = \sqrt{16} = 4$ and the
right side is $6 - 2 = 4$, but,
$x = -1$ does not check since the left
side is $\sqrt{-1 + 10} = \sqrt{9} = 3$ while the
right side is $-1 - 2 = -3$.

23. $\sqrt{10 + x} - x = 10$
$\sqrt{10 + x} = x + 10$ Square each side
$$(\sqrt{10 + x})^2 = (x + 10)^2$$
$$10 + x = x^2 + 20x + 100$$
$$0 = x^2 + 19x + 90$$
$$0 = (x + 10)(x + 9)$$
$$x + 10 = 0, \quad x + 9 = 0$$
$$x = -10, \qquad x = -9 \quad \text{and both}$$
$$\text{values check}$$

25. $\sqrt{x + 8} - 1 = 2x$
$\sqrt{x + 8} = 2x + 1$ Square each side
$$(\sqrt{x + 8})^2 = (2x + 1)^2$$
$$x + 8 = 4x^2 + 4x + 1$$
$$0 = 4x^2 + 3x - 7$$
$$0 = (4x + 7)(x - 1)$$
$$4x + 7 = 0, \quad x - 1 = 0$$
$$4x = -7, \qquad x = 1$$
$$x = -\frac{7}{4}, \qquad x = 1$$

Here $x = 1$ checks since the left side is $\sqrt{1 + 8} - 1 = \sqrt{9} - 1 = 2$ and the

right side is $2(1) = 2$, but $x = -\frac{7}{4}$ does not check since the left side is

$\sqrt{-\frac{7}{4} + 8} - 1 = \sqrt{\frac{25}{4}} - 1 = \frac{5}{2} - 1 = \frac{3}{2}$ while the right side is $2\left(-\frac{7}{4}\right) = -\frac{7}{2}$.

27. $\sqrt{2 - t} - t = 10$

$\sqrt{2 - t} = t + 10$ Square each side

$(\sqrt{2 - t})^2 = (t + 10)^2$

$2 - t = t^2 + 20t + 100$

$0 = t^2 + 21t + 98$

$0 = (t + 7)(t + 14)$

$t + 7 = 0,\ t + 14 = 0$

$t = -7,\qquad t = -14$

Here $t = -7$ checks since the left side is $\sqrt{2 - (-7)} - (-7) = \sqrt{9} + 7 = 3 + 7 = 10$ and the right side is 10, but $t = -14$ does not check since the left side is $\sqrt{2 - (-14)} - (-14) = \sqrt{16} + 14 = 4 + 14 = 18$ while the right side is 10.

31. $\sqrt{v - 3} - v = -5$

$\sqrt{v - 3} = v - 5$ Square each side

$(\sqrt{v - 3})^2 = (v - 5)^2$

$v - 3 = v^2 - 10v + 25$

$0 = v^2 - 11v + 28$

$0 = (v - 7)(v - 4)$

$v - 7 = 0,\ v - 4 = 0$

$v = 7,\qquad v = 4$

Here $v = 7$ checks, but $v = 4$ does not check since the left side is $\sqrt{4 - 3} - 4 = \sqrt{1} - 4 = 1 - 3 = -3$ while the right side is -5.

35. $\sqrt[3]{3x^2 + 4x - 1} = \sqrt[3]{3x^2 + 7}$

Cube each side

$(\sqrt[3]{3x^2 + 4x - 1})^3 = (\sqrt[3]{3x^2 + 7})^3$

$3x^2 + 4x - 1 = 3x^2 + 7$

$4x - 1 = 7$

$4x = 8$

$x = 2$ and this value checks

39. $\sqrt[3]{3x - 1} + 1 = x$

$\sqrt[3]{3x - 1} = x - 1$ Cube each side

$(\sqrt[3]{3x - 1})^3 = (x - 1)^3$

$3x - 1 = x^3 - 3x^2 + 3x - 1$

$0 = x^3 - 3x^2$

$0 = x^2(x - 3)$

$x^2 = 0,\ x - 3 = 0$

$x = 0,\qquad x = 3$ and both values check

29. $\sqrt{2r + 3} + 6 = r$

$\sqrt{2r + 3} = r - 6$ Square each side

$(\sqrt{2r + 3})^2 = (r - 6)^2$

$2r + 3 = r^2 - 12r + 36$

$0 = r^2 - 14r + 33$

$0 = (r - 11)(r - 3)$

$r - 11 = 0,\ r - 3 = 0$

$r = 11,\qquad r = 3$

Here $r = 11$ checks since the left side is $\sqrt{2 \cdot 11 + 3} + 6 = \sqrt{25} + 6 = 5 + 6 = 11$ and the right side is 11, but $x = 3$ does not check since the left side is $\sqrt{2 \cdot 3 + 3} + 6 = \sqrt{9} + 6 = 3 + 6 = 9$ while the right side is 3.

33. $\sqrt[3]{x - 1} = 3$ Cube each side

$(\sqrt[3]{x - 1})^3 = 3^3$

$x - 1 = 27$

$x = 28$

and this value checks since the left side is $\sqrt[3]{28 - 1} = \sqrt[3]{27} = 3$ and the right side is 3

37. $\sqrt[3]{x^2 + 2x} = -1$ Cube each side

$(\sqrt[3]{x^2 + 2x})^3 = (-1)^3$

$x^2 + 2x = -1$

$x^2 + 2x + 1 = 0$

$(x + 1)(x + 1) = 0$

$x + 1 = 0$

$x = -1$ and this value checks

41. $\sqrt{9-x} + \sqrt{x+8} = 3$

$\sqrt{9-x} = 3 - \sqrt{x+8}$ Square each side

$(\sqrt{9-x})^2 = (3 - \sqrt{x+8})^2$

$9 - x = 9 - 6\sqrt{x+8} + x + 8$

$9 - x = 17 + x - 6\sqrt{x+8}$

$-8 - 2x = -6\sqrt{x+8}$ Now $\div$ by -2:

$4 + x = 3\sqrt{x+8}$ Square again

$(4+x)^2 = [3\sqrt{x+8}]^2$

$16 + 8x + x^2 = 9(x+8)$

$16 + 8x + x^2 = 9x + 72$

$x^2 - x - 56 = 0$

$(x-8)(x+7) = 0$

$x - 8 = 0, \quad x + 7 = 0$

$x = 8, \qquad x = -7$ and **neither** value checks since when $x = 8$, the left side is $\sqrt{9-8} + \sqrt{8+8} = \sqrt{1} + \sqrt{16} = 1 + 4 = 5$ and the right side is 3. Also, when $x = -7$, the left side is $\sqrt{9-(-7)} + \sqrt{-7+8} = \sqrt{16} + \sqrt{1} = 4 + 1 = 5$ and the right side is 3. Thus the solution set is $\emptyset$.

43. $\sqrt{2x-3} - \sqrt{x+2} = 1$

$\sqrt{2x-3} = 1 + \sqrt{x+2}$ Square each side

$(\sqrt{2x-3})^2 = (1 + \sqrt{x+2})^2$

$2x - 3 = 1 + 2\sqrt{x+2} + x + 2$

$2x - 3 = 3 + x + 2\sqrt{x+2}$

$x - 6 = 2\sqrt{x+2}$ Square again

$(x-6)^2 = [2\sqrt{x+2}]^2$

$x^2 - 12x + 36 = 4(x+2)$

$x^2 - 12x + 36 = 4x + 8$

$x^2 - 16x + 28 = 0$

$(x-14)(x-2) = 0$

$x - 14 = 0, \quad x - 2 = 0$

$x = 14, \qquad x = 2$ Here $x = 14$ checks since the left side is $\sqrt{2(14)-3} - \sqrt{14+2} = \sqrt{25} - \sqrt{16} = 5 - 4 = 1$ and the right side is 1, but $x = 2$ does not check since the left side is $\sqrt{2(2)-3} - \sqrt{2+2} = \sqrt{1} - \sqrt{4} = 1 - 2 = -1$ while the right side is 1.

45. $\sqrt{2x+2} + \sqrt{2x+6} = 4$ Square each side

$\left(\sqrt{2x+2} + \sqrt{2x+6}\right)^2 = 4^2$

$2x + 2 + \sqrt{2x+6} = 16$

$\sqrt{2x+6} = 14 - 2x$ Square again

$(\sqrt{2x+6})^2 = (14-2x)^2$

$2x + 6 = 196 - 56x + 4x^2$

$0 = 190 - 58x + 4x^2$ Divide by 2:

$$0 = 95 - 29x + 2x^2$$
$$0 = (5 - x)(19 - 2x) = 0$$
$$5 - x = 0, \quad 19 - 2x = 0$$
$$x = 5, \qquad x = \frac{19}{2}$$

Here $x = 5$ checks since the left side is $\sqrt{10 + 2 + \sqrt{10 + 6}} = \sqrt{10 + 2 + 4} = \sqrt{16} = 4$ and the right side is 4, but $x = \frac{19}{2}$ does not check since the left side is $\sqrt{19 + 2 + \sqrt{19 + 6}} = \sqrt{19 + 2 + 5} = \sqrt{26}$ while the right side is 4.

47. $\sqrt{11 - x} - \sqrt{x + 6} = 3$

$\sqrt{11 - x} = 3 + \sqrt{x + 6}$ Square each side

$$(\sqrt{11 - x})^2 = (3 + \sqrt{x + 6})^2$$
$$11 - x = 9 + 6\sqrt{x + 6} + x + 6$$
$$11 - x = 15 + x + 6\sqrt{x + 6}$$
$$-4 - 2x = 6\sqrt{x + 6} \qquad \text{Divide by 2:}$$
$$-2 - x = 3\sqrt{x + 6} \qquad \text{Square again}$$
$$(-2 - x)^2 = [3\sqrt{x + 6}]^2$$
$$4 + 4x + x^2 = 9(x + 6)$$
$$4 + 4x + x^2 = 9x + 54$$
$$x^2 - 5x - 50 = 0$$
$$(x - 10)(x + 5) = 0$$
$$x - 10 = 0, \ x + 5 = 0$$
$$x = 10, \qquad x = -5 \quad \text{Here } x = -5 \text{ checks since the left side is}$$

$\sqrt{11 - (-5)} - \sqrt{-5 + 6} = \sqrt{16} - \sqrt{1} = 4 - 1 = 3$ and the right side is 3, but $x = 10$ does not check since the left side is $\sqrt{11 - 10} - \sqrt{10 + 6} = \sqrt{1} - \sqrt{16} = 1 - 4 = -3$ while the right side is 3.

49. $\sqrt{2x - 3} - \sqrt{x + 7} + 2 = 0$

$\sqrt{2x - 3} = \sqrt{x + 7} - 2$ Square each side

$$(\sqrt{2x - 3})^2 = (\sqrt{x + 7} - 2)^2$$
$$2x - 3 = x + 7 - 4\sqrt{x + 7} + 4$$
$$2x - 3 = x + 11 - 4\sqrt{x + 7}$$
$$x - 14 = -4\sqrt{x + 7} \qquad \text{Square again}$$
$$(x - 14)^2 = [-4\sqrt{x + 7}]^2$$
$$x^2 - 28x + 196 = 16(x + 7)$$
$$x^2 - 28x + 196 = 16x + 112$$
$$x^2 - 44x + 84 = 0$$
$$(x - 2)(x - 42) = 0$$
$$x - 2 = 0, \ x - 42 = 0$$
$$x = 2, \qquad x = 42$$

Here $x = 2$ checks since the left side is $\sqrt{2 \cdot 2 - 3} - \sqrt{2 + 7} + 2 = \sqrt{1} - \sqrt{9} + 2 = 1 - 3 + 2 = 0$ and the right side is 0, but $x = 42$ does not check since the left side is $\sqrt{2 \cdot 42 - 3} - \sqrt{42 + 7} + 2 = \sqrt{81} - \sqrt{49} + 2 = 9 - 7 + 2 = 4$ while the right side is 0.

51. $\sqrt{5 + \sqrt{x + 1}} + 1 = \sqrt{x + 1}$

$\sqrt{5 + \sqrt{x + 1}} = \sqrt{x + 1} - 1$ Square each side

$\left(\sqrt{5 + \sqrt{x + 1}}\right)^2 = (\sqrt{x + 1} - 1)^2$

$5 + \sqrt{x + 1} = x + 1 - 2\sqrt{x + 1} + 1$

$5 + \sqrt{x + 1} = x + 2 - 2\sqrt{x + 1}$

$3\sqrt{x + 1} = x - 3$ Square again

$[3\sqrt{x + 1}]^2 = (x - 3)^2$

$9(x + 1) = x^2 - 6x + 9$

$9x + 9 = x^2 - 6x + 9$

$0 = x^2 - 15x$

$0 = x(x - 15)$

$x = 0, \quad x - 15 = 0$

$x = 0, \qquad x = 15$

Here $x = 15$ checks since the left side is $\sqrt{5 + \sqrt{15 + 1}} + 1 = \sqrt{5 + \sqrt{16}} + 1 = \sqrt{5 + 4} + 1 = \sqrt{9} + 1 = 3 + 1 = 4$ and the right side is $\sqrt{15 + 1} = \sqrt{16} = 4$, but $x = 0$ does not check since the left side is $\sqrt{5 + \sqrt{0 + 1}} + 1 = \sqrt{5 + 1} + 1 = \sqrt{6} + 1$ while the right side is $\sqrt{0 + 1} = \sqrt{1} = 1$.

53. Let r be the rate the car was traveling initially. The time to complete the distance of 660 miles is $\frac{660}{r}$ hours. When the rate is increased by 5 mph, the new rate is $r + 5$ and the new time is $\frac{660}{r + 5}$ hours. The equation is

$\frac{660}{r} = \frac{660}{r + 5} + 1$ Multiply by $r(r + 5)$

$\frac{660}{\cancel{r}} \overset{1}{[\cancel{r}(r + 5)]} = \frac{660}{\cancel{r + 5}} \overset{1}{[r\cancel{(r + 5)}]} + 1[r(r + 5)]$

$660(r + 5) = 660r + r(r + 5)$

$660r + 3300 = 660r + r^2 + 5r$

$0 = r^2 + 5r - 3300$

$0 = (r - 55)(r + 60)$

$r - 55 = 0, \quad r + 60 = 0$

$r = 55, \qquad r = -60$

Since r must be positive, the only meaningful value is $r = 55$ mph.

55. Let x be the speed of the boat in still water. The 54 minutes becomes $\frac{54}{60}$ hours or $\frac{9}{10}$ hour. The following table (as was done in Section 2.2) appears as

	Rate	Time	Distance
upstream:	$x - 3$	$\frac{12}{x - 3}$	12
downstream:	$x + 3$	$\frac{12}{x + 3}$	12

Since the sum of the two times is $\frac{9}{10}$ hr, the equation is $\frac{12}{x - 3} + \frac{12}{x + 3} = \frac{9}{10}$.

Now multiply by $10(x + 3)(x - 3)$ to get

$$12[10(x + 3)] + 12[10(x - 3)] = 9(x + 3)(x - 3)$$
$$120x + 360 + 120x - 360 = 9(x^2 - 9)$$
$$240x = 9x^2 - 81$$
$$0 = 9x^2 - 240x - 81 \quad \text{Now divide by 3 to have}$$
$$0 = 3x^2 - 80x - 27$$
$$0 = (3x + 1)(x - 27)$$
$$0 = 3x + 1, \quad 0 = x - 27$$
$$-\frac{1}{3} = x, \qquad x = 27 \text{ mph}$$

The only meaningful value is 27 mph for the speed of the boat.

57. Consider $ax^2 + bx + c = 0$. The discriminant is $b^2 - 4ac$. If $b^2 - 4ac$ is a nonperfect square, then $\sqrt{b^2 - 4ac}$ is an irrational value and

$$x = \frac{-b \pm \sqrt{b^2 - 4ac}}{2a} \text{ produces two irrational values.}$$

If $b^2 - 4ac$ is a perfect square, then $b^2 - 4ac$ is a whole number and

$$x = \frac{-b \pm \sqrt{b^2 - 4ac}}{2a} \text{ produces two rational values}$$

59. The cube of a negative number is always a negative number. In addition, the cube root of a negative number is a negative number.

Exercises 2.7

1. $x^2 - x - 12 > 0$
$(x - 4)(x + 3) > 0$

factors $\begin{cases} x - 4 & ---------0++++ \\ x + 3 & ---0+++++++++ \end{cases}$

Product: $+++0----0++++$

The solution is $(-\infty, -3) \cup (4, \infty)$ and the graph is

3. $x^2 - 2x - 3 \leq 0$
$(x - 3)(x + 1) \leq 0$

factors $\begin{cases} x - 3 & ---------0++++ \\ x + 1 & ---0+++++++++ \end{cases}$

Product: $+++0----0++++$

The solution is $[-1, 3]$ and the graph is

5. $-x^2 + 4x - 3 \leq 0$
$x^2 - 4x + 3 \geq 0$
$(x - 3)(x - 1) \geq 0$

factors $\begin{cases} x - 3 & ---------0++++ \\ x - 1 & ---0+++++++++ \end{cases}$

Product: $+++0----0++++$

The solution is $(-\infty, 1] \cup [3, \infty)$ and the graph is

7.
$$-2x^2 + 6 < 0$$
$$2x^2 - 6 > 0$$
$$2(x^2 - 3) > 0$$
$$2(x + \sqrt{3})(x - \sqrt{3}) > 0$$

factors $\begin{cases} x + \sqrt{3} & ---0+++++++++ \\ x - \sqrt{3} & --------0++++ \end{cases}$

Product: $+++0----0++++$

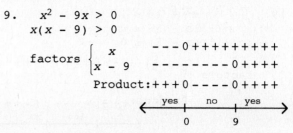

The solution is $(-\infty, -\sqrt{3}) \cup (\sqrt{3}, \infty)$

and the graph is

9.
$$x^2 - 9x > 0$$
$$x(x - 9) > 0$$

factors $\begin{cases} x & ---0+++++++++ \\ x - 9 & --------0++++ \end{cases}$

Product: $+++0----0++++$

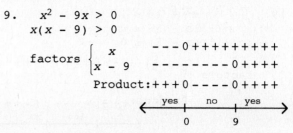

The solution is $(-\infty, 0) \cup (9, \infty)$

and the graph is

11.
$$8x - 2x^2 \geq 0$$
$$-2x^2 + 8x \geq 0$$
$$2x^2 - 8x \leq 0$$
$$2x(x - 4) \leq 0$$

factors $\begin{cases} x & ---0+++++++++ \\ x - 4 & --------0++++ \end{cases}$

Product: $+++0----0++++$

The solution is $[0, 4]$

and the graph is

13.
$$-2x^2 + 4 < 0$$
$$2x^2 - 4 > 0$$
$$2(x + \sqrt{2})(x - \sqrt{2}) > 0$$

factors $\begin{cases} x + \sqrt{2} & ---0+++++++++ \\ x - \sqrt{2} & --------0++++ \end{cases}$

Product: $+++0----0++++$

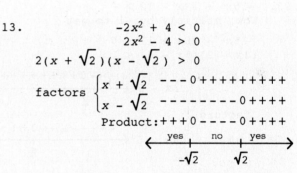

The solution is $(-\infty, -\sqrt{2}) \cup (\sqrt{2}, \infty)$

and the graph is

15.
$$2x - x^2 \leq 0$$
$$-x^2 + 2x \leq 0$$
$$x^2 - 2x \geq 0$$
$$x(x - 2) \geq 0$$

factors $\begin{cases} x & ---0+++++++++ \\ x - 2 & --------0++++ \end{cases}$

Product: $+++0----0++++$

The solution is $(-\infty, 0] \cup [2, \infty)$

and the graph is

17.
$$x^2 + 6x + 16 < 8$$
$$x^2 + 6x + 8 < 0$$
$$(x + 4)(x + 2) < 0$$

factors $\begin{cases} x + 4 & ---0+++++++++ \\ x + 2 & --------0++++ \end{cases}$

Product: $+++0----0++++$

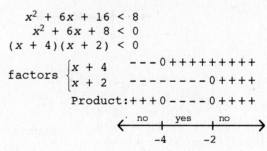

The solution is $(-4, -2)$

and the graph is

19.
$$z^2 + 2z \leq 15$$
$$z^2 + 2z - 15 \leq 0$$
$$(z + 5)(z - 3) \leq 0$$

factors $\begin{cases} z + 5 \quad ---0+++++++++ \\ z - 3 \quad ---------0++++ \end{cases}$

Product: $+++0----0++++$

```
      no   |  yes  |   no
<---------+-------+--------->
          -5      3
```

The solution is $[-5, 3]$

and the graph is

```
<----[=======]---->
     -5      3
```

21.
$$3w^2 + 13w - 10 > 0$$
$$(3w - 2)(w + 5) > 0$$

factors $\begin{cases} 3w - 2 \quad ---------0++++ \\ w + 5 \quad ---0+++++++++ \end{cases}$

Product: $+++0----0++++$

```
      yes  |  no   |  yes
<---------+-------+--------->
          -5      2/3
```

The solution is $(-\infty, -5) \cup \left(\dfrac{2}{3}, \infty\right)$

and the graph is

```
<===)----(===>
   -5    2/3
```

23.
$$-15x^2 + 28x - 12 > 0$$
Now, multiply by -1 to get

$$15x^2 - 28x + 12 < 0$$
$$(3x - 2)(5x - 6) < 0$$

$3x - 2$
```
     ---0+++++++++
<--------+--------->
         2/3
```

factors

$5x - 6$
```
     ---------0++++
<-----------+------->
            6/5
```

Product:
```
     +++0----0++++
      no  | yes | no
<--------+-----+------->
         2/3   6/5
```

The solution is $\left(\dfrac{2}{3}, \dfrac{6}{5}\right)$ and the graph is

```
<-----(=====)----->
      2/3   6/5
```

25.
$$3z^2 > 4z$$
$$3z^2 - 4z > 0$$
$$z(3z - 4) > 0$$

factors

z
```
     ---0+++++++++
<--------+--------->
         0
```

$3z - 4$
```
     ---------0++++
<-----------+------->
            4/3
```

Product:
```
     +++0----0++++
      yes | no  | yes
<--------+-----+------->
         0     4/3
```

The solution is $(-\infty, 0) \cup \left(\dfrac{4}{3}, \infty\right)$ and

the graph is

```
<===)----(===>
    0    4/3
```

27.
$$r - 6r^2 > -35$$
$$-6r^2 + r + 35 > 0$$
$$6r^2 - r - 35 < 0$$
$$(3r + 7)(2r - 5) < 0$$

factors

$3r + 7$
```
     ---0+++++++++
<--------+--------->
        -7/3
```

$2r - 5$
```
     ---------0++++
<-----------+------->
            5/2
```

Product:
```
     +++0----0++++
      no  | yes | no
<--------+-----+------->
        -7/3   5/2
```

The solution is $\left(-\dfrac{7}{3}, \dfrac{5}{2}\right)$ and the

graph is

```
<-----(=====)----->
     -7/3   5/2
```

29.
$$21 - 4x^2 > -5x$$
$$-4x^2 + 5x + 21 > 0$$
$$4x^2 - 5x - 21 < 0$$
$$(4x + 7)(x - 3) < 0$$

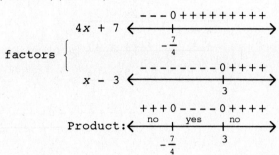

The solution is $\left(-\frac{7}{4}, \; 3\right)$ and the graph is

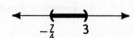

31. $-4s^2 - 4s + 1 \geq 0$
$4s^2 + 4s - 1 \leq 0$

For the zeros, $s = \dfrac{-4 \pm \sqrt{4^2 - 4(4)(-1)}}{2(4)} = \dfrac{-4 \pm \sqrt{16 + 16}}{8} = \dfrac{-4 \pm \sqrt{32}}{8}$

$= \dfrac{-4 \pm 4\sqrt{2}}{8} = \dfrac{4(-1 \pm \sqrt{2})}{2 \cdot 4} = \dfrac{-1 \pm \sqrt{2}}{2}$

The factors are $x - \left(\dfrac{-1 + \sqrt{2}}{2}\right)$ and $x - \left(\dfrac{-1 - \sqrt{2}}{2}\right)$. The number lines are

$x - \left(\dfrac{-1 - \sqrt{2}}{2}\right)$

```
          ---0+++++++++
        <---|---------->
          -1 - √2
          -------
             2
```

$x - \left(\dfrac{-1 + \sqrt{2}}{2}\right)$

```
          --------0++++
        <---------|---->
              -1 + √2
              -------
                 2
```

Product:
```
        +++0----0++++
         no   yes   no
        <---|----|---->
         -1 - √2  -1 + √2
         -------  -------
            2        2
```

The solution is $\left[\dfrac{-1 - \sqrt{2}}{2}, \; \dfrac{-1 + \sqrt{2}}{2}\right]$ and the graph is

```
        <---[----]---->
          -1 - √2  -1 + √2
          -------  -------
             2        2
```

33. $(x - 1)(x - 2)(x - 3) > 0$

$$x - 1 \qquad \text{-- 0 + + + + + + + + + +}$$
at 1

$$x - 2 \qquad \text{- - - - - - 0 + + + + +}$$
at 2

$$x - 3 \qquad \text{- - - - - - - - - 0 + +}$$
at 3

Product: -- 0 + + + 0 - - - 0 + + (no, yes, no, yes) at 1, 2, 3

The solution is $(1, 2) \cup (3, \infty)$ and the graph is (number line with open circles at 1, 2, 3)

35. $\dfrac{x + 1}{2x - 1} > 3$

$$\frac{x + 1}{2x - 1} - 3 > 0$$

$$\frac{x + 1}{2x - 1} - \frac{3(2x - 1)}{2x - 1} > 0$$

$$\frac{x + 1 - 3(2x - 1)}{2x - 1} > 0$$

$$\frac{-5x + 4}{2x - 1} > 0$$

$$-5x + 4 \qquad \text{+ + + + + + + + 0 - - - -}$$
at $\frac{4}{5}$

$$2x - 1 \qquad \text{- - - 0 + + + + + + + + +}$$
at $\frac{1}{2}$

Quotient: - - - u + + + + 0 - - - - (no, yes, no) at $\frac{1}{2}$, $\frac{4}{5}$

The solution is $\left(\dfrac{1}{2}, \dfrac{4}{5}\right)$ and the graph is (number line, open at $\frac{1}{2}$ and $\frac{4}{5}$)

37. $\dfrac{3x - 1}{x + 2} < 1$

$$\frac{3x - 1}{x + 2} - 1 < 0$$

$$\frac{3x - 1}{x + 2} - \frac{1(x + 2)}{x + 2} < 0$$

$$\frac{3x - 1 - (x + 2)}{x + 2} < 0$$

$$\frac{2x - 3}{x + 2} < 0$$

$$2x - 3 \qquad \text{- - - - - - - 0 + + + +}$$
at $\frac{3}{2}$

$$x + 2 \qquad \text{- - - 0 + + + + + + + + +}$$
at -2

Quotient: + + + u - - - 0 + + + + (no, yes, no) at -2, $\frac{3}{2}$

The solution is $\left(-2, \dfrac{3}{2}\right)$ and the graph is (number line, open at -2 and $\frac{3}{2}$)

39. $\dfrac{w + 1}{w + 3} \geq 2$

$$\frac{w + 1}{w + 3} - 2 \geq 0$$

$$\frac{w + 1}{w + 3} - \frac{2(w + 3)}{w + 3} \geq 0$$

$$\frac{w + 1 - 2(w + 3)}{w + 3} \geq 0$$

$$\frac{-w - 5}{w + 3} \geq 0$$

$$-w - 5 \qquad \text{+ + + 0 - - - - - - - - -}$$
at -5

$$w + 3 \qquad \text{- - - - - - - - 0 + + + +}$$
at -3

Quotient: - - - 0 + + + + u - - - - (no, yes, no) at -5, -3

The solution is $[-5, -3)$ and the graph is (number line, closed at -5, open at -3)

66

41. $\dfrac{7 - z}{(z - 2)(z - 3)} < 0$

$7 - z$ $\longleftarrow$ $+ + + + + + + + + + + + 0 - - -$
     7

$z - 2$ $\longleftarrow$ $- - - 0 + + + + + + + + + + +$
     2

$z - 3$ $\longleftarrow$ $- - - - - - - 0 + + + + + + + +$
     3

Quotient: $+ + + u - - - u + + + + + 0 - - -$
$\longleftarrow \underset{2}{\overset{no}{|}} \underset{3}{\overset{yes}{|}} \underset{7}{\overset{no}{\quad}} \overset{yes}{|} \longrightarrow$

The solution is $(2, 3) \cup (7, \infty)$ and
the graph is $\longleftarrow$ (—) —($\longrightarrow$
                2 3   7

43. $\dfrac{2}{w + 2} \geq \dfrac{1}{2w + 1}$

$\dfrac{2}{w + 2} - \dfrac{1}{2w + 1} \geq 0$

$\dfrac{2(2w + 1) - (w + 2)}{(w + 2)(2w + 1)} \geq 0$

$\dfrac{3w}{(w + 2)(2w + 1)} \geq 0$

$3w$ $\longleftarrow$ $- - - - - - - - - - - - - 0 + + +$
                                   0

$w + 2$ $\longleftarrow$ $- - - 0 + + + + + + + + + + + +$
           -2

$2w + 1$ $\longleftarrow$ $- - - - - - - 0 + + + + + + + +$
             $-\frac{1}{2}$

Quotient: $- - - u + + + u - - - - - 0 + + +$
$\longleftarrow \underset{-2}{\overset{no}{|}} \underset{-\frac{1}{2}}{\overset{yes}{|}} \underset{0}{\overset{no}{\quad}} \overset{yes}{|} \longrightarrow$

The solution is $\left(-2, -\dfrac{1}{2}\right) \cup [0, \infty)$ and
the graph is $\longleftarrow$ (——) [$\longrightarrow$
              -2 $-\frac{1}{2}$ 0

45. $\dfrac{5}{2w + 3} \geq \dfrac{-5}{w}$

$\dfrac{5}{2w + 3} + \dfrac{5}{w} \geq 0$

$\dfrac{5(w) + 5(2w + 3)}{(2w + 3)(w)} \geq 0$

$\dfrac{15w + 15}{w(2w + 3)} \geq 0$

$\dfrac{15(w + 1)}{w(2w + 3)} \geq 0$

$w + 1$ $\longleftarrow$ $- - - - - - - 0 + + + + + + + +$
             -1

w $\longleftarrow$ $- - - - - - - - - - - - - 0 + + +$
                                 0

$2w + 3$ $\longleftarrow$ $- - - 0 + + + + + + + + + + + +$
           $-\frac{3}{2}$

Quotient: $- - - u + + + 0 - - - - - u + + +$
$\longleftarrow \underset{-\frac{3}{2}}{\overset{no}{|}} \underset{-1}{\overset{yes}{|}} \underset{0}{\overset{no}{\quad}} \overset{yes}{|} \longrightarrow$

The solution is $\left(-\dfrac{3}{2}, -1\right] \cup (0, \infty)$ and the graph is $\longleftarrow$ (—] —($\longrightarrow$
                                                                    $-\frac{3}{2}$ -1 0

47. $\pi x^2 + 3\sqrt{2}\, x > \sqrt{15}$

$\pi x^2 + 3\sqrt{2}\, x - \sqrt{15} > 0$

Let's use the quadratic formula to solve $\pi x^2 + 3\sqrt{2}\, x - \sqrt{15} = 0$

$$x = \frac{-3\sqrt{2} \pm \sqrt{(3\sqrt{2})^2 - 4(\pi)(-\sqrt{15})}}{2\pi}$$

$$\approx \frac{-4.2426 \pm \sqrt{18 + 48.6693}}{6.2832}$$

$$= \frac{-4.2426 \pm \sqrt{66.6693}}{6.2832}$$

$$= \frac{-4.2426 \pm 8.1652}{6.2832}$$

$$x = \frac{-4.2426 - 8.1652}{6.2832} = -1.97, \quad x = \frac{-4.2426 + 8.1652}{6.2832} = 0.62$$

The factors are $(x + 1.97)$ and $(x - 0.62)$ so that the inequality is $(x + 1.97)(x - 0.62) > 0$.

```
                 - - - - - - - - 0 + + + +
  x - 0.62 <---------------------+------------->
                                0.62

              - - - 0 + + + + + + + + +
  x + 1.97 <-------+------------------------>
                 -1.97

              + + + 0 - - - - 0 + + + +
                yes     no     yes
  Product:  <------+---------+------------>
                 -1.97      0.62
```

The solution is $(-\infty, -1.97) \cup (0.62, \infty)$.

49. $3\pi x^2 + 0.0537 \le -3.96x$

$3\pi x^2 + 3.96x + 0.0537 \le 0$

Let's use the quadratic formula to solve
$3\pi x^2 + 3.96x + 0.0537 = 0$

$$x = \frac{-3.96 \pm \sqrt{(3.96)^2 - 4(3\pi)(0.0537)}}{2(3\pi)}$$

$$= \frac{-3.96 \pm \sqrt{15.6816 - 2.0244}}{18.8496}$$

$$= \frac{-3.96 \pm \sqrt{13.6572}}{18.8496}$$

$$= \frac{-3.96 \pm 3.6956}{18.8496}$$

$$x = \frac{-3.96 + 3.6956}{18.8496} = -0.01, \quad x = \frac{-3.96 - 3.6956}{18.8496} = -0.41$$

The factors are $(x + 0.01)$ and $(x + 0.41)$ so that the inequality is $(x + 0.01)(x + 0.41) \le 0$

```
                 - - - - - - - - 0 + + + +
  x + 0.01 <---------------------+------------->
                                -0.01

              - - - 0 + + + + + + + + +
  x + 0.41 <-------+------------------------>
                 -0.41

              + + + 0 - - - - 0 + + + +
                no      yes     no
  Product:  <------+---------+------------>
                 -0.41     -0.01
```

The solution is $[-0.41, -0.01]$.

51. Let x be the width. Then $2x$ is the length and

$$\begin{aligned} \text{Area} &\geq 1800 \\ x(2x) &\geq 1800 \\ 2x^2 &\geq 1800 \\ x^2 &\geq 900 \\ x &\geq 30 \text{ feet} \end{aligned}$$

Thus, the width ≥ 30 feet and the length $\geq 2(30) = 60$ feet.

53.
$$\begin{aligned} 128t - 16t^2 &\geq 192 \\ 128t - 16t^2 - 192 &\geq 0 \\ 16t^2 - 128t + 192 &\leq 0 \\ 16(t - 2)(t - 6) &\leq 0 \end{aligned}$$

$$t - 2 \quad \xleftarrow{\quad \overset{\displaystyle ---0+++++++++}{\underset{2}{|}} \quad}$$

$$t - 6 \quad \xleftarrow{\quad \overset{\displaystyle --------0++++}{\underset{6}{|}} \quad}$$

$$\xleftarrow{\quad \overset{\displaystyle +++0----0++++}{\underset{2 \qquad 6}{\text{no} \quad \text{yes} \quad \text{no}}} \quad}$$

Product:

The height is 192 feet or more when $2 \leq t \leq 6$.

55. The second part comes from both factors being negative so that the product is positive.

CHAPTER REVIEW

1.
$$\begin{aligned} 6x + 5 &= 4 \\ 6x &= 4 - 5 \\ 6x &= -1 \\ \frac{6x}{6} &= \frac{-1}{6} \\ x &= -\frac{1}{6} \end{aligned}$$

2.
$$\begin{aligned} 7x + 4(3 - 2x) &= 5 + x \\ 7x + 12 - 8x &= 5 + x \\ 12 - x &= 5 + x \\ 7 - x &= x \\ 7 &= 2x \\ \frac{7}{2} &= x \end{aligned}$$

3.
$$\begin{aligned} (3x - 1)(4x + 2) &= (12x - 1)(x + 2) \\ 12x^2 - 4x + 6x - 2 &= 12x^2 - x + 24x - 2 \\ 12x^2 + 2x - 2 &= 12x^2 + 23x - 2 \\ 2x - 2 &= 23x - 2 \\ 2x &= 23x \\ -21x &= 0 \\ x &= 0 \end{aligned}$$

4.
$$\frac{3}{x + 1} + 5 = \frac{4x}{x + 1}$$
$$\left[\frac{3}{x + 1} + 5 = \frac{4x}{x + 1} \right] (x + 1)$$
$$\begin{aligned} 3 + 5(x + 1) &= 4x \\ 3 + 5x + 5 &= 4x \\ 8 + 5x &= 4x \\ 8 &= -x \\ -8 &= x \end{aligned}$$

5.
$$\left[\frac{3}{x - 1} + \frac{4}{x + 2} = \frac{5x}{(x - 1)(x + 2)} \right] (x - 1)(x + 2)$$
$$\begin{aligned} 3(x + 2) + 4(x - 1) &= 5x \\ 3x + 6 + 4x - 4 &= 5x \\ 7x + 2 &= 5x \\ 2 &= -2x \\ -1 &= x \end{aligned}$$

6. $|2x - 9| = 0$
is written as
$$\begin{aligned} 2x - 9 &= 0 \\ 2x &= 9 \\ x &= \frac{9}{2} \end{aligned}$$

7.
$$\begin{aligned} |3x + 4| &= 19 \\ 3x + 4 = 19 \text{ or } 3x + 4 &= -19 \\ 3x = 15 \text{ or } \qquad 3x &= -23 \\ x = 5 \text{ or } \qquad x &= -\frac{23}{3} \end{aligned}$$

8. $|x + 2| = -5$ has no solution since the left side is positive whereas the right side is negative.

9. $|7x - 1| = |4x + 3|$

$7x - 1 = 4x + 3$ or $7x - 1 = -(4x + 3)$

$7x = 4x + 4$ or $7x - 1 = -4x - 3$

$3x = 4$ or $7x = -4x - 2$

$x = \dfrac{4}{3}$ or $11x = -2$

$x = -\dfrac{2}{11}$

10. $\dfrac{|3 - 2x|}{|2 - x|} = 1$

$\dfrac{3 - 2x}{2 - x} = 1$ or $\dfrac{-(3 - 2x)}{2 - x} = 1$

$3 - 2x = 1(2 - x)$ or $-(3 - 2x) = 1(2 - x)$

$3 - 2x = 2 - x$ or $-3 + 2x = 2 - x$

$1 - 2x = -x$ or $2x = 5 - x$

$1 = x$ or $3x = 5$

$x = \dfrac{5}{3}$

11. $ax + by = bx + ay$

$by = bx - ax + ay$

$by - ay = bx - ax$

$y(b - a) = x(b - a)$

$$y = \frac{x\cancel{(b - a)}}{\cancel{b - a}} = x$$

12. $\dfrac{1}{x} + \dfrac{1}{y} + \dfrac{1}{z} = 1$

$\dfrac{1}{z} = 1 - \dfrac{1}{x} - \dfrac{1}{y}$

$\dfrac{1}{z} = \dfrac{xy - y - x}{xy}$

$z = \dfrac{xy}{xy - y - x}$

13. a) Let $S = 96$ to get

$|176 - 32t| = 96$

$176 - 32t = 96$ or $176 - 32t = -96$

$-32t = -80$ or $-32t = -272$

$t = \dfrac{-80}{-32} = \dfrac{5}{2}$ or $t = \dfrac{-272}{-32} = \dfrac{17}{2}$

b) The highest point occurs when $S = 0$:

$|176 - 32t| = 0$

$176 - 32t = 0$

$176 = 32$

$\dfrac{176}{32} = t$

$\dfrac{11}{2} = t$

14. Let x be the number of juniors. Then $2x - 2300$ is the number of sophomores and the equation is

$x + (2x - 2300) = 8650$

$3x - 2300 = 8650$

$3x = 10,950$

$x = \dfrac{10,950}{3} = 3650$

Thus, there are $x = 3650$ juniors and $2x - 2300 = 2(3650) - 2300 = 7300 - 2300 = 5000$ sophomores.

15. Let x be the short side. Then $x + 20$ is the long side and Perimeter = 240 gives

$2(x) + 2(x + 20) = 240$

$2x + 2x + 40 = 240$

$4x + 40 = 240$

$4x = 200$

$x = 50$ in

The short side is $x = 50$ in and the long side is $x + 20 = 50 + 20 = 70$ in.

16. Let x represent Beckie's present age. The equation is

$x + 4 = 3(x - 6)$

$x + 4 = 3x - 18$

$x + 22 = 3x$

$22 = 2x$

$11 = x$ Beckie is now 11 years old.

17. Let x be the amount invested at 8%. Then $36,000 - x$ is the amount invested at 9% and the table is

Principal	Rate	Time	Interest
x	0.08	1	$0.08x$
$36,000 - x$	0.09	1	$0.09(36,000 - x)$

The equation is
$$0.08x + 0.09(36,000 - x) = 3090$$
$$0.08x + 3240 - 0.09x = 3090$$
$$-0.01x + 3240 = 3090$$
$$-0.01x = -150$$
$$x = \frac{-150}{-0.01} = \$15,000$$

Thus, $x = \$15,000$ was invested at 8% and $36,000 - x = \$36,000 - \$15,000 = \$21,000$ was invested at 9%.

18. Let x be the amount invested at 17%. The table is

Principal	Rate	Time	Interest
4000	15% = 0.15	1 yr	4000(0.15)
x	17% = 0.17	1 yr	$x(0.17)$

The equation is
$$4000(0.15) + x(0.17) = 1144$$
$$600 + 0.17x = 1144$$
$$0.17x = 544$$
$$x = \frac{544}{0.17} = \$3200$$

19. Let x be the number of adult tickets. Then $x + 14$ is the number of children tickets and the table is

Tickets	Price per ticket	Revenue
x	7.50	$7.50x$
$x + 14$	5.80	$5.80(x + 14)$

The equation is
$$7.50x + 5.80(x + 14) = 1836.80$$
$$7.50x + 5.80x + 81.20 = 1836.80$$
$$13.30x + 81.20 = 1836.80$$
$$13.30x = 1755.60$$
$$x = \frac{1755.60}{13.30} = 132$$

Thus, $x = 132$ adult tickets and $x + 14 = 132 + 14 = 146$ children tickets were sold.

20. Let x be the amount of pure acid needed. The table is

Amount	Concentration	Pure Acid
5	2% = 0.02	5(0.02)
x	100% = 1.00	$x(1.00)$
$x + 5$	25% = 0.25	$(x + 5)(0.25)$

The equation is
$$5(0.02) + x(1.00) = (x + 5)(0.25)$$
$$0.10 + x = 0.25x + 1.25$$
$$x = 0.25x + 1.15$$
$$0.75x = 1.15$$
$$x = \frac{1.15}{0.75} = \frac{115}{75} = \frac{23}{15} \text{ or } 1\frac{8}{15} \text{ liters}$$

21. Let x be the amount (in grams) of 30% alloy to be used. The table is

Mixture	Percent silver	Quantity	Amount of pure silver
Original	15%	500	0.15(500)
Added	30%	x	0.30(x)
Final	25%	500 + x	0.25(500 + x)

The equation is
$$0.15(500) + 0.30x = 0.25(500 + x)$$
$$75 + 0.30x = 125 + 0.25x$$
$$0.30x = 50 + 0.25x$$
$$0.05x = 50$$
$$x = \frac{50}{0.05} = 1000 \text{ grams}$$

22. Let x be the amount of 8-8-8 fertilizer needed. The table is

Amount	Concentration	Pure Nitrogen
x	8% = 0.08	x(0.08)
700	5% = 0.05	700(0.05)
700 + x	6% = 0.06	(700 + x)(0.06)

The equation is
$$x(0.08) + 700(0.05) = (700 + x)(0.06)$$
$$0.08x + 35 = 42 + 0.06x$$
$$0.08x = 7 + 0.06x$$
$$0.02x = 7$$
$$x = \frac{7}{0.02} = \frac{700}{2} = 350 \text{ pounds}$$

23. Let r be the rate of the current. The table is

Travel	Rate	Time	Distance
Upstream	7 - r	5	5(7 - r)
Downstream	7 + r	2	2(7 + r)

The equation is
$$5(7 - r) = 2(7 + r)$$
$$35 - 5r = 14 + 2r$$
$$21 - 5r = 2r$$
$$21 = 7r$$
$$3 = r$$
The rate of the current is 3 miles per hour.

24. Let x be the rate of the bike. The table is

	Time	Rate	Distance
Stella:	$\frac{5}{6}$ hr	4 mph	$\left(\frac{5}{6}\right)(4)$ miles
Mother:	$\frac{1}{3}$ hr	x	$\left(\frac{1}{3}\right)(x)$ miles

The equation is $\frac{1}{3}x = \frac{20}{6}$

$$x = \left(\frac{20}{6}\right)(3) = \frac{60}{6} = 10 \text{ mph}$$

25. Let x be the time required for Jane to mow the yard working alone.

In 1 hour, Jane completes $\frac{1}{x}$ of the job.

In 1 hour, Ron completes $\frac{1}{4}$ of the job.

In 1 hour, both complete $\frac{1}{1} = 1$ of the job.

The equation is $\frac{1}{x} + \frac{1}{4} = 1$ Multiply by $4x$: $\left[\frac{1}{x} + \frac{1}{4} = 1\right]4x$

$$x + 4 = 4x$$
$$4 = 3x$$
$$\frac{4}{3} = x$$

Thus, $x = \frac{4}{3}$ hr $= 1\frac{1}{3}$ hr $= 1$ hr 20 min is the time required by Jane to mow the yard.

26. $7x - 3 < 4$ 27. $5y + 7 < 7y + 3$ 28. $5z - 11 \geq 7z - 3$
       $7x < 7$ $5y + 4 < 7y$ $5z - 8 \geq 7z$
       $x < \frac{7}{7}$ $4 < 2y$ $-8 \geq 2z$
       $x < 1$ which is $(-\infty, 1)$ $\frac{4}{2} < y$ $-\frac{8}{2} \geq z$
                                        $2 < y$ which is $(2, \infty)$ $-4 \geq z$ which is $(-\infty, -4]$

29. $4t - 21 < 9t + 14$ 30. $3x + 5 > 4$ and $-5x + 6 < 4x - 1$
       $4t - 35 < 9t$ $3x > -1$ and $6 < 9x - 1$
       $-35 < 5t$ $x > -\frac{1}{3}$ and $7 < 9x$
       $-\frac{35}{5} < t$ $x > -\frac{1}{3}$ and $\frac{7}{9} < x$
       $-7 < t$ which is $(-7, \infty)$

$$\frac{7}{9} < x \text{ which is } \left(\frac{7}{9}, \infty\right)$$

31. $13 > 5 - 2t \geq 7$
       $8 > -2t \geq 2$
       $-4 < t \leq -1$ (Remember to reverse the inequality when ÷ by a negative number.)
    which is $(-4, -1]$

32. $-10 < 2 - 3y \leq 11$ 33. $3x - 5 < 4$ or $7x + 2 > 5$
     $-12 < -3y \leq 9$ $3x < 9$ or $7x > 3$
       $4 > y \geq -3$ which is $[-3, 4)$ $x < 3$ or $x > \frac{7}{3}$

                                      When combined, this is the entire real number
                                      line which is $(-\infty, \infty)$.

34. $3(2 - u) < 4u + 27$ or $3(u - 2) > 7(u + 2)$
       $6 - 3u < 4u + 27$ or $3u - 6 > 7u + 14$
           $6 < 7u + 27$ or $-6 > 4u + 14$
         $-21 < 7u$ or $-20 > 4u$
           $-3 < u$ or $-5 > u$
    This gives $(-\infty, -5) \cup (-3, \infty)$ and the graph is

$$\xleftarrow{\hspace{1cm}}\ )\ \ (\ \xrightarrow{\hspace{1cm}}$$
$$\ \ \ \ \ -5\ -3$$

73

35. $|5t - 6| \leq 3$ becomes
$$-3 \leq 5t - 6 \leq 3$$
$$3 \leq 5t \leq 9$$
$$\frac{3}{5} \leq t \leq \frac{9}{5}$$

This gives $\left[\frac{3}{5}, \frac{9}{5}\right]$ and the graph is

36. $\dfrac{|2x - 7|}{3} < 5$

which is $|2x - 7| < 15$
$$-15 < 2x - 7 < 15$$
$$-8 < 2x < 22$$
$$-4 < x < 11$$

This gives $(-4, 11)$ and the graph is

37. $\left|\dfrac{3x - 8}{2}\right| \geq 2$ becomes $\quad \dfrac{3x - 8}{2} \geq 2 \quad$ or $\quad \dfrac{3x - 8}{2} \leq -2$

$$3x - 8 \geq 4 \quad \text{or} \quad 3x - 8 \leq -4$$
$$3x \geq 12 \quad \text{or} \quad 3x \leq 4$$
$$x \geq 4 \quad \text{or} \quad x \leq \frac{4}{3}$$

This gives $[4, \infty) \cup \left(-\infty, \dfrac{4}{3}\right]$ and the graph is

38. $|5 - 6y| \geq 5$ which is $5 - 6y \geq 5$ or $5 - 6y \leq -5$
$$-6y \geq 0 \quad \text{or} \quad -6y \leq -10$$
$$y \leq 0 \quad \text{or} \quad y \geq \frac{-10}{-6}$$
$$y \leq 0 \quad \text{or} \quad y \geq \frac{5}{3}$$

This gives $(-\infty, 0] \cup \left[\dfrac{5}{3}, \infty\right]$ and the graph is

39.
$$3x^2 + 7x = 6$$
$$3x^2 + 7x - 6 = 0$$
$$(3x - 2)(x + 3) = 0$$
$$3x - 2 = 0 \text{ or } x + 3 = 0$$
$$3x = 2 \text{ or } \quad x = -3$$
$$x = \frac{2}{3}, \quad \quad x = -3$$

40.
$$2z^2 + 5z = 3$$
$$2z^2 + 5z - 3 = 0$$
$$(2z - 1)(z + 3) = 0$$
$$2z - 1 = 0, \; z + 3 = 0$$
$$2z = 1, \quad \quad z = -3$$
$$z = \frac{1}{2}, \quad \quad z = -3$$

41.
$$5t^2 - 7t - 6 = 0$$
$$(5t + 3)(t - 2) = 0$$
$$5t + 3 = 0 \text{ or } t - 2 = 0$$
$$5t = -3 \text{ or } \quad t = 2$$
$$t = -\frac{3}{5}, \quad \quad t = 2$$

42.
$$3v(6v - 5) - 25 = 0$$
$$18v^2 - 15v - 25 = 0$$
$$(3v - 5)(6v + 5) = 0$$
$$3v - 5 = 0, \; 6v + 5 = 0$$
$$3v = 5, \quad \quad 6v = -5$$
$$v = \frac{5}{3}, \quad \quad v = -\frac{5}{6}$$

43.
$$x^2 - 45 = 4x$$
$$x^2 - 4x - 45 = 0$$
$$x^2 - 4x = 45$$
$$x^2 - 4x + 4 = 45 + 4$$
$$(x - 2)^2 = 49$$
$$x - 2 = \pm\sqrt{49}$$
$$x - 2 = \pm 7$$
$$x = 2 \pm 7 \text{ which gives } x = 2 + 7 = 9$$
$$\text{or } x = 2 - 7 = -5$$

44.
$$z^2 + 4 = -2 - 4z$$
$$z^2 + 4z + 4 = -2$$
$$z^2 + 4z = -6$$
$$z^2 + 4z + \underline{4} = -6 + \underline{4}$$
$$(z + 2)^2 = -2$$
$$z + 2 = \pm\sqrt{-2}$$
$$z + 2 = \pm i\sqrt{2}$$
$$z = -2 \pm i\sqrt{2}$$

45.
$$2t(t - 3) + 5 = 0$$
$$2t^2 - 6t + 5 = 0$$
$$2t^2 - 6t = -5$$
$$t^2 - 3t = -\frac{5}{2}$$
$$t^2 - 3t + \frac{9}{4} = -\frac{5}{2} + \frac{9}{4}$$
$$\left(t - \frac{3}{2}\right)^2 = -\frac{1}{4}$$
$$t - \frac{3}{2} = \pm\sqrt{-\frac{1}{4}}$$
$$t - \frac{3}{2} = \pm\frac{1}{2}i$$
$$t = \frac{3}{2} \pm \frac{1}{2}i = \frac{3 \pm i}{2}$$

46.
$$1 + 12w + 2w^2 = 0$$
$$2w^2 + 12w = -1$$
$$w^2 + 6w = -\frac{1}{2}$$
$$w^2 + 6w + 9 = -\frac{1}{2} + 9$$
$$(w + 3)^2 = \frac{17}{2}$$
$$(w + 3)^2 = \pm\sqrt{\frac{17}{2} \cdot \frac{2}{2}}$$
$$(w + 3)^2 = \pm\frac{\sqrt{34}}{2}$$
$$w = -3 \pm \frac{\sqrt{34}}{2} = \frac{-6 \pm \sqrt{34}}{2}$$

47.
$$0 = 1 - 4r^2 - 8r$$
$$4r^2 + 8r = 1$$
$$r^2 + 2r = \frac{1}{4}$$
$$r^2 + 2r + \underline{1} = \frac{1}{4} + \underline{1}$$
$$(r + 1)^2 = \frac{5}{4}$$
$$r + 1 = \pm\sqrt{\frac{5}{4}}$$
$$r + 1 = \pm\frac{\sqrt{5}}{2}$$
$$r = -1 \pm \frac{\sqrt{5}}{2} = \frac{-2}{2} \pm \frac{\sqrt{5}}{2} = \frac{-2 \pm \sqrt{5}}{2}$$

48.
$$(2t - 1)^2 - 3(2t - 1) - 10 = 0$$
$$4t^2 - 4t + 1 - 6t + 3 - 10 = 0$$
$$4t^2 - 10t - 6 = 0$$
$$2t^2 - 5t - 3 = 0$$
$$(2t + 1)(t - 3) = 0$$
$$2t + 1 = 0, \quad t - 3 = 0$$
$$t = -\frac{1}{2}, \qquad t = 3$$

Another solution is the following:
Let $z = 2t - 1$. Then the equation is
$$z^2 - 3z - 10 = 0$$
$$(z - 5)(z + 2) = 0$$
$$z - 5 = 0, \quad z + 2 = 0$$
$$z = 5, \qquad z = -2$$

Since $z = 2t - 1$, then $z = 5$ becomes
$$2t - 1 = 5$$
$$2t = 6$$
$$t = 3$$
and $z = -2$ becomes
$$2t - 1 = -2$$
$$2t = -1$$
$$t = -\frac{1}{2}$$

49. Let $x = z^2$. Then $x^2 = z^4$ and the equation $9z^4 - 10z^2 + 1 = 0$ becomes

$9x^2 - 10x + 1 = 0$

$(9x - 1)(x - 1) = 0$

$9x - 1 = 0$ or $x - 1 = 0$

$x = \dfrac{1}{9}$ or $\qquad x = 1$

Now substitute z^2 for x to get $x = \dfrac{1}{9}$ is $z^2 = \dfrac{1}{9}$ or $z = \pm\sqrt{\dfrac{1}{9}}$ or $z = \pm\dfrac{1}{3}$

$x = 1$ is $z^2 = 1$ or $z = \pm\sqrt{1}$ or $z = \pm 1$

50.

$9y^4 + 4 = 13y^2$

$9y^4 - 13y^2 + 4 = 0$

Let $x = y^2$ so that $x^2 = y^4$ and the equation becomes

$9x^2 - 13x + 4 = 0$

$(9x - 4)(x - 1) = 0$

$9x - 4 = 0, \quad x - 1 = 0$ or $\quad x = \dfrac{4}{9}, \quad x = 1$

Since $x = y^2$, then $x = \dfrac{4}{9}$ becomes $\dfrac{4}{9} = y^2$ or $y = \pm\sqrt{\dfrac{4}{9}} = \pm\sqrt{\dfrac{2}{3}}$

and $x = 1$ becomes $1 = y^2$ or $y = \pm\sqrt{1} = \pm 1$

51. Let $y = \sqrt{x}$. Then $y^2 = x$ and the equation $7 - 15\sqrt{x} + 2x = 0$ becomes

$7 - 15y + 2y^2 = 0$

$(2y - 1)(y - 7) = 0$

$2y - 1 = 0$ or $y - 7 = 0$

$y = \dfrac{1}{2}$ or $\qquad y = 7$

Now substitute $\sqrt{x}$ for y to get $y = \dfrac{1}{2}$ is $\sqrt{x} = \dfrac{1}{2}$ or $x = \dfrac{1}{4}$

$y = 7$ is $\sqrt{x} = 7$ or $x = 49$

52. $y^2 - 2y - 5 = 0$

$$y = \frac{-(-2) \pm \sqrt{(-2)^2 - 4(1)(-5)}}{2(1)} = \frac{2 \pm \sqrt{4 + 20}}{2} = \frac{2 \pm \sqrt{24}}{2}$$

$$= \frac{2 \pm 2\sqrt{6}}{2} = \frac{2(1 \pm \sqrt{6})}{2} = 1 \pm \sqrt{6}$$

53. $2 - r - 2r^2 = 0$

$-2r^2 - r + 2 = 0$

$2r^2 + r - 2 = 0$

$$x = \frac{-1 \pm \sqrt{1^2 - 4(2)(-2)}}{2(2)} = \frac{-1 \pm \sqrt{1 + 16}}{4} = \frac{-1 \pm \sqrt{17}}{4}$$

54. $2t^2 = -5 - 2t$

$2t^2 + 2t + 5 = 0$

$$t = \frac{-2 \pm \sqrt{2^2 - 4(2)(5)}}{2(2)} = \frac{-2 \pm \sqrt{4 - 40}}{4} = \frac{-2 \pm \sqrt{-36}}{4}$$

$$= \frac{-2 \pm 6i}{4} = \frac{2(-1 \pm 3i)}{4} = \frac{-1 \pm 3i}{2}$$

55. $5x(x - 1) = -1 - 2x$

$5x^2 - 5x = -1 - 2x$

$5x^2 - 3x = -1$

$5x^2 - 3x + 1 = 0$

$$x = \frac{-(-3) \pm \sqrt{(-3)^2 - 4(5)(1)}}{2(5)} = \frac{3 \pm \sqrt{9 - 20}}{10} = \frac{3 \pm \sqrt{-11}}{10} = \frac{3 \pm i\sqrt{11}}{10}$$

56. $b^2 - 4ac = (-7)^2 - 4(1)(-10) = 49 + 40 = 89 > 0$.
    The roots are two distinct real numbers.

57. $\quad\quad 2t^2 + 6 = 5t$
    $2t^2 - 5t + 6 = 0$
    Now, $b^2 - 4ac = (-5)^2 - 4(2)(6) = 25 - 48 = -23 < 0$
    The roots are two nonreal complex numbers that are conjugates.

58. $\quad\quad 2y^2 - 3y = 4$
    $2y^2 - 3y - 4 = 0$
    $b^2 - 4ac = (-3)^2 - 4(2)(-4) = 9 + 32 = 41 > 0$
    The roots are two distinct real numbers.

59. $z(z - 1) + 4 = 0$
    $\quad z^2 - z + 4 = 0$
    Now, $b^2 - 4ac = (-1)^2 - 4(1)(4) = 1 - 16 = -15 < 0$
    The roots are two nonreal complex numbers that are conjugates.

60. $\quad\quad 4x^2 - 12x = -9$
    $4x^2 - 12x + 9 = 0$
    $b^2 - 4ac = (-12)^2 - 4(4)(9) = 144 - 144 = 0$
    The roots are two equal real numbers.

61. Let $P = 40$ in the equation $P = 30x - x^2 - 160$ to get
    $$\begin{aligned} 40 &= 30x - x^2 - 160 \\ x^2 + 40 &= 30x - 160 \\ x^2 - 30x + 40 &= -160 \\ x^2 - 30x + 200 &= 0 \\ (x - 20)(x - 10) &= 0 \\ x - 20 = 0, \quad x - 10 &= 0 \\ x = 20, \quad\quad x &= 10 \end{aligned}$$

62. $D = 22 + 6p - 0.5p^2$. Let $D = 38$ to get
    $38 = 22 + 6p - 0.5p^2$
    or $\quad 0.5p^2 - 6p + 16 = 0$ Now multiply by 2 to obtain
    or $\quad\quad p^2 - 12p + 32 = 0$
    $\quad\quad (p - 4)(p - 8) = 0$
    $\quad p - 4 = 0, \quad p - 8 = 0$
    $\quad\quad p = 4, \quad\quad p = 8$
    The price can be $4.00 or $8.00.

63. Let $P = 144$ in the equation $P = 24n - n^2$ to get
    $$\begin{aligned} 144 &= 24n - n^2 \\ n^2 + 144 &= 24n \\ n^2 - 24n + 144 &= 0 \\ (n - 12)^2 &= 0 \\ n - 12 &= 0 \\ n &= 12 \end{aligned}$$

64. $D = 18 + 5p - 0.5p^2$. Let $D = 6$ to get
    $6 = 18 + 5p - 0.5p^2$
    $\quad 0.5p^2 - 5p - 12 = 0$ Now, multiply by 2 to obtain
    $\quad\quad p^2 - 10p - 24 = 0$
    $\quad (p - 12)(p + 2) = 0$
    $p - 12 = 0, \quad p + 2 = 0$
    $\quad\quad p = 12, \quad\quad p = -2$
    The only meaningful value is $p = \$12$ per pound.

65. Let x be the width. Then the height is $\left(1 + \dfrac{1}{3}\right)x = \dfrac{4}{3}x$.

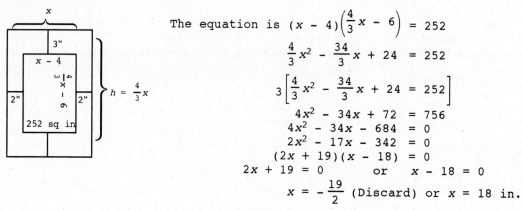

The equation is $(x - 4)\left(\dfrac{4}{3}x - 6\right) = 252$

$$\dfrac{4}{3}x^2 - \dfrac{34}{3}x + 24 = 252$$

$$3\left[\dfrac{4}{3}x^2 - \dfrac{34}{3}x + 24 = 252\right]$$

$$4x^2 - 34x + 72 = 756$$

$$4x^2 - 34x - 684 = 0$$

$$2x^2 - 17x - 342 = 0$$

$$(2x + 19)(x - 18) = 0$$

$$2x + 19 = 0 \qquad \text{or} \qquad x - 18 = 0$$

$$x = -\dfrac{19}{2} \text{ (Discard) or } x = 18 \text{ in.}$$

Thus, the width is $x = 18$ in and the length is $\dfrac{4}{3}x = \dfrac{4}{3}(18) = 24$ in.

66. Let x be the width of the plowed area. Then the unplowed part has dimensions $100 - 2x$ by $80 - 2x$. The unplowed part is 40% of 100×80 or $0.40(100 \times 80)$ or 4800.

The equation is
$$(100 - 2x)(80 - 2x) = 4800$$
$$4x^2 - 360x + 8000 = 4800$$
$$4x^2 - 360x + 3200 = 0 \quad \text{Now} \div \text{ by 4 to obtain}$$
$$x^2 - 90x + 800 = 0$$
$$(x - 10)(x - 80) = 0$$
$$x - 10 = 0, \ x - 80 = 0$$
$$x = 10, \qquad x = 80 \quad \text{We discard } x = 80 \text{ and conclude}$$
$$\text{that } x = 10 \text{ feet.}$$

The width of the plowed area is 10 feet.

67. $\left[3r + 1 - \dfrac{20}{3r + 2} = 0\right](3r + 2)$

$$(3r + 1)(3r + 2) - 20 = 0$$
$$9r^2 + 9r + 2 - 20 = 0$$
$$9r^2 + 9r - 18 = 0$$
$$r^2 + r - 2 = 0$$
$$(r + 2)(r - 1) = 0$$
$$r + 2 = 0, \ r - 1 = 0$$
$$r = -2, \qquad r = 1$$

68. $y - \dfrac{6}{2y - 3} = 2$

$$-\dfrac{6}{2y - 3} = 2 - y \quad \text{Now multiply by } 2y - 3$$

$$-\dfrac{6}{\cancel{2y - 3}} \cdot \cancel{(2y - 3)} = (2 - y)(2y - 3)$$

$$-6 = -2y^2 + 7y - 6$$
$$2y^2 - 7y = 0$$
$$y(2y - 7) = 0$$
$$y = 0, \ 2y - 7 = 0$$
$$y = 0, \qquad y = \dfrac{7}{2}$$

69. $\left[\dfrac{8}{2z - 5} + 7 = 2z\right](2z - 5)$

$8 + 7(2z - 5) = 2z(2z - 5)$
$8 + 14z - 35 = 4z^2 - 10z$
$14z - 27 = 4z^2 - 10z$
$0 = 4z^2 - 24z + 27$
$0 = (2z - 3)(2z - 9)$
$2z - 3 = 0, \ 2z - 9 = 0$
$z = \dfrac{3}{2}, \qquad z = \dfrac{9}{2}$

70. $t + 2 - \dfrac{4t}{2t + 3} = \dfrac{2t + 9}{2t + 3}$ Multiply by $2t + 3$

$(t + 2)(2t + 3) - \dfrac{4t}{\cancel{2t + 3}}^{\,1} \cdot \cancel{(2t + 3)}^{\,1} = \dfrac{2t + 9}{\cancel{2t + 3}}^{\,1} \cdot \cancel{(2t + 3)}^{\,1}$

$2t^2 + 7t + 6 - 4t = 2t + 9$
$2t^2 + 3t + 6 = 2t + 9$
$2t^2 + t - 3 = 0$
$(2t + 3)(t - 1) = 0$
$2t + 3 = 0, \ t - 1 = 0$
$t = -\dfrac{3}{2}, \qquad t = 1$

Only $t = 1$ works since $t = -\dfrac{3}{2}$ does **not** check due to division by 0.

71. $\sqrt{3x + 1} = x - 1$
$(\sqrt{3x + 1})^2 = (x - 1)^2$
$3x + 1 = x^2 - 2x + 1$
$0 = x^2 - 5x$
$0 = x(x - 5)$
$x = 0, \ x - 5 = 0$
$x = 0, \qquad x = 5$
Here, $x = 0$ does **not** check
whereas $x = 5$ checks.

72. $\sqrt{2q + 4} = 2q - 8$
$(\sqrt{2q + 4})^2 = (2q - 8)^2$
$2q + 4 = 4q^2 - 32q + 64$
$0 = 4q^2 - 34q + 60$ Now $\div$ by 2
$0 = 2q^2 - 17q + 30$
$0 = (q - 6)(2q - 5)$
$q - 6 = 0, \ 2q - 5 = 0$
$q = 6, \qquad q = \dfrac{5}{2}$

Here $q = 6$ checks, but $q = \dfrac{5}{2}$ does **not** check since the left side is

$\sqrt{2q + 4} = \sqrt{2\left(\dfrac{5}{2}\right) + 4} = \sqrt{5 + 4} = \sqrt{9} = 3$

while the right side is

$2q - 8 = 2\left(\dfrac{5}{2}\right) - 8 = 5 - 8 = -3.$

73. $3t = 5 - \sqrt{7 - 3t}$
$3t - 5 = -\sqrt{7 - 3t}$
$(3t - 5)^2 = (-\sqrt{7 - 3t})^2$
$9t^2 - 30t + 25 = 7 - 3t$
$9t^2 - 27t + 18 = 0$
$t^2 - 3t + 2 = 0$
$(t - 1)(t - 2) = 0$
$t - 1 = 0, \ t - 2 = 0$
$t = 1, \qquad t = 2$
Here, $t = 2$ does **not** check
whereas $t = 1$ checks.

74.
$$\sqrt{5y + 1} = y + 1$$
$$(\sqrt{5y + 1})^2 = (y + 1)^2$$
$$5y + 1 = y^2 + 2y + 1$$
$$0 = y^2 - 3y$$
$$0 = y(y - 3)$$
$$y = 0, \quad y - 3 = 0$$
$$y = 0, \qquad y = 3$$
Here both values check.

75.
$$\sqrt{2z - 1} + z = 2$$
$$\sqrt{2z - 1} = 2 - z$$
$$(\sqrt{2z - 1})^2 = (2 - z)^2$$
$$2z - 1 = 4 - 4z + z^2$$
$$0 = 5 - 6z + z^2$$
$$0 = z^2 - 6z + 5$$
$$0 = (z - 5)(z - 1)$$
$$z - 5 = 0, \quad z - 1 = 0$$
$$z = 5, \qquad z = 1$$
Here, $z = 5$ does **not** check whereas $z = 1$ checks.

76.
$$\sqrt{2r - 1} = 1 + \sqrt{r - 1}$$
$$(\sqrt{2r - 1})^2 = (1 + \sqrt{r - 1})^2$$
$$2r - 1 = 1 + 2\sqrt{r - 1} + r - 1$$
$$2r - 1 = 2\sqrt{r - 1} + r$$
$$r - 1 = 2\sqrt{r - 1}$$
$$(r - 1)^2 = [2\sqrt{r - 1}]^2$$
$$r^2 - 2r + 1 = 4(r - 1)$$
$$r^2 - 2r + 1 = 4r - 4$$
$$r^2 - 6r + 5 = 0$$
$$(r - 5)(r - 1) = 0$$
$$r - 5 = 0, \quad r - 1 = 0$$
$$r = 5, \qquad r = 1$$
and both values check

77.
$$\sqrt{2z - 1} - \sqrt{z + 3} = 1$$
$$\sqrt{2z - 1} = 1 + \sqrt{z + 3}$$
$$(\sqrt{2z - 1})^2 = (1 + \sqrt{z + 3})^2$$
$$2z - 1 = 1 + 2\sqrt{z + 3} + z + 3$$
$$2z - 1 = 4 + 2\sqrt{z + 3} + z$$
$$z - 5 = 2\sqrt{z + 3}$$
$$(z - 5)^2 = [2\sqrt{z + 3}]^2$$
$$z^2 - 10z + 25 = 4(z + 3)$$
$$z^2 - 10z + 25 = 4z + 12$$
$$z^2 - 14z + 13 = 0$$
$$(z - 13)(z - 1) = 0$$
$$z - 13 = 0, \quad z - 1 = 0$$
$$z = 13, \qquad z = 1$$
Here, $z = 1$ does **not** check whereas $z = 13$ checks.

78.
$$\sqrt{2x + 11} - 2 - \sqrt{x + 2} = 0$$
$$\sqrt{2x + 11} = 2 + \sqrt{x + 2}$$
$$(\sqrt{2x + 11})^2 = (2 + \sqrt{x + 2})^2$$
$$2x + 11 = 4 + 4\sqrt{x + 2} + x + 2$$
$$2x + 11 = 6 + 4\sqrt{x + 2} + x$$
$$x + 5 = 4\sqrt{x + 2}$$
$$(x + 5)^2 = [4\sqrt{x + 2}]^2$$
$$x^2 + 10x + 25 = 16(x + 2)$$
$$x^2 + 10x + 25 = 16x + 32$$
$$x^2 - 6x - 7 = 0$$
$$(x - 7)(x + 1) = 0$$
$$x - 7 = 0, \quad x + 1 = 0$$
$$x = 7, \qquad x = -1$$
and both values check

79.
$$w^2 + 6w + 8 < 0$$
$$(w + 4)(w + 2) < 0$$

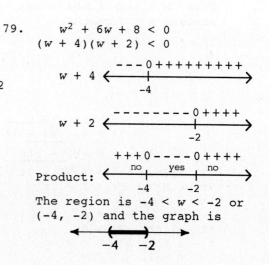

The region is $-4 < w < -2$ or $(-4, -2)$ and the graph is

80.
$$2t^2 \geq 10 - t$$
$$2t^2 + t - 10 \geq 0$$
$$(2t + 5)(t - 2) \geq 0$$

$$2t + 5 \quad \xleftarrow{\qquad \underset{-\frac{5}{2}}{|} \quad} \text{---0+++++++}\xrightarrow{\qquad}$$

$$t - 2 \quad \xleftarrow{\qquad} \text{--------0++++}\underset{2}{|}\xrightarrow{\qquad}$$

$$\text{Product:} \quad \xleftarrow{\quad \underset{\substack{| \\ -\frac{5}{2}}}{\text{yes}} \quad \underset{\substack{| \\ 2}}{\text{no}} \quad \text{yes}\quad}$$
+++0----0++++

The region is $t \leq -\dfrac{5}{2}$ along with

$t \geq 2$ or $\left(-\infty, -\dfrac{5}{2}\right] \cup [2, \infty)$ and the

graph is

81.
$$5t < 2 - 12t^2$$
$$12t^2 + 5t - 2 < 0$$
$$(3t + 2)(4t - 1) < 0$$
$$3t + 2 = 0 \quad \text{or} \quad 4t - 1 = 0$$
$$3t = -2 \quad \text{or} \qquad 4t = 1$$
$$t = -\frac{2}{3} \quad \text{or} \qquad t = \frac{1}{4}$$

$$3t + 2 \quad \xleftarrow{\qquad} \text{---0+++++++++}\underset{-\frac{2}{3}}{|}\xrightarrow{\qquad}$$

$$4t - 1 \quad \xleftarrow{\qquad} \text{--------0++++}\underset{\frac{1}{4}}{|}\xrightarrow{\qquad}$$

$$\text{Product:} \quad \xleftarrow{\quad \underset{\substack{| \\ -\frac{2}{3}}}{\text{no}} \quad \underset{\substack{| \\ \frac{1}{4}}}{\text{yes}} \quad \text{no}\quad}$$
+++0----0++++

The region is $-\dfrac{2}{3} < t < \dfrac{1}{4}$ or $\left(-\dfrac{2}{3}, \dfrac{1}{4}\right)$

and the graph is

82.
$$x^2 + 3x + 2 > 6$$
$$x^2 + 3x - 4 > 0$$
$$(x + 4)(x - 1) > 0$$

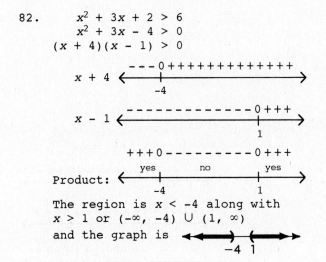

$$x + 4 \quad \xleftarrow{\qquad} \text{---0+++++++++++++}\underset{-4}{|}\xrightarrow{\qquad}$$

$$x - 1 \quad \xleftarrow{\qquad} \text{-------------0+++}\underset{1}{|}\xrightarrow{\qquad}$$

$$\text{Product:} \quad \xleftarrow{\quad \underset{\substack{| \\ -4}}{\text{yes}} \quad \underset{\substack{| \\ 1}}{\text{no}} \quad \text{yes}\quad}$$
+++0---------0+++

The region is $x < -4$ along with
$x > 1$ or $(-\infty, -4) \cup (1, \infty)$
and the graph is

83.
$$2y^3 + 5y^2 > 3y$$
$$2y^3 + 5y^2 - 3y > 0$$
$$y(2y - 1)(y + 3) > 0$$
$$y = 0 \text{ or } 2y - 1 = 0 \text{ or } y + 3 = 0$$
$$y = 0 \text{ or} \qquad y = \frac{1}{2} \text{ or} \qquad y = -3$$

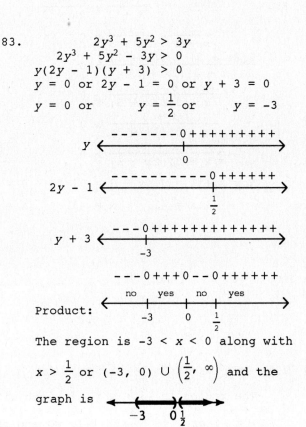

$$y \quad \xleftarrow{\qquad} \text{---------0+++++++++}\underset{0}{|}\xrightarrow{\qquad}$$

$$2y - 1 \quad \xleftarrow{\qquad} \text{----------0++++++}\underset{\frac{1}{2}}{|}\xrightarrow{\qquad}$$

$$y + 3 \quad \xleftarrow{\qquad} \text{---0++++++++++++++}\underset{-3}{|}\xrightarrow{\qquad}$$

$$\text{Product:} \quad \xleftarrow{\quad \underset{\substack{| \\ -3}}{\text{no}} \quad \underset{\substack{| \\ 0}}{\text{yes}} \quad \underset{\substack{| \\ \frac{1}{2}}}{\text{no}} \quad \text{yes}\quad}$$
---0+++0--0+++++

The region is $-3 < x < 0$ along with

$x > \dfrac{1}{2}$ or $(-3, 0) \cup \left(\dfrac{1}{2}, \infty\right)$ and the

graph is

84.
$$\frac{2x + 5}{x + 3} \leq 1$$

$$\frac{2x + 5}{x + 3} - 1 \leq 0$$

$$\frac{2x + 5}{x + 3} - \frac{x + 3}{x + 3} \leq 0$$

$$\frac{2x + 5 - (x + 3)}{x + 3} < 0$$

$$\frac{x + 2}{x + 3} < 0$$

$x + 2$
$$\xleftarrow{\hspace{1cm}} \begin{array}{c} -------0+++++++ \\ \mid \\ -2 \end{array} \xrightarrow{\hspace{1cm}}$$

$x + 3$
$$\xleftarrow{\hspace{1cm}} \begin{array}{c} ---0++++++++++++ \\ \mid \\ -3 \end{array} \xrightarrow{\hspace{1cm}}$$

Quotient:
$$\xleftarrow{\hspace{1cm}} \begin{array}{c} +++u----0+++++++ \\ \mid \quad\quad \mid \\ -3 \quad\quad -2 \end{array} \xrightarrow{\hspace{1cm}}$$

The region is $-3 < x \leq -2$ or $(-3, -2]$ and the graph is

85.
$$\frac{2}{x + 3} \geq \frac{1}{2x + 3}$$

$$\frac{2}{x + 3} - \frac{1}{2x + 3} \geq 0$$

$$\frac{2(2x + 3) - 1(x + 3)}{(x + 3)(2x + 3)} \geq 0$$

$$\frac{4x + 6 - x - 3}{(x + 3)(2x + 3)} \geq 0$$

$$\frac{3x + 3}{(x + 3)(2x + 3)} \geq 0$$

$$\frac{3(x + 1)}{(x + 3)(2x + 3)} \geq 0$$

Now $x + 1 = 0$, $x + 3 = 0$, $2x + 3 = 0$

$\quad\quad x = -1, \quad\quad x = -3, \quad\quad x = -\frac{3}{2}$

$x + 1$
$$\xleftarrow{\hspace{1cm}} \begin{array}{c} -------------0+++ \\ \mid \\ -1 \end{array} \xrightarrow{\hspace{1cm}}$$

$x + 3$
$$\xleftarrow{\hspace{1cm}} \begin{array}{c} ---0+++++++++++++ \\ \mid \\ -3 \end{array} \xrightarrow{\hspace{1cm}}$$

$2x + 3$
$$\xleftarrow{\hspace{1cm}} \begin{array}{c} --------0++++++++ \\ \mid \\ -\frac{3}{2} \end{array} \xrightarrow{\hspace{1cm}}$$

Quotient:
$$\xleftarrow{\hspace{1cm}} \begin{array}{c} ---u++++u----0+++ \\ \mid \quad\quad \mid \quad\quad \mid \\ -3 \quad -\frac{3}{2} \quad -1 \end{array} \xrightarrow{\hspace{1cm}}$$

The region is $-3 < x < -\frac{3}{2}$ along with
$x \geq -1$ or $\left(-3, -\frac{3}{2}\right) \cup [-1, \infty)$ and the graph is

Exercises 3.1

1. $y = 2x - 3$
    Let $x = 0$ to get
    $\quad y = 2 \cdot 0 - 3 = 0 - 3 = -3$
    for the y intercept
    Let $y = 0$ to get
    $\quad 0 = 2x - 3$
    $\quad 3 = 2x$
    $\quad \dfrac{3}{2} = x$ for the x intercept

    The graph is

    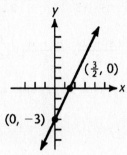

3. $4x = 5.$ Solve for x to have
    $\quad x = \dfrac{5}{4}$ and this is the x intercept
    There is **no** y intercept.

    The graph is

    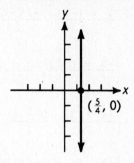

5. $2x = 3y$
    Let $y = 0$ to get
    $\quad 2x = 3 \cdot 0$
    $\quad 2x = 0$
    $\quad\ \ x = 0$
    The x and y intercepts are 0.

    The graph is

    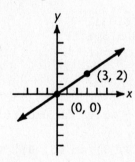

7. $2x + 5y = 7.$ Let $x = 0$ to obtain
    $\quad 2 \cdot 0 + 5y = 7$
    $\quad\ \ \ 0 + 5y = 7$
    $\quad\qquad\ 5y = 7$
    $\quad\qquad\ \ y = \dfrac{7}{5}$ for the y intercept
    Let $y = 0$:
    $\quad 2x + 5 \cdot 0 = 7$
    $\quad\ \ 2x + 0 = 7$
    $\quad\qquad 2x = 7$
    $\quad\qquad\ x = \dfrac{7}{2}$ for the x intercept

    The graph is

    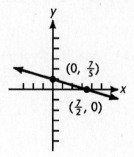

9. $y = -3$ is a horizontal
    line so that -3 is the y
    intercept. There is no x
    intercept. The graph is

    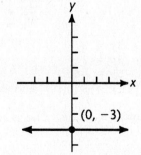

11. $m = \dfrac{3 - (-1)}{-1 - 2} = \dfrac{3 + 1}{-3} = \dfrac{4}{-3} = -\dfrac{4}{3}$ 13. $m = \dfrac{2 - 0}{3 - 0} = \dfrac{2}{3}$

15. $m = \dfrac{3 - (-1)}{3 - 3} = \dfrac{3 + 1}{0} = \dfrac{4}{0}$ which is undefined due to division by 0

17. $m = \dfrac{5 - 5}{2 - (-4)} = \dfrac{0}{2 + 4} = \dfrac{0}{6} = 0$ 19. $m = \dfrac{5 - 1}{7 - (-1)} = \dfrac{4}{7 + 1} = \dfrac{4}{8} = \dfrac{1}{2}$

21. $\left(\dfrac{1 + (-3)}{2}, \dfrac{-2 + 1}{2}\right) = \left(\dfrac{-2}{2}, \dfrac{-1}{2}\right) = \left(-1, -\dfrac{1}{2}\right)$

23. $\left(\dfrac{-3 + (-5)}{2}, \dfrac{2 + 2}{2}\right) = \left(\dfrac{-8}{2}, \dfrac{4}{2}\right) = (-4, 2)$ 25. $\left(\dfrac{2 + 1}{2}, \dfrac{7 + (-3)}{2}\right) = \left(\dfrac{3}{2}, \dfrac{4}{2}\right) = \left(\dfrac{3}{2}, 2\right)$

27. $\left(\dfrac{-\pi + 3\pi}{2}, \dfrac{0 + 0}{2}\right) = \left(\dfrac{2\pi}{2}, \dfrac{0}{2}\right) = (\pi, 0)$ 29. $\left(\dfrac{a + b}{2}, \dfrac{b + a}{2}\right)$ or $\left(\dfrac{a + b}{2}, \dfrac{a + b}{2}\right)$

31. Let (x, y) represent the other point. Then, the representation is
$\left(\dfrac{x + 2}{2}, \dfrac{y + 8}{2}\right) = (1, 5)$.

Now $\dfrac{x + 2}{2} = 1$ gives $x + 2 = 2$ or $x = 0$

and $\dfrac{y + 8}{2} = 5$ gives $y + 8 = 10$ or $y = 2$. The point is $(0, 2)$.

33. Let $y = 0$ to obtain $3x = 6$ or $x = 2$. One point is $(2, 0)$. Now let $x = 6$ to
get $\quad 3 \cdot 6 - 4y = 6$
$\quad\quad\quad 18 - 4y = 6$
$\quad\quad\quad\quad -4y = -12$
$\quad\quad\quad\quad\quad y = 3$ This is the point $(6, 3)$.
The two points are $(2, 0)$ and $(6, 3)$. The slope is
$\quad m = \dfrac{3 - 0}{6 - 2} = \dfrac{3}{4}$

35. $x + 4 = 0$ can be written as $x = -4$. Two points on this line could be $(-4, 0)$
or $(-4, 3)$. The slope is $m = \dfrac{3 - 0}{-4 - (-4)} = \dfrac{3}{0}$ which is undefined.

37. $y = 3$. Two points on this line could be $(0, 3)$ or $(5, 3)$. The slope is
$\quad m = \dfrac{3 - 3}{5 - 0} = \dfrac{0}{5} = 0$

39. $7x - 5y = 3$. Let $y = 5$ to obtain $\quad 7x - 5 \cdot 5 = 3$
$\quad\quad\quad\quad\quad\quad\quad\quad\quad\quad 7x - 25 = 3$
$\quad\quad\quad\quad\quad\quad\quad\quad\quad\quad\quad\quad 7x = 28$
$\quad\quad\quad\quad\quad\quad\quad\quad\quad\quad\quad\quad\quad x = 4$ and the point is $(4, 5)$

Now, let $x = 9$ to get $7 \cdot 9 - 5y = 3$
$\quad\quad\quad\quad\quad\quad\quad\quad 63 - 5y = 3$
$\quad\quad\quad\quad\quad\quad\quad\quad\quad -5y = -60$
$\quad\quad\quad\quad\quad\quad\quad\quad\quad\quad y = 12$ and the point is $(9, 12)$

The two points are $(4, 5)$ and $(9, 12)$. The slope is $m = \dfrac{12 - 5}{9 - 4} = \dfrac{7}{5}$.

41. $2y = 7x - 5$. Let $x = 1$ to obtain
$$2y = 7 \cdot 1 - 5$$
$$2y = 7 - 5$$
$$2y = 2$$
$$y = 1$$
and the point is $(1, 1)$

Now, let $x = 3$ to get
$$2y = 7 \cdot 3 - 5$$
$$2y = 21 - 5$$
$$2y = 16$$
$$y = 8$$
and the point is $(3, 8)$

The two points are $(1, 1)$ and $(3, 8)$. The slope is
$$m = \frac{8 - 1}{3 - 1} = \frac{7}{2}$$

43. $5x + 2y = 10$. Let $y = 10$ to obtain
$$5x + 2 \cdot 0 = 10$$
$$5x + 0 = 10$$
$$5x = 10$$
$$x = 2$$
and the point is $(2, 0)$

Now, let $x = 0$ to get
$$5 \cdot 0 + 2y = 10$$
$$0 + 2y = 10$$
$$2y = 10$$
$$y = 5$$
and the point is $(0, 5)$

The two points are $(0, 5)$ and $(2, 0)$. The slope is
$$m = \frac{0 - 5}{2 - 0} = -\frac{5}{2}$$

45. $x = 2y + 6$. Let $y = 0$ to obtain
$$x = 2 \cdot 0 + 6$$
$$x = 0 + 6$$
$$x = 6 \quad \text{and the point is } (6, 0)$$

Now, let $y = 3$ to get
$$x = 2 \cdot 3 + 6$$
$$x = 6 + 6$$
$$x = 12 \quad \text{and the point is } (12, 3)$$

The two points are $(6, 0)$ and $(12, 3)$. The slope is $m = \dfrac{3 - 0}{12 - 6} = \dfrac{3}{6} = \dfrac{1}{2}$.

47. The slope of the line joining $(1, y)$ to $(2, 3)$ is $\dfrac{3 - y}{2 - 1} = \dfrac{3 - y}{1} = 3 - y$.

Set this value equal to 5 and then solve:
$$3 - y = 5$$
$$-y = 2$$
$$y = -2$$

49. The slope of the line joining $(0, 0)$ to $(1, 2)$ is $m = \dfrac{2 - 0}{1 - 0} = 2$ and the

slope of the line joining $(0, 0)$ to $(3, y)$ is $m = \dfrac{y - 0}{3 - 0} = \dfrac{y}{3}$.

Equate these two quantities: $\dfrac{y}{3} = 2$ or $y = 6$

51. The slope of the line joining $(-1, 3)$ to $(-2, 0)$ is $m = \dfrac{0 - 3}{-2 - (-1)} = \dfrac{-3}{-1} = 3$.

The slope of the line joining $(2, 0)$ to $(3, 3)$ is $m = \dfrac{3 - 0}{3 - 2} = \dfrac{3}{1} = 3$.

The slope of the line joining $(-1, 3)$ to $(3, 3)$ is $m = \dfrac{3 - 3}{3 - (-1)} = \dfrac{0}{4} = 0$.

The slope of the line joining $(-2, 0)$ to $(2, 0)$ is $m = \dfrac{0 - 0}{2 - (-2)} = \dfrac{0}{4} = 0$.

Since the opposite sides have the same slopes, the opposite sides are parallel and a parallelogram exists.

53. The slope of the line joining $(-2, 8)$ to $(-6, 4)$ is $m = \dfrac{4 - 8}{-6 - (-2)} = \dfrac{-4}{-4} = 1$.

The slope of the line joining $(-4, 1)$ to $(2, 2)$ is $m = \dfrac{2 - 1}{2 - (-4)} = \dfrac{1}{6}$.

Since the slopes are different, opposite sides are **not** parallel and no parallelogram exists.

55.

x	y
1	0
4	3
4	-3

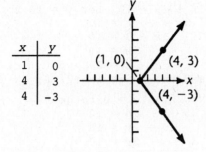

57.

x	y
0	-1
1	0
1	-2

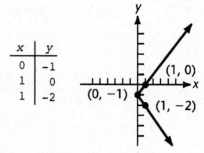

59.

x	y
-4	0
0	2
0	-2

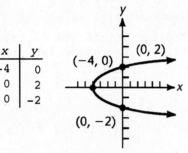

61.

x	C
0	9,000
10,000	16,000

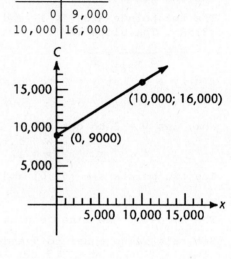

63.

x	P
0	750
12	0

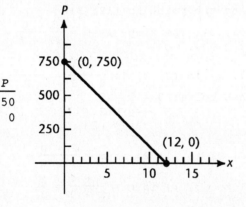

65. Figure d   67. Figure c   69. Figure b   71. 3.01   73. -4.41

75. The phrase *'no slope'* could refer to zero slope (equation is $y = 7$) or to undefined slope (equation is $x = 3$).

77. All y values for the points must be the same since
$$\frac{y_2 - y_1}{x_2 - x_1} = 0 \text{ means that } y_2 - y_1 = 0 \text{ or } y_2 = y_1 .$$
When plotted, all such points produce a straight line.

Exercises 3.2

1. $D = \mathcal{R}, R = \mathcal{R}$ 3. $D = [4, \infty), R = (-\infty, 0]$ 5. $D = \mathcal{R}, R = [-3, \infty)$

7. $D = (-\infty, 1) \cup (1, \infty), R = (-\infty, 0) \cup (0, \infty)$ 9. $D = \mathcal{R}, R = [0, \infty)$

11. $D = \mathcal{R}, R = [3, \infty)$

13. Is a function since for each value of x, there is one value for y.

15. Not a function since when x is 2, there are two values for y. They are $y = 1$ and $y = 2$.

17. Not a function since every point on the straight line has an x value of 3 and all the y values are different.

19. Is a function since for every value of x, there is one value for y. That is, for any value of x, the y value is $x - 1$.

21. Is a function since for every value of x, there is one value for y. That is, for any value of x, the y value is $|x|$.

23. $f(3) = 4 \cdot 3 - 2 = 12 - 2 = 10$ 25. $g(4) = 4^2 = 16$

27. $f(b) = 4 \cdot b - 2 = 4b - 2$

29. $f(-x) = 4(-x) - 2 = -4x - 2$ 31. $g(x + 1) = (x + 1)^2 = x^2 + 2x + 1$

33. $f(3) = 2 \cdot 3 = 6$ 35. $g(-3) = (-3)^2 - (-3) = 9 + 3 = 12$ 37. $f(0) = 2 \cdot 0 = 0$

39. $f(-x) = 2(-x) = -2x$ 41. $f(x + h) = 2(x + h) = 2x + 2h$

43. $g(x + h) - g(x) = [(x + h)^2 - (x + h)] - [x^2 - x]$
$$= x^2 + 2xh + h^2 - x - h - x^2 + x = 2xh + h^2 - h$$

45. $\dfrac{f(x + h) - f(x)}{h} = \dfrac{[3(x + h) - 4] - [3x - 4]}{h} = \dfrac{3x + 3h - 4 - 3x + 4}{h} = \dfrac{3h}{h} = 3$

47. $\dfrac{f(x + h) - f(x)}{h} = \dfrac{[2(x + h)^2 + 1] - [2x^2 + 1]}{h} = \dfrac{2x^2 + 4xh + 2h^2 + 1 - 2x^2 - 1}{h}$

$$= \dfrac{4xh + 2h^2}{h} = \dfrac{h(4x + 2h)}{h} = 4x + 2h$$

49. $\dfrac{f(x + h) - f(x)}{h} = \dfrac{[3(x + h)^2 - 2(x + h) + 4] - [3x^2 - 2x + 4]}{h}$

$$= \dfrac{3x^2 + 6xh + 3h^2 - 2x - 2h + 4 - 3x^2 + 2x - 4}{h} = \dfrac{6xh + 3h^2 - 2h}{h}$$

$$= \dfrac{h(6x + 3h - 2)}{h} = 6x + 3h - 2$$

51. $\dfrac{f(x + h) - f(x)}{h} = \dfrac{[2(x + h)^3 + 1] - [2x^3 + 1]}{h}$

$$= \dfrac{2x^3 + 6x^2h + 6xh^2 + 2h^3 + 1 - 2x^3 - 1}{h}$$

$$= \dfrac{6x^2h + 6xh^2 + 2h^3}{h} = \dfrac{h(6x^2 + 6xh + 2h^2)}{h} = 6x^2 + 6xh + 2h^2$$

53. $f(x) = 2x^5 - 3x^4 + 5x - 2$
   $f(3.6) = 2(3.6)^5 - 3(3.6)^4 + 5(3.6) - 2$
          $= 1209.32352 - 503.8848 + 18 - 2$
          $= 721.43872$ which rounds to 721.439

55. $\dfrac{f(x + h) - f(x)}{h} = \dfrac{[(x + h)^3 + 4(x + h)^2 - 2] - [x^3 + 4x^2 - 2]}{h}$

$$= \dfrac{x^3 + 3x^2h + 3xh^2 + h^3 + 4x^2 + 8xh + 4h^2 - 2 - x^3 - 4x^2 + 2}{h}$$

$$= \dfrac{3x^2h + 3xh^2 + h^3 + 8xh + 4h^2}{h} = 3x^2 + 3xh + h^2 + 8x + 4h$$

Now let $x = -1$ and $h = 0.01$ to obtain

$3(-1)^2 + 3(-1)(0.01) + (0.01)^2 + 8(-1) + 4(0.01)$

$\quad = 3 - 0.03 + 0.0001 - 8 + 0.04$

$\quad = -4.9899$ which rounds to -4.99

For problems 57—67, the appropriate graphs are given below.

57.

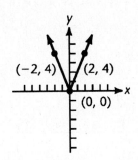

59.

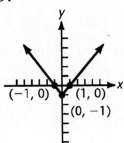

61.

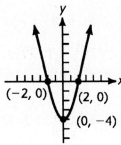

63.

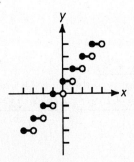

65.

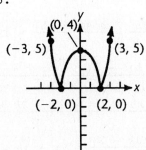

67.

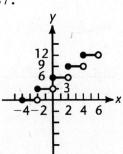

69. The graph is **not** a function since for $x > -2$ every vertical line intersects the graph in two points.

71. The graph is a function since for $-3 \le x \le 3$ every vertical line intersects the graph in one point.

73. The graph is a function since every vertical line intersects the graph in one point.

75. The graph is **not** a function since for $x < -1$ or $x > 1$, every vertical line intersects the graph in two points.

77. This is the graph of the charges for miles traveled:

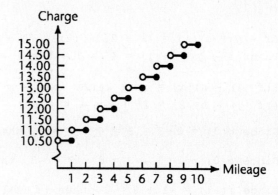

79. a) Total amount of fence used is represented by $\ell + \ell + w + w + w = 2\ell + 3w$.
    The equation is $2\ell + 3w = 120$
    $$2\ell = 120 - 3w$$
    $$\ell = \frac{120 - 3w}{2}$$
    $$\ell = 60 - \frac{3}{2}w$$

    b) $A = (\text{length})(\text{width}) = \left(60 - \frac{3}{2}w\right)(w) = 60w - \frac{3}{2}w^2$

81. The box appears

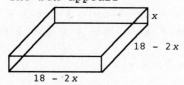

$$\begin{aligned} V = (\text{length})(\text{width})(\text{height}) &= (18 - 2x)(18 - 2x)(x) \\ &= [2(9 - x)][2(9 - x)](x) \\ &= 4(9 - x)^2(x) \\ &= 4x(9 - x)^2 \end{aligned}$$

83. The definition of a function states: a function f with domain D is a correspondence such that for each element of D, there is one and only one associated element y in a second set of real numbers. Therefore, each x-y pair forms an ordered pair such that no two distinct ordered pairs have the same x value. But, this is precisely the alternate definition of a function.

Exercises 3.3

1. $(f + g)(x) = (x + 2) + (2x - 3)$
   $$= 3x - 1$$
   Then $(f + g)(3) = 3 \cdot 3 - 1 = 9 - 1 = 8$

3. $(h - g)(x) = x^2 - (2x - 3)$
   $$= x^2 - 2x + 3$$
   Then $(h - g)(-1) = (-1)^2 - 2(-1) + 3$
   $$= 1 + 2 + 3 = 6$$

5. $(f \cdot h)(x) = (x + 2)(x^2) = x^3 + 2x^2$
   Then $(f \cdot h)(-3) = (-3)^3 + 2(-3)^2$
   $$= -27 + 18 = -9$$

7. $(f \cdot f)(x) = (x + 2)(x + 2) = (x + 2)^2$
   Then $(f \cdot f)(0) = (0 + 2)^2 = 2^2 = 4$

9. $\left(\dfrac{f}{g}\right)(x) = \dfrac{x + 2}{2x - 3}$

   Then $\left(\dfrac{f}{g}\right)(1) = \dfrac{1 + 2}{2 \cdot 1 - 3} = \dfrac{3}{-1} = -3$

11. $\left(\dfrac{h}{g}\right)(x) = \dfrac{x^2}{2x - 3}$

    Then $\left(\dfrac{h}{g}\right)(0) = \dfrac{0^2}{2 \cdot 0 - 3} = \dfrac{0}{-3} = 0$

13. $(f - g + h)(x) = (x + 2) - (2x - 3) + x^2 = x + 2 - 2x + 3 + x^2 = x^2 - x + 5$
Then $(f - g + h)(1) = 1^2 - 1 + 5 = 1 - 1 + 5 = 5$

15. $((f \cdot g) - h)(x) = (x + 2)(2x - 3) - x^2 = 2x^2 + x - 6 - x^2 = x^2 + x - 6$
$((f \cdot g) - h)(2) = 2^2 + 2 - 6 = 4 + 2 - 6 = 0$

17. Since $g(3) = 2 \cdot 3 - 3 = 6 - 3 = 3$, then $(f \circ g)(3) = f(g(3)) = f(3) = 3 + 2 = 5$

19. Since $g(0) = 2 \cdot 0 - 3 = 0 - 3 = -3$, then $(h \circ g)(0) = h(g(0)) = h(-3) = (-3)^2 = 9$

21. Since $f(-1) = -1 + 2 = 1$, then $(f \circ f)(-1) = f(f(-1)) = f(1) = 1 + 2 = 3$

23. Since $h(-2) = (-2)^2 = 4$ and $g(4) = 2 \cdot 4 - 3 = 8 - 3 = 5$, then
$((f \circ g) \circ h)(-2) = ((f \circ g)(h(-2)) = ((f \circ g)(4) = f(g(4)) = f(5) = 5 + 2 = 7$

25. $(f + g)(x) = (x - 2) + (2x + 1) = 3x - 1$ for $x \in \mathcal{R}$
$(f - g)(x) = (x - 2) - (2x + 1) = x - 2 - 2x - 1 = -x - 3$ for $x \in \mathcal{R}$
$(f \cdot g)(x) = (x - 2)(2x + 1) = 2x^2 - 3x - 2$ for $x \in \mathcal{R}$
$\left(\dfrac{f}{g}\right)(x) = \dfrac{x - 2}{2x + 1}$ for $x \in \mathcal{R}$ with $x \neq -\dfrac{1}{2}$

27. $(f + g)(x) = (4x - 1) + x(x - 1) = 4x - 1 + x^2 - x = x^2 + 3x - 1$ for $x \in \mathcal{R}$
$(f - g)(x) = (4x - 1) - x(x - 1) = 4x - 1 - x^2 + x = -x^2 + 5x - 1$ for $x \in \mathcal{R}$
$(f \cdot g)(x) = (4x - 1)[x(x - 1)] = (4x - 1)(x^2 - x) = 4x^3 - 5x^2 + x$ for $x \in \mathcal{R}$
$\left(\dfrac{f}{g}\right)(x) = \dfrac{4x - 1}{x(x - 1)} = \dfrac{4x - 1}{x^2 - x}$ for $x \in \mathcal{R}$ with $x \neq 0$, $x \neq 1$

29. $(f + g)(x) = \sqrt{x} + 2x$ for $x \in [0, \infty)$
$(f - g)(x) = \sqrt{x} - 2x$ for $x \in [0, \infty)$
$(f \cdot g)(x) = \sqrt{x}(2x) = 2x\sqrt{x}$ for $x \in [0, \infty)$
$\left(\dfrac{f}{g}\right)(x) = \dfrac{\sqrt{x}}{2x}$ for $x \in (0, \infty)$

31. $(f + g)(x) = 2x + 1 + \sqrt{x + 2}$, $x \in [-2, \infty)$
$(f - g)(x) = 2x + 1 - \sqrt{x + 2}$, $x \in [-2, \infty)$
$(f \cdot g)(x) = (2x + 1)\sqrt{x + 2}$, $x \in [-2, \infty)$
$\left(\dfrac{f}{g}\right)(x) = \dfrac{(2x + 1)}{\sqrt{x + 2}}$, $x \in (-2, \infty)$

33. Observe that $D_f = [2, \infty)$ and $D_g = (-\infty, 3]$ so that $D_f \cap D_g = [2, 3]$
$(f + g)(x) = \sqrt{x - 2} + \sqrt{3 - x}$, $x \in [2, 3]$
$(f - g)(x) = \sqrt{x - 2} - \sqrt{3 - x}$, $x \in [2, 3]$
$(f \cdot g)(x) = \sqrt{5x - x^2 - 6}$, $x \in [2, 3]$
$\left(\dfrac{f}{g}\right)(x) = \dfrac{\sqrt{x - 2}}{\sqrt{3 - x}}$, $x \in [2, 3)$

35. Observe that $D_f = [-1, \infty)$ and $D_g = [-6, \infty)$ so that $D_f \cap D_g = [-1, \infty)$

$(f + g)(x) = \sqrt{x + 1} + \sqrt{x + 6}, \ x \in [-1, \infty)$

$(f - g)(x) = \sqrt{x + 1} - \sqrt{x + 6}, \ x \in [-1, \infty)$

$(f \cdot g)(x) = \sqrt{x^2 + 7x + 6}, \ x \in [-1, \infty)$

$\left(\dfrac{f}{g}\right)(x) = \dfrac{\sqrt{x + 1}}{\sqrt{x + 6}}, \ x \in [-1, \infty)$

37. $(f + g)(x) = (x - 4) + (x^2 - 5x + 6) = x^2 - 4x + 2$ for $x \in \mathcal{R}$

$(f - g)(x) = (x - 4) - (x^2 - 5x + 6) = x - 4 - x^2 + 5x - 6$
$$= -x^2 + 6x - 10 \text{ for } x \in \mathcal{R}$$

$(f \cdot g)(x) = (x - 4)(x^2 - 5x + 6) = x^3 - 9x^2 + 26x - 24$ for $x \in \mathcal{R}$

$\left(\dfrac{f}{g}\right)(x) = \dfrac{x - 4}{x^2 - 5x + 6}$ or $\dfrac{x - 4}{(x - 3)(x - 2)}$ for $x \in \mathcal{R}$ with $x \neq 3, \ x \neq 2$

39. $(f + g)(x) = (6x^2 + 2x - 5) + (x^2 + 3x - 4) = 7x^2 + 5x - 9$ for $x \in \mathcal{R}$

$(f - g)(x) = (6x^2 + 2x - 5) - (x^2 + 3x - 4) = 6x^2 + 2x - 5 - x^2 - 3x + 4$
$$= 5x^2 - x - 1 \text{ for } x \in \mathcal{R}$$

$(f \cdot g)(x) = (6x^2 + 2x - 5)(x^2 + 3x - 4)$
$$= 6x^2(x^2 + 3x - 4) + 2x(x^2 + 3x - 4) - 5(x^2 + 3x - 4)$$
$$= 6x^4 + 18x^3 - 24x^2 + 2x^3 + 6x^2 - 8x - 5x^2 - 15x + 20$$
$$= 6x^4 + 20x^3 - 23x^2 - 23x + 20 \text{ for } x \in \mathcal{R}$$

$\left(\dfrac{f}{g}\right)(x) = \dfrac{6x^2 + 2x - 5}{x^2 + 3x - 4}$ or $\dfrac{6x^2 + 2x - 5}{(x + 4)(x - 1)}$ for $x \in \mathcal{R}$ with $x \neq -4, \ x \neq 1$

41. $(f \circ g)(x) = f(g(x)) = f(x^2) = x^2 + 1$ for $x \in \mathcal{R}$

$(g \circ f)(x) = g(f(x)) = g(x + 1) = (x + 1)^2$ for $x \in \mathcal{R}$

43. $(f \circ g)(x) = f(g(x)) = f(x - 1) = (x - 1)^{10}$ for $x \in \mathcal{R}$

$(g \circ f)(x) = g(f(x)) = g(x^{10}) = x^{10} - 1$ for $x \in \mathcal{R}$

45. $(f \circ g)(x) = f(g(x)) = f(x + 3) = 2(x + 3)^2 + (x + 3)$
$$= 2(x + 3)^2 + x + 3 \text{ for } x \in \mathcal{R}$$

$(g \circ f)(x) = g(f(x)) = g(2x^2 + x) = (2x^2 + x) + 3 = 2x^2 + x + 3$ for $x \in \mathcal{R}$

47. $(f \circ g)(x) = f(g(x)) = f(x - 3) = \sqrt{x - 3}$ for $x \in [3, \infty)$

$(g \circ f)(x) = g(f(x)) = g(\sqrt{x}) = \sqrt{x} - 3$ for $x \in [0, \infty)$

49. $(f \circ g)(x) = f(g(x)) = f(1 - x) = \dfrac{1}{1 - x}$ for $x \in \mathcal{R}$ with $x \neq 1$

$(g \circ f)(x) = g(f(x)) = g\left(\dfrac{1}{x}\right) = 1 - \dfrac{1}{x}$ for $x \in \mathcal{R}$ with $x \neq 0$

51. $(f \circ g)(x) = f(g(x)) = f(x - 3) = \sqrt{x - 3 + 2} = \sqrt{x - 1}$ for $x \in [1, \infty)$

$(g \circ f)(x) = g(f(x)) = g(\sqrt{x + 2}) = \sqrt{x + 2} - 3$ for $x \in [-2, \infty)$

53. $(f \circ g)(x) = f(g(x)) = f\left(\dfrac{x + 5}{3}\right) = 3\left(\dfrac{x + 5}{3}\right) - 5 = x + 5 - 5 = x$ for $x \in \mathcal{R}$

$(g \circ f)(x) = g(f(x)) = g(3x - 5) = \dfrac{(3x - 5) + 5}{3} = \dfrac{3x}{3} = x$ for $x \in \mathcal{R}$

55. $(f \circ g)(x) = f(g(x)) = f\left(\dfrac{1 - x}{x}\right) = \dfrac{1}{\dfrac{1 - x}{x} + 1} = \dfrac{1(x)}{\left(\dfrac{1 - x}{x} + 1\right)x} = \dfrac{x}{1 - x + x} = \dfrac{x}{1}$

$\qquad\qquad\qquad\qquad\qquad\qquad\qquad\qquad\qquad\qquad\qquad\qquad\qquad = x$ for $x \in \mathcal{R}$

$\quad (g \circ f)(x) = g(f(x)) = g\left(\dfrac{1}{x + 1}\right) = \dfrac{1 - \dfrac{1}{x + 1}}{\dfrac{1}{x + 1}} = \dfrac{\left(1 - \dfrac{1}{x + 1}\right)(x + 1)}{\left(\dfrac{1}{x + 1}\right)(x + 1)} = \dfrac{x + 1 - x}{1}$

$\qquad\qquad\qquad\qquad\qquad\qquad\qquad\qquad\qquad\qquad\qquad\qquad\qquad\qquad = x$ for $x \in \mathcal{R}$

57. $(f \circ g)(x) = f(g(x)) = f(x^2 + 1) = \sqrt{(x^2 + 1) - 1} = \sqrt{x^2} = |x|$ for $x \in \mathcal{R}$

$\quad (g \circ f)(x) = g(f(x)) = g(\sqrt{x - 1}) = (\sqrt{x - 1})^2 + 1 = x - 1 + 1$ for $x \in [1, \infty)$

59. Several outcomes are possible. One solution is $f(x) = x - 1$ and $g(x) = x^3$.

61. Several outcomes are possible. One solution is $f(x) = \sqrt{x}$ and $g(x) = x + 3$.

63. Several outcomes are possible. One solution is $f(x) = x^{50}$ and $g(x) = 2x - 9$.

65. Several outcomes are possible. One solution is $f(x) = x^3$ and $g(x) = \dfrac{1}{x + 3}$

67. a) $Y_1 = \sqrt{x - 2}$ 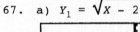 $\qquad\qquad Y_2 = \sqrt{|x - 2|}$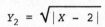

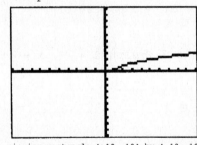 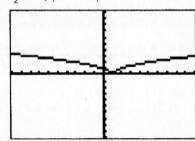

viewing rectangle [-10, 10] by [-10, 10]   viewing rectangle [-10, 10] by [-10, 10]

b) The graph of Y_1 consists of the right half of the graph of Y_2.

69. Several answers are possible. Here are a few suggestions.

a) Suppose you are traveling in a car at x miles per hour. The total stopping distance, $S(x)$, consists of two different distances: distance traveled during the reaction time, $D_r(x)$ and the distance traveled while actually braking, $D_b(x)$. Thus, $S(x) = D_r(x) + D_b(x)$.

b) A manufacturer produces and sells x units of a commodity. The cost, $C(x)$, and the revenue $R(x)$, both depend on x. The profit function, $P(x)$, is given by $P(x) = R(x) - C(x)$.

c) A revenue function, $R(x)$, depends on demand, $D(x)$, for an item as well as the price per item, $p(x)$. The equation is $R(x) = D(x) \cdot p(x)$.

d) In ecomomics, the responsiveness of demand to price fluctuations is known as elasticity of demand, $E(x)$, which is defined as the quotient of the relative change in demand, $D(x)$, and the relative change in price, $p(x)$. The equation is $E(x) = \dfrac{D(x)}{p(x)}$.

e) A stone is thrown into a lake causing ripples to form in the shape of concentric circles. The radius of the circles, $r(t)$, depends on the amount of time which has elapsed since the stone was thrown. The area of the circles depends on the radius, $A(r)$. Since the radius depends upon t, the area can also be described as a function which depends upon t. The equation is $A(t) = (A \circ r)(t) = A(r(t))$.

Exercises 3.4

1. Use the Point-Slope Form to obtain
    $$y - 1 = -1[x - (-1)]$$
    $$y - 1 = -1(x + 1)$$
    $$y - 1 = -x - 1$$
    $$y + x = 0$$
    or $x + y = 0$
    The graph is

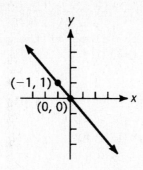

3. Use the Slope-Intercept Form to obtain
    $$y = 2x + 4$$
    $$0 = 2x - y + 4$$
    The graph is

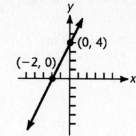

5. Use the Point-Slope Form to obtain
    $$y - 5 = 0[x - (-7)]$$
    $$y - 5 = 0$$
    or $y = 5$

    The graph is shown at the right.

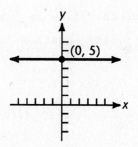

7. $y = \dfrac{1}{2}x + 3$
    Now multiply by 2:
    $$2y = x + 6$$
    $$0 = x - 2y + 6$$
    $$-6 = x - 2y$$

9. $m = \dfrac{3 - 4}{4 - 3} = \dfrac{-1}{1} = -1$. Using the point (3, 4),
    the equation is
    $$y - 4 = -1(x - 3)$$
    $$y - 4 = -x + 3$$
    $$y = -x + 7$$
    $$x + y = 7$$

11. Since the x intercept is 7, the
    corresponding point is (7, 0).
    The equation is
    $$y - 0 = 4(x - 7)$$
    $$y = 4x - 28$$
    $$y + 28 = 4x$$
    $$28 = 4x - y$$

13. The two points corresponding to the
    intercepts are (2, 0) and (0, -1).
    Now, $m = \dfrac{-1 - 0}{0 - 2} = \dfrac{-1}{-2} = \dfrac{1}{2}$ and using
    -1 for the y intercept, the equation
    is
    $$y = \frac{1}{2}x - 1$$
    $$2y = x - 2$$
    $$2y + 2 = x$$
    $$2 = x - 2y$$

15. The two points are (0, 3) and (-1, -1) so that $m = \dfrac{-1 - 3}{-1 - 0} = \dfrac{-4}{-1} = 4$ and the
    equation is $y = 4x + 3$
    $$y - 3 = 4x$$
    $$-3 = 4x - y$$

17. If there is **no** x intercept, the line must be horizontal and its equation is $y = 3$.

19. If the slope **does not exist**, then the line must be vertical and the equation is $x = 5$.

21. Start with $3x - y = 6$ and solve for y:
$$3x - y = 6$$
$$-y = -3x + 6$$
$$y = 3x - 6$$
The slope of this line is 3 and, thus, the slope of the parallel line is also 3. Its equation is
$$y - 3 = 3[x - (-2)]$$
$$y - 3 = 3x + 6$$
$$y - 9 = 3x$$
$$-9 = 3x - y$$

23. Start with $5x = 6y - 1$ and solve for y:
$$5x = 6y - 1$$
$$5x + 1 = 6y$$
$$\frac{5}{6}x + \frac{1}{6} = y$$
The slope of this line is $\frac{5}{6}$ and, thus, the slope of the perpendicular line is $-\frac{1}{\frac{5}{6}} = -\frac{6}{5}$. Its equation is $y - (-3) = -\frac{6}{5}[x - (-5)]$
$$y + 3 = -\frac{6}{5}(x + 5) \quad \text{Now multiply by 5:}$$
$$5(y + 3) = -6(x + 5)$$
$$5y + 15 = -6x - 30$$
$$5y = -6x - 45$$
$$6x + 5y = -45$$

25. For the equation, $y = x$, the slope is 1. The slope of the perpendicular line is $-\frac{1}{1} = -1$ and its equation is $y - 0 = -1(x - 0)$
$$y = -x$$
$$x + y = 0$$

27. Start with $5x - 7y = 5$ and solve for y:
$$-7y = -5x + 5$$
$$y = \frac{5}{7}x - \frac{5}{7}$$
The slope of this line is $\frac{5}{7}$ and, thus, the slope of the parallel line is also $\frac{5}{7}$. Its equation is
$$y = \frac{5}{7}x - 1$$
$$7y = 5x - 7$$
$$7y + 7 = 5x$$
$$7 = 5x - 7y$$

29. The two points are $(0, 3)$ and $(3, 0)$ so that $m = \frac{0 - 3}{3 - 0} = \frac{-3}{3} = -1$ and the equation is $y = -x + 3$ where 3 is the y intercept.

31. The two points are $(0, 0)$ and $(1, 4)$ so that $m = \frac{4 - 0}{1 - 0} = \frac{4}{1} = 4$ and the equation is $y = 4x + 0$ or $y = 4x$ where 0 is the y intercept.

33. The equation of the horizontal line is $y = 4$.

For problems 35—41, the graphs are given below.

35.   37.   39.   41.

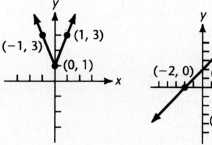

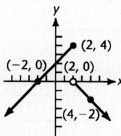

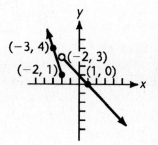

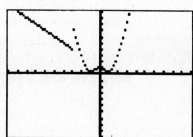

viewing rectangle is [-10, 10] by [-10, 10]

43. The slope of the line joining $(-1, 3)$ and $(4, 7)$ is $m_1 = \dfrac{7 - 3}{4 - (-1)} = \dfrac{4}{5}$ and

the slope of the line joining $(4, 7)$ and $(6, 10)$ is $m_2 = \dfrac{10 - 7}{6 - 4} = \dfrac{3}{2}$. Since
the slope values are different $(m_1 \neq m_2)$ the points do **not** fall on a
straight line.

45. The slope of the line joining $(-1, 5)$ and $(1, 2)$ is $m_1 = \dfrac{2 - 5}{1 - (-1)} = \dfrac{-3}{2} = -\dfrac{3}{2}$.

The slope of the line joining $(-1, 5)$ and $(-2, 0)$ is $m_2 = \dfrac{0 - 5}{-2 - (-1)} = \dfrac{-5}{-1} = 5$.

The slope of the line joining $(1, 2)$ to $(-2, 0)$ is $\dfrac{0 - 2}{-2 - 1} = \dfrac{-2}{-3} = \dfrac{2}{3}$.

Since $m_1 m_3 = \left(-\dfrac{3}{2}\right)\left(\dfrac{2}{3}\right) = -1$, then $m_1 \perp m_3$ and a right triangle exists.

47. The slope of the line joining $(0, 6)$ to $(-3, 0)$ is $m_1 = \dfrac{0 - 6}{-3 - 0} = \dfrac{-6}{-3} = 2$. The

slope of the line joining $(0, 6)$ to $(3, 0)$ is $\dfrac{0 - 6}{3 - 0} = \dfrac{-6}{3} = -2$. The slope of

the line joining $(-3, 0)$ to $(3, 0)$ is $\dfrac{0 - 0}{3 - (-3)} = \dfrac{0}{6} = 0$. Since the product of

any two slopes is **not** -1, no right triangle exists.

49. Let x be the number of hours the plumber works. The equation is
$f(x) = 30 + 28x$ for $x \in [0, 8]$.

51. The two points are $(0, 18{,}000)$ and $(6, 3{,}000)$ so that the slope is
$m = \dfrac{3{,}000 - 18{,}000}{6 - 3} = \dfrac{-15{,}000}{6} = -2{,}500$. Since the y intercept is 18,000, the
equation is $f(x) = 18{,}000 - 2{,}500x$ for $x \in [0, 6]$ where x represents years.

53. Let r represent the riser. Then $\dfrac{r}{12} = \dfrac{7}{9}$

$$\text{or} \quad 9r = 7 \cdot 12$$
$$9r = 84$$
$$r = \dfrac{84}{9} = 9\dfrac{3}{9} = 9\dfrac{1}{3} \text{ inches}$$

55. The change in altitude is 3402 - 3349 = 53 ft. Let x be the distance (horizontal) covered by the truck on the incline. Then

$$\frac{53}{x} = \frac{4}{100}$$

or $\quad 4x = 5300$

$$x = \frac{5300}{4} = 1325 \text{ ft}$$

Now, the distance covered by the truck is $\sqrt{(1325)^2 + (53)^2}$

$$= \sqrt{1,755,625 + 2809}$$

$$= \sqrt{1,758,434} \approx 1326 \text{ ft}$$

57. Since neither of the points is on an axis, we do not know the y intercept. Therefore, the slope-intercept form is not very useful here. In order to use the standard form of the equation of a straight line, we must set up and then solve a system of two equations (one equation for each point) but we would have three unknowns—the paramenters a, b, and c. Therefore, the standard form is not very useful also. However, we can easily calculate the slope using the two given points. Thus, using the slope and one of the two points, it is possible to determine the equation making use of the point-slope form. Therefore, the point-slope form is the easiest one to use.

Exercises 3.5

1. Use $y = ax^2 + bx + c$ with $a = 1$, $b = 0$, and $c = -4$.
   Then the x coordinate of the vertex is $x = -\frac{b}{2a} = -\frac{0}{2 \cdot 1}$
   = 0. The corresponding y value is $y = 0^2 - 4 = -4$ and
   the vertex is (0, -4). The axis of symmetry is $x = -\frac{b}{2a}$
   = 0. Since $a = 1 > 0$, the parabola opens upward. Two
   other points are (-2, 0) and (2, 0) and the graph is
   shown at the right.

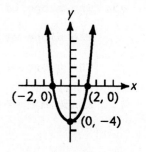

3. Use $y = a(x - h)^2 + k$ with $a = \frac{2}{3}$, $h = 2$ and $k = -1$.
   Then the vertex is $(h, k) = (2, -1)$ and the axis of
   symmetry is $x = h = 2$. Since $a = \frac{2}{3} > 0$, the parabola
   opens upward. Two other points are (-1, 5) and (5, 5)
   and the graph is shown at the right.

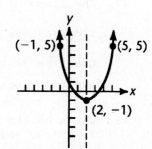

5. Use $y = ax^2 + bx + c$ with $a = -1$, $b = 4$, and $c = 0$. The
   x coordinate of the vertex is $x = -\frac{b}{2a} = -\frac{4}{2(-1)} = \frac{-4}{-2} = 2$.
   The corresponding y value is $y = 4(2) - 2^2 = 8 - 4 = 4$
   and the vertex is (2, 4). The axis of symmetry is
   $x = -\frac{b}{2a} = 2$. Since $a = -1 < 0$, the parabola opens
   downward. Two other points are (0, 0) and (4, 0) and the
   graph is shown at the right.

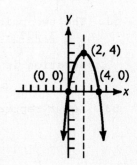

7. Use $y = ax^2 + bx + c$ with $a = 1$, $b = 8$, and
$c = 13$. The x coordinate of the vertex is
$x = -\dfrac{b}{2a} = -\dfrac{8}{2 \cdot 1} = -\dfrac{8}{2} = -4$. The corresponding
y value is $y = (-4)^2 + 8(-4) + 13 =$
$16 - 32 + 13 = -3$ and the vertex is $(-4, -3)$.
The axis of symmetry is $x = -\dfrac{b}{2a} = -4$. Since
$a = 1 > 0$, the parabola opens upward. Two
other points are $(-6, 1)$ and $(-2, 1)$ and the
graph is shown at the right.

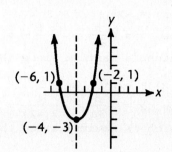

9. Use $y = ax^2 + bx + c$ with $a = -1$, $b = -4$, and
$c = -1$. The x coordinate of the vertex is
$x = -\dfrac{b}{2a} = -\dfrac{(-4)}{2(-1)} = -\dfrac{-4}{-2} = -2$. The corresponding
y value is $y = -(-2)^2 - 4(-2) - 1 = -4 + 8 - 1$
$= 3$ and the vertex is $(-2, 3)$. The axis of
symmetry is $x = -\dfrac{b}{2a} = -2$. Since $a = -1 < 0$, the
parabola opens downward. Two other points are
$(0, -1)$ and $(-4, -1)$ and the graph is shown at
the right.

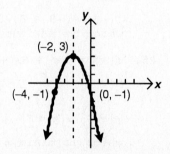

11. Use $y = ax^2 + bx + c$ with $a = 3$, $b = 12$, and
$c = 17$. The x coordinate of the vertex is
$x = -\dfrac{b}{2a} = -\dfrac{12}{2 \cdot 3} = -\dfrac{12}{6} = -2$ and the corresponding
y value is $y = 3(-2)^2 + 12(-2) + 17 =$
$12 - 24 + 17 = 5$ and the vertex is $(-2, 5)$.
The axis of symmetry is $x = -\dfrac{b}{2a} = -2$.
Since $a = 3 > 0$, the parabola opens upward. Four
other points are $(-4, 17)$, $(-3, 8)$, $(-1, 8)$, and
$(0, 17)$ and the graph is shown at the right.

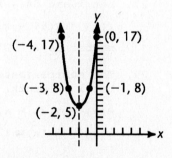

13. Use $y = ax^2 + bx + c$ with $a = -2$, $b = 4$, and
$c = -5$. The x coordinate of the vertex is
$x = -\dfrac{b}{2a} = -\dfrac{4}{2(-2)} = -\dfrac{4}{-4} = 1$. The corresponding y
value is $y = -2(1)^2 + 4(1) - 5 = -2 + 4 - 5 = -3$
and the vertex is $(1, -3)$. The axis of symmetry
is $x = -\dfrac{b}{2a} = 1$. Since $a = -2 < 0$, the parabola
opens downward. Two other points are $(0, -5)$ and
$(2, -5)$ and the graph is shown at the right.

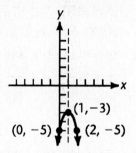

15. $f(x) = x^2 - 8x + 5$
Now, $x = -\dfrac{b}{2a} = -\dfrac{-8}{2 \cdot 1} = -\dfrac{-8}{2} = 4$ and $f(4) = 4^2 - 8 \cdot 4 + 5 = 16 - 32 + 5 = -11$
which is minimum since the parabola opens upward.

17. $f(x) = x^2 + 10x - 20$
Now, $x = -\dfrac{b}{2a} = -\dfrac{10}{2 \cdot 1} = -\dfrac{10}{2} = -5$ and $f(-5) = (-5)^2 + 10(-5) - 20$
$= 25 - 50 - 20 = -45$ which is minimum since the parabola opens upward.

19. $f(x) = -x^2 + 6x + 3$

Now, $x = -\dfrac{b}{2a} = -\dfrac{6}{2(-1)} = -\dfrac{6}{-2} = 3$ and $f(3) = -3^2 + 6 \cdot 3 + 3 = -9 + 18 + 3 = 12$

which is maximum since the parabola opens downward.

21. $f(x) = -2x^2 + 4x - 3$

Now, $x = -\dfrac{b}{2a} = -\dfrac{4}{2(-2)} = -\dfrac{4}{-4} = 1$ and $f(1) = -2(1)^2 + 4 \cdot 1 - 3 = -2 + 4 - 3 = -1$

which is maximum since the parabola opens downward.

23. $f(x) = -9 - 6x - x^2$

Now $x = -\dfrac{b}{2a} = -\dfrac{-6}{2(-1)} = -\dfrac{-6}{-2} = -3$ and

$f(-3) = -9 - 6(-3) - (-3)^2 = -9 + 18 - 9 = 0$

which is maximum since the parabola opens downward.

25. $f(x) = 3x^2 + 9x + 14$

Now $x = -\dfrac{b}{2a} = -\dfrac{9}{2(3)} = -\dfrac{9}{6} = -\dfrac{3}{2}$ and

$f\left(-\dfrac{3}{2}\right) = 3\left(-\dfrac{3}{2}\right)^2 + 9\left(-\dfrac{3}{2}\right) + 14 = \dfrac{27}{4} - \dfrac{27}{2} + 14 = \dfrac{27}{4} - \dfrac{54}{4} + \dfrac{56}{4} = \dfrac{29}{4}$

which is minimum since the parabola opens upward.

27. Let x and $54 - x$ represent the two numbers. The function is

$f(x) = x(54 - x) = 54x - x^2$. Now $x = -\dfrac{b}{2a} = -\dfrac{54}{2(-1)} = -\dfrac{54}{-2} = 27$. Thus, one

number is $x = 27$ and the other number is $54 - x = 54 - 27 = 27$.

29. The semiperimeter is 24 ft (to represent the length and width).
Let x = length. Then the width is $24 - x$ and the function for area is

$A(x) = x(24 - x) = 24x - x^2$. Now, $x = -\dfrac{b}{2a} = -\dfrac{24}{2(-1)} = -\dfrac{24}{-2} = 12$. Thus, the

length is $x = 12$ ft and the width is $24 - x = 24 - 12 = 12$ ft.

31. $C = 0.001x^2 - 0.5x + 800$

$x = -\dfrac{b}{2a} = -\dfrac{-0.5}{2(0.001)} = -\dfrac{-0.5}{0.002} = 250$

Then $C = 0.001(250)^2 - 0.5(250) + 800 = 62.50 - 125 + 800 = \$737.50$.

33. $S = -16t^2 + 64t$

$t = -\dfrac{b}{2a} = -\dfrac{64}{2(-16)} = -\dfrac{64}{-32} = 2$ seconds

Then, $S = -16(2)^2 + 64(2) = -64 + 128 = 64$ ft.

35. The dog pen appears as width is x
length is $40 - 2x$

Area = $A = x(40 - 2x)$
$A = 40x - 2x^2$

Now $x = -\dfrac{b}{2a} = -\dfrac{40}{2(-2)} = -\dfrac{40}{-4} = 10$ ft.

The width is $x = 10$ ft and the length is $40 - 2x = 40 - 2 \cdot 10 = 20$ ft.

37. The figure looks like

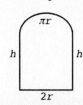

The perimeter is $h + h + \pi r + 2r = 2h + \pi r + 2r$.
Since the perimeter is fixed at 20, the equation becomes
$2h + \pi r + 2r = 20$. Now solve for h:

$$2h = 20 - \pi r - 2r$$

$$h = \frac{20 - \pi r - 2r}{2}$$

The area of the region is the area of the rectangle + area of semicircle

$$A = h(2r) + \frac{1}{2}\pi r^2$$

$$A = \left(\frac{20 - \pi r - 2r}{2}\right)(2r) + \frac{1}{2}\pi r^2$$

$$A = (20 - \pi r - 2r)r + \frac{1}{2}\pi r^2$$

$$A = 20r - \pi r^2 - 2r^2 + \frac{1}{2}\pi r^2$$

$$A = 20r - 2r^2 - \frac{1}{2}\pi r^2$$

$$= 20r - \left(2 + \frac{1}{2}\pi\right)r^2$$

Now, maximum area occurs for $r = -\dfrac{b}{2a} = -\dfrac{20}{2\left[-\left(2 + \frac{1}{2}\pi\right)\right]} = \dfrac{10}{2 + \frac{1}{2}\pi} = \dfrac{20}{4 + \pi}$ ft.

Since $r = \dfrac{20}{4 + \pi}$, substitute into $2h + \pi r + 2r = 20$ to get

$$2h + \pi\left(\frac{20}{4 + \pi}\right) + 2\left(\frac{20}{4 + \pi}\right) = 20$$

$$2h = 20 - \frac{20\pi}{4 + \pi} - \frac{40}{4 + \pi} = \frac{20(4 + \pi) - 20\pi - 40}{4 + \pi} = \frac{40}{4 + \pi}$$

Then, $\quad h = \dfrac{1}{2}\left(\dfrac{40}{4 + \pi}\right) = \dfrac{20}{4 + \pi}$ ft.

39. Using the graphics calculator the x intercepts are $x = 1.80$ and $x = 2.20$. The vertex is $(2, -0.04)$.

41. The graph is

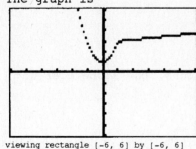

viewing rectangle [-6, 6] by [-6, 6]

43.

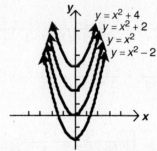

If $k > 0$, the graph of $y = x^2$ is shifted up k units.
If $k < 0$, the graph of $y = x^2$ is shifted down k units.

45.

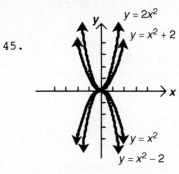

If $a < 0$, the graph of $y = ax^2$ is reflected about the x axis.

CHAPTER REVIEW

1. No x intercept;
   y intercept 4

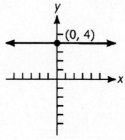

2. x intercept -2;
   No y intercept

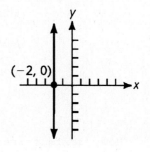

3. $2x - 5y = 10$

x intercept: Let $y = 0$, $\quad 2x - 5y = 10$
$$2x - 5(0) = 10$$
$$2x = 10$$
$$x = 5$$

The point is $(5, 0)$.

y intercept: Let $x = 0$, $\quad 2x - 5y = 10$
$$2(0) - 5y = 10$$
$$-5y = 10$$
$$y = -2$$

The point is $(0, -2)$ and the graph is

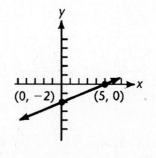

4. $m = \dfrac{y_2 - y_1}{x_2 - x_1} = \dfrac{3 - 3}{5 - (-2)} = \dfrac{0}{7} = 0$

5. $m = \dfrac{y_2 - y_1}{x_2 - x_1} = \dfrac{1 - 3}{-3 - (-2)} = \dfrac{-2}{-3 + 2} = \dfrac{-2}{-1} = 2$

6. $m = \dfrac{y_2 - y_1}{x_2 - x_1} = \dfrac{5 - (-7)}{-1 - (2)} = \dfrac{5 + 7}{-1 - 2} = \dfrac{12}{-3} = -4$

7. $m = \dfrac{y_2 - y_1}{x_2 - x_1} = \dfrac{-4 - 6}{5 - 5} = \dfrac{-10}{0}$ which is undefined. Thus, the slope does **not** exist.

8. midpoint $= \left(\dfrac{x_1 + x_2}{2}, \dfrac{y_1 + y_2}{2}\right) = \left(\dfrac{5 + 3}{2}, \dfrac{6 + 2}{2}\right) = \left(\dfrac{8}{2}, \dfrac{8}{2}\right) = (4, 4)$

9. midpoint $= \left(\dfrac{x_1 + x_2}{2}, \dfrac{y_1 + y_2}{2}\right) = \left(\dfrac{-3 + 7}{2}, \dfrac{8 + 2}{2}\right) = \left(\dfrac{4}{2}, \dfrac{10}{2}\right) = (2, 5)$

10. midpoint $= \left(\dfrac{x_1 + x_2}{2}, \dfrac{y_1 + y_2}{2}\right) = \left(\dfrac{0 + 7}{2}, \dfrac{-3 + 5}{2}\right) = \left(\dfrac{7}{2}, \dfrac{2}{2}\right) = \left(\dfrac{7}{2}, 1\right)$

11. midpoint $= \left(\dfrac{x_1 + x_2}{2}, \dfrac{y_1 + y_2}{2}\right) = \left(\dfrac{-4 + (-1)}{2}, \dfrac{6 + (-3)}{2}\right) = \left(\dfrac{-5}{2}, \dfrac{3}{2}\right)$ or $\left(-\dfrac{5}{2}, \dfrac{3}{2}\right)$

12. $3x - y = 10$ can be written as $y = 3x + 10$.
    When $x = 0$, then $y = 3(0) + 10 = 0 + 10 = 10$ and the point is $(0, 10)$. Also,
    when $x = 1$, then $y = 3(1) + 10 = 3 + 10 = 13$ and the second point is $(1, 13)$.
    Finally, $m = \dfrac{y_2 - y_1}{x_2 - x_1} = \dfrac{13 - 10}{1 - 0} = \dfrac{3}{1} = 3$.

13. Let $y = 0$. Then $x + 3y = 0$ becomes
    $$x + 3(0) = 0$$
    $$x + 0 = 0$$
    $$x = 0$$
    and the first point is $(0, 0)$.
    Now let $y = 1$ and the equation becomes
    $$x + 3(1) = 0$$
    $$x + 3 = 0$$
    $$x = -3$$
    and the second point is $(-3, 1)$.
    Finally, $m = \dfrac{y_2 - y_1}{x_2 - x_1} = \dfrac{1 - 0}{-3 - 0} = \dfrac{1}{-3}$ or $-\dfrac{1}{3}$.

14. $2x + 7y = 14$
    When $x = 0$, then the equation is
    $$2(0) + 7y = 14$$
    $$7y = 14$$
    $$y = 2$$
    and the point is $(0, 2)$.
    Also, when $y = 0$, then the equation is
    $$2x + 7(0) = 14$$
    $$2x = 14$$
    $$x = 7$$
    and the point is $(7, 0)$.
    Finally, $m = \dfrac{y_2 - y_1}{x_2 - x_1} = \dfrac{0 - 2}{7 - 0} = \dfrac{-2}{7} = -\dfrac{2}{7}$.

15. Let $y = 0$. Then $5x - 3y = 15$ becomes
    $$5x - 3(0) = 15$$
    $$5x - 0 = 15$$
    $$5x = 15$$
    $$x = 3$$
    and the first point is $(3, 0)$.
    Now let $x = 0$. Then $5x - 3y = 15$ becomes
    $$5(0) - 3y = 15$$
    $$0 - 3y = 15$$
    $$y = -\dfrac{15}{3} = -5$$
    and the second point is $(0, -5)$.
    Finally, $m = \dfrac{y_2 - y_1}{x_2 - x_1} = \dfrac{-5 - 0}{0 - 3} = \dfrac{-5}{-3} = \dfrac{5}{3}$.

16. $x = -y^2$ is a parabola opening to the left:

x	y
0	0
-1	1
-1	-1
-4	2
-4	-2

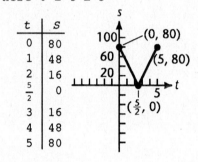

17.

y	x
2	0
1	1
0	2
3	1
4	2

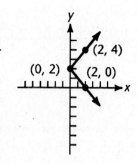

18. Here $0 \le t \le 5$

t	S
0	80
1	48
2	16
$\frac{5}{2}$	0
3	16
4	48
5	80

19. Here, 1900 corresponds to $t = 0$ and 1950 corresponds to $t = 50$. When $t = 0$, then $P = 80 + 1.5(0) = 80 + 0 = 80$(million).

Also, when $t = 50$, then
$$P = 80 + 1.5(50)$$
$$P = 80 + 75 = 155 \text{ million}$$
Thus, $0 \le t \le 50$ and $80 \le P \le 155$ and the graph is shown at the right.

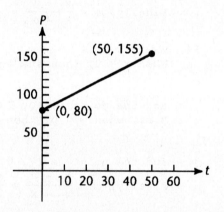

20. $D = \mathcal{R}$, $R = [1, \infty)$

21. For the domain, we need $x - 9 \ge 0$ or $x \ge 9$ which is $[9, \infty)$.
For the range, $y \ge 0$ which is $[0, \infty)$. That is, $D = [9, \infty)$, $R = [0, \infty)$.

22. $D = \mathcal{R}$, $R = [0, \infty)$

23. For the domain, we need $x - 4 \ne 0$ or $x \ne 4$. This is written as $(-\infty, 4) \cup (4, \infty)$. The range is $y \ne 0$ or $(-\infty, 0) \cup (0, \infty)$. That is, $D = (-\infty, 4) \cup (4, \infty)$, $R = (-\infty, 0) \cup (0, \infty)$.

24. $f(4) = (4)^2 - 2(4) = 16 - 8 = 8$

25. First, $g(9) = \sqrt{9} + 3 = 3 + 3 = 6$.
Then $(f \circ g)(9) = f(g(9)) = f(6) = 6^2 - 2 \cdot 6 = 36 - 12 = 24$

26. First $g(a) = \sqrt{a} + 3$. Then,

$$(f \circ g)(a) = f(g(a)) = f(\sqrt{a} + 3) = (\sqrt{a} + 3)^2 - 2(\sqrt{a} + 3)$$

$$= a + 6\sqrt{a} + 9 - 2\sqrt{a} - 6$$

$$= a + 4\sqrt{a} + 3$$

27. $f(x + h) - f(x) = [(x + h)^2 - 2(x + h)] - [x^2 - 2x]$

$$= (x^2 + 2xh + h^2 - 2x - 2h) - (x^2 - 2x)$$

$$= x^2 + 2xh + h^2 - 2x - 2h - x^2 + 2x$$

$$= 2xh + h^2 - 2h$$

$$= h(2x + h - 2)$$

28. The graph of $G(x) = \left[\!\left[\dfrac{x}{2} \right]\!\right]$ is a step function:

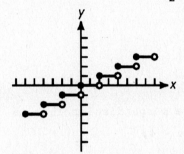

29. $(f + g)(x) = f(x) + g(x) = \sqrt{x - 3} + x - 4$ and the domain is $x \in [3, \infty)$
     $(f - g)(x) = f(x) - g(x) = \sqrt{x - 3} - (x - 4) = \sqrt{x - 3} - x + 4$ and the domain
     is $x \in [3, \infty)$

30. a) $(f \cdot g)(x) = f(x) \cdot g(x) = \sqrt{x - 3}\,(x - 4)$ for $x \in [3, \infty)$

     b) $\left(\dfrac{f}{g}\right)(x) = \dfrac{f(x)}{g(x)} = \dfrac{\sqrt{x - 3}}{x - 4}$ for $x \in [3, 4) \cup (4, \infty)$

31. Not a function since several vertical lines cut the graph in two points.

32. Is a function since vertical lines intersect the graph in one point.

33. Is a function since every vertical line cuts the graph in one point.

34. Is **not** a function since vertical lines near the y axis intersect the graph
     in two points.

35. Here $m = -2$ and the point is $(-1, 0)$.
     The equation is $y - 0 = -2[x - (-1)]$
     $$y = -2(x + 1)$$
     $$y = -2x - 2$$
     or $2x + y = -2$

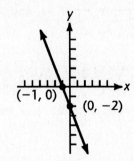

36. $m = \dfrac{y_2 - y_1}{x_2 - x_1} = \dfrac{4 - 1}{-3 - 1} = \dfrac{3}{-4} = -\dfrac{3}{4}$.　　　37. The two points are (4, 0) and (0, -3).

Using (1, 1) as the point, the equation is
$$m = \dfrac{y_2 - y_1}{x_2 - x_1} = \dfrac{-3 - 0}{0 - 4} = \dfrac{-3}{-4} = \dfrac{3}{4}$$

$$y - 1 = -\dfrac{3}{4}(x - 1)$$

The equation is $y = \dfrac{3}{4}x - 3$

$$4(y - 1) = -3(x - 1)$$
$$4y - 4 = -3x + 3$$
or $4y = 3x - 12$
$$3x + 4y - 4 = 3$$
or $4y - 3x = -12$
$$3x + 4y = 7$$
or $3x - 4y = 12$

38. Solve $5x - y = 10$ for y: $5x - y = 10$
$$-y = -5x + 10$$
$$y = 5x - 10$$
The slope of this line is 5 and, consequently, the slope of the parallel line is also 5. Using (-2, -1) as a point on the line, the equation of the parallel line is $y - (-1) = 5[x - (-2)]$
$$y + 1 = 5(x + 2)$$
$$y + 1 = 5x + 10$$
$$1 = 5x - y + 10$$
$$-9 = 5x - y$$

39. $4x + y = 9$
$$y = -4x + 9$$
and the slope of this line is -4. The slope of the perpendicular line is
$-\dfrac{1}{m} = -\dfrac{1}{-4} = \dfrac{1}{4}$ and its equation is $y - (-5) = \dfrac{1}{4}(x - 4)$
$$y + 5 = \dfrac{1}{4}(x - 4)$$
$$4y + 20 = x - 4$$
$$4y + 24 = x$$
$$24 = x - 4y$$

40. Since the slope does not exist, the line must be a vertical line whose equation is $x = a$. Now the line passes through the point (-2, 6) which means that $a = -2$ and the final equation is $x = -2$.

41. $\dfrac{y - 1}{3 - (-1)} = 2$ or $\dfrac{y - 1}{4} = 2$
$$y - 1 = 8$$
$$y = 9$$

42. The slope is $m = \dfrac{y_2 - y_1}{x_2 - x_1} = \dfrac{0 - 4}{2 - 0} = \dfrac{-4}{2} = -2$.

Since the y intercept is 4, the equation is $y = -2x + 4$ or $2x + y = 4$.

43. $m = \dfrac{y_2 - y_1}{x_2 - x_1} = \dfrac{5 - 1}{0 - 2} = \dfrac{4}{-2} = -2$ and the y intercept is 5 so that the equation

is $\qquad y = -2x + 5$
or $2x + y = 5$

44. The slope is $m = \dfrac{y_2 - y_1}{x_2 - x_1} = \dfrac{2 - 0}{1 - 0} = \dfrac{2}{1} = 2$.

Since the y intercept is 0, the equation is $y = 2x + 0$ or $0 = 2x - y$.

45. $m = \dfrac{y_2 - y_1}{x_2 - x_1} = \dfrac{1 - 0}{2 - 0} = \dfrac{1}{2}$ and the y intercept is 0 so that the equation is

$$y = \dfrac{1}{2} x$$

or $2y = x$

or $0 = x - 2y$

46. The slope is $m = \dfrac{y_2 - y_1}{x_2 - x_1} = \dfrac{1 - (-3)}{2 - 0} = \dfrac{1 + 3}{2} = \dfrac{4}{2} = 2.$

Since the y intercept is -3, the equation is $y = 2x - 3$ or $0 = 2x - y - 3$

or $3 = 2x - y$

47.

x	y
1	3
2	2
4	0
0	1
-1	-1

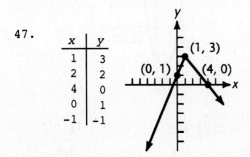

48. a) The data points are (0, 1800) and (15, 1200). Now,

$$m = \dfrac{1200 - 1800}{15 - 0} = \dfrac{-600}{15} = -40$$

Thus, the copier depreciated at a value of \$40 each month.

b) Using $m = -40$ and $b = 1800$ (the y intercept is 1800).
The linear equation is $V(t) = -40t + 1800$

49. The data points are (32, 800) and (44, 968)

$$m = \dfrac{968 - 800}{44 - 32} = \dfrac{168}{12} = 14$$

and using the point (32, 800), the equation is

$C - 800 = 14(x - 32)$

$C - 800 = 14x - 448$

$\quad\quad C = 14x + 352,$

or $C(x) = 14x + 352$

50. $y = 3(x - 1)^2 + 2$

Vertex: (1, 2)

Axis: $x = 1$

Two other points are (0, 5) and (2, 5) and the graph is

105

51. Vertex: $x = -\dfrac{b}{2a} = -\dfrac{4}{2(-1)} = 2$. Then $y = 4(2) - 2^2 = 8 - 4 = 4$ and the point is $(2, 4)$. Axis: $x = -\dfrac{b}{2a} = 2$ Two other points are $(0, 0)$ and $(4, 0)$ and the graph is

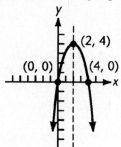

52. $y = x^2 + 2x - 3$

For the vertex, $x = -\dfrac{b}{2a} = -\dfrac{2}{2(1)} = -1$ and $y = (-1)^2 + 2(-1) - 3 = 1 - 2 - 3 = -4$ and the point is $(-1, -4)$.

Axis: $x = -\dfrac{b}{2a} = -1$

Four other points are $(-3, 0)$, $(-2, -3)$, $(0, -3)$, and $(1, 0)$ and the graph is

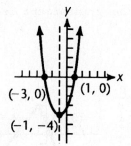

53. Vertex: $x = -\dfrac{b}{2a} = -\dfrac{3}{2(-2)} = \dfrac{3}{4}$. Then $y = -2\left(\dfrac{3}{4}\right)^2 + 3\left(\dfrac{3}{4}\right) + 1$
$$= -\dfrac{9}{8} + \dfrac{9}{4} + 1 = \dfrac{17}{8}$$
and the point is $\left(\dfrac{3}{4}, \dfrac{17}{8}\right)$.

Axis: $x = -\dfrac{b}{2a} = \dfrac{3}{4}$

Two other points are $(0, 1)$ and $\left(\dfrac{3}{2}, 1\right)$ and the graph is shown at the right.

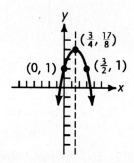

54. The extreme value is $f(3) = 2(3 - 3)^2 + 1 = 0 + 1 = 1$ and it is a minimum value.

55. $f(x) = 1 - (x - 2)^2 = -(x - 2)^2 + 1$
The extreme value occurs when $x = 2$ and it is $f(2) = -(2 - 2)^2 + 1 = 0 + 1 = 1$ which is a maximum value.

56. The extreme value occurs when $x = -\dfrac{b}{2a} = -\dfrac{(-4)}{2(1)} = -\dfrac{-4}{2} = 2$ and it is $f(2) = 2^2 - 4(2) + 2 = 4 - 8 + 2 = -2$ which is a minimum value.

57. $f(x) = 1 - 6x - x^2$
The extreme value occurs for
$$x = -\dfrac{b}{2a} = -\dfrac{-6}{2(-1)} = -3$$
and it is $f(-3) = 1 - 6(-3) - (-3)^2 = 1 + 18 - 9 = 10$.
Since the coefficient of x^2 is negative $f(-3) = 10$ is a maximum value.

58. The extreme value occurs when $x = -\dfrac{b}{2a} = -\dfrac{-4}{2(-2)} = -\dfrac{-4}{-4} = -1$ and it is $f(-1) = -2(-1)^2 - 4(-1) + 1 = -2 + 4 + 1 = 3$ which is a maximum value.

59. $f(x) = 2x^2 - 8x + 13$
    The extreme value occurs for
    $$x = -\frac{b}{2a} = -\frac{-8}{2(2)} = 2$$
    and it is $f(2) = 2(2)^2 - 8(2) + 13 = 8 - 16 + 13 = 5$.
    Since the coefficient of x^2 is positive, $f(2) = 5$ is a minimum value.

60. $P(x) = 0.8x - 0.001x^2$
    The maximum value occurs for
    $$x = -\frac{b}{2a} = -\frac{0.8}{2(-0.001)} = -\frac{0.8}{-0.002} = 400 \text{ units}$$
    and the value is $P(400) = 0.8(400) - 0.001(400)^2$
    $$= 320 - 160$$
    $$= 160 \text{ dollars}$$

61. $h(t) = 48t - 16t^2$
    The maximum value occurs for
    $$t = -\frac{b}{2a} = -\frac{48}{2(-16)} = -\frac{48}{-32} = \frac{3}{2} = 1.5 \text{ seconds}$$
    and the value is $h\left(\frac{3}{2}\right) = 48\left(\frac{3}{2}\right) - 16\left(\frac{3}{2}\right)^2 = 72 - 36 = 36 \text{ ft.}$

62. The minimum value occurs for $x = -\frac{b}{2a} = -\frac{-60}{2(3)} = -\frac{-60}{6} = 10$ lawnmowers and the
    value is $C(10) = 1500 - 60(10) + 3(10)^2$
    $$= 1500 - 600 + 300$$
    $$= 1200 \text{ dollars}$$

63. $R(x) = 40x - x^2 - 150 = -x^2 + 40x - 150$
    The maximum value occurs for
    $$x = -\frac{b}{2a} = -\frac{40}{2(-1)} = 20 \text{ bird houses}$$
    and the value is $R(20) = -(20)^2 + 40(20) - 150 = -400 + 800 - 150 = 250$
    dollars.

64. The maximum value occurs for $n = -\frac{b}{2a} = -\frac{80}{2(-0.02)} = -\frac{80}{-0.04} = 2000$ anchors
    and the value is $P(2000) = 80(2000) - 0.02(2000)^2$
    $$= 160,000 - 80,000$$
    $$= 80,000 \text{ dollars}$$

CHAPTER 4   GRAPHING TECHNIQUES

Exercises 4.1

1. $d = \sqrt{(4 - 0)^2 + (3 - 0)^2} = \sqrt{4^2 + 3^2} = \sqrt{16 + 9} = \sqrt{25} = 5$

3. $d = \sqrt{(3 - 0)^2 + [1 - (-3)]^2} = \sqrt{3^2 + 4^2} = \sqrt{9 + 16} = \sqrt{25} = 5$

5. $d = \sqrt{[11 - (-9)]^2 + (1 - 22)^2} = \sqrt{(20)^2 + (-21)^2} = \sqrt{400 + 441} = \sqrt{841} = 29$

7. $d = \sqrt{[-2 - (2)]^2 + [-9 - (-9)]^2} = \sqrt{(-4)^2 + (-18)^2} = \sqrt{16 + 324} = \sqrt{340} = \sqrt{4 \cdot 85}$
    $= 2\sqrt{85}$

9. Since $y = 10$ is a horizontal line, the shortest distance is along the vertical line from $(3, 6)$ to $(3, 10)$.
    $$d = \sqrt{(3 - 3)^2 + (10 - 6)^2} = \sqrt{0^2 + 4^2} = \sqrt{16} = 4$$

11. The distance is $10 - f(x)$.

13. It is a circle with center at $(0, 0)$ and $r = \sqrt{16} = 4$.

15. It is a circle with center at $(0, -4)$ and $r = \sqrt{\dfrac{3}{25}} = \dfrac{\sqrt{3}}{5}$.

17. Use the method of completing the square
    $$x^2 + 2x + \underline{} + y^2 + 6y + \underline{} = 0 + \underline{} + \underline{}$$
    $$x^2 + 2x + \underline{1} + y^2 + 6y + \underline{9} = 0 + \underline{1} + 9$$

    $(x + 1)^2 + (y + 3)^2 = 10$ Here $h = -1$, $k = -3$, and $r^2 = 10$ so that it is a circle with center at $(-1, -3)$ and $r = \sqrt{10}$.

19. $x^2 + y^2 = 3x + 4y - 4$
    $x^2 - 3x + y^2 - 4y = -4$
    Use the method of completing the square:
    $$x^2 - 3x + \underline{} + y^2 - 4y + \underline{} = -4 + \underline{} + \underline{}$$
    $$x^2 - 3x + \underline{\dfrac{9}{4}} + y^2 - 4y + \underline{4} = -4 + \underline{\dfrac{9}{4}} + \underline{4}$$

    $\left(x - \dfrac{3}{2}\right)^2 + (y - 2)^2 = \dfrac{9}{4}$ Here $h = \dfrac{3}{2}$, $y = 2$, and $r^2 = \dfrac{9}{4}$ so that it is a circle with center at $\left(\dfrac{3}{2}, 2\right)$ and $r = \sqrt{\dfrac{9}{4}} = \dfrac{3}{2}$.

21. $x^2 + y^2 = 10x - 6y - 36$
    $x^2 - 10x + y^2 + 6y = -36$
    Use the method of completing the square
    $$x^2 - 10x + \underline{} + y^2 + 6y + \underline{} = -36 + \underline{} + \underline{}$$
    $$x^2 - 10x + \underline{25} + y^2 + 6y + \underline{9} = -36 + \underline{25} + \underline{9}$$
    $(x - 5)^2 + (y + 3)^2 = -2$ Since $r^2 = -2 < 0$, no circle exists.

23. $x^2 + y^2 - 4x + 16y + 68 = 0$
    $x^2 - 4x + y^2 + 16y + 68 = 0$
    Use the method of completing the square
    $$x^2 - 4x + \underline{} + y^2 + 16y + \underline{} = -68 + \underline{} + \underline{}$$
    $$x^2 - 4x + \underline{4} + y^2 + 16y + \underline{64} = -68 + \underline{4} + \underline{64}$$

    $(x - 2)^2 + (y + 8)^2 = 0$ Here $h = 2$, $k = -8$, and $r^2 = 0$ so that it is a circle with center at $(2, -8)$ and $r = 0$. It is a point-circle.

25. Use $h = 0$, $k = 1$, and $r = 2$ to obtain $(x - 0)^2 + (y - 1)^2 = 2^2$
    or $x^2 + (y - 1)^2 = 4$

27. Here $r = \sqrt{(5 - 2)^2 + [1 - (-2)]^2} = \sqrt{3^2 + 3^2} = \sqrt{9 + 9} = \sqrt{18}$. Also, use $h = 2$ and $k = -2$ to obtain $(x - 2)^2 + [y - (-2)]^2 = (\sqrt{18})^2$
    or $(x - 2)^2 + (y + 2)^2 = 18$

29. Center is the midpoint of the diameter: $\left(\dfrac{0 + 6}{2}, \dfrac{4 + (-2)}{2}\right) = \left(\dfrac{6}{2}, \dfrac{2}{2}\right) = (3, 1)$. The radius is the distance from $(3, 1)$ to $(0, 4) = \sqrt{(0 - 3)^2 + (4 - 1)^2} = \sqrt{9 + 9} = \sqrt{18}$. Use $h = 3$, $k = 1$, and $r = \sqrt{18}$ to obtain
    $(x - 3)^2 + (y - 1)^2 = (\sqrt{18})^2$
    or $(x - 3)^2 + (y - 1)^2 = 18$

31. $x^2 + y^2 = 4$ is a circle with center at $(0, 0)$ and $r = \sqrt{4} = 2$.

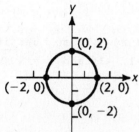

33. $x^2 - 4x + y^2 = 21$
    $x^2 - 4x + \underline{4} + y^2 = 21 + \underline{4}$
    $(x - 2)^2 + y^2 = 25$
    It is a circle with center at $(2, 0)$ and $r = \sqrt{25} = 5$.

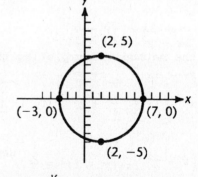

35. $x^2 + y^2 = 10x$
    $x^2 - 10x + y^2 = 0$
    $x^2 - 10x + \underline{25} + y^2 = 0 + 25$
    $(x - 5)^2 + y^2 = 25$
    It is a circle with center at $(5, 0)$ and $r = \sqrt{25} = 5$.

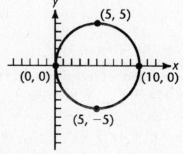

37. $x = 12 - 3(y + 2)^2$
$x = -3(y + 2)^2 + 12$
It is a parabola with vertex at
(12, -2) and $a = -3$, $k = -2$, and
$h = 12$. Two other points are
obtained by letting $y = 0$ to get
$x = 12 - 3(0 + 2)^2 = 12 - 3(4) = 0$.
The point is (0, 0). Also, when
$y = -4$, then $x = 12 - 3(-4 + 2)^2 =$
$12 - 3(-2)^2 = 12 - 3(4) = 12 - 12 =$
0. The point is (0, -4).

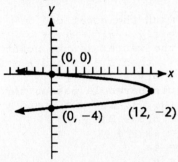

39. $x = 3y^2 + 6y - 20$
It is a parabola which opens to the
right. For the vertex, $y = -\dfrac{b}{2a}$
$= -\dfrac{6}{2 \cdot 3} = -\dfrac{6}{6} = -1$ and the x value is
$x = 3(-1)^2 + 6(-1) - 20 = 3 - 6 - 20$
$= -23$. The vertex is (-23, -1).
Two other points are (-20, 0) and
(-20, -2) which are obtained by
letting $y = 0$ to get
$x = 3(0)^2 + 6(0) - 20 = -20$ and by
letting $y = -2$ to get
$x = 3(-2)^2 + 6(-2) - 20$
$= 3(4) - 12 - 20 = -20$.

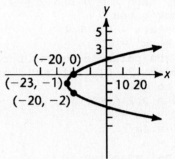

41. $y = \sqrt{x - 3}$ Here $y \geq 0$.
$y^2 = x - 3$
$y^2 + 3 = x$ which is the upper part
of a parabola since $y \geq 0$. Now
$a = 1$, $k = 0$, and $h = 3$ so that
the vertex is (3, 0). For another
point, when $y = 1$, then $x = 4$. The
point is (4, 1).

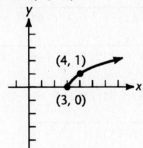

43. $x = \sqrt{y - 2}$ Here $x \geq 0$.
$x^2 = y - 2$
$x^2 + 2 = y$ which is the right side of
a parabola. Now $a = 1$, $h = 0$, and
$k = 2$. The vertex is $(h, k) = (0, 2)$.
For another point, let $x = 1$ to get
$y = 1^2 + 2 = 1 + 2 = 3$. The point is
(1, 3).

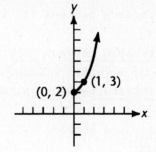

45. $x = 2 - 2\sqrt{y - 1}$ Here $y \geq 1$ and $x \leq 2$.
$$x - 2 = -2\sqrt{y - 1}$$
$$(x - 2)^2 = [-2\sqrt{y - 1}]^2$$
$$(x - 2)^2 = 4(y - 1)$$
$$(x - 2)^2 = 4y - 4$$
$$(x - 2)^2 + 4 = 4y$$
$\dfrac{1}{4}(x - 2)^2 + 1 = y$ which is the left side of a parabola since $x \leq 2$.

111

Now $a = \dfrac{1}{4}$, $h = 2$, and $k = 1$ so that the vertex
is $(h, k) = (2, 1)$. For another point, let $x = 0$
to get

$$\frac{1}{4}(0 - 2)^2 + 1 = y$$

$$\frac{1}{4}(4) + 1 = y$$

$$2 = y$$

and the point is $(0, 2)$

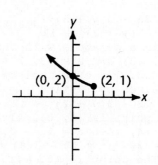

47. Start with the equation
$$y = a(x - h)^2 + k$$
Since the vertex is the point
$(0, 4)$, then $h = 0$ and $k = 4$.
This now gives
$$y = a(x - 0)^2 + 4$$
$$y = ax^2 + 4$$
Since the parabola passes through
the point $(2, 0)$, let $x = 2$ and
$y = 0$ to get
$$0 = a(2)^2 + 4$$
$$0 = 4a + 4$$
$$-4 = 4a$$
$$-1 = a$$
The final equation is
$$y = -x^2 + 4$$

49. Start with the equation
$$x = a(y - k)^2 + h$$
Since the vertex is the point
$(-4, 2)$, then $h = -4$ and $k = 2$.
$$x = a(y - 2)^2 - 4$$
Since the parabola passes through the
point $(0, 0)$, let $x = 0$ and $y = 0$
$$0 = a(0 - 2)^2 - 4$$
$$0 = 4a - 4$$
$$4 = 4a$$
$$1 = a$$
The final equation is
$$x = (y - 2)^2 - 4$$

51. The slope of the line joining $(2, 2)$ to $(-2, -2)$ is $m_1 = \dfrac{-2 - 2}{-2 - 2} = \dfrac{-4}{-4} = 1$.

The slope of the line joining $(2, 2)$ and $(-1, 5)$ is $m_2 = \dfrac{5 - 2}{-1 - 2} = \dfrac{3}{-3} = -1$.

Since one slope is the negative reciprocal of the other slope, the two lines
are $\perp$ and the right angle occurs at the point $(2, 2)$.

53. The distance between $(6, -9)$ and $(9, 1)$ is $\sqrt{(9 - 6)^2 + [1 - (-9)]^2} = \sqrt{3^2 + 10^2}$
$= \sqrt{109}$.

The distance between $(-1, 4)$ and $(-4, -6)$ is $\sqrt{[-4 - (-1)]^2 + (-6 - 4)^2}$
$= \sqrt{(-3)^2 + (-10)^2} = \sqrt{109}$.

The distance between $(6, -9)$ and $(-4, -6)$ is $\sqrt{(-4 - 6)^2 + [-6 - (-9)]^2}$
$= \sqrt{(-10)^2 + 3^2} = \sqrt{109}$.

The distance between $(-1, 4)$ and $(9, 1)$ is $\sqrt{[9 - (-1)]^2 + (1 - 4)^2} = \sqrt{10^2 + (-3)^2}$
$= \sqrt{109}$.

The distance between $(9, 1)$ and $(-4, -6)$ is $\sqrt{(-4 - 9)^2 + (-6 - 1)^2}$
$= \sqrt{(-13)^2 + (-7)^2} = \sqrt{218}$.

The distance between $(6, -9)$ and $(-1, 4)$ is $\sqrt{(-1 - 6)^2 + [4 - (-9)]^2}$
$= \sqrt{(-7)^2 + (13)^2} = \sqrt{218}$.

It is a square since all four sides have the same length ($\sqrt{109}$) and the two
diagonals have the same length ($\sqrt{218}$).

55. Set up a coordinate system:

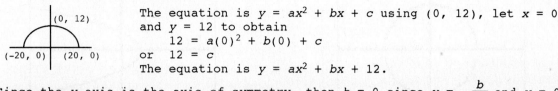

The equation is $y = ax^2 + bx + c$ using $(0, 12)$, let $x = 0$ and $y = 12$ to obtain

$$12 = a(0)^2 + b(0) + c$$

or $12 = c$

The equation is $y = ax^2 + bx + 12$.

Since the y axis is the axis of symmetry, then $b = 0$ since $y = -\dfrac{b}{2a}$ and $y = 0$ produce $-\dfrac{b}{2a} = 0$ or $b = 0$. The equation is now $y = ax^2 + 12$.

Finally, to find a, use the point $(20, 0)$ with $x = 20$ and $y = 0$ to get

$$0 = a(20)^2 + 12$$
$$0 = 400a + 12$$
$$-12 = 400a$$
$$\frac{-12}{400} = a \quad \text{or} \quad a = -\frac{3}{100}$$

The final equation is $y = -\dfrac{3}{100}x^2 + 12$.

Now let $x = 10$ to obtain

$$y = -\frac{3}{100}(10)^2 + 12$$
$$= -3 + 12$$
$$= 9 \text{ meters}$$

57. Solve this problem in the same manner as problem 55.
The coordinate system is:

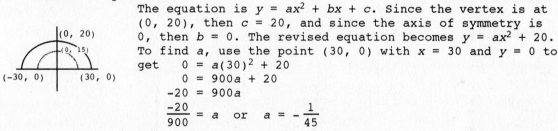

The equation is $y = ax^2 + bx + c$. Since the vertex is at $(0, 20)$, then $c = 20$, and since the axis of symmetry is 0, then $b = 0$. The revised equation becomes $y = ax^2 + 20$.

To find a, use the point $(30, 0)$ with $x = 30$ and $y = 0$ to get

$$0 = a(30)^2 + 20$$
$$0 = 900a + 20$$
$$-20 = 900a$$
$$\frac{-20}{900} = a \quad \text{or} \quad a = -\frac{1}{45}$$

The final equation is $y = -\dfrac{1}{45}x^2 + 20$.

For the second reflector, the equation is $y = -\dfrac{1}{45}x^2 + 15$ where 20 from the original equation is changed to 15. To find the length of the opening, set $y = 0$ and solve for x:

$$0 = -\frac{1}{45}x^2 + 15$$
$$-15 = -\frac{1}{45}x^2$$
$$15 \cdot 45 = x^2 \quad \text{so that } x = \pm\sqrt{15 \cdot 45} \pm \sqrt{(15)^2 3} = \pm 15\sqrt{3}.$$

The intercepts are $(-15\sqrt{3}, 0)$ and $(15\sqrt{3}, 0)$ and the distance between them is $30\sqrt{3}$ cm.

59.

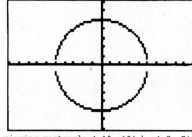

viewing rectangle [-10, 10] by [-7, 7]

61.

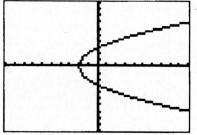

viewing rectangle [-10, 10] by [-10, 10]

63.

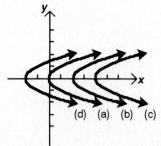

If $h > 0$, the graph of $x = y^2$ is shifted to the right by h units. If $h < 0$, the graph of $x = y^2$ is shifted to the left $|h|$ units.

65.

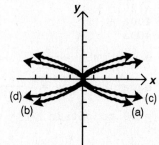

If $a < 0$, the graph of $x = ay^2$ is reflected about the x axis.

Exercises 4.2

For exercises 1—7, the graphs are given below.

1. a)

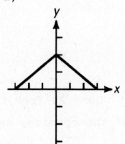

b)

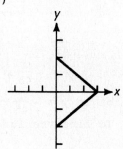

c)

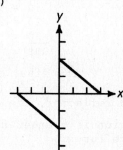

3. a)

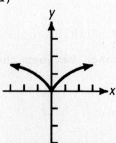

b)

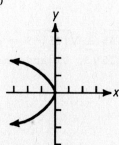

c)

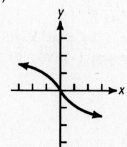

5. a) 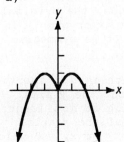   b) 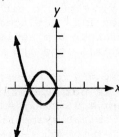   c)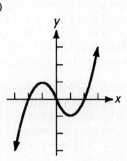

7. a) 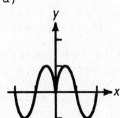   b) 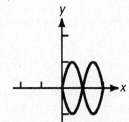   c)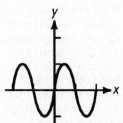

9. $y = |x|$: Replace x with $-x$ to get $y = |-x| = |x|$ which is identical to the original equation. It is symmetric with respect to the y axis.
   Replace y with $-y$ to get $-y = |x|$ which is **not** identical to the original equation. It is not symmetric with respect to the x axis.
   Replace x with $-x$ and y with $-y$ to get $-y = |-x|$ or $-y = |x|$ which is **not** identical to the original equation. It is not symmetric about the origin.

11. $|y| = |x|$: Replace x with $-x$ to get $|y| = |-x|$ or $|y| = |x|$ which is identical to the original equation. It is symmetric with respect to the y axis.
    Replace y with $-y$ to get $|-y| = |x|$ or $|y| = |x|$ which is identical to the original equation. It is symmetric with respect to the x axis.
    Replace x with $-x$ and y with $-y$ to get $|-y| = |-x|$ or $|y| = |x|$ which is identical to the original equation. It is symmetric about the origin.

13. $y^2 = x + 1$: Replace x with $-x$ to get $y^2 = -x + 1$ which is **not** identical to the original equation. It is **not** symmetric about the y axis.
    Replace y with $-y$ to get $(-y)^2 = x + 1$ or $y^2 = x + 1$ which is identical to the original equation. It is symmetric about the x axis.
    Replace x with $-x$ and y with $-y$ to get $(-y)^2 = -x + 1$ or $y^2 = -x + 1$ which is **not** identical to the original equation. It is **not** symmetric about the origin.

15. $y = x$: Replace x with $-x$ to get $y = -x$ which is **not** identical to the original equation. It is not symmetric about the y axis.
    Replace y with $-y$ to get $-y = x$ which is **not** identical to the original equation. It is not symmetric about the x axis.
    Replace x with $-x$ and y with $-y$ to get $-y = -x$ or $y = x$ which is identical to the original equation. It is symmetric about the origin.

17. $x^2 + y^2 = 25$: Here when you replace x by $-x$, the $(-x)^2 = x^2$ and when y is replaced by $-y$, then $(-y)^2 = y^2$. In either case, the original equation is obtained when x is replaced with $-x$ or when y is replaced with $-y$ or when both are replaced. It is symmetric about the y axis, the x axis, and the origin.

19. $xy = 9$: Replace x with $-x$ to get $(-x)y = 9$ or $-xy = 9$ which is **not** identical to the original equation. It is **not** symmetric about the y axis. Replace y with $-y$ to get $x(-y) = 9$ or $-xy = 9$ which is **not** identical to the original equation. It is not symmetric about the x axis. Replace x with $-x$ and y with $-y$ to get $(-x)(-y) = 9$ or $xy = 9$ which is identical to the original equation. It is symmetric about the origin.

21. $y = x^4 - 4x^2$: Replace x with $-x$ to get $y = (-x)^4 - 4(-x)^2$ or $y = x^4 - 4x^2$ which is identical to the original equation. It is symmetric about the y axis.
Replace y with $-y$ to get $-y = x^4 - 4x^2$ which is **not** identical to the original equation. It is not symmetric about the x axis. Replace x with $-x$ and y with $-y$ to get $-y = (-x)^4 - 4(-x)^2$ or $-y = x^4 - 4x^2$ which is **not** identical to the original equation. It is not symmetric about the origin.

23. a) $y = 2f(x)$ stretches $y = f(x)$ by a factor of 2.

    b) $y = f(x) - 2$ shifts $y = f(x)$ down by 2 units.

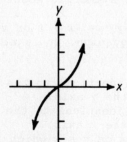

25. a) $y = f(x - 1)$ shifts $y = f(x)$ to the right by 1 unit.

    b) $y = -f(x)$ reflects $y = f(x)$ about the x axis.

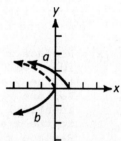

27. $y = \frac{1}{2}x^3$ stretches (compresses) $y = x^3$ by a factor of $\frac{1}{2}$.

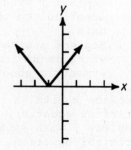

29. $y = |x + 1|$ shifts $y = |x|$ one unit to the left.

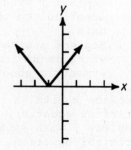

31. $y = (x - 1)^3$ shifts $y = x^3$ one unit to the right.

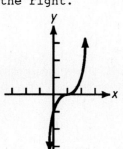

33. $y = x^3 - 1$ shifts $y = x^3$ down one unit.

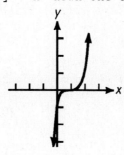

35. $y = (x + 1)^3 - 2$ shifts $y = x^3$ one unit to the left and down two units.

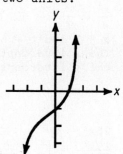

For exercises 37—41, the graphics calculator obtains the desired results.

37.

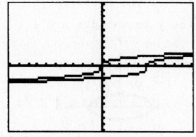

viewing rectangle [-10, 10] by [-10, 10]

The original graph has been shifted to the right 5 units.

39.

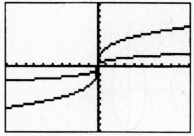

viewing rectangle [-10, 10] by [-10, 10]

The original graph has been stretched by a factor of 3, that is, each y coordinate has been multiplied by 3.

41.

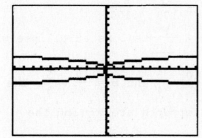

viewing rectangle [-10, 10] by [-10, 10]

The original graph has been reflected through the y axis, that is, each point on the graph has moved to the point that is symmetric with respect to the y axis.

43. For your discussion consider several different cases where x is replaced by $-x$.

45. There are several functions. Consider $f(x) = x$ or $f(x) = -x$.

Exercises 4.3

1. $\frac{x^2}{4} + \frac{y^2}{9} = 1$ is an ellipse with
   $a = 2$ and $b = 3$.
   The x intercepts are ±2 and
   the y intercepts are ±3.

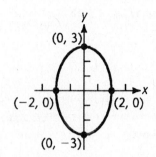

3. $25x^2 + 4y^2 = 100$ Divide by 100:
   $$\frac{25x^2}{100} + \frac{4y^2}{100} = \frac{100}{100}$$
   $\frac{x^2}{4} + \frac{y^2}{25} = 1$ is an ellipse with
   $a = 2$ and $b = 5$. The x intercepts
   are ±2 and the y intercepts are
   ±5.

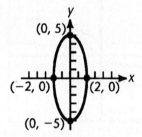

5. $4x^2 + 25y^2 = 25$ Divide by 25:
   $$\frac{4x^2}{25} + \frac{25y^2}{25} = \frac{25}{25}$$
   $\frac{x^2}{\frac{25}{4}} + \frac{y^2}{1} = 1$ is an ellipse with $a = \frac{5}{2}$
   and $b = 1$. The x intercepts are $\pm\frac{5}{2}$
   and the y intercepts are ±1.

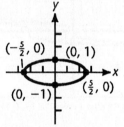

7. $\frac{4x^2}{9} + \frac{9y^2}{16} = 1$ can be written as
   $\frac{x^2}{\frac{9}{4}} + \frac{y^2}{\frac{16}{9}} = 1$ is an equation of an
   ellipse with $a = \frac{3}{2}$ and $b = \frac{4}{3}$.
   The x intercepts are $\pm\frac{3}{2}$ and the
   y intercepts are $\pm\frac{4}{3}$.

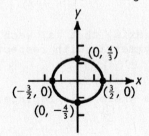

9. $64x^2 + 36y^2 = 100$ Divide by 100:
   $$\frac{64x^2}{100} + \frac{36y^2}{100} = \frac{100}{100}$$
   $\frac{x^2}{\frac{100}{64}} + \frac{y^2}{\frac{100}{36}} = 1$ is an ellipse with
   $a = \frac{10}{8} = \frac{5}{4}$ and $b = \frac{10}{6} = \frac{5}{3}$. The
   x intercepts are $\pm\frac{5}{4}$ and the
   y intercepts are $\pm\frac{5}{3}$.

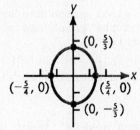

11. $2x^2 + y^2 = 8$ Divide by 8:

$$\frac{2x^2}{8} + \frac{y^2}{8} = \frac{8}{8}$$

$\frac{x^2}{4} + \frac{y^2}{8} = 1$ is an ellipse with $a = 2$ and

$b = \sqrt{8} = \sqrt{4 \cdot 2} = 2\sqrt{2}$. The x intercepts are

± 2 and the y intercepts are $\pm 2\sqrt{2}$.

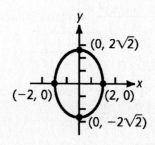

13. $x^2 + 4y^2 = 0$ holds true only if
$x = 0$ and $y = 0$. Thus, the graph
is the point $(0, 0)$.

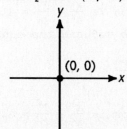

15. $2y = \sqrt{16 - x^2}$ Notice that $y \geq 0$

$$(2y)^2 = (\sqrt{16 - x^2})^2$$
$$4y^2 = 16 - x^2$$
$x^2 + 4y^2 = 16$ Now $\div$ by 16:
$$\frac{x^2}{16} + \frac{4y^2}{16} = \frac{16}{16}$$

$\frac{x^2}{16} + \frac{y^2}{4} = 1$ is an ellipse with $a = 4$

and $b = 2$. Since $y \geq 0$, we want the
upper half of the ellipse.

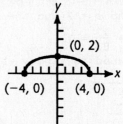

17. $5x = -\sqrt{9 - 4y^2}$ Notice that $x \leq 0$

$$(5x)^2 = [-\sqrt{9 - 4y^2}]^2$$
$$25x^2 = 9 - 4y^2$$
$25x^2 + 4y^2 = 9$ Now $\div$ by 9:
$$\frac{25x^2}{9} + \frac{4y^2}{9} = \frac{9}{9}$$

$\frac{x^2}{\frac{9}{25}} + \frac{y^2}{\frac{9}{4}} = 1$ is an ellipse with

$a = \frac{3}{5}$ and $b = \frac{3}{2}$. Since $x \leq 0$, we

want the left half of the ellipse.

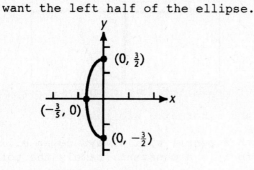

19. $y = -\sqrt{16 - 9x^2}$ Notice that $y \leq 0$

$$(y)^2 = [-\sqrt{16 - 9x^2}]^2$$
$$y^2 = 16 - 9x^2$$
$9x^2 + y^2 = 16$ Now $\div$ by 16
$$\frac{9x^2}{16} + \frac{y^2}{16} = \frac{16}{16}$$

$\frac{x^2}{\frac{16}{9}} + \frac{y^2}{16} = 1$ is an ellipse with $a = \frac{4}{3}$

and $b = 4$. Since $y \leq 0$, we want the
bottom half of the ellipse.

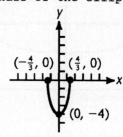

21. $x = \dfrac{\sqrt{25 - 16y^2}}{2}$ Notice that $x \geq 0$

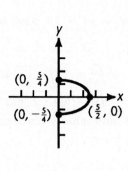

$$2x = \sqrt{25 - 16y^2}$$
$$(2x)^2 = (\sqrt{25 - 16y^2})^2$$
$$4x^2 = 25 - 16y^2$$
$$4x^2 + 16y^2 = 25 \quad \text{Now} \div \text{by 25:}$$
$$\frac{4x^2}{25} + \frac{16y^2}{25} = \frac{25}{25}$$
$$\frac{x^2}{\frac{25}{4}} + \frac{y^2}{\frac{25}{16}} = 1 \text{ is an ellipse with } a = \frac{5}{2} \text{ and } b = \frac{5}{4}.$$

Since $x \geq 0$, we want the right half of an ellipse.

23. The equation is $\dfrac{x^2}{a^2} + \dfrac{y^2}{b^2} = 1$. But $a = 4$ and $b = 2$ and the equation becomes
$$\frac{x^2}{16} + \frac{y^2}{4} = 1$$

25. The equation is $\dfrac{x^2}{a^2} + \dfrac{y^2}{b^2} = 1$. But $a = 2$ and $b = 6$ and the equation becomes
$$\frac{x^2}{4} + \frac{y^2}{36} = 1$$

27. The equation is $\dfrac{x^2}{a^2} + \dfrac{y^2}{b^2} = 1$. But $a = 6$ and $b = 9$ and the equation is
$$\frac{x^2}{36} + \frac{y^2}{81} = 1$$
Now we need to find y when $x = 3$.
$$\frac{9}{36} + \frac{y^2}{81} = 1$$
$$\frac{1}{4} + \frac{y^2}{81} = 1$$
$$\frac{y^2}{81} = 1 - \frac{1}{4}$$
$$\frac{y^2}{81} = \frac{3}{4} \quad \text{Now multiply by 81:}$$
$$y^2 = \frac{3}{4} \cdot 81$$
$$y = \sqrt{\frac{3 \cdot 81}{4}} = \frac{9}{2}\sqrt{3} \text{ ft}$$

29. The equation is $\dfrac{x^2}{a^2} + \dfrac{y^2}{b^2} = 1$. But $a = 3$ and $b = 2$ and the equation is
$$\frac{x^2}{9} + \frac{y^2}{4} = 1$$
Now we need to find x when $y = 1$
$$\frac{x^2}{9} + \frac{1}{4} = 1$$
$$\frac{x^2}{9} = 1 - \frac{1}{4}$$
$$\frac{x^2}{9} = \frac{3}{4} \quad \text{Now multiply by 9:}$$
$$x^2 = \frac{3}{4} \cdot 9$$
$$x = \pm\sqrt{\frac{3 \cdot 9}{4}} = \pm\frac{3}{2}\sqrt{3} \text{ ft}$$
The total distance across is $3\sqrt{3}$ ft.

31.
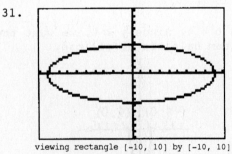
viewing rectangle [-10, 10] by [-10, 10]

33.
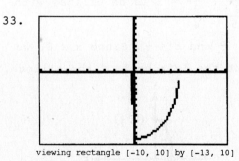
viewing rectangle [-10, 10] by [-13, 10]

35. The graphs are identical except major and minor axes are reversed.

37. The curve that results from tracing with a pencil will always be an ellipse since the sum of the distances to the foci is a constant, namely the total length of the string.

Exercises 4.4

1. $\dfrac{x^2}{9} - \dfrac{y^2}{16} = 1$

   $a = 3$ and $b = 4$ so that the vertices are $(\pm a, 0)$ or $(\pm 3, 0)$ and the asymptotes are

   $y = \pm\dfrac{b}{a}x = \pm\dfrac{4}{3}x.$

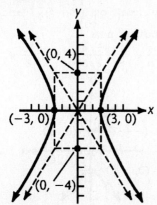

3. $\dfrac{y^2}{4} - \dfrac{x^2}{25} = 1$

   $a = 2$ and $b = 5$ so that the vertices are $(0, \pm a)$ or $(0, \pm 2)$ and the asymptotes are $y = \pm\dfrac{a}{b}x = \pm\dfrac{2}{5}x.$

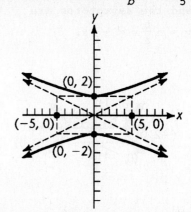

5. $y^2 - x^2 = 9$ Divide by 9:

   $\dfrac{y^2}{9} - \dfrac{x^2}{9} = 1$

   $a = 3$ and $b = 3$ so that the vertices are $(0, \pm a)$ or $(0, \pm 3)$ and the asymptotes are

   $y = \pm\dfrac{a}{b}x = \pm\dfrac{3}{3}x.$

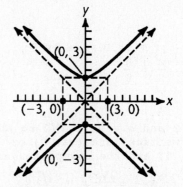

7. $4x^2 - 9y^2 = 36$ Divide by 36:

   $\dfrac{4x^2}{36} - \dfrac{9y^2}{36} = \dfrac{36}{36}$

   $\dfrac{x^2}{9} - \dfrac{y^2}{4} = 1$

   $a = 3$ and $b = 2$ so that the vertices are $(\pm a, 0)$ or $(\pm 3, 0)$ and the asymptotes are

   $y = \pm\dfrac{b}{a}x = \pm\dfrac{2}{3}x.$

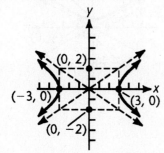

9. $4x^2 - 25y^2 = 0$ holds only when

   $$4x^2 = 25y^2$$
   $$x^2 = \dfrac{25}{4}y^2$$
   $$x = \pm\dfrac{5}{2}y$$

   which are 2 straight lines given by

   $x = \dfrac{5}{2}y$ and $x = -\dfrac{5}{2}y.$

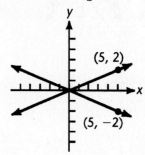

11. $x^2 - 4y^2 = 4$ Divide by 4:

$$\frac{x^2}{4} - \frac{4y^2}{4} = \frac{4}{4}$$

$$\frac{x^2}{4} - \frac{y^2}{1} = 1$$

$a = 2$ and $b = 1$ so that the vertices are $(\pm a, 0)$ or $(\pm 2, 0)$ and the asymptotes are

$$y = \pm \frac{b}{a} x = \pm \frac{1}{2} x.$$

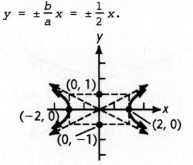

13. $\dfrac{16y^2}{25} - \dfrac{9x^2}{4} = 1$ can be written as

$$\frac{y^2}{\frac{25}{16}} - \frac{x^2}{\frac{4}{9}} = 1$$

$a = \dfrac{5}{4}$ and $b = \dfrac{2}{3}$ so that the vertices

are $(0, \pm a)$ or $\left(0, \pm \dfrac{5}{4}\right)$ and the

asymptotes are

$$y = \pm \frac{a}{b} x = \pm \frac{5/4}{2/3} x = \pm \frac{15}{8} x.$$

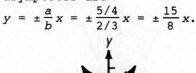
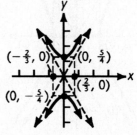

15. $2x^2 - y^2 = 8$ Divide by 8:

$$\frac{2x^2}{8} - \frac{y^2}{8} = \frac{8}{8}$$

$$\frac{x^2}{4} - \frac{y^2}{8} = 1$$

$a = 2$ and $b = \sqrt{8} = 2\sqrt{2}$ so that the vertices are $(\pm a, 0)$ or $(\pm 2, 0)$ and the asymptotes are

$$y = \pm \frac{b}{a} x = \pm \frac{2\sqrt{2}}{2} x = \pm \sqrt{2}\, x.$$

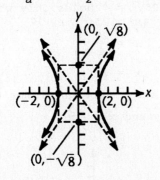

17. $5y = \sqrt{9 + 4x^2}$ Observe that $y \geq 0$

$$(5y)^2 = (\sqrt{9 + 4x^2})$$
$$25y^2 = 9 + 4x^2$$

$25y^2 - 4x^2 = 9$ Divide by 9:

$$\frac{25y^2}{9} - \frac{4x^2}{9} = \frac{9}{9} \text{ or } \frac{y^2}{\frac{9}{25}} - \frac{x^2}{\frac{9}{4}} = 1$$

$a = \dfrac{3}{5}$ and $b = \dfrac{3}{2}$ so that the vertex is

$(0, a)$ or $\left(0, \dfrac{3}{5}\right)$ since $y \geq 0$ and the

asymptotes are $y = \pm \dfrac{a}{b} x = \pm \dfrac{3/5}{3/2} x = \pm \dfrac{2}{5} x.$

The graph is is the upper part of the hyperbola.

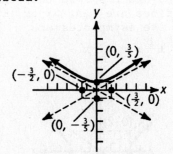

19. $2x = \sqrt{16 + y^2}$ Observe the $x \geq 0$

      $(2x)^2 = (\sqrt{16 + y^2})^2$

         $4x^2 = 16 + y^2$

  $4x^2 - y^2 = 16$ Divide by 16:

  $\dfrac{x^2}{4} - \dfrac{y^2}{16} = 1$

Here $a = 2$ and $b = 4$ so that the vertex is
$(a, 0)$ or $(2, 0)$ and the asymptotes are

$y = \pm\dfrac{b}{a}x = \pm\dfrac{4}{2}x = \pm 2x$. The graph is the

right side of the hyperbola.

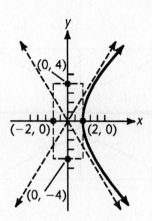

21. The equation is $\dfrac{x^2}{a^2} - \dfrac{y^2}{b^2} = 1$. Since the vertices are $(2, 0)$, and $(-2, 0)$ we see

    that $a = 2$ so that $a^2 = 4$ and the equation becomes $\dfrac{x^2}{4} - \dfrac{y^2}{b^2} = 1$. To find b, we

    need to make use of the asymptotes. The asymptotes are given by $y = \pm\dfrac{b}{a}x = \dfrac{b}{2}x$

    where $a = 2$. Now the point $(2, 3)$ is on the line $y = \dfrac{b}{2}x$. Substitute

    $x = 2$ and $y = 3$ to obtain

        $y = \dfrac{b}{2}x$

        $3 = \dfrac{b}{2} \cdot 2$

        $3 = b$

    Thus, the final equation is $\dfrac{x^2}{4} - \dfrac{y^2}{9} = 1$.

23. The equation is $\dfrac{y^2}{a^2} - \dfrac{x^2}{b^2} = 1$. Since the vertices are $(0, 2)$ and $(0, -2)$ we see

    that $a = 2$ so that $a^2 = 4$ and the equation becomes $\dfrac{y^2}{4} - \dfrac{x^2}{b^2} = 1$. To find b, we

    need to make use of the asymptotes. The asymptotes are given by $y = \pm\dfrac{a}{b}x =$

    $\pm\dfrac{2}{b}x$ where a is 2. Now the point $(2, 2)$ is on the line $y = \dfrac{2}{b}x$. Substitute

    $x = 2$ and $y = 2$ to obtain

        $2 = \dfrac{2}{b} \cdot 2$

        $2 = \dfrac{4}{b}$

      $2b = 4$

       $b = 2$

    Thus, the final equation is $\dfrac{y^2}{4} - \dfrac{x^2}{4} = 1$.

For problems 25 and 27, the graphics calculator produces the following graphs.

25.

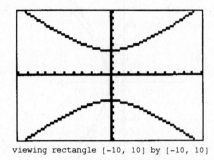

viewing rectangle [-10, 10] by [-10, 10]

27.

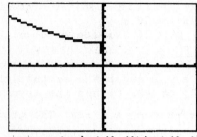

viewing rectangle [-10, 10] by [-10, 10]

29. Both graphs are hyperbolas with vertices along the x axis: the first at $(\pm p, 0)$ and the second at $(\pm q, 0)$. The asymptotes of the first graph are $y = \pm\dfrac{q}{p}x$ while the asymptotes for the second graph are $y = \pm\dfrac{p}{q}x$.

Exercises 4.5

1. The center is $(h, k) = (2, 1)$. The vertices are $(h, k \pm b) = (2, 1 \pm 3)$ which are $(2, 4)$ and $(2, -2)$. The graph is an ellipse.

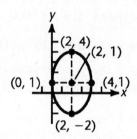

3. $\dfrac{(x - 1)^2}{9} - \dfrac{(y - 2)^2}{25} = 1$

The center is $(1, 2)$ and the vertices are $(h \pm a, k) = (1 \pm 3, 2)$ which are $(4, 2)$ and $(-2, 2)$. The asymptotes go through the point $(1, 2)$ with slope of $\pm\dfrac{b}{a} = \pm\dfrac{5}{3}$. The graph is a hyperbola.

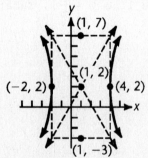

5. $\dfrac{(y + 3)^2}{4} - \dfrac{(x - 2)^2}{1} = 1$

The center is $(2, -3)$ and the vertices are $(h, k \pm a) = (2, -3 \pm 2)$ which are $(2, -1)$ and $(2, -5)$. The asymptotes go through the point $(2, -3)$ with a slope of $\pm\dfrac{a}{b} = \pm\dfrac{2}{1} = \pm 2$. The graph is a hyperbola.

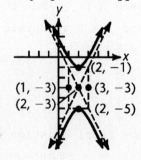

7. $4(x + 2)^2 - 9(y + 2)^2 = 36$

Divide by 36:

$$\frac{4(x + 2)^2}{36} - \frac{9(y + 2)^2}{36} = 1$$

$$\frac{(x + 2)^2}{9} - \frac{(y + 2)^2}{4} = 1$$

The center is $(-2, -2)$ and the vertices are $(h \pm a, k) = (-2 \pm 3, -2)$ which are $(1, -2)$ and $(-5, -2)$. The asymptotes go through $(-2, -2)$ with a slope of $\pm \frac{b}{a} = \pm \frac{2}{3}$. The graph is a hyperbola.

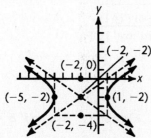

9. $25(x + 1)^2 + 4(y - 3)^2 = 100$

Divide by 100:

$$\frac{25(x + 1)^2}{100} + \frac{4(y - 3)^2}{100} = 1$$

$$\frac{(x + 1)^2}{4} + \frac{(y - 3)^2}{25} = 1$$

The center is $(-1, 3)$ and the vertices are $(h, k \pm b) = (-1, 3 \pm 5)$ which are $(-1, 8)$ and $(-1, -2)$. The graph is an ellipse.

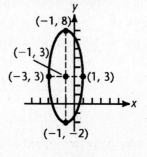

11. $9(x + 2)^2 + (y - 3)^2 = 81$ Divide by 81:

$$\frac{9(x + 2)^2}{81} + \frac{(y - 3)^2}{81} = 1$$

$$\frac{(x + 2)^2}{9} + \frac{(y - 3)^2}{81} = 1$$

The center is $(-2, 3)$ and the vertices are $(h, k \pm b) = (-2, 3 \pm 9)$ which are $(-2, 12)$ and $(-2, -6)$. The graph is an ellipse.

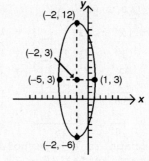

13. $4x^2 + 9y^2 - 8x - 36y + 4 = 0$

$$4x^2 - 8x + 9y^2 - 36y = -4$$

$$4(x^2 - 2x + \underline{}) + 9(y^2 - 4y + \underline{}) = -4 + \underline{} + \underline{}$$

$$4(x^2 - 2x + \underline{1}) + 9(y^2 - 4y + \underline{4}) = -4 + \underline{4(1)} + \underline{9(4)}$$

$$4(x - 1)^2 + 9(y - 2)^2 = 36 \quad \text{Now divide by 36:}$$

$$\frac{(x - 1)^2}{9} + \frac{(y - 2)^2}{4} = 1$$

The graph is an ellipse with center at $(1, 2)$. Also $a = 3$ and $b = 2$.

15. $9x^2 - y^2 - 36x + 2y + 44 = 0$

$$9x^2 - 36x - y^2 + 2y = -44$$

$$9(x^2 - 4x + \underline{}) - (y^2 - 2y + \underline{}) = -44 + \underline{} + \underline{}$$

$$9(x^2 - 4x + \underline{4}) - (y^2 - 2y + \underline{1}) = -44 + \underline{9(4)} + \underline{(-1)(1)}$$

$$9(x - 2)^2 - (y - 1)^2 = -9 \quad \text{Now divide by -9:}$$

$$\frac{9(x - 2)^2}{-9} - \frac{(y - 1)^2}{-9} = 1$$

$$-\frac{(x - 2)^2}{1} + \frac{(y - 1)^2}{9} = 1$$

$$\text{or} \quad \frac{(y - 1)^2}{9} - \frac{(x - 2)^2}{1} = 1$$

The graph is a hyperbola with the center at $(2, 1)$. Also, $a = 3$ and $b = 1$.

17. $9x^2 + 16y^2 - 54x + 64y + 1 = 0$

$9x^2 - 54x + 16y^2 + 64y = -1$

$9(x^2 - 6x + \underline{\quad}) + 16(y^2 + 4y + \underline{\quad}) = -1 + \underline{\quad} + \underline{\quad}$

$9(x^2 - 6x + \underline{9}) + 16(y^2 + 4y + \underline{4}) = -1 + 81 + 64$

$9(x - 3)^2 + 16(y + 2)^2 = 144$

$\dfrac{9(x - 3)^2}{144} + \dfrac{16(y + 2)^2}{144} = 1, \quad \dfrac{(x - 3)^2}{16} + \dfrac{(y + 2)^2}{9} = 1$

The graph is an ellipse with center at (3, -2). Also, $a = 4$ and $b = 3$.

19. $x^2 - 4y^2 + 6x + 32y = 155$

$x^2 + 6x - 4y^2 + 32y = 155$

$x^2 + 6x + \underline{\quad} - 4(y^2 - 8y + \underline{\quad}) = 155 + \underline{\quad} + \underline{\qquad}$

$x^2 + 6x + \underline{9} - 4(y^2 - 8y + \underline{16}) = 155 + \underline{9} + \underline{(-4)(16)}$

$(x + 3)^2 - 4(y - 4)^2 = 100$ Now divide by 100:

$\dfrac{(x + 3)^2}{100} - \dfrac{4(y - 4)^2}{100} = 1$

$\dfrac{(x + 3)^2}{100} - \dfrac{(y - 4)^2}{25} = 1$

The graph is a hyperbola with center at (-3, 4). Also, $a = 10$ and $b = 5$.

21. $4x^2 - y^2 + 2y = 17$

$4x^2 - (y^2 - 2y + \underline{\quad}) = 17 + \underline{\qquad}$

$4x^2 - (y^2 - 2y + \underline{1}) = 17 + \underline{(-1)(1)}$

$4x^2 - (y - 1)^2 = 16$ Now divide by 16:

$\dfrac{4x^2}{16} - \dfrac{(y - 1)^2}{16} = 1$

$\dfrac{x^2}{4} - \dfrac{(y - 1)^2}{16} = 1$

The graph is a hyperbola with center at (0, 1) and vertices at (0 ± 2, 1) which are (2, 1) and (-2, 1). Through (0, 1) the asymptotes have slopes of $\pm\dfrac{b}{a} = \pm\dfrac{4}{2} = \pm 2$.

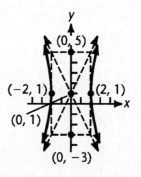

23. $4x^2 - 8x + y^2 + 6y = 3$

$4(x^2 - 2x + \underline{\quad}) + y^2 + 6y + \underline{\quad} = 3 + \underline{\qquad} + \underline{\quad}$

$4(x^2 - 2x + \underline{1}) + y^2 + 6y + \underline{9} = 3 + \underline{(4)(1)} + \underline{9}$

$4(x - 1)^2 + (y + 3)^2 = 16$ Now divide by 16:

$\dfrac{4(x - 1)^2}{16} + \dfrac{(y + 3)^2}{16} = 1$

$\dfrac{(x - 1)^2}{4} + \dfrac{(y + 3)^2}{16} = 1$

The graph is an ellipse with center at (1, -3) and vertices at (1, -3 ± 4) which are (1, 1) and (1, -7).

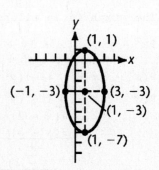

25. $x^2 + 9y^2 + 6x = 27$

$x^2 + 6x + \underline{} + 9y^2 = 27 + \underline{}$

$x^2 + 6x + \underline{9} + 9y^2 = 27 + \underline{9}$

$\qquad (x + 3)^2 + 9y^2 = 36 \quad$ Now divide by 36:

$\qquad \dfrac{(x + 3)^2}{36} + \dfrac{9y^2}{36} = 1$

$\qquad \dfrac{(x + 3)^2}{36} + \dfrac{y^2}{9} = 1$

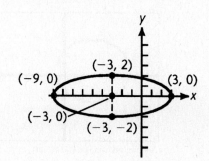

The graph is an ellipse with center at $(-3, 0)$ and vertices at $(-3 \pm 6, 0)$ which are $(3, 0)$ and $(-9, 0)$.

27. $4y^2 - 9x^2 + 8y - 36x = 68$

$-9x^2 - 36x + 4y^2 + 8y = 68$

$-9(x^2 + 4x + \underline{}) + 4(y^2 + 2y + \underline{}) = 68 + \underline{} + \underline{}$

$-9(x^2 + 4x + \underline{4}) + 4(y^2 + 2y + \underline{1}) = 68 + \underline{(-9)(4)} + \underline{(4)(1)}$

$-9(x + 2)^2 + 4(y + 1)^2 = 36 \quad$ Now divide by 36:

$\dfrac{-9(x + 2)^2}{36} + \dfrac{4(y + 1)^2}{36} = 1$

$-\dfrac{(x + 2)^2}{4} + \dfrac{(y + 1)^2}{9} = 1$

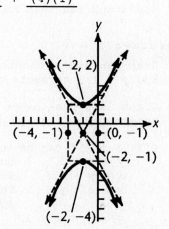

The graph is a hyperbola with center at $(-2, -1)$ and vertices at $(-2, -1 \pm 3)$ which are $(-2, 2)$ and $(-2, -4)$. Through $(-2, -1)$ the asymptote have slopes of $\pm \dfrac{a}{b} = \pm \dfrac{3}{2}$.

29. $h = 0$, $k = 0$, $a = 3$, and $b = 2$

The equation $\dfrac{(x - h)^2}{a^2} + \dfrac{(y - k)^2}{b^2} = 1$ becomes $\dfrac{x^2}{9} + \dfrac{y^2}{4} = 1$

31. $h = 0$, $k = 0$, $a = 1$, and $b = 4$

The equation $\dfrac{(x - h)^2}{a^2} + \dfrac{(y - k)^2}{b^2} = 1$ becomes $\dfrac{x^2}{1} + \dfrac{y^2}{16} = 1$ or $x^2 + \dfrac{y^2}{16} = 1$.

33. $h = 1$, $k = 0$, $a = 3 - 1 = 2$, and $b = 1 - 0 = 1$

The equation $\dfrac{(x - h)^2}{a^2} + \dfrac{(y - k)^2}{b^2} = 1$ becomes $\dfrac{(x - 1)^2}{4} + \dfrac{y^2}{1} = 1$ or

$\dfrac{(x - 1)^2}{4} + y^2 = 1$.

35. $h = 2$, $k = 3$, $a = 2 - 0 = 2$, and $b = 6 - 3 = 3$.

The equation $\dfrac{(x - h)^2}{a^2} + \dfrac{(y - k)^2}{b^2} = 1$ becomes $\dfrac{(x - 2)^2}{4} + \dfrac{(y - 3)^2}{9} = 1$.

37.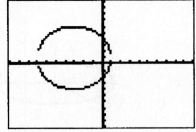

39.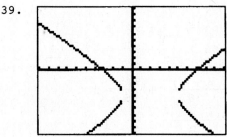

viewing rectangle [-10, 10] by [-10, 10]   viewing rectangle [-10, 10] by [-10, 10]

41. Determine the center of the ellipse (h, k) and the values for a and b. If $a > b$, locate the endpoints of the minor axis by moving b units up or down from the center to obtain $(h, k \pm b)$. Locate the vertices by moving a units left or right from the center to have $(h \pm a, b)$.

    If $a < b$, locate the endpoints of the minor axis by moving a units left or right from the center to obtain $(h \pm a, k)$. Locate the vertices by moving b units up or down from the center to have $(h, k \pm b)$. With the ends located, the graph can then be drawn.

Exercises 4.6

1. Graph is **not** a one-to-one function since any horizontal line between $y = 0$ and $y = 1$ intersects the graph in three distinct points.

3. Graph is a one-to-one function since every horizontal line (except $y = 0$) intersects the graph in one point.

5. $g(x) = x + 3$ can be written as
   $$y = x + 3 \qquad \text{Interchanging } x \text{ and } y \text{ gives}$$
   $$x = y + 3 \qquad \text{Now solve for } y:$$
   $$x - 3 = y \qquad \text{This is the inverse function}$$
   $$\text{and is written as}$$
   $$g^{-1}(x) = x - 3.$$
   The graphs of $g(x)$ and $g^{-1}(x)$ are straight lines with g passing through $(-3, 0)$ and $(0, 3)$ and g^{-1} passing through $(0, -3)$ and $(3, 0)$.

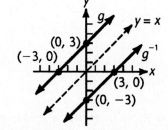

7. $g(x) = \dfrac{4}{3}x - 4$ can be written as

   $$y = \frac{4}{3}x - 4 \qquad \text{Interchanging } x \text{ and } y \text{ gives}$$

   $$x = \frac{4}{3}y - 4 \qquad \text{Now solve for } y:$$

   $$x + 4 = \frac{4}{3}y$$

   $$\frac{3}{4}(x + 4) = \frac{3}{4}\left[\frac{4}{3}y\right]$$

   $$\frac{3}{4}x + 3 = y \qquad \text{This is the inverse function}$$
   $$\text{and is written as}$$

   $$g^{-1}(x) = \frac{3}{4}x + 3.$$
   The graphs of $g(x)$ and $g^{-1}(x)$ are straight lines with g passing through $(3, 0)$ and $(0, -4)$ and g^{-1} passing through $(0, 3)$ and $(-4, 0)$.

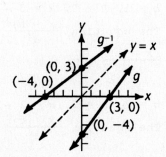

9. $g(x) = 8 - x^3$ can be written as
    $y = 8 - x^3$ Interchanging x and y gives
    $x = 8 - y^3$ Now solve for y:
    $x - 8 = -y^3$

    $\sqrt[3]{8 - x} = y$ This is the inverse function
                      and is written as

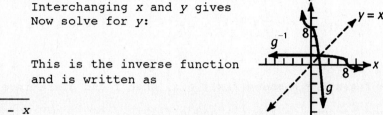

    $g^{-1}(x) = \sqrt[3]{8 - x}$
    The points $(0, 8)$, $(1, 7)$, $(2, 0)$, $(3, -19)$ lie
    on the graph of g and the points $(8, 0)$, $(7, 1)$,
    $(0, 2)$, $(-19, 3)$ lie on the graph of g^{-1}.

11. $2x - y = 4$
    Solve for y: $2x - y = 4$
                    $-y = -2x + 4$
                     $y = 2x - 4$
    and we write $f(x) = 2x - 4$ which is one-to-one since the graph is a non-horizontal straight line. A defining equation for f^{-1} is obtained by interchanging x and y values from the original equation.
          $2y - x = 4$
             $2y = x + 4$
               $y = \frac{1}{2}x + 2$ and we write $f^{-1}(x) = \frac{1}{2}x + 2$

    Now $f(f^{-1}(x)) = f\left(\frac{1}{2}x + 2\right) = 2\left(\frac{1}{2}x + 2\right) - 4 = x + 4 - 4 = x,$

         $f^{-1}(f(x)) = f^{-1}(2x - 4) = \frac{1}{2}(2x - 4) + 2 = x - 2 + 2 = x.$

13. $y = x^2 + 2$ can be written as $f(x) = x^2 + 2$ which is **not** one-to-one
    since $f(1) = 1^2 + 2 = 1 + 2 = 3$ and
                $f(-1) = (-1)^2 + 2 = 1 + 2 = 3$
    That is, $y = 3$ when $x = 1$ or $x = -1$.

15. $y = (x - 1)^2$ can be written as $f(x) = (x - 1)^2$ which is **not** one-to-one
    since $f(0) = (0 - 1)^2 = 1$
        and $f(2) = (2 - 1)^2 = 1$
    That is, $y = 1$ when $x = 0$ or $x = 2$.

17. $y = x^3 - 1$ can be written as $f(x) = x^3 - 1$ which is a one-to-one function.
    For every value of y, there is precisely one value for x. Also, every
    horizontal line intersects the graph in one point. A defining equation for
    f^{-1} is obtained by interchanging x and y values from the original equation.
            $x = y^3 - 1$
          $x + 1 = y^3$

       $\sqrt[3]{x + 1} = y$ and we write $f^{-1}(x) = \sqrt[3]{x + 1}$

    Now, $f(f^{-1}(x)) = f(\sqrt[3]{x + 1}) = (\sqrt[3]{x + 1})^3 - 1 = x + 1 - 1 = x$

        $f^{-1}(f(x)) = f^{-1}(x^3 - 1) = \sqrt[3]{x^3 - 1 + 1} = \sqrt[3]{x^3} = x$

19. We need to compute $f(g(x))$ and $g(f(x))$.

$$f(g(x)) = f\left(-\frac{4}{3}x + 4\right) = -\frac{3}{4}\left(-\frac{4}{3}x + 4\right) + 3 = x - 3 + 3 = x$$

and $g(f(x)) = g\left(-\frac{3}{4}x + 3\right) = -\frac{4}{3}\left(-\frac{3}{4}x + 3\right) + 4 = x - 4 + 4 = x$

Since $f(g(x)) = x$ and $g(f(x)) = x$, then f and g are inverse functions.

21. We need to compute $f(g(x))$ and $g(f(x))$

$$f(g(x)) = f\left(\frac{x + 9}{3}\right) = 3\left(\frac{x + 9}{3}\right) - 9 = x + 9 - 9 = x$$

and $g(f(x)) = g(3x - 9) = \dfrac{3x - 9 + 9}{3} = \dfrac{3x}{3} = x$

Since $f(g(x)) = x$ and $g(f(x)) = x$, then f and g are inverse functions.

23. We need to compute $f(g(x))$ and $g(f(x))$

$$f(g(x)) = f\left(\frac{1}{x - 2}\right) = \dfrac{2\left(\dfrac{1}{x - 2}\right) + 1}{\dfrac{1}{x - 2}} = \dfrac{\dfrac{2}{x - 2} + 1}{\dfrac{1}{x - 2}} \cdot \dfrac{(x - 2)}{(x - 2)} = \dfrac{2 + x - 2}{1} = x$$

and $g(f(x)) = g\left(\dfrac{2x + 1}{x}\right) = \dfrac{1}{\dfrac{2x + 1}{x} - 2} = \dfrac{1}{\dfrac{2x + 1}{x} - \dfrac{2x}{x}} = \dfrac{1}{\dfrac{1}{x}} = x$

Since $f(g(x)) = x$ and $g(f(x)) = x$, then f and g are inverse functions.

25. We need to compute $f(g(x))$ and $g(f(x))$

$$f(g(x)) = f\left(\frac{x + 1}{2x}\right) = \dfrac{2\left(\dfrac{x + 1}{2x}\right)}{\dfrac{x + 1}{2x} + 1} = \dfrac{\dfrac{x + 1}{x}}{\dfrac{x + 1}{2x} + 1} = \dfrac{\dfrac{x + 1}{x}}{\dfrac{x + 1}{2x} + 1\left(\dfrac{2x}{2x}\right)}$$

$$= \dfrac{\dfrac{x + 1}{x}}{\dfrac{3x + 1}{2x}} = \dfrac{x + 1}{\cancel{x}_1} \cdot \dfrac{2\cancel{x}}{3x + 1} = \dfrac{2(x + 1)}{3x + 1}$$

Since $f(g(x)) \neq x$, then f and g are **not** inverse functions.

27. We need to compute $f(g(x))$ and $g(f(x))$

$f(g(x)) = f(x^2 + 1) = \sqrt{x^2 + 1 - 1} = \sqrt{x^2} = x$ (since $x \geq 0$)

and $g(f(x)) = g(\sqrt{x - 1}) = (\sqrt{x - 1})^2 + 1 = x - 1 + 1 = x$

Since $f(g(x)) = x$ and $g(f(x)) = x$, then f and g are inverse functions.

29. We need to compute $f(g(x))$ and $g(f(x))$.

$f(g(x)) = f(x^2 - 2) = \sqrt{x^2 - 2 + 2} = \sqrt{x^2} = |x|$

Since $f(g(x)) \neq x$, then f and g are **not** inverse functions.

Also, observe that $g(x) = x^2 - 2$ is **not** a one-to-one function

since $g(2) = 2^2 - 2 = 4 - 2 = 2$

and $g(-2) = (-2)^2 - 2 = 4 - 2 = 2$

31. $f^{-1}(x)$ represents the inverse of $f(x)$ whereas $[f(x)]^{-1}$ means $\dfrac{1}{f(x)}$ the reciprocal of $f(x)$.

33. The inverse of $f \circ g$ is $g^{-1} \circ f^{-1}$

Exercises 4.7

1. $\dfrac{3 \text{ lbs}}{4 \text{ lbs}} = \dfrac{3}{4}$ 3. $\dfrac{12 \text{ m}}{150 \text{ cm}} = \dfrac{12(100) \text{ cm}}{150 \text{ cm}} = \dfrac{1200}{150} = \dfrac{8}{1}$ 5. $\dfrac{90 \text{ boys}}{30 \text{ girls}} = \dfrac{3 \text{ boys}}{1 \text{ girl}}$

7. $44 \text{ ft/sec} = \dfrac{44 \overset{1}{\cancel{\text{ft}}}}{1 \cancel{\text{sec}}} \cdot \dfrac{1 \text{ mile}}{5,280 \underset{1}{\cancel{\text{ft}}}} \cdot \dfrac{60 \overset{1}{\cancel{\text{sec}}}}{1 \underset{1}{\cancel{\text{min}}}} \cdot \dfrac{60 \overset{1}{\cancel{\text{min}}}}{1 \text{ hr}} = \dfrac{(44)(1)(60)(60) \text{ miles}}{(1)(5,280)(1)(1) \text{ hr}} \approx 30 \text{ mph}$

9. $48 \text{ oz/in}^2 = \dfrac{48 \overset{1}{\cancel{\text{oz}}}}{1 \cancel{\text{in}^2}} \cdot \dfrac{1 \text{ lb}}{16 \underset{1}{\cancel{\text{oz}}}} \cdot \dfrac{144 \overset{1}{\cancel{\text{in}^2}}}{1 \text{ ft}^2} = \dfrac{(48)(1)(144) \text{ lb}}{(1)(16)(1) \text{ ft}^2} = 432 \text{ lb/ft}^2$

11. $20 \text{ miles/gal} = \dfrac{20 \text{ miles}}{1 \underset{1}{\cancel{\text{gal}}}} \cdot \dfrac{1 \overset{1}{\cancel{\text{gal}}}}{4 \text{ qt}} = \dfrac{(20)(1) \text{ miles}}{(1)(4) \text{ qt}} = 5 \text{ miles/qt}$

13. Let x be the miles traveled by the car. The proportion is
$$\frac{60 \text{ miles}}{2.5 \text{ gal}} = \frac{x \text{ miles}}{9 \text{ gal}}$$
The equation is $\dfrac{60}{2.5} = \dfrac{x}{9}$
Cross multiply: $2.5x = (60)(9)$
$$x = \frac{(60)(9)}{2.5} = \frac{540}{2.5} = 216 \text{ miles}$$

15. Let x be the number of pints of oil. But x pints $= \dfrac{x}{80}$ gallons and the

proportion is $\dfrac{50 \text{ gal of gas}}{1 \text{ gal of oil}} = \dfrac{6 \text{ gal of gas}}{\dfrac{x}{8} \text{ gal of oil}}$

The equation is $\dfrac{50}{1} = \dfrac{6}{\dfrac{x}{8}}$

Cross multiply: $\dfrac{50x}{8} = 6$

$$x = \left(\frac{8}{50}\right)(6) = \frac{48}{50} = \frac{24}{25} \text{ pts of oil}$$

17. Let x be the amount of money received by one boy. Then $40 - x$ is the amount
received by the other boy. The proportion is
$$\frac{5}{3} = \frac{x}{40 - x}$$
Cross multiply: $5(40 - x) = 3x$
$$200 - 5x = 3x$$
$$200 = 8x$$
$$\frac{200}{8} = x \text{ or } x = \$25$$
The other boy receives $40 - 25 = \$15$.

19. $y = kx^2 \rightarrow 12 = k(2)^2$
$$12 = 4k$$
$$3 = k$$
The equation is $y = 3x^2$.
Now let $x = 5$ to get
$$y = 3(5)^2$$
$$y = 3(25) = 75$$

21. $y = \dfrac{k}{t^2} \rightarrow 2 = \dfrac{k}{2^2}$
$$2 = \dfrac{k}{4}$$
$$8 = k$$
The equation is $y = \dfrac{8}{t^2}$.
Now let $t = 4$ to obtain
$$y = \dfrac{8}{4^2} = \dfrac{8}{16} = \dfrac{1}{2}$$

23. $y = \dfrac{kz}{w} \rightarrow \dfrac{1}{3} = \dfrac{k\left(\frac{1}{6}\right)}{18}$
$$\dfrac{1}{3} = \dfrac{k}{108}$$
$$\dfrac{108}{3} = k \text{ or } k = 36$$
The equation is $y = \dfrac{36z}{w}$.
Now let $z = 1$ and $w = 4$ to get
$$y = \dfrac{36(1)}{4} = 9$$

25. $t^2 = \dfrac{k}{z^3} \rightarrow (4)^2 = \dfrac{k}{\left(\frac{1}{2}\right)^3}$
$$16 = \dfrac{k}{\frac{1}{8}}$$
$$16\left(\dfrac{1}{8}\right) = k \text{ or } k = 2$$
The equation is $t^2 = \dfrac{2}{z^3}$.
Now let $z = 2$ to obtain
$$t^2 = \dfrac{2}{2^3} = \dfrac{2}{8} = \dfrac{1}{4}$$
and $\quad t = \pm\sqrt{\dfrac{1}{4}} = \pm\dfrac{1}{2}$

27. $x = \dfrac{kyt^3}{4z - 3} \rightarrow 2 = \dfrac{k(-4)(-2)^3}{4(1) - 3}$
$$2 = \dfrac{32k}{1}$$
$$\dfrac{2}{32} = k \text{ or } k = \dfrac{1}{16}$$
The equation is $x = \dfrac{\frac{1}{16}yt^3}{4z - 3} = \dfrac{yt^3}{16(4z - 3)}$.
Now let $y = 7$, $z = -1$, and $t = 4$ to get
$$x = \dfrac{7(4)^3}{16[4(-1) - 3]} = \dfrac{7 \cdot 64}{16(-7)} = -4$$

29. $f = kx \rightarrow 30 = k(3)$
$$10 = k$$
The equation is $f = 10x$.
Now let $x = 16 - 12 = 4$ to get
$$f = 10(4) = 40 \text{ lbs}$$

31. $p = kd \rightarrow 1250 = k(20)$
$$\dfrac{1250}{20} = k$$
$$62.5 = k$$
The equation is $p = 62.5d$.
Now let $d = 50$ to have
$$p = 62.5(50)$$
$$= 3125 \text{ lb/ft}^2$$

33. $C = kx^2 \rightarrow 450 = k(15)^2$
$$450 = 225k$$
$$\dfrac{450}{225} = k$$
$$2 = k$$
The equation is $C = 2x^2$.
Now let $x = 80$ to obtain
$$C = 2(80)^2$$
$$= 2(6400)$$
$$= \$12,800$$

35. $t = k\sqrt{x} \rightarrow 2.5 = k\sqrt{4}$
$$2.5 = 2k$$
$$\frac{2.5}{2} = k \text{ or } k = 1.25$$

The equation is $t = 1.25\sqrt{x}$.
Now let $x = 2.25$ to have
$$t = 1.25\sqrt{2.25}$$
$$t = 1.25(1.5)$$
$$= 1.875 \text{ sec}$$

37. $d = kt^2 \rightarrow 144 = d(3)^2$
$$144 = 9d$$
$$\frac{144}{9} = d \text{ or } d = 16$$

The equation is $d = 16t^2$.
Now let $t = 10$ to get
$$d = 16(10)^2$$
$$= 16(100)$$
$$= 1600 \text{ ft}$$

39. $h = \dfrac{kV}{r^2} \rightarrow 32 = \dfrac{k\left(\frac{512\pi}{3}\right)}{(4)^2}$
$$32 = \frac{k(512\pi)}{48}$$
$$\frac{(32)(48)}{512\pi} = k \text{ or } k = \frac{3}{\pi}$$

The equation is $h = \dfrac{\left(\frac{3}{\pi}\right)V}{r^2} = \dfrac{3V}{\pi r^2}$.
Now let $V = 6\pi$ and $r = 3$ to get
$$h = \frac{3(6\pi)}{\pi(3)^2} = \frac{18\pi}{9\pi} = 2 \text{ cm}$$

41. $S = \dfrac{kA}{r} \rightarrow 20 = \dfrac{k(400\pi)}{20}$
$$20 = \frac{\overset{20}{\cancel{400}}\pi k}{\underset{1}{\cancel{20}}}$$
$$\frac{1}{\pi} = k$$

The equation is $S = \dfrac{\left(\frac{1}{\pi}\right)A}{r} = \dfrac{A}{\pi r}$.
Now let $A = 12$ and $r = 2$ to get
$$S = \frac{12}{\pi \cdot 2} = \frac{6}{\pi} \text{ inches}$$

43. $t = \dfrac{k}{p} \rightarrow 12 = \dfrac{k}{5}$
$$60 = k$$
The equation is $t = \dfrac{60}{p}$.
Now let $t = 10$ to get
$$10 = \frac{60}{p}$$
$$10p = 60$$
$$p = \frac{60}{10} = 6 \text{ workers}$$

45. $y = kx \rightarrow \dfrac{1}{k}y = x$

Since $\dfrac{1}{k}$ is a constant, we have
that x varies directly as y.

47. If z varies jointly as x and y, z may **not** vary directly as y. Suppose $k > 0$. If z varies directly with y, then z should increase as y increases. But if x decreases as y increases, the effect of x on z may be greater than the effect of y on z, causing z to decrease rather than increase.

CHAPTER REVIEW

1. $d = \sqrt{(x_2 - x_1)^2 + (y_2 - y_1)^2} = \sqrt{(3 - 7)^2 + (-1 - 2)^2} = \sqrt{(-4)^2 + (-3)^2} = \sqrt{16 + 9}$
$$= \sqrt{25} = 5$$

2. $d = \sqrt{(x_2 - x_1)^2 + (y_2 - y_1)^2} = \sqrt{(1 - 3)^2 + [0 - (-2)]^2} = \sqrt{(-2)^2 + (2)^2} = \sqrt{4 + 4}$
$$= \sqrt{8} = \sqrt{4 \cdot 2} = \sqrt{4}\sqrt{2} = 2\sqrt{2}$$

3. $d = \sqrt{(x_2 - x_1)^2 + (y_2 - y_1)^2} = \sqrt{[4 - (-2)]^2 + [0 - (-3)]^2} = \sqrt{6^2 + 3^2} = \sqrt{36 + 9}$
$$= \sqrt{45} = \sqrt{9 \cdot 5} = 3\sqrt{5}$$

4. $d = \sqrt{(x_2 - x_1)^2 + (y_2 - y_1)^2} = \sqrt{[-1 - (-1)]^2 + (-1 - 7)^2} = \sqrt{0^2 + (-8)^2}$

$= \sqrt{0 + 64} = \sqrt{64} = 8$

5. Graph is a circle with center at (3, -5) and $r = 4$.

6. $x^2 + y^2 + 2x + 4y = 4$

$x^2 + 2x + y^2 + 4y = 4$

$x^2 + 2x + \underline{} + y^2 + 4y + \underline{} = 4 + \underline{} + \underline{}$

$x^2 + 2x + \underline{1} + y^2 + 4y + \underline{4} = 4 + \underline{1} + \underline{4}$

$(x + 1)^2 + (y + 2)^2 = 9$

It is a circle with center at (-1, -2) and $r = \sqrt{9} = 3$.

7. $x^2 - 12x + y^2 = 6y - 50$

$x^2 - 12x + y^2 - 6y = -50$

$x^2 - 12x + \underline{36} + y^2 - 6y + \underline{9} = -50 + \underline{36} + \underline{9}$

$(x - 6)^2 + (y - 3)^2 = -5$

This is **not** a circle due to the -5 on the right side.

8. $(x - h)^2 + (y - k)^2 = r^2$

$(x - 2)^2 + [y - (-3)]^2 = 5^2$

$(x - 2)^2 + (y + 3)^2 = 25$

9. $r = \sqrt{[-1 - (-4)]^2 + [-7 - (-3)]^2} = \sqrt{3^2 + (-4)^2} = \sqrt{9 + 16} = \sqrt{25} = 5$

The equation is $[x - (-3)]^2 + [y - (-4)]^2 = 5^2$

or $(x + 3)^2 + (y + 4)^2 = 25$

10. The center is the midpoint of the line joining (-3, -4) and (1, 2):

$\left(\dfrac{x_1 + x_2}{2}, \dfrac{y_1 + y_2}{2}\right) = \left(\dfrac{-3 + 1}{2}, \dfrac{-4 + 2}{2}\right) = \left(\dfrac{-2}{2}, \dfrac{-2}{2}\right) = (-1, -1)$

Now the radius is the distance from the radius to one endpoint. Let's use (1, 2) as the endpoint.

$d = \sqrt{(-1 - 1)^2 + (-1 - 2)^2} = \sqrt{(-2)^2 + (-3)^2} = \sqrt{4 + 9} = \sqrt{13}$

The equation is $[x - (-1)]^2 + [y - (-1)]^2 = (\sqrt{13})^2$

$(x + 1)^2 + (y + 1)^2 = 13$

11. The center is the midpoint of the diameter:

$\left(\dfrac{x_1 + x_2}{2}, \dfrac{y_1 + y_2}{2}\right) = \left(\dfrac{4 + (-2)}{2}, \dfrac{-2 + (-4)}{2}\right) = (1, -3).$

The radius is the distance from (1, -3) to one endpoint of the diameter. Here we will use (4, -2).

$r = \sqrt{(4 - 1)^2 + [-2 - (-3)]^2} = \sqrt{3^2 + 1^2} = \sqrt{9 + 1} = \sqrt{10}$

The equation is $(x - 1)^2 + [y - (-3)]^2 = (\sqrt{10})^2$

$(x - 1)^2 + (y + 3)^2 = 10$

12. a)   b)   c)

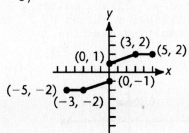

13. x axis: Replace y with $-y$ to get: $(-y)^2 = x^2$
$$y^2 = x^2$$ which is identical to the original equation

Thus, it is symmetric with respect to the x axis.

y axis: Replace x with $-x$ to get $y^2 = (-x)^2$
$$y^2 = x^2$$ which is identical to the original equation

Thus, it is symmetric with respect to the y axis.

Origin: Replace x with $-x$ and y with $-y$ to get $(-y)^2 = (-x)^2$
$$y^2 = x^2$$ which is identical to the original equation

Thus, it is symmetric with respect to the origin. The graph is

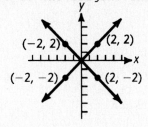

14. $y = |x| - 1$

y axis: Replace x with $-x$: $y = |x| - 1$
$$y = |-x| - 1$$
$$y = |x| - 1$$ which is identical to the original equation

It is symmetric about the y axis.

x axis: Replace y with $-y$: $y = |x| - 1$
$$-y = |x| - 1$$ which is **not** identical to the original equation

Origin: Replace y with $-y$ and x with $-x$: $y = |x| - 1$
$$-y = |-x| - 1$$
$$-y = |x| - 1$$ which is **not** identical to the original equation

It is **not** symmetric with respect to the x axis or the origin.

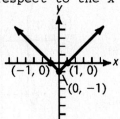

135

15. For $y = 2f(x)$, multiply all original y values by 2. The resulting graph is

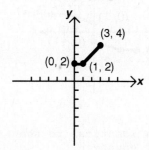

16. For $y = f(x) + 1$, increase all y values by 1 unit and shift the graph upward 1 unit.

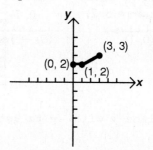

17. For $y = f(x + 1)$, shift the original graph 1 unit to the left. The resulting graph is

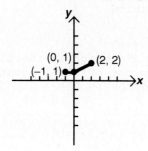

18. For $y = -f(x)$, change all original y values by $-y$. The graph is reflected about the x axis.

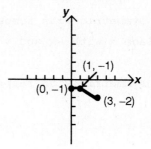

19. $y = 4 - (x - 2)^2 = -(x - 2)^2 + 4$
The vertex is $(2, 4)$. It is a parabola opening downward.

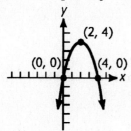

20. $4x^2 + 9y^2 = 144$

$$\frac{4x^2}{144} + \frac{9y^2}{144} = 1$$

$$\frac{x^2}{36} + \frac{y^2}{16} = 1$$

The graph is an ellipse with $a = 6$ and $b = 4$.

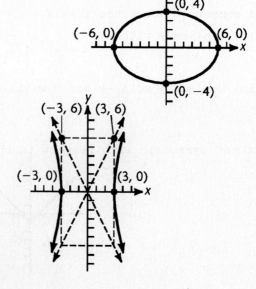

21. $4x^2 - y^2 = 36$

$$\frac{4x^2}{36} - \frac{y^2}{36} = 1$$

$$\frac{x^2}{9} - \frac{y^2}{36} = 1$$

It is a hyperbola with $a = 3$ and $b = 6$. Now vertices: $(\pm a, 0) = (\pm 3, 0)$;

asymptotes: $y = \pm\dfrac{b}{a}x = \pm\dfrac{6}{3}x = \pm 2x$.

22. $25y^2 - 4x^2 = 100$

$\dfrac{25y^2}{100} - \dfrac{4x^2}{100} = \dfrac{100}{100}$

$\dfrac{y^2}{4} - \dfrac{x^2}{25} = 1$

The graph is a hyperbola with $a = 2$ and $b = 5$. The vertices are $(0, \pm a)$ or $(0, \pm 2)$. Asymptotes: $y = \pm\dfrac{a}{b}x = \pm\dfrac{2}{5}x$.

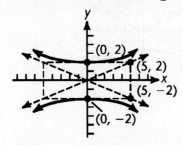

23. $x = -2y^2 + 8y - 6$

Vertex: $y = -\dfrac{b}{2a} = -\dfrac{8}{2(-2)} = 2$

and $x = -2(2)^2 + 8(2) - 6$
$= -8 + 16 - 6 = 2$.

The point is $(2, 2)$. It is a parabola opening to the left.

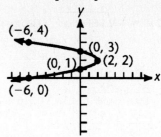

24. $2y = \sqrt{1 - 4x^2}$ Notice that $y \geq 0$

$(2y)^2 = (\sqrt{1 - 4x^2})^2$

$4y^2 = 1 - 4x^2$

$4x^2 + 4y^2 = 1$

$4(x^2 + y^2) = 1$

$x^2 + y^2 = \dfrac{1}{4}$

This is an equation for a circle with $r = \sqrt{\dfrac{1}{4}} = \dfrac{1}{2}$. Since $y \geq 0$, the graph is an upper semicircle.

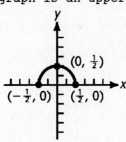

25. $\dfrac{4x^2}{9} + \dfrac{9y^2}{16} = 1$

$\dfrac{x^2}{\frac{9}{4}} + \dfrac{y^2}{\frac{16}{9}} = 1$

The graph is an ellipse with $a = \dfrac{3}{2}$ and $b = \dfrac{4}{3}$.

x intercepts: $(\pm a, 0) = \left(\pm\dfrac{3}{2}, 0\right)$;

y intercepts: $(0, \pm b) = \left(0, \pm\dfrac{4}{3}\right)$

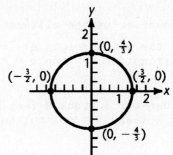

26. $3x = -\sqrt{16 - 9y^2}$ Notice that $x \leq 0$

$(3x)^2 = (-\sqrt{16 - 9y^2})^2$

$9x^2 = 16 - 9y^2$

$9x^2 + 9y^2 = 16$

$9(x^2 + y^2) = 16$

$x^2 + y^2 = \dfrac{16}{9}$

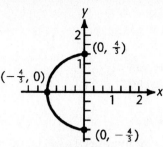

26. (cont.)

This is an equation for a circle with

$r = \sqrt{\dfrac{16}{9}} = \dfrac{4}{3}$. Since $x \leq 0$, the graph

is the left semicircle.

27. $2y = -\sqrt{16 + x^2}$ (Note: $y < 0$)

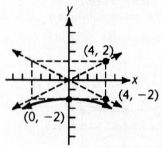

$$(2y)^2 = (-\sqrt{16 + x^2})^2$$
$$4y^2 = 16 + x^2$$
$$-x^2 + 4y^2 = 16$$
$$-\dfrac{x^2}{16} + \dfrac{4y^2}{16} = 1$$
$$-\dfrac{x^2}{16} + \dfrac{y^2}{4} = 1$$

It is a hyperbola with $b = 4$ and $a = 2$.

Vertex: $(0, -a) = (0, -2)$ Asymptotes: $y = \pm\dfrac{a}{b}x = \pm\dfrac{2}{4}x = \pm\dfrac{1}{2}x$

Since $y < 0$, only the lower part is graphed.

28. $(x + 3)^2 - 4(y - 2)^2 = 16$

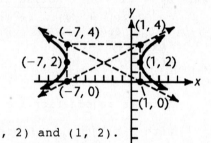

$$\dfrac{(x + 3)^2}{16} - \dfrac{4(y - 2)^2}{16} = \dfrac{16}{16}$$
$$\dfrac{(x + 3)^2}{16} - \dfrac{(y - 2)^2}{4} = 1$$

It is a hyperbola. Center is at $(-3, 2)$ with
$h = -3$ and $k = 2$. Since $a = 4$ and $b = 2$, the
vertices are $(h \pm a, b)$ or $(-3 \pm 4, 2)$ to get $(-7, 2)$ and $(1, 2)$.

29. $x^2 - 2x + y^2 = 12y - 38$
Based upon the solution to problem 7, this is expressed $(x - 1)^2 + (y - 6)^2 = -1$ and there is **no** graph due to -1 on the right side.

30. The equation of the ellipse is $\dfrac{x^2}{a^2} + \dfrac{y^2}{b^2} = 1$

Since the x intercepts are ± 6, then $a = 6$ or $a^2 = 36$. Also, since the
y intercept is 8, then $b = 8$ or $b^2 = 64$. The equation becomes

$$\dfrac{x^2}{36} + \dfrac{y^2}{64} = 1$$

In order for a truck 8 feet wide and 6 feet high to pass through the arch,
let $x = 4$ and $y = 6$ into the equation:

$$\dfrac{x^2}{36} + \dfrac{y^2}{64} \overset{?}{=} 1$$
$$\dfrac{4^2}{36} + \dfrac{6^2}{64} \overset{?}{=} 1$$
$$\dfrac{16}{36} + \dfrac{36}{64} \overset{?}{=} 1$$
$$0.4444 + 0.5625 \overset{?}{=} 1$$
$$1.0069 > 1$$

Since 1.0069 is greater than 1, then the point is outside the ellipse, the
truck will not make it through the arch.

31. It is **not** a one-to-one function since horizontal lines above the x axis cut the graph in two points.

32. The graph is **not** a function since vertical lines between $x = 0$ and $x = -2$ intersect the graph in two points. Hence, it cannot be a one-to-one function.

33. It is a one-to-one function since every horizontal line cuts the graph in one point.

34. The graph is a one-to-one function since vertical and horizontal lines intersect the graph in one point.

35. $2x - 3y = 6$
$$-3y = -2x + 6$$
$$y = \frac{-2x + 6}{-3} = \frac{2}{3}x - 2$$

This is the representation for $f(x)$: $f(x) = \frac{2}{3}x - 2$.

For the inverse, interchage x and y in the equation $2x - 3y = 6$ to obtain
$$2y - 3x = 6$$
$$2y = 3x + 6$$
$$y = \frac{3x + 6}{2} = \frac{3}{2}x + 3$$

This is the representation for $f^{-1}(x)$: $f^{-1}(x) = \frac{3}{2}x + 3$.

36. $y = x^2 - 4$ can be expressed as $f(x) = x^2 - 4$.
Since the graph is a parabola it is **not** a one-to-one function and therefore has no inverse which is a function.

37. $y - x^2 = 1$
$$y = x^2 + 1$$
or $f(x) = x^2 + 1$
Since this is **not** a one-to-one function, no inverse exists.

38. $(f \circ g)(x) = f(g(x)) = f\left(\dfrac{1}{2x - 1}\right) = 2\left(\dfrac{1}{2x - 1}\right) - 1 = \dfrac{2}{2x - 1} - 1$
$$= \frac{2}{2x - 1} - \frac{2x - 1}{2x - 1} = \frac{-2x + 3}{2x - 1} \neq x$$
Since $(f \circ g)(x) \neq x$, f and g are not inverses of each other.

39. $(f \circ g)(x) = f(g(x)) = f\left(\dfrac{3}{2}x - 9\right) = \dfrac{2}{3}\left(\dfrac{3}{2}x - 9\right) + 6 = x - 6 + 6 = x$

$(g \circ f)(x) = g(f(x)) = g\left(\dfrac{2}{3}x + 6\right) = \dfrac{3}{2}\left(\dfrac{2}{3}x + 6\right) - 9 = x + 9 - 9 = x$

Since $(f \circ g)(x) = (g \circ f)(x) = x$, then f and g are inverse functions.

40.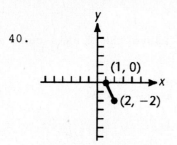

41.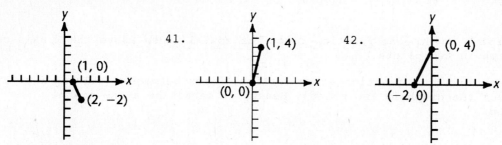

42.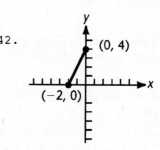

43. cement: 4 cu meters
    sand: $2(4) = 8$ cu meters
    gravel: $3(4) = 12$ cu meters
    sum: $4 + 8 + 12 = 24$ cu meters

Another solution is as follows. Let x be the amount of concrete. Now $\frac{1}{6}$ of the amount is cement. The equation is $\frac{1}{6}x = 4$ or $x = 24$ cu meters.

44. 45 miles per hour $= \dfrac{45 \; \overset{1}{\cancel{\text{miles}}}}{1 \; \cancel{\text{hour}}} \cdot \dfrac{1 \; \cancel{\text{hour}}}{3600 \; \text{sec}} \cdot \dfrac{5280 \; \text{feet}}{\underset{1}{\cancel{\text{mile}}}}$

$= \dfrac{(45)(5280)}{3600}$ ft/sec $= 66$ ft/sec

45. $y = \dfrac{kt}{z^2}$ Let $y = 20$, $t = 15$, and $z = 3$ to have

$20 = \dfrac{k(15)}{3^2}$ or $20 = \dfrac{k(15)}{9}$

$\dfrac{9(20)}{15} = k$

$12 = k$

The equation is $y = \dfrac{12t}{z^2}$.

Now let $t = 9$ and $z = 6$ to get

$y = \dfrac{12(9)}{6^2} = \dfrac{108}{36} = 3$

46. $x = \dfrac{kyz}{w^2}$ Let $x = 40$, $y = 12$, $z = 6$, and $w = 3$ to have

$40 = \dfrac{k(12)(6)}{3^2}$ or $40 = \dfrac{k(72)}{9}$

$\dfrac{9(40)}{72} = k$ or $k = 5$

The equation is $x = \dfrac{5yz}{w^2}$.

Now let $y = 4$, $z = 10$, and $w = 5$.

$x = \dfrac{5(4)(10)}{5^2} = \dfrac{200}{25} = 8$

47. $I = \dfrac{kE}{R}$ Let $I = 22$, $E = 154$, and $R = 7$ to have

$22 = \dfrac{k(154)}{7}$ or $154 = k(154)$

or $1 = k$

The equation is $I = \dfrac{E}{R}$.

Now let $E = 220$ and $R = 5$ to get

$I = \dfrac{220}{5} = 44$ amps

48. $D = \dfrac{k(T - 75)}{d}$ Let $D = 350$, $T = 90$, and $d = 3$ to have

$350 = \dfrac{k(90 - 75)}{3} = \dfrac{k(15)}{3} = 5k$

$\dfrac{350}{5} = k$

$70 = k$

The equation is $D = \dfrac{70(T - 75)}{d}$

Now let $T = 95$ and $d = 2$ to have

$D = \dfrac{70(95 - 75)}{2} = \dfrac{70(20)}{2} = 700$

CHAPTER 5 POLYNOMIALS AND RATIONAL FUNCTIONS

Exercises 5.1

1.

$$
\begin{array}{r}
x^2 - 3x + 3 \\
x - 1 \overline{\smash{\big)}\, x^3 - 4x^2 + 6x - 3} \\
\underline{x^3 - \ x^2} \\
-3x^2 + 6x \\
\underline{-3x^2 + 3x} \\
3x - 3 \\
\underline{3x - 3} \\
0
\end{array}
$$

Thus, $\dfrac{x^3 - 4x^2 + 6x - 3}{x - 1} = x^2 - 3x + 3 + \dfrac{0}{x - 1}$

3.

$$
\begin{array}{r}
x^2 - 3x + 2 \\
2x - 5 \overline{\smash{\big)}\, 2x^3 - 11x^2 + 19x - 10} \\
\underline{2x^3 - \ 5x^2} \\
-6x^2 + 19x \\
\underline{-6x^2 + 15x} \\
4x - 10 \\
\underline{4x - 10} \\
0
\end{array}
$$

Thus, $\dfrac{2x^3 - 11x^2 + 19x - 10}{2x - 5} = x^2 - 3x + 2 + \dfrac{0}{2x - 5}$

5.

$$
\begin{array}{r}
x^2 - x + 4 \\
3x - 2 \overline{\smash{\big)}\, 3x^3 - 5x^2 + 14x + 3} \\
\underline{3x^3 - 2x^2} \\
-3x^2 + 14x \\
\underline{-3x^2 + 2x} \\
12x + 3 \\
\underline{12x - 8} \\
11
\end{array}
$$

Thus, $\dfrac{3x^3 - 5x^2 + 14x + 3}{3x - 2} = x^2 - x + 4 + \dfrac{11}{3x - 2}$

7.

$$
\begin{array}{r}
2x^2 - x + 3 \\
x^2 + 2x - 2 \overline{\smash{\big)}\, 2x^4 + 3x^3 - 3x^2 + 5x - 8} \\
\underline{2x^4 + 4x^3 - 4x^2} \\
-x^3 + \ x^2 + 5x \\
\underline{-x^3 - 2x^2 + 2x} \\
3x^2 + 3x - 8 \\
\underline{3x^2 + 6x - 6} \\
-3x - 2
\end{array}
$$

Thus, $\dfrac{2x^4 + 3x^3 - 3x^2 + 5x - 8}{x^2 + 2x - 2}$

$\qquad = 2x^2 - x + 3 + \dfrac{-3x - 2}{x^2 + 2x - 2}$

9.
$$
2x^2 + 3x - 2 \overline{\smash{\big)}\, \begin{aligned} & 2x^2 - 3x + 9 \\ & 4x^4 + 0x^3 + 5x^2 - 7x + 3 \end{aligned}}
$$

$$
\underline{4x^4 + 6x^3 - 4x^2}
$$
$$
-6x^3 + 9x^2 - 7x
$$
$$
\underline{-6x^3 - 9x^2 + 6x}
$$
$$
18x^2 - 13x + 3
$$
$$
\underline{18x^2 + 27x - 18}
$$
$$
-40x + 21
$$

Thus, $\dfrac{4x^4 + 5x^2 - 7x + 3}{2x^2 + 3x - 2}$

$$= 2x^2 - 3x + 9 + \dfrac{-40x + 21}{2x^2 + 3x - 2}$$

11.
$$
2 \,\underline{\big|}\, \begin{array}{rrrr} 2 & -4 & +5 & -1 \\ & 4 & 0 & 10 \\ \hline 2 & 0 & 5 & 9 \end{array}
$$

Quotient   Remainder

Thus, $\dfrac{2x^3 - 4x^2 + 5x - 1}{x - 2} = 2x^2 + 5 + \dfrac{9}{x - 2}$

13.
$$
-3 \,\underline{\big|}\, \begin{array}{rrrr} -3 & +0 & +2 & -75 \\ & 9 & -27 & 75 \\ \hline -3 & 9 & -25 & 0 \end{array}
$$

Quotient   Remainder

Thus, $\dfrac{-3x^3 + 2x - 75}{x + 3} = -3x^2 + 9x - 25 + \dfrac{0}{x + 3}$

15.
$$
3 \,\underline{\big|}\, \begin{array}{rrrrrr} 1 & -1 & -7 & +0 & +0 & +24 \\ & 3 & 6 & -3 & -9 & -27 \\ \hline 1 & 2 & -1 & -3 & -9 & -3 \end{array}
$$

Quotient   Remainder

Thus, $\dfrac{x^5 - x^4 - 7x^3 + 24}{x - 3}$

$$= x^4 + 2x^3 - x^2 - 3x - 9 + \dfrac{-3}{x - 3}$$

17.
$$
-1 \,\underline{\big|}\, \begin{array}{rrrr} 4 & 0 & 1 & 7 \\ & -4 & 4 & -5 \\ \hline 4 & -4 & 5 & 2 \end{array}
$$

Quotient   Remainder

Thus, $\dfrac{4x^3 + x + 7}{x + 1} = 4x^2 - 4x + 5 + \dfrac{2}{x + 1}$

19.
$$
-1 \,\underline{\big|}\, \begin{array}{rrrr} 4 & 0 & -3 & -2 \\ & -4 & 4 & -1 \\ \hline 4 & -4 & 1 & -3 \end{array}
$$

Quotient   Remainder

Thus, $4m^3 - 3m - 2 = (m + 1)(4m^2 - 4m + 1) + (-3)$

21.
$$
x^2 + 3x + 2 \overline{\smash{\big)}\, \begin{aligned} & 3x^3 - 6x^2 + 2x - 4 \\ & 3x^5 + 3x^4 - 10x^3 - 10x^2 - 8x - 8 \end{aligned}}
$$

$$
\underline{3x^5 + 9x^4 + 6x^3}
$$
$$
-6x^4 - 16x^3 - 10x^2
$$
$$
\underline{-6x^4 - 18x^3 - 12x^2}
$$
$$
2x^3 + 2x^2 - 8x
$$
$$
\underline{2x^3 + 6x^2 + 4x}
$$
$$
-4x^2 - 12x - 8
$$
$$
\underline{-4x^2 - 12x - 8}
$$
$$
0
$$

Thus,
$$3x^5 + 3x^4 - 10x^3 - 10x^2 - 8x - 8 =$$
$$(x^2 + 3x + 2)(3x^3 - 6x^2 + 2x - 4) + 0$$

23.

$$
\begin{array}{r}
3k + 2r \\
6k - 5r \overline{\smash{\big)}\ 18k^2 - 3rk - 10r^2} \\
\underline{18k^2 - 15rk} \\
12rk - 10r^2 \\
\underline{12rk - 10r^2} \\
0
\end{array}
$$

Thus, $18k^2 - 3rk - 10r^2 = (6k - 5r)(3k + 2r) + 0$

25.

$$
\begin{array}{r}
x^2 - 3ax + 2a^2 \\
4x^2 - ax + a^2 \overline{\smash{\big)}\ 4x^4 - 13ax^3 + 12a^2x^2 - 5a^3x + 2a^4} \\
\underline{4x^4 - ax^3 + a^2x^2} \\
-12ax^3 + 11a^2x^2 - 5a^3x \\
\underline{-12ax^3 + 3a^2x^2 - 3a^3x} \\
8a^2x^2 - 2a^3x + 2a^4 \\
\underline{8a^2x^2 - 2a^3x + 2a^4} \\
0
\end{array}
$$

Thus, $4x^4 - 13ax^3 + 12a^2x^2 - 5a^3x + 2a^4 = (4x^2 - ax + a^2)(x^2 - 3ax + 2a^2) + 0$

27.

$$
\begin{array}{r|rrrr}
-3 & 1 & 0 & 0 & -27 \\
   &   & -3 & 9 & -27 \\
\hline
   & 1 & -3 & 9 & -54
\end{array}
$$

$\underbrace{}_{\text{Quotient}}$ $\underbrace{}_{\text{Remainder}}$

Thus, $x^3 - 27$
$= (x + 3)(x^2 - 3x + 9) + (-54)$

29.

$$
\begin{array}{r|rrr}
3 & 1 & -7 & 12 \\
  &   & 3 & -12 \\
\hline
  & 1 & -4 & 0
\end{array}
$$

$\underbrace{}_{\text{Quotient}}$ $\underbrace{}_{\text{Remainder}}$

Thus, $Q(x) = x - 4$ and $r = 0$

31.

$$
\begin{array}{r|rrrr}
2 & 2 & -1 & -2 & 2 \\
  &   & 4 & 6 & 8 \\
\hline
  & 2 & 3 & 4 & 10
\end{array}
$$

$\underbrace{}_{\text{Quotient}}$ $\underbrace{}_{\text{Remainder}}$

Thus, $Q(x) = 2x^2 + 3x + 4$ and $r = 10$

33.

$$
\begin{array}{r|rrrrrr}
2 & 1 & -4 & 0 & -1 & 0 & 2 \\
  &   & 2 & -4 & -8 & -18 & -36 \\
\hline
  & 1 & -2 & -4 & -9 & -18 & -34
\end{array}
$$

$\underbrace{}_{\text{Quotient}}$ $\underbrace{}_{\text{Remainder}}$

Thus, $Q(x) = x^4 - 2x^3 - 4x^2 - 9x - 18$
and $r = -34$

35.

$$
\begin{array}{r|rrrrr}
-\frac{1}{2} & 4 & 0 & -1 & 6 & -2 \\
             &   & -2 & 1 & 0 & -3 \\
\hline
             & 4 & -2 & 0 & 6 & -5
\end{array}
$$

$\underbrace{}_{\text{Quotient}}$ $\underbrace{}_{\text{Remainder}}$

Thus, $Q(x) = 4x^3 - 2x^2 + 6$
and $r = -5$

37.

$$
\begin{array}{r|rrrr}
-i & 1 & 3 & 1 & 3 \\
   &   & -i & -3i - 1 & -3 \\
\hline
   & 1 & 3 - i & -3i & 0
\end{array}
$$

$\underbrace{}_{\text{Quotient}}$ $\underbrace{}_{\text{Remainder}}$

Thus, $Q(x) = x^2 + (3 - i)x - 3i$
and $r = 0$

39.
$$3i \overline{\smash{\big)}\ \begin{array}{cccc} -i & 4 & 0 & 5 \end{array}}$$
$$\begin{array}{cccc} & 3 & 21i & -63 \end{array}$$
$$\begin{array}{cccc} -i & 7 & 21i & -58 \end{array}$$

$\underbrace{}_{\text{Quotient}}$ $\underbrace{}_{\text{Remainder}}$

Thus, $Q(x) = -ix^2 + 7x + 21i$
and $r = -58$

41.
$$-2 \overline{\smash{\big)}\ \begin{array}{ccccc} 1 & 0 & -8 & -8 & 1 \end{array}}$$
$$\begin{array}{ccccc} & -2 & 4 & 8 & 0 \end{array}$$
$$\begin{array}{ccccc} 1 & -2 & -4 & 0 & 1 \end{array}$$

Here, $r = 1$

43.
$$1 \overline{\smash{\big)}\ \begin{array}{ccccc} 1 & -1 & 2 & 3 & 0 & 4 \end{array}}$$
$$\begin{array}{ccccc} & 1 & 0 & 2 & 5 & 5 \end{array}$$
$$\begin{array}{ccccc} 1 & 0 & 2 & 5 & 5 & 9 \end{array}$$

Here, $r = 9$

45.
$$2i \overline{\smash{\big)}\ \begin{array}{ccccc} 1 & -i & 2 & 4i & 8 \end{array}}$$
$$\begin{array}{ccccc} & 2i & -2 & 0 & -8 \end{array}$$
$$\begin{array}{ccccc} 1 & i & 0 & 4i & 0 \end{array}$$

Here, $r = 0$

47.
$$2+i \overline{\smash{\big)}\ \begin{array}{cccc} 2 & -9 & 14 & -5 \end{array}}$$
$$\begin{array}{cccc} & 4+2i & -12-i & 5 \end{array}$$
$$\begin{array}{cccc} 2 & -5+2i & 2-i & 0 \end{array}$$

Here, $r = 0$

49.
$$-a \overline{\smash{\big)}\ \begin{array}{cccc} 1 & 0 & 0 & 0 & -a^4 \end{array}}$$
$$\begin{array}{cccc} & -a & a^2 & -a^3 & a^4 \end{array}$$
$$\begin{array}{cccc} 1 & -a & a^2 & -a^3 & 0 \end{array}$$

Here, $r = 0$

51.
$$-2 \overline{\smash{\big)}\ \begin{array}{ccccccc} 1 & 0 & 0 & 0 & 0 & 30 & 0 & k \end{array}}$$
$$\begin{array}{ccccccc} & -2 & 4 & -8 & 16 & -32 & 4 & -8 \end{array}$$
$$\begin{array}{ccccccc} 1 & -2 & 4 & -8 & 16 & -2 & 4 & k-8 \end{array}$$

$\underbrace{}_{\text{Remainder}}$

To have the remainder 0, we must have $k - 8 = 0$ or $k = 8$

53.
$$-2 \overline{\smash{\big)}\ \begin{array}{cccc} -1 & k & -5 & -20 \end{array}}$$
$$\begin{array}{cccc} & 2 & -2k-4 & 4k+18 \end{array}$$
$$\begin{array}{cccc} -1 & k+2 & -2k-9 & 4k-2 \end{array}$$

To have the remainder positive, we must have $4k - 2 > 0$
$$4k > 2$$
$$k > \frac{2}{4}$$
$$k > \frac{1}{2}$$

55.
$$1.01 \overline{\smash{\big)}\ \begin{array}{cccccc} 1 & 0 & 1 & -11 & 0 & 43 \end{array}}$$
$$\begin{array}{cccccc} & 1.01 & 1.0201 & 2.0403 & -9.0493 & -9.1398 \end{array}$$
$$\begin{array}{cccccc} 1 & 1.01 & 2.0201 & -8.9597 & -9.0493 & 33.8602 \end{array}$$

The remainder is 33.8602 which rounds to 33.86

57.
$$-0.91 \overline{\smash{\big)}\ \begin{array}{ccccccc} -3 & 0 & 1 & -5.1 & 0 & -1 & -2.9 \end{array}}$$
$$\begin{array}{ccccccc} & 2.73 & -2.4843 & 1.350713 & 3.411851 & -3.104785 & 3.735354 \end{array}$$
$$\begin{array}{ccccccc} -3 & 2.73 & -1.4843 & -3.749287 & 3.411851 & -4.104785 & 0.835354 \end{array}$$

The remainder is 0.835354 which rounds to 0.84

59. Use dividend = (quotient)(divisor) + remainder to observe that for (quotient)(divisor) the exponents are added in the multiplication process. When multiplied, their degrees are added where the sum must be equal to the degree of the dividend.

Exercises 5.2

1. $r = P(2) = 2^4 - 2(2)^3 + 2(2^2) - 1$
$= 16 - 16 + 8 - 1$
$= 7$

3. $r = P(1) = 4(1)^5 - 3(1)^2 + 5(1) - 1$
$= 4 - 3 + 5 - 1$
$= 5$

5. $r = P(-3) = (-3)^3 + 2(-3)^2 - 5(-3) + 1$
$= -27 + 18 + 15 + 1$
$= 7$

7. $r = P(\sqrt{2}) = 2(\sqrt{2})^3 - 2(\sqrt{2})^2 + 4$
$= 4\sqrt{2} - 4 + 4$
$= 4\sqrt{2}$

9. $r = P(i) = (i)^2 - 3(i) + 1$
$= i^2 - 3i + 1$
$= -1 - 3i + 1$
$= -3i$

11. $r = P(-1) = (-1)^{1023} - 3(-1)^{15} + 7$
$= -1 - 3(-1) + 7$
$= -1 + 3 + 7$
$= 9$

13.
```
-1| 3   -5    1    -1     7
        -3    8    -9    10
   ------------------------
     3   -8    9   -10    17
```
Here, $P(-1) = r = 17$

15.
```
-2| -1    0    1     5     0     0    -2
          2   -4     6   -22    44   -88
   -------------------------------------
    -1    2   -3    11   -22    44   -90
```
Here, $P(-2) = r = -90$

17.
```
1/2| 4    6    0   -2
          2    4    2
    -----------------
     4    8    4    0
```
Here, $P\left(\frac{1}{2}\right) = r = 0$

19.
```
√2| 4     0      -2       6
        4√2     8      6√2
   ------------------------
    4   4√2     6    6 + 6√2
```
Here, $P(\sqrt{2}) = r = 6 + 6\sqrt{2}$

21.
```
1 - i| 3     0          -2         1          -5
           3 - 3i     0 - 6i    -8 - 4i    -11 + 3i
      --------------------------------------------
       3   3 - 3i    -2 - 6i    -7 - 4i    -16 + 3i
```
Here, $P(1 - i) = r = -16 + 3i$

23.
```
1 + 2i| 1      0          2          -3           0
            1 + 2i     -3 + 4i    -9 + 2i    -16 - 22i
       ---------------------------------------------
        1    1 + 2i    -1 + 4i   -12 + 2i    -16 - 22i
```
Here, $P(1 + 2i) = -16 - 22i$

25. $P(-1) = (-1)^2 - 3(-1) + 4$
$= 1 + 3 + 4$
$= 8$

Since $P(-1) \neq 0$, $D(x) = x + 1$ is **not** a factor of $P(x)$.

27. $P(1) = (1)^{19} - (1)^{17} + (1)^2 - 1$
$= 1 - 1 + 1 - 1$
$= 0$

Since $P(1) = 0$, $D(x) = x - 1$ is a factor of $P(x)$.

29. $P(-2) = (-2)^4 - 3(-2)^3 + 7(-2) - 1$
$= 16 + 24 - 14 - 1$
$= 25$

Since $P(-2) \neq 0$, $D(x) = x + 2$ is **not** a factor of $P(x)$.

31. $P(i) = i^3 + 5(i)^2 + i + 5$
$= -i - 5 + i + 5$
$= 0$

Since $P(i) = 0$, $D(x) = x - i$ is a factor of $P(x)$.

33. $P(1 + 2i) = (1 + 2i)^2 - (3 - i)(1 + 2i) + 8 + i$
$= -3 + 4i + (-5 - 5i) + 8 + i$
$= 0$

Since $P(1 + 2i) = 0$, $D(x) = x - (1 + 2i)$ is a factor of $P(x)$.

35.
$$2 \overline{\smash{\big)}\ 1 \quad -1 \quad -1 \quad -5}$$

	2	2	2
1	1	1	-3

Since $P(2) = r = -3$, then $c = 2$ is **not** a zero of $P(x)$.

37.
$$-\frac{1}{2} \overline{\smash{\big)}\ 8 \quad 0 \quad 2 \quad -4 \quad -3}$$

	-4	2	-2	3
8	-4	4	-6	0

Since $P\left(-\frac{1}{2}\right) = r = 0$, then $c = -\frac{1}{2}$ is a zero of $P(x)$.

39.
$$\frac{3}{4} \overline{\smash{\big)}\ 4 \quad 5 \quad -10 \quad 4}$$

	3	6	-3
4	8	-4	1

Since $P\left(\frac{3}{4}\right) = r = 1$, then $c = \frac{3}{4}$ is **not** a zero of $P(x)$.

41.
$$1 + i \overline{\smash{\big)}\ 1 \quad -3 \quad 4 \quad -2}$$

	$1 + i$	$-3 - i$	2
1	$-2 + i$	$1 - i$	0

Since $P(1 + i) = 0$, then $c = 1 + i$ is a zero of $P(x)$.

43. Since 3 is a zero, then $x - 3$ is a factor of $P(x)$. To find the quotient when $P(x)$ is divided by $x - 3$, use synthetic division:

$$3 \overline{\smash{\big)}\ 1 \quad -3 \quad 1 \quad -3}$$

	3	0	3
1	0	1	0

Quotient

The quotient is $x^2 + 1$ and the equation can be written as
$(x - 3)(x^2 + 1) = 0$
$x - 3 = 0, \quad x^2 + 1 = 0$
$x = 3, \qquad x^2 = -1$
$x = 3, \qquad x = \pm i$

45. Since -7 is a zero, then $x + 7$ is a factor of $P(x)$. To find the quotient when $P(x)$ is divided by $x + 7$, use synthetic division:

$$-7 \overline{\smash{\big)}\ 2 \quad 17 \quad 19 \quad -14}$$

	-14	-21	14
2	3	-2	0

Quotient

The quotient is $2x^2 + 3x - 2$ and the equation can be written as

$$(x + 7)(2x^2 + 3x - 2) = 0$$
$$(x + 7)(2x - 1)(x + 2) = 0$$
$$x + 7 = 0, \quad 2x - 1 = 0, \quad x + 2 = 0$$
$$x = -7, \qquad x = \frac{1}{2}, \qquad x = -2$$

47.

```
-2 | 1    -k        3          7k
   |       -2      4 + 2k    -14 - 4k
   ------------------------------------
     1  -2 - k    7 + 2k    -14 + 3k      Here, r = -14 + 3k
```

Since $P(x)$ is divisible by $x + 2$, we want $r = 0$ or $-14 + 3k = 0$

$$3k = 14$$
$$k = \frac{14}{3}$$

49.

$$P(1) = 1^3 + 3(1)^2 - 3(1) - 9$$
$$= 1 + 3 - 3 - 9$$
$$= -8$$

and $P(2) = 2^3 + 3(2)^2 - 3(2) - 9$
$$= 8 + 12 - 6 - 9$$
$$= 5$$

Since $P(1) = -5$ (a negative number) and $P(2) = 5$ (a positive number) then $P(x)$ must have a zero between 1 and 2.

51. To compute $P(3)$ and $P(4)$, use synthetic division:

```
3 | 1   1   -11   -12   -12          4 | 1   1   -11   -12   -12
  |     3    12     3   -27            |     4    20    36    96
  --------------------------------     --------------------------------
    1   4    1    -9   -39 = P(3)        1   5    9    24    84 = P(4)
```

Since $P(3) = -39$ (a negative number) and $P(4) = 84$ (a positive number) then $P(x)$ must have a zero between 3 and 4.

53.

$$P(0) = 2(0)^4 + 6(0)^3 + 5(0)^2 - 3(0) - 3$$
$$= 0 + 0 + 0 + 0 - 3$$
$$= -3$$
$$P(1) = 2(1)^4 + 6(1)^3 + 5(1)^2 - 3(1) - 3$$
$$= 2 + 6 + 5 - 3 - 3$$
$$= 7$$

Since $P(0) = -3$ (a negative number) and $P(1) = 7$ (a positive number) then $P(x)$ has a zero between 0 and 1.

55. Since $P(1) = 1^4 + 2(1)^3 + 2(1)^2 + 1 - 1$
$$= 1 + 2 + 2 + 1 - 1$$
$$= 5 \text{ (a positive number)}$$

and $P(0) = 0^4 + 2(0)^3 + 2(0)^2 + 0 - 1$
$$= 0 + 0 + 0 + 0 - 1$$
$$= -1 \text{ (a negative number)}$$

Then $P(x)$ must have a zero between $x = 0$ and $x = 1$.

57. Since $P(2) = 2^4 + 2(2)^3 - 6(2) - 9$
$= 16 + 16 - 12 - 9$
$= 11$ (a positive number)
and $P(1) = 1^4 + 2(1)^3 - 6(1) - 9$
$= 1 + 2 - 6 - 9$
$= -12$ (a negative number)

Since $P(2) > 0$ and $P(1) < 0$, then $P(x)$ has a zero between 1 and 2.

Also, $P(-2) = (-2)^4 + 2(-2)^3 - 6(-2) - 9$
$= 16 - 16 + 12 - 9$
$= 3$ (a positive number)
$P(-1) = (-1)^4 + 2(-1)^3 - 6(-1) - 9$
$= 1 - 2 + 6 - 9$
$= -4$ (a negative number)

Since $P(-2) > 0$ and $P(-1) < 0$, then $P(x)$ has a zero between -2 and -1.

59. If $x - c$ is a factor of $P(x)$, then $P(c) = 0$ and the graph crosses the x axis at the point $(c, 0)$. That is, c is the x intercept.

Exercises 5.3

1. $(x - 1)(x + 2)^2 = 0$
$x - 1 = 0$, $(x + 2)^2 = 0$
$x - 1 = 0$ or $x = 1$ of multiplicity 1
$(x + 2)^2 = 0$ or $x = -2$ of multiplicity 2
Degree of $P(x)$ is 3

3. $-(x + 3)^2(4x - 3)^3 = 0$
$(x + 3)^2 = 0$, $(4x - 3)^3 = 0$
$(x + 3)^2 = 0$ or $x = -3$ of multiplicity 2
$(4x - 3)^3 = 0$ or $4x = 3$
or $x = \frac{3}{4}$ of multiplicity 3
Degree of $P(x)$ is 5

5. $x^2(x^2 + 1)(x^2 - 1) = 0$
$x^2(x + i)(x - i)(x + 1)(x - 1) = 0$
$x^2 = 0$ or $x = 0$ of multiplicity 2
$x + i = 0$ or $x = -i$ ⎤
$x - i = 0$ or $x = i$ ⎥ multiplicity 1
$x + 1 = 0$ or $x = -1$ ⎥
$x - 1 = 0$ or $x = 1$ ⎦
Degree of $P(x)$ is 6

7. The factors are
$(x - 3)^2$ and $x - (-1) = x + 1$
The polynomial is
$P(x) = (x - 3)^2(x + 1)$

9. The factors are
x^5 and $[x - (-2)]^3 = (x + 2)^3$
The polynomial is
$P(x) = x^5(x + 2)^3$

11. The factors are $(x - 1)$, $(x + 1)$, and $(x - 2)$.
The polynomial is $P(x) = a(x - 1)(x + 1)(x - 2)$.
Since $P(0) = 4$, then $a(0 - 1)(0 + 1)(0 - 2) = 4$
$+2a = 4$
$a = 2$
and the final form of the polynomial is $P(x) = 2(x - 1)(x + 1)(x - 2)$.

13. $P(x) = x^3 - 4x^2 + 5x + 1$

                           2 variations: 0 or 2 positive real zeros

  $P(-x) = (-x)^3 - 4(-x)^2 + 5(-x) + 1$

        $= -x^3 - 4x^2 - 5x + 1$

                        1 variation: 1 negative real zero

The possibilities are (1): 2 positive and 1 negative real zeros
                        (2): 1 negative and 2 nonreal complex zeros

15. $P(x) = 2x^4 + 3x^3 - 2x + 1$

                           2 variations: 0 or 2 positive real zeros

  $P(-x) = 2(-x)^4 + 3(-x)^3 - 2(-x) + 1$

        $= 2x^4 - 3x^3 + 2x + 1$

                         2 variations: 0 or 2 negative real zeros

The possibilities are (1): 2 positive and 2 negative zeros
                        (2): 2 positive and 2 nonreal complex zeros
                        (3): 2 negative and 2 nonreal complex zeros
                        (4): 4 nonreal complex zeros

17. $P(x) = 2x^4 - x^3 + 3x - 1$

                           3 variations: 1 or 3 positive real zeros

  $P(-4) = 2(-x)^4 - (-x)^3 + 3(-x) - 1$

        $= 2x^4 + x^3 - 3x - 1$

                         1 variation: 1 negative real zero

The possibilities are (1): 3 positive and 1 negative zeros
                        (2): 1 positive, 1 negative, and 2 nonreal complex zeros

19. $P(x) = x^6 + x^3 + 2x + 3$

                           0 variations: **no** positive real zeros

  $P(-x) = (-x)^6 + (-x)^3 + 2(-x) + 3$

        $= x^6 - x^3 - 2x + 3$

                         2 variations: 0 or 2 negative real zeros

The possibilities are (1): 2 negative and 4 nonreal complex zeros
                        (2): 6 nonreal complex zeros

21. $P(x) = 2x^5 - 5x^4 - 12x^3 + 3x^2 - 14x + 8$

                           4 variations: 0, 2, or 4 positive real zeros

  $P(-x) = 2(-x)^5 - 5(-x)^4 - 12(-x)^3 + 3(-x)^2 - 14(-x) + 8$

        $= -2x^5 - 5x^4 + 12x^3 + 3x^2 + 14x + 8$

                         1 variation: 1 negative real zero

The possibilities are (1): 4 positive and 1 negative zeros
                        (2): 2 positive, 1 negative, and 2 nonreal complex zeros
                        (3): 1 negative and 4 nonreal complex zeros

23. $P(x) = x^7 + x^6 - 4x^5 - 4x^4 + x^3 + x^2 + 6x + 6$

                           2 variations: 0 or 2 positive real zeros

  $P(-x) = (-x)^7 + (-x)^6 - 4(-x)^5 - 4(-x)^4 + (-x)^3 + (-x)^2 + 6(-x) + 6$

        $= -x^7 + x^6 + x^5 - 4x^4 - x^3 + x^2 - 6x + 6$

                         5 variations: 1, 3, or 5 negative real zeros

The possibilities are (1): 2 positive and 5 negative zeros
(2): 2 positive, 3 negative, and 2 nonreal complex zeros
(3): 2 positive, 1 negative, and 4 nonreal complex zeros
(4): 5 negative and 2 nonreal complex zeros
(5): 3 negative and 4 nonreal complex zeros
(6): 1 negative and 6 nonreal complex zeros

25. $P(x) = x^6 + 4x^4 + x^2 + 5$

0 variations: 0 positive real zeros

$P(-x) = (-x)^6 + 4(-x)^4 + (-x)^2 + 5$
$= x^6 + 4x^4 + x^2 + 5$

0 variations: 0 negative real zeros

The only possibility is 6 nonreal complex zeros.

27. Since $-3i$ is a zero, $x - (-3i) = (x + 3i)$ is a factor. Also, its conjugate is $3i$ and the corresponding factor is $x - 3i$. Now we have
$$x^2 + 9 = 0$$
$$(x + 3i)(x - 3i) = 0$$
$$x + 3i = 0, \ x - 3i = 0$$
$$x = -3i, \quad x = 3i$$
Both zeros are $3i, -3i$.

29. Since -2 and $1 + 2i$ are the zeros, then $x - (-2) = x + 2$ and $x - (1 + 2i) = x - 1 - 2i$ are two of the three factors. Also, the conjugate of $1 + 2i$ is $1 - 2i$ and the factor is $x - (1 - 2i) = x - 1 + 2i$. Now that all three factors are known, we have:
$$x^3 + x + 10 = 0$$
$$(x + 2)(x - 1 - 2i)(x - 1 + 2i) = 0$$
$$x + 2 = 0, \ x - 1 - 2i = 0, \ x - 1 + 2i = 0$$
$$x = -2, \quad x = 1 + 2i, \quad x = 1 - 2i$$
The zeros are $-2, 1 + 2i, 1 - 2i$.

An easier solution might be the following. The 3rd degree polynomial has 3 zeros. Since two of them are -2 and $1 + 2i$, the 3rd zero must be the conjugate of $1 + 2i$ which is $1 - 2i$. The three zeros are $-2, 1 + 2i, 1 - 2i$.

31. Since $-2i$ and $4i$ are two zeros, the factors are $x - (-2i) = x + 2i$ and $x - 4i$. Also, the conjugate of $-2i$ is $2i$ and the corresponding factor is $(x - 2i)$. Finally, the conjugate of $4i$ is $-4i$ and the corresponding factor is $x + 4i$. The polynomial equation becomes:
$$x^4 + 20x^2 + 64 = 0$$
$$(x + 2i)(x - 2i)(x + 4i)(x - 4i) = 0$$
$$x + 2i = 0, \ x - 2i = 0, \ x + 4i = 0, \ x - 4i = 0$$
$$x = -2i, \quad x = 2i, \quad x = -4i, \quad x = 4i$$
The zeros are $\pm 2i, \pm 4i$.

An easier solution might be the following. The 4th degree polynomial has 4 zeros. Since two of them are $-2i$ and $4i$, the other two must be their conjugates which are $2i$ and $-4i$. The four zeros are $\pm 2i, \pm 4i$.

33. Since $-2i$ is one zero, its conjugate, $2i$, must be another zero. The first two factors are $x + 2i$ and $x - 2i$. Their product is $(x + 2i)(x - 2i) = x^2 + 4$. Since $P(x)$ is a third degree equation, the last zero can be obtained through the last factor which is obtained by dividing $x^3 + 4x^2 + 4x + 16$ by $x^2 + 4$.

$$
\begin{array}{r}
x + 4 \\
x^2 + 0x + 4 \overline{\smash{\big)}\ x^3 + 4x^2 + 4x + 16} \\
\underline{x^3 + 0x^2 + 4x} \\
4x^2 + 0x + 16 \\
\underline{4x^2 + 0x + 16} \\
0
\end{array}
$$

Thus, the 3 factors are $x + 2i$, $x - 2i$, and $x + 4$. The polynomial equation becomes
$$x^3 + 4x^2 + 4x + 16 = 0$$
$$(x + 2i)(x - 2i)(x + 4) = 0$$
$$x + 2i = 0,\ x - 2i = 0,\ x + 4 = 0$$
$$x = -2i,\qquad x = 2i,\qquad x = -4$$

The zeros are $\pm 2i$ and -4.

35. Since $2 + i$ is a zero, its conjugate, $2 - i$, must be another zero. The first two factors are $x - (2 + i)$ and $x - (2 - i)$.
Their product is
$$[x - (2 + i)][x - (2 - i)]$$
$$= (x - 2 - i)(x - 2 + i)$$
$$= x^2 - 4x + 5$$
Since $P(x)$ is a third degree equation, the last zero is obtained from the last factor which is obtained by dividing $x^3 - 2x^2 - 3x + 10$ by $x^2 - 4x + 5$.

$$
\begin{array}{r}
x + 2 \\
x^2 - 4x + 5 \overline{\smash{\big)}\ x^3 - 2x^2 - 3x + 10} \\
\underline{x^3 - 4x^2 + 5x} \\
2x^2 - 8x + 10 \\
\underline{2x^2 - 8x + 10} \\
0
\end{array}
$$

Thus, the 3 factors are $x - 2 - i$, $x - 2 + i$, and $x + 2$. The polynomial equation becomes
$$x^3 - 2x^2 - 3x + 10 = 0$$
$$(x - 2 - i)(x - 2 + i)(x + 2) = 0$$
$$x - 2 - i = 0,\ x - 2 + i = 0,\ x + 2 = 0$$
$$x = 2 + i,\qquad x = 2 - i,\ x = -2$$

The zeros are $2 + i$, $2 - i$, and -2.

37. Since $-3i$ is a zero, its conjugate, $3i$, is also a zero. The first two factors are $x + 3i$ and $x - 3i$. Their product is $(x + 3i)(x - 3i) = x^2 - (3i)^2 = x^2 - 9(-1) = x^2 + 9$. To obtain the remaining factor(s), we must divide $P(x)$ by $x^2 + 9$.

$$
\begin{array}{r}
x^2 + x + 1 \\
x^2 + 9 \overline{\smash{\big)}\ x^4 + x^3 + 10x^2 + 9x + 9} \\
\underline{x^4 \qquad\ + 9x^2} \\
x^3 + x^2 + 9x \\
\underline{x^3 + \qquad 9x} \\
x^2 + 0x + 9 \\
\underline{x^2 \qquad + 9} \\
0
\end{array}
$$

Thus, the polynomial equation becomes $x^4 + x^3 + 10x^2 + 9x + 9 = 0$
$$(x + 3i)(x - 3i)(x^2 + x + 1) = 0$$

Since $x^2 + x + 1$ is **not** factorable, set each factor equal to 0:
$x + 3i = 0$, $x - 3i = 0$, $x^2 + x + 1 = 0$

$$x = -3i, \quad x = 3i, \quad x = \frac{-1 \pm \sqrt{1^2 - 4(1)(1)}}{2 \cdot 1} = \frac{-1 \pm \sqrt{-3}}{2} = \frac{-1 \pm \sqrt{3}\,i}{2}$$

where the quadratic formula was used to solve the quadratic equation. The zeros
are $-3i$, $3i$, and $\dfrac{-1 \pm \sqrt{3}\,i}{2}$.

39. Since $1 + i$ is a zero, then its conjugate, $1 - i$, is also a zero. The first
two factors are $x - (1 + i)$ and $x - (1 - i)$.
Their product is $[x - (1 + i)][x - (1 - i)] = (x - 1 - i)(x - 1 + i)$
$= x^2 - 2x + 2$. To obtain the remaining factor(s), we must divide
$P(x)$ by $x^2 - 2x + 2$.

$$
\begin{array}{r}
x^2 - 1 \\
x^2 - 2x + 2\overline{\smash{\big)}\,x^4 - 2x^3 + x^2 + 2x - 2} \\
\underline{x^4 - 2x^3 + 2x^2 } \\
-x^2 + 2x - 2 \\
\underline{-x^2 + 2x - 2} \\
0
\end{array}
$$

Thus, the polynomial equation becomes
$$x^4 - 2x^3 + x^2 + 2x - 2 = 0$$
$$(x - 1 - i)(x - 1 + i)(x^2 - 1) = 0$$
$$(x - 1 - i)(x - 1 + i)(x - 1)(x + 1) = 0$$
$$x - 1 - i = 0, \ x - 1 + i = 0, \ x - 1 = 0, \ x + 1 = 0$$
$$x = 1 + i, \quad x = 1 - i, \ x = 1, \quad x = -1$$

The zeros are $1 + i$, $1 - i$, 1, and -1

41. Since $1 + i$ is a zero of multiplicity 2, then its conjugate, $1 - i$, must be
a zero of multiplicity 2. The first four factors are $[x - (1 + i)]^2$ and
$[x - (1 - i)]^2$ which can be expressed as $(x - 1 - i)^2$ and $(x - 1 + i)^2$. Their
product is $(x - 1 - i)^2(x - 1 + i)^2 = x^4 - 4x^3 + 8x^2 - 8x + 4$. To obtain the
last factor, divide $P(x)$ by $x^4 - 4x^3 + 8x^2 - 8x + 4$ to get

$$
\begin{array}{r}
x - 2 \\
x^4 - 4x^3 + 8x^2 - 8x + 4\overline{\smash{\big)}\,x^5 - 6x^4 + 16x^3 - 24x^2 + 20x - 8} \\
\underline{x^5 - 4x^4 + 8x^3 - 8x^2 + 4x } \\
-2x^4 + 8x^3 - 16x^2 + 16x - 8 \\
\underline{-2x^4 + 8x^3 - 16x^2 + 16x - 8} \\
0
\end{array}
$$

The polynomial equation becomes
$$x^5 - 6x^4 + 16x^3 - 24x^2 + 20x - 8 = 0$$
$$(x - 1 - i)^2(x - 1 + i)^2(x - 2) = 0$$
$$(x - 1 - i)^2 = 0, \ (x - 1 + i)^2 = 0, \ x - 2 = 0$$
$$x = 1 + i \text{ (multiplicity 2)}, \ x = 1 - i \text{ (multiplicity 2)}, \ x = 2$$

The zeros are $1 + i$, and $1 - i$ each of multiplicity 2, and 2.

43. The factors are $x - 3$ and $x + 5$. The polynomial is $P(x) = (x - 3)(x + 5)$
$$= x^2 + 2x - 15$$

Also, $Q(x) = (x - 3)(x + 5)$
$$= x^2 + 2x - 15$$

45. For $P(x)$, the factors are $x - 2i$ and $x + i$ so that
$$P(x) = (x - 2i)(x + i) = x^2 - ix + 2$$
For $Q(x)$, the zeros are $2i$, $-2i$ (the conjugate of $2i$), $-i$ and i (the conjugate of $-i$).
The factors are $x - 2i$, $x + 2i$, $x + i$, and $x - i$ so that
$$\begin{aligned} Q(x) &= (x - 2i)(x + 2i)(x + i)(x - i) \\ &= (x^2 + 4)(x^2 + 1) \\ &= x^4 + 5x + 4 \end{aligned}$$

47. For $P(x)$, the factors are $x - 3$ and $x - (2 - i) = x - 2 + i$ so that
$$P(x) = (x - 3)(x - 2 + i) = x^2 + (-5 + i)x + 6 - 3i$$
For $Q(x)$, the zeros are 3, $2 - i$, and $2 + i$ (the conjugate of $2 - i$). The factors are $x - 3$, $x - (2 - i) = x - 2 + i$, and $x - (2 + i) = x - 2 - i$ so that
$$\begin{aligned} Q(x) &= (x - 3)(x - 2 + i)(x - 2 - i) \\ &= (x - 3)(x^2 - 4x + 5) \\ &= x^3 - 7x^2 + 17x - 15 \end{aligned}$$

49. For $P(x)$, the factors are $x - 3$, $x - (1 - i) = x - 1 + i$ and $x - (3 + 2i) = x - 3 - 2i$ so that $P(x) = (x - 3)(x - 1 + i)(x - 3 - 2i)$
$$= x^3 - (7 + i)x^2 + (17 + 2i)x - 15 + 3i$$
For $Q(x)$, the zeros are 3, $1 - i$, $1 + i$ (the conjugate of $1 - i$), $3 + 2i$, and $3 - 2i$ (the conjugate of $3 + 2i$).
The factors are $x - 3$, $x - (1 - i) = x - 1 + i$
$$x - (1 + i) = x - 1 - i$$
$$x - (3 + 2i) = x - 3 - 2i$$
$$x - (3 - 2i) = x - 3 + 2i \quad \text{so that}$$

$$\begin{aligned} Q(x) &= (x - 3)(x - 1 + i)(x - 1 - i)(x - 3 - 2i)(x - 3 + 2i) \\ &= (x - 3)(x^2 - 2x + 2)(x^2 - 6x + 13) \\ &= x^5 - 11x^4 + 51x^3 - 119x^2 + 140x - 78 \end{aligned}$$

51. If all the coefficients are positive, then there is **no** variation in signs which means that there are **no** positive real zeros.

53. If the degree is odd, there can be only an even number of nonreal complex zeros which means there has to be an odd number (at least one) of real zeros.

Exercises 5.4

1. a) To show that $x = 1$ is an upper bound, use synthetic division

```
1| 1   0   0   7  -1   3
        1   1   1   8   7
   ─────────────────────────
   1   1   1   8   7  10
```
→ Since all values are positive, 1 is an upper bound.

b) Use synthetic division

```
-3| 1   0   0    7   -1     3
        -3   9  -27   60  -177
    ───────────────────────────
    1  -3   9  -20   59  -174
```
→ Since the values have alternate signs, -3 is a lower bound.

3. $P(x) = 2x^3 + x^2 - 10x - 4$

```
2 | 2   1   -10   -4          3 | 2   1   -10   -4
  |     4    10    0            |     6    21    33
   ─────────────────            ─────────────────
    2   5    0    -4             2   7   11    29
```

2 is **not** an upper bound since some of the values are negative

3 is an upper bound since all values are positive

```
-2 | 2   1   -10   -4         -3 | 2   1   -10   -4
   |    -4     6    8            |    -6    15   -15
    ─────────────────            ─────────────────
     2  -3   -4     4             2  -5    5    -19
```

-2 is **not** a lower bound since values do not alternate signs

-3 is a lower bound since values alternate signs

5. $P(x) = 2x^3 - 3x^2 + 8x - 13$

```
1 | 2   -3   8   -13       2 | 2   -3   8   -13       -1 | 2   -3    8   -13
  |      2  -1     7         |      4   2    20           |     -2    5   -13
   ──────────────────        ──────────────────           ───────────────────
    2   -1   7    -6          2    1  10     7             2   -5   13   -26
```

1 is **not** an upper bound

2 is an upper bound

-1 is a lower bound for the negative numbers

7. $P(x) = x^4 + x^3 + 2x^2 + 4x - 9$

```
1 | 1   1   2   4   -9        2 | 1   1   2    4   -9
  |     1   1   3    7          |     2   6   16   40
   ──────────────────           ──────────────────────
    1   1   3   7   -2           1   3   8   20   31
```

1 is **not** an upper bound

2 is an upper bound

```
-2 | 1   1   2   4   -9       -3 | 1   1    2    4   -9
   |    -2   2  -8    8          |    -3    6  -24   60
    ──────────────────           ──────────────────────
     1  -1   4  -4   -1           1  -2    8  -20   51
```

-2 is **not** a lower bound

-3 is a lower bound

9. $P(x) = x^4 + x^3 - 5x^2 + x - 5$

```
1 | 1   1   -5    1   -5       2 | 1   1   -5   1   -5
  |     1   2    -3   -2         |     2   6    2    6
   ────────────────────          ──────────────────────
    1   2  -3   -2   -7           1   3   1    3    1
```

1 is **not** an upper bound

2 is an upper bound

```
-2 | 1   1   -5    1    -5      -3 | 1   1   -5    1   -5
   |    -2   2    6   -14          |    -3   6   -3    6
    ─────────────────────          ─────────────────────
     1  -1  -3    7   -19           1  -2   1   -2    1
```

-2 is **not** a lower bound

-3 is a lower bound

11. Rational zeros would be the factors of 3 divided by the factors of 1. The list is {±1, ±3}.

13. Rational zeros would be the factors of 8 divided by the factors of 1. The list is {±1, ±2, ±4, ±8}.

15. Rational zeros would be the factors of 12 divided by the factors of 2. The list is $\left\{\pm 1,\ \pm 2,\ \pm 3,\ \pm 4,\ \pm 6,\ \pm 12,\ \pm\dfrac{1}{2},\ \pm\dfrac{3}{2}\right\}$.

17. Rational zeros would be the factors of 2 divided by the factors of 2. $\left\{\pm 1,\ \pm 2,\ \pm\dfrac{1}{2}\right\}$

19. Rational zeros are the factors of 6 divided by the factors of 4. The list is $\left\{\pm 1,\ \pm 2,\ \pm 3,\ \pm 6,\ \pm\dfrac{1}{2},\ \pm\dfrac{3}{2},\ \pm\dfrac{1}{4},\ \pm\dfrac{3}{4}\right\}$

21. $x^3 - 3x^2 + 4x - 12 = 0$ This can be solved by first factoring the left side.
$$x^2(x - 3) + 4(x - 3) = 0$$
$$(x - 3)(x^2 + 4) = 0$$
$$(x + 3)(x + 2i)(x - 2i) = 0$$
$$x + 3 = 0,\ x + 2i = 0,\ x - 2i = 0$$
$$x = -3,\qquad x = -2i,\qquad x = 2i$$

The zeros are -3, $-2i$, and $2i$.

23. $2x^3 + 7x^2 + 2x - 3 = 0$

The list for rational zeros is $\left\{\pm 1,\ \pm 3,\ \pm\dfrac{1}{2},\ \pm\dfrac{3}{2}\right\}$. Let's try -1 as a zero using synthetic division.

```
-1| 2   7    2   -3
  |    -2   -5    3
   ─────────────────
    2   5   -3    0  ← Remainder
```

Quotient   which is $2x^2 + 5x - 3$

Thus, $x + 1$ is a factor and the original equation becomes
$$2x^3 + 7x^2 + 2x - 3 = 0$$
$$(x + 1)(2x^2 + 5x - 3) = 0$$
$$(x + 1)(2x - 1)(x + 3) = 0$$
$$x + 1 = 0,\ 2x - 1 = 0,\ x + 3 = 0$$
$$x = -1,\qquad x = \frac{1}{2},\qquad x = -3$$

The zeros are -1, $\dfrac{1}{2}$, and -3.

25. $3x^3 - 5x^2 - 4 = 0$

The list for rational zeros is $\left\{\pm 1,\ \pm 2,\ \pm 4,\ \pm\dfrac{1}{3},\ \pm\dfrac{2}{3},\ \pm\dfrac{4}{3}\right\}$. Let's try 2 as a zero using synthetic division.

```
2| 3  -5   0   -4
 |      6   2    4
  ─────────────────
   3   1   2    0  ← Remainder
```

Quotient   which is $3x^2 + x + 2$

Thus, $x - 2$ is a factor and the original equation becomes
$$3x^3 - 5x^2 - 4 = 0$$
$$(x - 2)(3x^2 + x + 2) = 0$$
$$x - 2 = 0,\ 3x^2 + x + 2 = 0$$

$$x = 2, \qquad x = \frac{-1 \pm \sqrt{1^2 - 4(3)(2)}}{2 \cdot 3}$$

$$= \frac{-1 \pm \sqrt{-23}}{6} = \frac{-1 \pm \sqrt{23}\, i}{6}$$

The zeros are 2, $\dfrac{-1 \pm \sqrt{23}\, i}{6}$.

27. $x^4 - x^3 - 2x^2 + 6x - 4 = 0$

The list of rational zeros is $\{\pm 1, \pm 2, \pm 4\}$. Let's try 1 as a rational zero.

```
1│1  -1  -2   6  -4
  │    1   0  -2   4
  ─────────────────────
   1   0  -2   4   0  ← Remainder
   └──────────┘
     Quotient      which is x³ - 2x + 4
```

Thus, $x^4 - x^3 - 2x^2 + 6x - 4 = 0$ becomes $(x - 1)(x^3 - 2x + 4) = 0$.

Now let's try -2 as a rational zero for $x^3 - 2x + 4 = 0$.

```
-2│1   0  -2   4
  │    -2   4  -4
  ─────────────────
   1  -2   2   0  ← Remainder
   └───────┘
    Quotient   which is x² - 2x + 2
```

The final factorization is:
$$(x - 1)(x + 2)(x^2 - 2x + 2) = 0$$
$$x - 1 = 0, \; x + 2 = 0, \; x^2 - 2x + 2 = 0$$
$$x = 1, \qquad x = -2, \; x^2 - 2x + 2 = 0$$

$$x = \frac{-(-2) \pm \sqrt{(-2)^2 - 4(1)(2)}}{2 \cdot 1}$$

$$= \frac{2 \pm \sqrt{-4}}{2} = \frac{2 \pm 2i}{2} = \frac{2(1 \pm i)}{2} = 1 \pm i$$

The zeros are 1, -2, $1 \pm i$.

29. $2x^4 - x^3 - 13x^2 + 5x + 15 = 0$

The list for rational zeros is $\left\{\pm 1, \; \pm 3, \; \pm 5, \; \pm 15, \; \pm \dfrac{1}{2}, \; \pm \dfrac{3}{2}, \; \pm \dfrac{5}{2}, \; \pm \dfrac{15}{2}\right\}$. Let's try -1 as a rational zero.

```
-1│2  -1  -13   5   15
  │    -2    3  10  -15
  ──────────────────────
   2  -3  -10  15    0  ← Remainder
   └──────────────┘
      Quotient which is 2x³ - 3x² - 10x + 15
```

Thus, $2x^4 - x^3 - 13x^2 + 5x + 15 = 0$ becomes $(x + 1)(2x^3 - 3x^2 - 10x + 15) = 0$.

Now, let's try $\dfrac{3}{2}$ as a rational zero for $2x^3 - 3x^2 - 10x + 15 = 0$.

```
3/2│2  -3  -10   15
   │     3    0  -15
   ──────────────────
    2   0  -10    0  ← Remainder
    └──────────┘
     Quotient   which is 2x² - 10
```

The final factorization is:

$$(x + 1)\left(x - \frac{3}{2}\right)(2x^2 - 10) = 0$$

$$(x + 2)\left(x - \frac{3}{2}\right)(2)(x + \sqrt{5})(x - \sqrt{5}) = 0$$

$$x + 2 = 0, \quad x - \frac{3}{2} = 0, \quad x + \sqrt{5} = 0, \quad x - \sqrt{5} = 0$$

$$x = -2, \quad x = \frac{3}{2}, \quad x = -\sqrt{5}, \quad x = \sqrt{5}$$

The zeros are -2, $\frac{3}{2}$, $\pm\sqrt{5}$.

31. $2x^4 + 5x^3 - 7x^2 - 10x + 6 = 0$

The list for rational zeros is $\left\{\pm 1, \pm 2, \pm 3, \pm 6, \pm\frac{1}{2}, \pm\frac{3}{2}\right\}$. Let's try -3 as a rational zero.

```
-3 | 2    5    -7   -10    6
   |     -6     3    12   -6
   ───────────────────────────
     2   -1    -4     2    0  ← Remainder
```

Quotient which is $2x^3 - x^2 - 4x + 2$

Thus, $2x^4 + 5x^3 - 7x^2 - 10x + 6 = 0$ becomes $(x + 3)(2x^3 - x^2 - 4x + 2) = 0$.

Now, let's try $\frac{1}{2}$ as a rational zero for $2x^3 - x^2 - 4x + 2 = 0$.

```
1/2 | 2   -1   -4    2
    |      1    0   -2
    ──────────────────
      2    0   -4    0  ← Remainder
```

Quotient which is $2x^2 - 4$

The final factorization is:

$$(x + 3)\left(x - \frac{1}{2}\right)(2x^2 - 4) = 0$$

$$(x + 3)\left(x - \frac{1}{2}\right)(2)(x + \sqrt{2})(x - \sqrt{2}) = 0$$

$$x + 3 = 0, \quad x - \frac{1}{2} = 0, \quad x + \sqrt{2} = 0, \quad x - \sqrt{2} = 0$$

$$x = -3, \quad x = \frac{1}{2}, \quad x = -\sqrt{2}, \quad x = \sqrt{2}$$

The zeros are -3, $\frac{1}{2}$, $\pm\sqrt{2}$.

33. The list for rational zeros is $\{\pm 1, \pm 2, \pm 3, \pm 6\}$. None of these values produces a remainder of 0 upon using synthetic division. Thus, there are **no** rational zeros.

35. The list for rational zeros is $\{\pm1, \pm2, \pm4, \pm8\}$. Now 1 is a rational zero since a remainder of 0 is obtained:

$$1\overline{)\begin{array}{rrrr} 1 & 1 & 2 & 4 & -8 \\ & 1 & 2 & 4 & 8 \\ \hline 1 & 2 & 4 & 8 & 0 \end{array}} \leftarrow \text{Remainder}$$

$\underbrace{\qquad\qquad}$

Quotient which is $x^3 + 2x^2 + 4x + 8$

Also, -2 is a rational zero since a remainder of 0 is obtained:

$$-2\overline{)\begin{array}{rrrr} 1 & 2 & 4 & 8 \\ & -2 & 0 & -8 \\ \hline 1 & 0 & 4 & 0 \end{array}} \leftarrow \text{Remainder}$$

The rational zeros are 1 and -2. The last quotient is $x^2 + 4$ and the equation $x^2 + 4 = 0$ has no rational zeros.

37. The list for rational zeros is $\left\{\pm1, \pm2, \pm\dfrac{1}{3}, \pm\dfrac{2}{3}\right\}$.

Now, -2 is a rational zero since a remainder of 0 is obtained:

$$-2\overline{)\begin{array}{rrr} 3 & 8 & 3 & -2 \\ & -6 & -4 & 2 \\ \hline 3 & 2 & -1 & 0 \end{array}} \leftarrow \text{Remainder}$$

$\underbrace{\qquad\qquad}$

Quotient   which is $3x^2 + 2x - 1$

The original equation, $3x^3 + 8x^2 + 3x - 2$ can be written as:

$$(x + 2)(3x^2 + 2x - 1) = 0$$
$$(x + 2)(3x - 1)(x + 1) = 0$$
$$x + 2 = 0, \quad 3x - 1 = 0, \quad x + 1 = 0$$
$$x = -2, \qquad x = \frac{1}{3}, \qquad x = -1$$

The rational zeros are $-2, \dfrac{1}{3}, -1$.

39. The list for rational zeros is $\left\{-1, -3, -5, -15, -\dfrac{1}{2}, -\dfrac{3}{2}, -\dfrac{5}{2}, -\dfrac{15}{2}\right\}$. Here, only negative values are considered since $P(x) = 2x^5 + 17x^2 + 38x + 15$ has **no** variation in signs. Now, -5 is a rational zero since a remainder of 0 is obtained:

$$-5\overline{)\begin{array}{rrr} 2 & 17 & 38 & 15 \\ & -10 & -35 & -15 \\ \hline 2 & 7 & 3 & 0 \end{array}} \leftarrow \text{Remainder}$$

$\underbrace{\qquad\qquad}$

Quotient   which is $2x^2 + 7x + 3$

Thus, the original equation, $2x^3 + 17x^2 + 38x + 15$ can be written as:
$$(x + 5)(2x^2 + 7x + 3) = 0$$
$$(x + 5)(2x + 1)(x + 3) = 0$$
$$x + 5 = 0, \; 2x + 1 = 0, \; x + 3 = 0$$
$$x = -5, \qquad x = -\frac{1}{2}, \qquad x = -3$$

The rational zeros are -5, $-\frac{1}{2}$, -3.

41. The list for possible rational zeros is $\left\{\pm 1, \; \pm 2, \; \pm 3, \; \pm 6, \; \pm\frac{1}{2}, \; \pm\frac{3}{2}\right\}$.

   2 is a rational zero since a remainder of 0 is obtained:

```
2 | 2  -1  -4  -1  -6
   |     4   6   4   6
     ─────────────────
     2   3   2   3   0  ← Remainder
     └──────────┬──────┘
       Quotient which is 2x³ + 3x² + 2x + 3
```

   Now $-\frac{3}{2}$ is also a rational zero:

```
-3/2 | 2   3   2   3
     |    -3   0  -3
       ──────────────
       2   0   2   0  ← Remainder
       └──────┬───────┘
     Quotient is 2x² + 2 and the equation 2x² + 2 = 0 has no rational zeros.
```

   The rational zeros are 2 and $-\frac{3}{2}$.

43. The list of possible rational zeros is $\left\{-1, \; -2, \; -3, \; -4, \; -6, \; -12, \; -\frac{1}{2}, \; -\frac{3}{2}\right\}$.

   Here only negative values are considered since $P(x) = 2x^4 + 3x^3 + 2x^2 + 11x + 12$ has no variation in signs.

   $-\frac{3}{2}$ is a rational zero since a remainder of 0 is obtained:

```
-3/2 | 2   3   2   11   12
     |    -3   0   -3  -12
       ──────────────────
       2   0   2    8    0  ← Remainder
       └──────┬──────────┘
       Quotient which is 2x³ + 2x + 8
```

   There are no additional rational roots for $2x^3 + 2x + 8 = 0$.

45. The list of possible rational zeros is $\left\{\pm1,\ \pm2,\ \pm3,\ \pm6,\ \pm\dfrac{1}{2},\ \pm\dfrac{3}{2}\right\}$.

2 is a rational zero since a remainder of 0 is obtained:

```
2│2  -1  -8   3   5  -6
 │    4   6  -4  -2   6
 └─────────────────────
  2   3  -2  -1   3   0  ← Remainder
```

Quotient   which is $2x^4 + 3x^3 - 2x^2 - x + 3$, and the equation $2x^4 + 3x^3 - 2x^2 - x + 3 = 0$ has $-\dfrac{3}{2}$ as a zero since a remainder of 0 is obtained:

```
-3/2│2   3  -2  -1   3
    │   -3   0   3  -3
    └──────────────────
     2   0  -2   2   0  ← Remainder
```

Quotient is $2x^3 - 2x + 2$, and the resulting equation is:

$2x^3 - 2x + 2 = 0$ Now ÷ by 2 to have
$x^3 - x + 1 = 0$

The list of possible rational zeros is $\{\pm1\}$ and neither one results in a remainder of 0 using synthetic division. Thus, the only rational zeros are 2 and $-\dfrac{3}{2}$.

47. $P(x) = 3x^4 + 5x^2 + 6$. Since there is **no** variation in signs, there are **no** positive rational zeros.

Now replace x by $-x$ to obtain
$P(-x) = 3(-x)^4 + 5(-x)^2 + 6 = 3x^4 + 5x^2 + 6$
which has **no** variation in signs. Thus, there are **no** negative rational zeros also. Consequently, there are **no** rational zeros for $P(x) = 3x^4 + 5x^2 + 6$.

49. Let $P(x) = x^2 - 3$. The list of possible rational zeros is $\{\pm1, \pm3\}$. However, **none** of these results in a remainder of 0 using synthetic division. Thus, there are **no** rational zeros for $P(x)$.
If $P(x) = 0$, then $x^2 - 3 = 0$
$$x^2 = 3$$
$$x = \pm\sqrt{3}$$
Consequently, $\sqrt{3}$ as well as $-\sqrt{3}$ cannot be rational.

51. The volume of the box is $\ell \cdot w \cdot h$ or
$$V(x) = (3 - 2x)(3 - 2x)(x)$$
$$= 9x - 12x^2 + 4x^3, \text{ where } x \text{ represents the side of a square being cut out of each corner.}$$
Now set $V(x) = 2$ to obtain:
$$2 = 9x - 12x^2 + 4x^3$$
or $\quad 0 = -2 + 9x - 12x^2 + 4x^3$

Now $\dfrac{1}{2}$ is a rational root since 0 is obtained using synthetic division. Thus, $x = \dfrac{1}{2}$ and the corner to be cut out is $\dfrac{1}{2}$ meter by $\dfrac{1}{2}$ meter.

$$\frac{1}{2}\overline{)\begin{array}{rrrr} 4 & -12 & 9 & -2 \end{array}}$$

$$\begin{array}{rrr} & 2 & -5 & 2 \end{array}$$

$$\underbrace{\begin{array}{rrr} 4 & -10 & 4 \end{array}}\quad 0 \leftarrow \text{Remainder}$$

The quotient is $4x^2 - 10x + 4 = 0$. Now $\div$ by 2 to get:
$$2x^2 - 5x + 2 = 0$$
$$(2x - 1)(x - 2) = 0$$
$$2x - 1 = 0, \quad x - 2 = 0$$
$$x = \frac{1}{2}, \qquad x = 2$$

The zeros are $\frac{1}{2}$ (of multiplicity 2) and 2. However, discard 2 since it is impossible to cut more than $\frac{3}{2}$ meter from any corner. The final answer is $\frac{1}{2}$ meter by $\frac{1}{2}$ meter.

53. The volume of the box is Volume = $\ell \cdot w \cdot h$ or $V = x \cdot x \cdot h = x^2 h$ where x is the length of each side of the base and h is the height. Since the volume is fixed at 4, this equation becomes
$$4 = x^2 h$$
But the relationship between x and h is obtained by using the fact that the surface area is 12 sq m. That is,
$$\text{base} + 4 \text{ sides} = 12$$
$$\text{or} \qquad x^2 + 4xh = 12$$
$$4xh = 12 - x^2$$
$$h = \frac{12 - x^2}{4x}$$
Now the equation $4 = x^2 h$ becomes
$$4 = x^2 \left(\frac{12 - x^2}{4x} \right)$$
$$4 = \frac{x(12 - x^2)}{4} \quad \text{Multiply by 4 to get:}$$
$$\text{or} \qquad 16 = x(12 - x^2)$$
$$16 = 12x - x^3$$
$$x^3 - 12x + 16 = 0$$

The list of possible rational zeros is $\pm 1, \pm 2, \pm 4, \pm 8, \pm 16$. Let's try 2.

$$2\overline{)\begin{array}{rrrr} 1 & 0 & -12 & 16 \end{array}}$$

$$\begin{array}{rrr} & 2 & 4 & -16 \end{array}$$

$$\begin{array}{rrrr} 1 & 2 & -8 & 0 \leftarrow \text{Remainder} \end{array}$$

Since the remainder is 0, $x = 2$ is the desired value and the volume of 4 cu meters is obtained when $x = 2$ meters.

55. The equation is $\dfrac{55\pi}{4} = \dfrac{\pi}{4}(36h - h^3)$ Multiply by 4 to have
$$55\pi = \pi(36h - h^3) \quad \text{Now divide by } \pi \text{ to get}$$
$$55 = 36h - h^3$$
$$\text{or} \quad h^3 - 36h + 55 = 0$$

The list of possible rational zeros is ±1, ±5, ±11, ±55. Let's try 5.

$$\begin{array}{r|rrrr} 5 & 1 & 0 & -36 & 55 \\ & & 5 & 25 & -55 \\ \hline & 1 & 5 & -11 & 0 \leftarrow \text{Remainder} \end{array}$$

Quotient is $x^2 + 5x - 11$ and the resulting equation is

$x^2 + 5x - 11 = 0$ Use the quadratic formula to get

$$x = \frac{-5 \pm \sqrt{(5)^2 - 4(1)(-11)}}{2(1)} = \frac{-5 \pm \sqrt{25 + 44}}{2} = \frac{-5 \pm \sqrt{69}}{2}$$

The only meaningful value is $\dfrac{-5 + \sqrt{69}}{2}$. Thus, the two values of h are $h = 5$

and $h = \dfrac{-5 + \sqrt{69}}{2}$.

57. The equation is $1347 = x^3 - 20x^2 + 500x$
    or $x^3 - 20x^2 + 500x - 1347 = 0$
    The list of possible rational zeros is ±1, ±3, ±449, ±1347. Let's try 3.

$$\begin{array}{r|rrrr} 3 & 1 & -20 & 500 & -1347 \\ & & 3 & -51 & 1347 \\ \hline & 1 & -17 & 449 & 0 \leftarrow \text{Remainder} \end{array}$$

Thus, $x = 3$ canoes will produce \$1347.

59. Possible rational zeros:

$$\left\{ \pm 1, \ \pm 3, \ \pm 5, \ \pm 15, \ \pm \frac{1}{2}, \ \pm \frac{3}{2}, \ \pm \frac{5}{2}, \ \pm \frac{15}{2}, \ \pm \frac{1}{4}, \right.$$

$$\left. \pm \frac{3}{4}, \ \pm \frac{5}{4}, \ \pm \frac{15}{4}, \ \pm \frac{1}{8}, \ \pm \frac{3}{8}, \ \pm \frac{5}{8}, \ \pm \frac{15}{8} \right\}$$

Using the graphics calculator, you should discover that the rational zeros
are $\dfrac{5}{4}$ and $-\dfrac{3}{2}$.

61. Possible rational zeros:

$$\left\{ \pm 1, \ \pm 2, \ \pm 5, \ \pm 10, \ \pm \frac{1}{3}, \ \pm \frac{2}{3}, \ \pm \frac{5}{3}, \ \pm \frac{10}{3}, \ \pm \frac{1}{5}, \ \pm \frac{2}{5}, \ \pm \frac{1}{15}, \ \pm \frac{2}{15} \right\}$$

Using the graphics calculator, you should discover that the rational zeros
are $\dfrac{5}{3}$ and $-\dfrac{1}{5}$.

63. When $x = 1$, then $x^n = 1^n = 1$ for n being any whole number. Then the
    polynomial, $P(x)$, results in the sum of the coefficients when $x = 1$. That is
    any term of the form $a_n x^n$ simply becomes a_n when $x = 1$ since $a_n x^n$ becomes
    $a_n(1)^n = a_n(1) = a_n$. If the sum of the coefficients is 0, then $x = 1$ is a
    zero.

Exercises 5.5

1. $P(0) = 2(0)^3 - 7(0)^2 + 0 + 3 = 3 > 0$
   $P(1) = 2(1)^3 - 7(1)^2 + 1 + 3 = -1 < 0$
   Since $P(0) > 0$ and $P(1) < 0$, a zero must occur between $x = 0$ and $x = 1$.

3. $P(1) = 1^4 + 2(1)^3 + (1)^2 - 5 = -1 < 0$
   $P(1.5) = 9.0625 > 0$ as seen below by synthetic division

   $$
   \begin{array}{r|rrrrr}
   1.5 & 1 & 2 & 1 & 0 & -5 \\
       &   & 1.5 & 5.25 & 9.375 & 14.0625 \\
   \hline
       & 1 & 3.5 & 6.25 & 9.375 & 9.0625
   \end{array}
   $$

   Since $P(1) < 0$ and $P(1.5) > 0$, a zero must occur between $x = 1$ and $x = 1.5$.

5. Several answers are possible. One such interval is [1, 1.5]:
   $P(1) = 2(1)^4 + 6(1)^3 + 1^2 - 9(1) - 6 = -6 < 0$
   and $P(1.5) = 13.125 > 0$

   $$
   \begin{array}{r|rrrrr}
   1.5 & 2 & 6 & 1 & -9 & -6 \\
       &   & 3 & 13.5 & 21.75 & 19.125 \\
   \hline
       & 2 & 9 & 14.5 & 12.75 & 13.125
   \end{array}
   $$

   Since $P(1) < 0$ and $P(1.5) > 0$, a zero must occur in the interval [1, 1.5].

7. Several answers are possible. One such interval is [-0.5, -0.4]:
   $P(-0.5) = (-0.5)^3 - 2(-0.5)^2 + (-0.5) + 1 = -0.125 - 0.5 - 0.5 + 1 = -0.125 < 0$
   and $P(-0.4) = (-0.4)^3 - 2(-0.4)^2 + (-0.4) + 1$
   $= -0.064 - 0.32 - 0.4 + 1 = 0.216 > 0$
   Since $P(-0.5) < 0$ and $P(-0.4) > 0$, a zero must occur in the interval [-0.5, -0.4].

Note: Starting with the solution to problem #9, all solutions involving synthetic division with decimals will be rounded off to 3 decimal places. If you carry out your solutions to more decimal places, your values will be slightly different.

9. Observe that $P(2.15) = 0.279 > 0$ and $P(2.25) = -0.156 < 0$, the zero is between 2.15 and 2.25 which, when rounded off to the nearest tenth would be 2.2.

   $$
   \begin{array}{r|rrrr}
   2.15 & 2 & -11 & 15 & -1 \\
        &   & 4.3 & -14.405 & 1.279 \\
   \hline
        & 2 & -6.7 & 0.595 & 0.279
   \end{array}
   \qquad
   \begin{array}{r|rrrr}
   2.25 & 2 & -11 & 15 & -1 \\
        &   & 4.5 & -14.625 & 0.844 \\
   \hline
        & 2 & -6.5 & 0.375 & -0.156
   \end{array}
   $$

11. Observe that $P(0.45) = 0.244 > 0$ and $P(0.55) = -0.481 < 0$, the zero is between 0.45 and 0.55 which, when rounded off to the nearest tenth would be 0.5.

   $$
   \begin{array}{r|rrrr}
   0.45 & 1 & 1 & -9 & 4 \\
        &   & 0.45 & 0.653 & -3.756 \\
   \hline
        & 1 & 1.45 & -8.347 & 0.244
   \end{array}
   \qquad
   \begin{array}{r|rrrr}
   0.55 & 1 & 1 & -9 & 4 \\
        &   & 0.55 & 0.853 & -4.481 \\
   \hline
        & 1 & 1.55 & -8.147 & -0.481
   \end{array}
   $$

13. Observe that $P(2.75) = 2.273 > 0$ and $P(2.65) = -3.581 < 0$, the zero is between 2.65 and 2.75 which, when rounded off to the nearest tenth would be 2.7.

2.75	2	-5	6	-22	7
		5.5	1.375	20.281	-4.727
	2	0.5	7.375	-1.719	2.273

2.65	2	-5	6	-22	7
		5.3	0.795	18.007	-10.581
	2	0.3	6.795	-3.993	-3.581

15. Observe that $P(0.35) = 0.239 > 0$ and $P(0.45) = -0.149 < 0$, the zero is between 0.35 and 0.45 which, when rounded off to the nearest tenth would be 0.4.

0.35	-1	-3	-1	1
		-0.35	-1.173	-0.761
	-1	-3.35	-2.173	0.239

0.45	-1	-3	-1	1
		-0.45	-1.553	-1.149
	-1	-3.45	-2.553	-0.149

17. Observe that $P(1.05) = -0.63 < 0$ and $P(1.15) = 0.237 > 0$, the zero is between 1.05 and 1.15 which, when rounded off to the nearest tenth would be 1.1.

1.05	2	2	-3	-2
		2.1	4.305	1.370
	2	4.1	1.305	-0.63

1.15	2	2	-3	-2
		2.3	4.945	2.237
	2	4.3	1.945	0.237

19. Observe that $P(1.25) = -1.581 < 0$ and $P(1.35) = 2.601 > 0$, the zero is between 1.25 and 1.35 which, when rounded off to the nearest tenth would be 1.3.

1.25	3	6	-2	-10	-5
		3.75	12.188	12.735	3.419
	3	9.75	10.188	2.735	-1.581

1.35	3	6	-2	-10	-5
		4.05	13.578	15.630	7.601
	3	10.05	11.578	5.630	2.601

21. Observe that $P(0.45) = -0.256 < 0$ and $P(0.55) = 0.019 > 0$. The zero is between 0.45 and 0.55 which, when rounded off to the nearest tenth would be 0.5.

0.45	1	1	1	-1
		0.45	0.653	0.744
	1	1.45	1.653	-0.256

0.55	1	1	1	-1
		0.55	0.853	1.019
	1	1.55	1.853	0.019

23. Observe that $P(-0.75) = 0.062 > 0$ and $P(-0.85) = -0.267 < 0$. The zero is between -0.75 and -0.85 which, when rounded off to the nearest tenth would be -0.8.

-0.75	4	4	2	1
		-3	-0.75	-0.938
	4	1	1.25	0.062

-0.85	4	4	2	1
		-3.4	-0.51	-1.267
	4	-0.6	1.49	-0.267

25. Observe that $P(1.05) = -0.587 < 0$ and $P(1.15) = 0.316 > 0$. The zero is between 1.05 and 1.15 which, when rounded off to the nearest tenth would be 1.1.

```
1.05 | 1    2        1       -5           1.15 | 1    2       1       -5
     |      1.05    3.203    4.413             |      1.15   3.623    5.316
       1    3.05    4.203   -0.587              1    3.15   4.623    0.316
```

27. Observe that $P(1.15) = -1.061 < 0$ and $P(1.25) = 0.391 > 0$. The zero is between 1.15 and 1.25 which, when rounded off to the nearest tenth would be 1.2.

```
1.15 | 1    3       3       -10           1.25 | 1    3       3       -10
     |      1.15    4.773    8.939             |      1.25    5.313   10.391
       1    4.15    7.773   -1.061              1    4.25    8.313    0.391
```

29. Observe that $P(-1.45) = 0.198 > 0$ and $P(-1.55) = -0.477 < 0$. The zero is between -1.45 and -1.55 which, when rounded off to the nearest tenth would be -1.5.

```
-1.45 | 1   -1      -3        1          -1.55 | 1   -1      -3        1
      |     -1.45    3.553   -0.802            |     -1.55    3.953   -1.477
        1   -2.45    0.553    0.198             1   -2.55    0.953   -0.477
```

For the next zero, observe that $P(0.25) = 0.203 > 0$ and $P(0.35) = -0.130 < 0$. The zero is between 0.25 and 0.35 which, when rounded off to the nearest tenth is 0.3.

```
0.25 | 1   -1      -3        1           0.35 | 1   -1      -3        1
     |      0.25   -0.188   -0.797            |      0.35   -0.228   -1.130
       1   -0.75   -3.188    0.203             1   -0.65   -3.228   -0.130
```

For the last zero, observe that $P(2.15) = -0.133 < 0$ and $P(2.25) = 0.579 > 0$. The zero is between 2.15 and 2.25 which, when rounded off to the nearest tenth is 2.2.

```
2.15 | 1   -1      -3        1           2.25 | 1   -1      -3        1
     |      2.15    2.473   -1.133            |      2.25    2.813   -0.421
       1    1.15   -0.527   -0.133             1    1.25   -0.187    0.579
```

The zeros are -1.5, 0.3, and 2.2.

31. Observe that $P(-1.45) = 0.008 > 0$ and $P(-1.55) = -0.543 < 0$. The zero is between -1.45 and -1.55 which, when rounded off to the nearest tenth is -1.5.

```
-1.45 | 2    2      -2       -1          -1.55 | 2    2      -2       -1
      |     -2.9    1.305    1.008             |     -3.1    1.705    0.457
        2   -0.9   -0.695    0.008              2   -1.1   -0.295   -0.543
```

For the next zero, observe that $P(-0.35) = -0.141 < 0$ and $P(-0.45) = 0.123 > 0$. The zero is between -0.35 and -0.45 which, when rounded off to the nearest tenth is -0.4.

-0.35	2	2	-2	-1
		-0.7	-0.455	0.859
	2	1.3	-2.455	-0.141

-0.45	2	2	-2	-1
		-0.9	-0.495	1.123
	2	1.1	-2.495	0.123

For the last zero, observe that $P(0.85) = -0.027 < 0$ and $P(0.95) = 0.620 > 0$. This zero is between 0.85 and 0.95 which, when rounded off to the nearest tenth is 0.9.

0.85	2	2	-2	-1
		1.7	3.145	0.973
	2	3.7	1.145	-0.027

0.95	2	2	-2	-1
		1.9	3.705	1.620
	2	3.9	1.705	0.620

The three zeros are -1.5, -0.4, and 0.9.

33. Observe that $P(-2.25) = 0.296 > 0$ and $P(-2.35) = -0.252 < 0$. The zero is between -2.25 and -2.35 which, when rounded off to the nearest tenth is -2.3.

-2.25	1	-1	-15	-17
		-2.25	7.313	17.296
	1	-3.25	-7.687	0.296

-2.35	1	-1	-15	-17
		-2.35	7.873	16.748
	1	-3.35	-7.127	-0.252

For the next zero, observe that $P(-1.45) = -0.402 < 0$ and $P(-1.55) = 0.123 > 0$. The zero is between -1.45 and -1.55 which, when rounded off to the nearest tenth is -1.5.

-1.45	1	-1	-15	-17
		-1.45	3.553	16.598
	1	-2.45	-11.447	-0.402

-1.55	1	-1	-15	-17
		-1.55	3.953	17.123
	1	-2.55	-11.047	0.123

For the last zero, observe that $P(4.75) = -3.638 < 0$ and $P(4.85) = 0.814 > 0$. This zero is between 4.75 and 4.85 which, when rounded off to the nearest tenth is 4.8.

4.75	1	-1	-15	-17
		4.75	17.813	13.362
	1	3.75	2.813	-3.638

4.85	1	-1	-15	-17
		4.85	18.673	17.814
	1	3.85	3.673	0.814

The three zeros are -2.3, -1.5, and 4.8.

35. Let $P(x) = 2x^4 - 8x^3 + x^2 + 4$ Here P has 2 variations in signs which means that it has 0 or 2 positive real zeros.

Now $P(-x) = 2(-x)^4 - 8(-x)^3 + (-x)^2 + 4$
$= 2x^4 + 8x^3 + x^2 + 4$

Here $P(-x)$ has **no** variation in signs which means that it has **no** negative real zeros.

Observe that $P(0.85) = 0.853 > 0$ and $P(0.95) = -0.327 < 0$. the first zero is between 0.85 and 0.95 which, when rounded off to the nearest tenth is 0.9.

0.85	2	-8	1	0	4
		1.7	-5.355	-3.702	-3.147
	2	-6.3	-4.355	-3.702	0.853

0.95	2	-8	1	0	4
		1.9	-5.795	-4.555	-4.327
	2	-6.1	-4.795	-4.555	-0.327

For the next zero, observe that $P(3.75) = -8.304 < 0$ and $P(3.85) = 1.702 > 0$. The zero is between 3.75 and 3.85 which, when rounded off to the nearest tenth is 3.8.

3.75	2	-8	1	0	4
		7.5	-1.875	-3.281	-12.304
	2	-0.5	-0.875	-3.281	-8.304

3.85	2	-8	1	0	4
		7.7	-1.155	-0.597	-2.298
	2	-0.3	-0.155	-0.597	1.702

The two positive real zeros are 0.9 and 3.8. There are no other real zeros.

37. Using a graphics calculator, the only real zero is 0.68.

39. Using a graphics calculator, the two real zeros are 1.00 and -1.22.

41. Using a graphics calculator, the real zeros are -1.00, 0.50, and 3.00.

43. Using a graphics calculator, the real zeros are -2.61, -0.58, and 2.70.

45. Not necessarily. It is possible for $P(x)$ to be negative for x between a and b which means there could be one or more zeros. For example, in the graph below, $P(a) > 0$ and $P(b) > 0$ yet though $P(x)$ has two zeros.

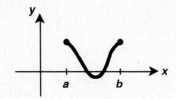

Exercises 5.6

1.

x	y
-3	-24
-2	0
-1	8
0	6
1	0
2	-4
3	0
4	18

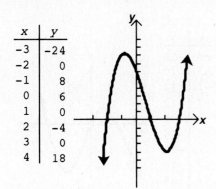

$P(x) = (x - 1)(x + 2)(x - 3)$
The zeros are 1, -2, and 3. Also,
$P(0) = (0 - 1)(0 + 2)(0 - 3) = 6$.

3.

x	y
-1	-4
0	0
$\frac{1}{2}$	$\frac{1}{8}$
1	0
2	2

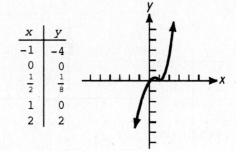

$P(x) = x(x - 1)^2$
The zeros are 0 and 1
(multiplicity 2). At $x = 1$, the
graph is tangent to the x axis.

5.

x	y
-2	-6
-1	0
0	-2
1	0
2	18

$P(x) = -2(x + 1)^2(1 - x)$
The zeros are -1 (multiplicity 2) and 0. At $x = -1$, the graph is tangent to the x axis. Also, $P(0) = -2(0 + 1)^2(1 - 0) = -2$.

7.

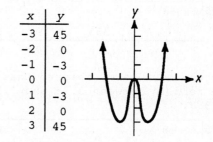

x	y
-3	45
-2	0
-1	-3
0	0
1	-3
2	0
3	45

$P(x) = x^2(x + 2)(x - 2)$
The zeros are -2, 2, and 0 (multiplicity 2). At $x = 0$, the graph is tangent to the x axis.

9.

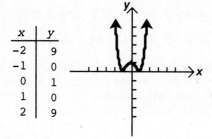

x	y
-2	9
-1	0
0	1
1	0
2	9

$P(x) = (x - 1)^2(x + 1)^2$
The zeros are 1 and -1 both of multiplicity 2. At $x = 1$ and $x = -1$, the graph is tangent to the x axis.
Also, $P(0) = (0 - 1)^2(0 + 1)^2 = 1$.

11.

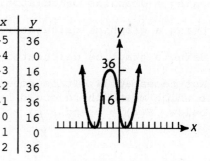

x	y
-5	36
-4	0
-3	16
-2	36
-1	36
0	16
1	0
2	36

$P(x) = (x - 1)^2(x + 4)^2$
The zeros are 1 (multiplicity 2) and -4 (multiplicity 2).
At $x = 1$ and $x = -4$, the graph is tangent to the x axis.
Also, $P(0) = (0 - 1)^2(0 + 4)^2 = 16$.

13.

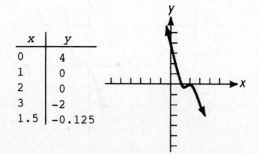

x	y
0	4
1	0
2	0
3	-2
1.5	-0.125

$$P(x) = (2 - x)(x^2 - 3x + 2)$$
$$= (2 - x)(x - 2)(x - 1)$$
$$= -(x - 2)^2(x - 1)$$

The zeros are 1 and 2 (multiplicity 2). At $x = 2$, the graph is tangent to the x axis.
Also, $P(0) = -(0 - 2)^2(0 - 1) = 4$.

15.

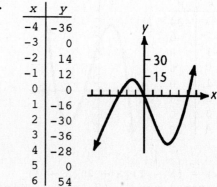

x	y
-4	-36
-3	0
-2	14
-1	12
0	0
1	-16
2	-30
3	-36
4	-28
5	0
6	54

$$P(x) = x^3 - 2x^2 - 15x$$
$$= x(x^2 - 2x - 15)$$
$$= x(x - 5)(x + 3)$$

The zeros are 0, 5, and -3.

17. The zeros are -1, 1, and 3. The factors are $x + 1$, $x - 1$, and $x - 3$. The polynomial is $P(x) = a(x + 1)(x - 1)(x - 3)$.
To find a, use the fact that $P(0) = 3$ to obtain:
$$3 = a(0 + 1)(0 - 1)(0 - 3)$$
$$3 = a(3)$$
$$1 = a$$
The final result is $P(x) = (x + 1)(x - 1)(x - 3)$.

19. The zeros are -1 and 2 of multiplicity 2. The factors are $x + 1$ and $(x - 2)^2$. The polynomial is $P(x) = a(x + 1)(x - 2)^2$.
To find a, use $P(0) = 8$ to obtain:
$$8 = a(0 + 1)(0 - 2)^2$$
$$8 = 4a$$
$$2 = a$$
The final result is $P(x) = 2(x + 1)(x - 2)^2$

21. The zeros are -1, 0, 1, and 3. The factors are $x + 1$, x, $x - 1$, and $x - 3$.
The polynomial is $P(x) = a(x + 1)(x)(x - 1)(x - 3)$.
To find a, use $P(2) = -6$ to obtain:
$$-6 = a(2 + 1)(2)(2 - 1)(2 - 3)$$
$$-6 = a(3)(2)(1)(-1)$$
$$-6 = a(-6)$$
$$1 = a$$
The final result is $P(x) = (x + 1)(x)(x - 1)(x - 3)$
or $P(x) = x(x + 1)(x - 1)(x - 3)$

23. The zeros are -1 of multiplicity 2 and 1 of multiplicity 2. The factors are $(x + 1)^2$ and $(x - 1)^2$. The polynomial is $P(x) = a(x + 1)^2(x - 1)^2$. To find a, use $P(0) = -3$ to obtain:
$$-3 = a(0 + 1)^2(0 - 1)^2$$
$$-3 = a(1)(1)$$
$$-3 = a(1)$$
$$-3 = a$$
The final result is $P(x) = -3(x + 1)^2(x - 1)^2$.

25. $P(x) = (x^2 + 1)(x + 1)(x - 1)$
The zeros are -1 and 1.

x	y
-2	15
-1	0
0	-1
1	0
2	15

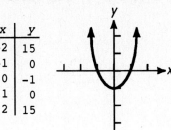

27. $P(x) = -2(x^2 + 1)(x^2 + x - 2)$
$= -2(x^2 + 1)(x + 2)(x - 1)$
The zeros are -2 and 1.

x	y
-2	0
-1	8
0	4
1	0
2	-40

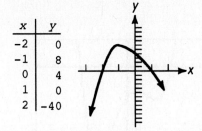

29. $P(x) = (x^2 + x + 1)(x - 2)$
     The only zero is 2.

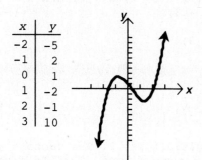

x	y
-2	-12
-1	-3
0	-2
1	-3
2	0
3	13

31. $y = x^3 - 3x + 4$
     The only zero is -2.2 rounded to the nearest tenth.

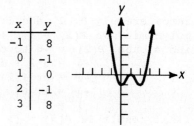

x	y
-3	-14
-2	2
-1	6
0	4
1	2
2	6

33. $y = x^3 - x^2 - 3x + 1$
     The zeros are 2.2, 0.3, and -1.5 rounded to the nearest tenth.

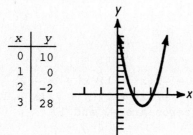

x	y
-2	-5
-1	2
0	1
1	-2
2	-1
3	10

35. $y = x^4 - 4x^3 + 4x^2 - 1$
     The zeros are -0.4, 1, and 2.4.

x	y
-1	8
0	-1
1	0
2	-1
3	8

37. $y = x^4 - 2x^3 + 3x^2 - 12x + 10$
     The zeros are 1 and 2.2 (to the nearest tenth).

x	y
0	10
1	0
2	-2
3	28

39.

viewing rectangle [-5, 5] by [-50, 5]

41.

viewing rectangle [-10, 10] by [-40, 10]

170

43. a) $R(x) = 8x^2 - 0.02x^3$
$$= 2x^2(4 - 0.01x), \text{ for } x \geq 0$$
$R(x) \geq 0$ when $2x^2(4 - 0.01x) \geq 0$
$$\text{or} \quad 4 - 0.01x \geq 0 \quad \text{since } x \geq 0$$
$$4 \geq 0.01x$$
$$400 \geq x$$
Thus, $x \geq 0$ and $x \leq 400$, so the interval is $0 \leq x \leq 400$ or $[0, 400]$

b) The graph is

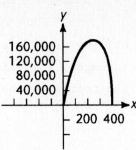

45. $V(x) = x(8 - 2x)(12 - 2x)$, for $x \geq 0$
In order for $V(x) \geq 0$, we need $x \geq 0$
$$8 - 2x \geq 0 \quad \text{or} \quad 4 \geq x$$
$$12 - 2x \geq 0 \quad \text{or} \quad 6 \geq x$$
The intersection of the inequalities is $x \geq 0$, $x \leq 4$, and $x \leq 6$ is $0 \leq x \leq 4$ or $[0, 4]$.

47. $(-\infty, 1.4)$ 49. $(-3.3, -0.7) \cup (2.1, \infty)$ 51. $(1.6, 3.9)$

53. Between $x = a$ and $x = b$, the graph could be

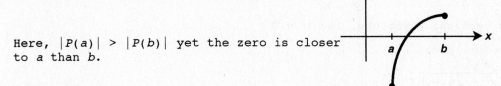

Here, $|P(a)| > |P(b)|$ yet the zero is closer to a than b.

Exercises 5.7

1. $f(x) = \dfrac{-1}{x + 3}$
For the vertical asymptote, let $x + 3 = 0$
$$x = -3$$
For the horizontal asymptote, $y = 0$ since (by the definition) $n < m$; that is $0 < 1$.

3. $f(x) = \dfrac{2x}{x(x + 3)} = \dfrac{2x}{x^2 + 3x}$
For the vertical asymptotes, let $x(x + 3) = 0$
$$x = 0, \quad x + 3 = 0$$
$$x = 0, \qquad x = -3$$
For the horizontal asymptote, $y = 0$ since $n < m$; that is $1 < 2$.

5. $f(x) = \dfrac{(3x + 1)(x - 2)}{(x + 2)^2} = \dfrac{3x^2 - 5x - 2}{x^2 + 4x + 4}$
For the vertical asymptote, let $(x + 2)^2 = 0$
$$x + 2 = 0$$
$$x = -2$$

For the horizontal asymptote, $y = \dfrac{a_n}{b_m} = \dfrac{3}{1} = 3$ since $n = m$; that is $2 = 2$.

7. $f(x) = \dfrac{x^2 - 1}{(x + 3)(2x - 3)} = \dfrac{x^2 - 1}{2x^2 + 3x - 9}$

    For the vertical asymptotes, let $(x + 3)(2x - 3) = 0$
    $$x + 3 = 0, \quad 2x - 3 = 0$$
    $$x = -3, \qquad x = \frac{3}{2}$$

    For the horizontal asymptote, $y = \dfrac{a_n}{b_m} = \dfrac{1}{2}$ since $n = m$; that is $2 = 2$.

9. $f(x) = \dfrac{x^2}{x + 2} = x - 2 + \dfrac{4}{x + 2}$ The oblique asymptote is $y = x - 2$ since when $|x|$ is large, the value of $\dfrac{4}{x + 2}$ is near zero and the points on the graph of f are close to the line $y = x - 2$.

    $$\begin{array}{r} x - 2 \\ x + 2 \overline{)\,x^2 + 0x + 0} \\ \underline{x^2 + 2x} \\ -2x + 0 \\ \underline{-2x - 4} \\ 4 \end{array}$$

    For the vertical asymptote, let $x + 2 = 0$
    $$x = -2$$

11. $f(x) = \dfrac{x^3 + 2x}{x^2} = x + \dfrac{2}{x}$ The oblique asymptote is $y = x$ since when $|x|$ is very large, the value of $\dfrac{2}{x}$ is near zero and the points on the graph of f are close to the line $y = x$.

    For the vertical asymptote, let $x^2 = 0$
    $$x = 0$$

13. $f(x) = \dfrac{x^2 - 1}{x^2 + 2x + 1}$

    For the vertical asymptote, let $x^2 + 2x + 1 = 0$
    $$(x + 1)^2 = 0$$
    $$x + 1 = 0$$
    $$x = -1$$

    For the horizontal asymptote, $y = \dfrac{a_n}{b_m} = \dfrac{1}{1} = 1$ where $n = m$; that is $2 = 2$.

15. $f(x) = \dfrac{x^2 - 5}{x + 5} = \dfrac{\overset{1}{\cancel{(x + 5)}}(x - 5)}{\underset{1}{\cancel{x + 5}}} = x - 5$ and there are **no** asymptotes.

17. $f(x) = \dfrac{2}{x + 4}$

    The vertical asymptote is $x + 4 = 0$ or $x = -4$.
    The horizontal asymptote is $y = 0$.

    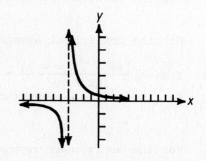

172

19. $f(x) = \dfrac{1}{x^2 - 3x} = \dfrac{1}{x(x - 3)}$

For the vertical asymptotes, let $x(x - 3) = 0$

$$x = 0, \quad x - 3 = 0$$
$$x = 0, \quad x = 3$$

The horizontal asymptote is $y = 0$.

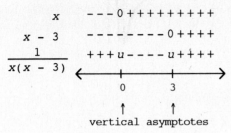

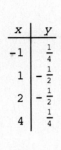

x	y
-1	$\frac{1}{4}$
1	$-\frac{1}{2}$
2	$-\frac{1}{2}$
4	$\frac{1}{4}$

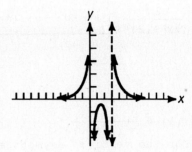

21. $f(x) = \dfrac{1}{(x + 1)^2}$

For the vertical asymptote, let $(x + 1)^2 = 0$

$$x + 1 = 0$$
$$x = -1$$

For the horizontal asymptote, $y = 0$.

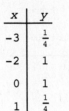

x	y
-3	$\frac{1}{4}$
-2	1
0	1
1	$\frac{1}{4}$

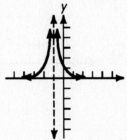

23. $f(x) = \dfrac{1}{x^2 + 4}$

There is no vertical asymptote since there is **no** value of x for which $x^2 + 4 = 0$. The horizontal asymptote is $y = 0$.

x	y
-2	$\frac{1}{8}$
-1	$\frac{1}{5}$
0	$\frac{1}{4}$
1	$\frac{1}{5}$
2	$\frac{1}{8}$

25. $f(x) = \dfrac{2x}{(x + 2)^2}$

For the vertical asymptote, let $(x + 2)^2 = 0$

$$x + 2 = 0$$
$$x = -2$$

173

For the horizontal asymptote, $y = 0$.

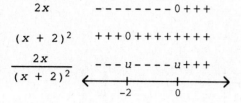

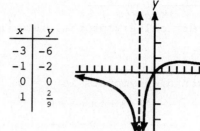

27. $f(x) = \dfrac{1}{x^2(x + 1)}$

For the vertical asymptotes, let $x^2(x + 1) = 0$
$$x^2 = 0, \quad x + 1 = 0$$
$$x = 0, \qquad x = -1$$

For the horizontal asymptote, $y = 0$.

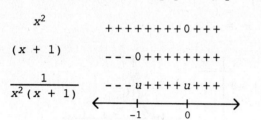

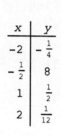

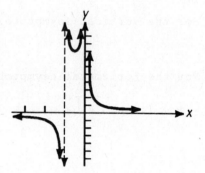

29. $f(x) = \dfrac{2x - 2}{x + 3} = \dfrac{2(x - 1)}{x + 3}$

For the vertical asymptote, let $x + 3 = 0$
$$x = -3$$

For the horizontal asymptote, $y = \dfrac{a_n}{b_m} = \dfrac{2}{1} = 2$. (Here, $m = n = 1$.)

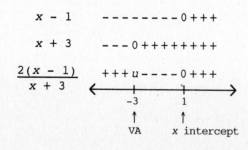

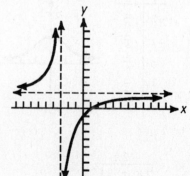

174

31. $f(x) = \dfrac{3x}{x + 2}$

For the vertical asymptote, let $x + 2 = 0$

$$x = -2$$

For the horizontal asymptote, $y = \dfrac{a_n}{b_m} = \dfrac{3}{1} = 3$. (Here, $m = n = 1$.)

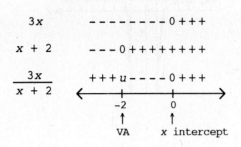
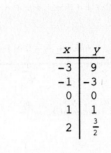
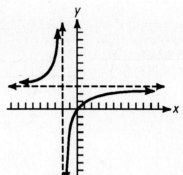

33. $f(x) = \dfrac{x^2 - x - 2}{x^2 - x} = \dfrac{(x - 2)(x + 1)}{x(x - 1)}$

For the vertical asymptotes, let $x(x - 1) = 0$

$$x = 0, \quad x - 1 = 0$$
$$x = 0, \qquad x = 1$$

For the horizontal asymptote, $y = \dfrac{a_n}{b_m} = \dfrac{1}{1} = 1$. (Here, $m = n = 2$.)

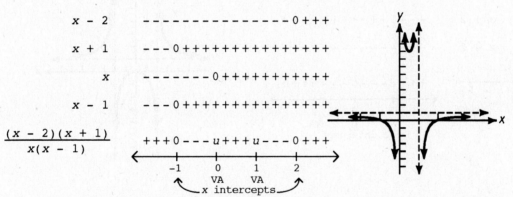

35. $f(x) = \dfrac{x^2 - 3}{2x^2 - x - 3} = \dfrac{(x + \sqrt{3})(x - \sqrt{3})}{(2x - 3)(x + 1)}$

For the vertical asymptotes, let $(2x - 3)(x + 1) = 0$

$$2x - 3 = 0, \quad x + 1 = 0$$
$$x = \dfrac{3}{2}, \qquad x = -1$$

For the horizontal asymptote, $y = \dfrac{a_n}{b_m} = \dfrac{1}{2}$. (Here, $m = n = 2$.)

175

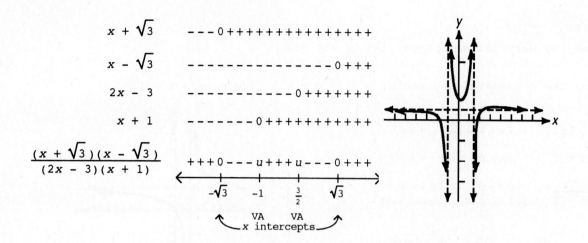

$x + \sqrt{3}$ $---0+++++++++++++$

$x - \sqrt{3}$ $---------------0+++$

$2x - 3$ $-----------0++++++$

$x + 1$ $-------0++++++++++$

$$\frac{(x + \sqrt{3})(x - \sqrt{3})}{(2x - 3)(x + 1)}$$ $+++0---u+++u---0+++$

$-\sqrt{3}$ -1 $\frac{3}{2}$ $\sqrt{3}$

VA   VA

x intercepts

37. $f(x) = \dfrac{x^2}{x^2 - 4} = \dfrac{x^2}{(x + 2)(x - 2)}$

For the vertical asymptotes, let $(x + 2)(x - 2) = 0$

$x + 2 = 0, \quad x - 2 = 0$

$x = -2, \qquad x = 2$

For the horizontal asymptote, $y = \dfrac{a_n}{b_m} = \dfrac{1}{1} = 1$. (Here, $m = n = 2$.)

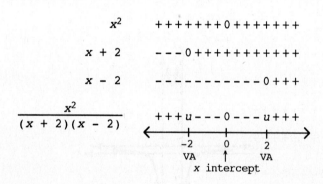

x^2 $+++++++0+++++++$

$x + 2$ $---0++++++++++++$

$x - 2$ $-------------0+++$

$$\frac{x^2}{(x + 2)(x - 2)}$$ $+++u---0---u+++$

-2 0 2

VA $\uparrow$ VA

x intercept

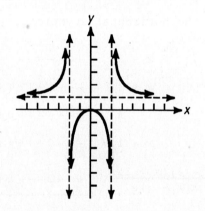

39. $f(x) = \dfrac{x^2 - 4}{x^2 - 1} = \dfrac{(x + 2)(x - 2)}{(x + 1)(x - 1)}$

For the vertical asymptotes, let $(x + 1)(x - 1) = 0$

$x + 1 = 0, \quad x - 1 = 0$

$x = -1, \qquad x = 1$

For the horizontal asymptote, $y = \dfrac{a_n}{b_m} = \dfrac{1}{1} = 1$. (Here, $m = n = 2$.)

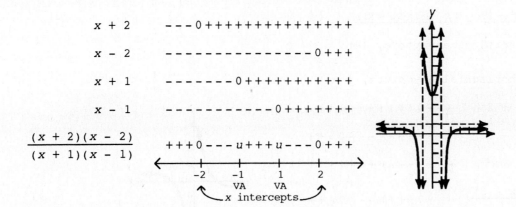

```
x + 2        - - - 0 + + + + + + + + + + + + + + +

x - 2        - - - - - - - - - - - - - - 0 + + +

x + 1        - - - - - - - 0 + + + + + + + + + +

x - 1        - - - - - - - - - - - 0 + + + + + +

(x + 2)(x - 2)   + + + 0 - - - u + + + u - - - 0 + + +
──────────────
(x + 1)(x - 1)   ←─────┬─────┬─────┬─────┬─────→
                      -2    -1     1     2
                            VA    VA
                        ↖           ↗
                         └ x intercepts ┘
```

41. $f(x) = \dfrac{x^2 - 1}{x^2(x + 2)} = \dfrac{(x + 1)(x - 1)}{x^2(x + 2)}$

    For the vertical asymptotes, let $x^2(x + 2) = 0$

    $x^2 = 0$, $x + 2 = 0$

    $x = 0$, $\quad x = -2$

    For the horizontal asymptote, $y = 0$.

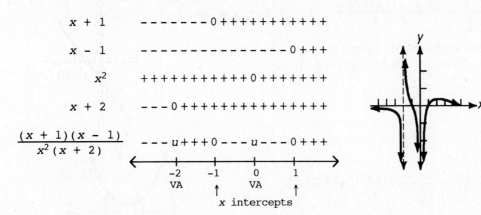

```
x + 1          - - - - - - - 0 + + + + + + + + + + +

x - 1          - - - - - - - - - - - - - - - 0 + + +

x²             + + + + + + + + + + + 0 + + + + + + +

x + 2          - - - 0 + + + + + + + + + + + + + + +

(x + 1)(x - 1)  - - - u + + + 0 - - - u - - - 0 + + +
──────────────
x²(x + 2)       ←────┬─────┬─────┬─────┬────→
                    -2    -1     0     1
                    VA          VA
                          ↑            ↑
                          x intercepts
```

43. $f(x) = \dfrac{x^2 + 4}{x - 2} = x + 2 + \dfrac{8}{x - 2}$

    For the vertical asymptote,
    let $\quad x - 2 = 0$
    $\quad\quad\quad x = 2$

    The oblique asymptote is $y = x + 2$

    since as $|x|$ becomes large, $\dfrac{8}{x - 2}$

    becomes 0.

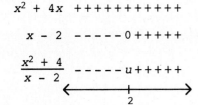

```
x² + 4x    + + + + + + + + + +

x - 2      - - - - - 0 + + + +

x² + 4     - - - - - u + + + +
─────
x - 2      ←──────┬──────→
                  2
```

x	y
0	-2
1	-5
2	u
3	13
4	10

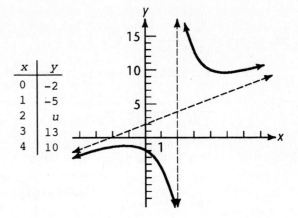

177

45. $f(x) = \dfrac{x^2 - 9}{x^3} = \dfrac{(x + 3)(x - 3)}{x^3}$

For the vertical asymptote, let $x^3 = 0$
$x = 0$

For the horizontal asymptote, $y = 0$

$x + 3$ $- - - 0 + + + + + + + + + +$

$x - 3$ $- - - - - - - - - - - - 0 + + +$

x^3 $- - - - - - - 0 + + + + + + +$

$\dfrac{(x + 3)(x - 3)}{x^3}$ $- - - 0 + + + u - - - 0 + + +$

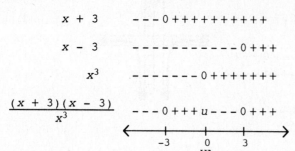

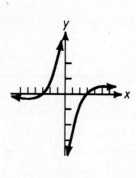

47. $f(x) = \dfrac{x^2 - 4}{x + 2} = \dfrac{(x + 2)(x - 2)}{x + 2} = x - 2$ for $x \neq -2$

The graph is a straight line, $y = x - 2$, except
for $x = -2$. Thus, a 'hole' occurs at $(-2, -4)$.

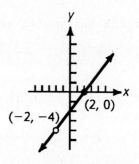

49. $f(x) = \dfrac{x^3 - 27}{x - 3} = \dfrac{(x - 3)(x^2 + 3x + 9)}{x - 3}$
$= x^2 + 3x + 9$ for $x \neq 3$

The graph is a parabola, $y = x^2 + 3x + 9$, except
for $x = 3$. Thus, a 'hole' occurs at $(3, 27)$.

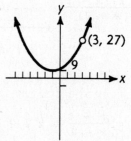

51. $D(x) = \dfrac{25x}{x + 12}$ for $x \geq 0$

The horizontal asymptote is $y = \dfrac{a_n}{b_m} = \dfrac{25}{1} = 25.$

x	y
0	0
1	$\frac{25}{13}$
10	$\frac{250}{22} \approx 11.4$
50	$\frac{1250}{62} \approx 20.2$
100	$\frac{2500}{112} \approx 22.3$

53. $C(x) = \dfrac{15x}{100 - x}$ for $0 \le x \le 100$

For the vertical asymptote, let $100 - x = 0$
$$x = 100$$

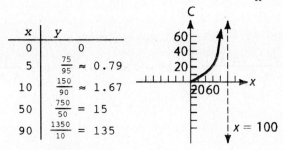

x	y
0	0
5	$\frac{75}{95} \approx 0.79$
10	$\frac{150}{90} \approx 1.67$
50	$\frac{750}{50} = 15$
90	$\frac{1350}{10} = 135$

55.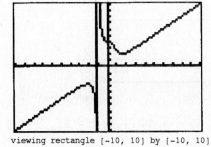

viewing rectangle [-10, 10] by [-10, 10]

57.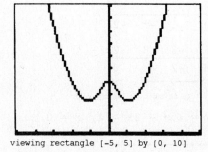

viewing rectangle [-5, 5] by [0, 10]

59. a)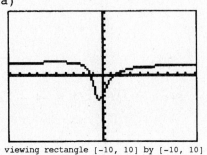

viewing rectangle [-10, 10] by [-10, 10]

b)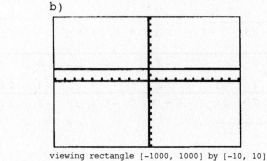

viewing rectangle [-1000, 1000] by [-10, 10]

61. If a rational function contains only even powers of x, then the graph of the function is symmetric about the y axis since $R(-x) = R(x)$.

CHAPTER REVIEW

1.
```
  3│ 2  -1   -4  -30
   │     6   15   33
   └────────────────
     2   5   11    3
```
Quotient Remainder

$$\dfrac{2x^3 - x^2 - 4x - 30}{x - 3} = 2x^2 + 5x + 11 + \dfrac{3}{x - 3}$$

$Q(x) = 2x^2 + 5x + 11, \ r = 3$

2.
```
-2 | 3   5   0   7
   |    -6   2  -4
   |_____
     3  -1   2   3
```
Quotient   Remainder

Thus, $\dfrac{3x^3 + 5x^2 + 7}{x + 2} = 3x^2 - x + 2 + \dfrac{3}{x + 2}$

$Q(x) = 3x^2 - x + 2, \ r = 3$

3.
```
-1 | 2   0   6   -2   1
   |    -2   2   -8  10
   |_____
     2  -2   8  -10  11
```
Quotient   Remainder

$\dfrac{2x^4 + 6x^2 - 2x + 1}{x + 1} = 2x^3 - 2x^2 + 8x - 10 + \dfrac{11}{x + 1}$

$Q(x) = 2x^3 - 2x^2 + 8x - 10, \ r = 11$

4.
```
-1/2 | 6   1   -5   -2
     |    -3    1    2
     |_____
       6  -2   -4    0
```
Quotient   Remainder

$Q(x) = 6x^2 - 2x - 4, \ r = 0$

5.
```
-i | 1     -1        3       -1      3
   |      -i    -1 + i    1 - 2i   -2
   |_____
     1  -1 - i   2 + i     -2i      1
```
Quotient   Remainder

$Q(x) = x^3 - (1 + i)x^2 + (2 + i)x - 2i, \ r = 1$

6.
```
2 | 4   5   -3   -40
  |     8   26    46
  |_____
    4  13   23     6   ← P(c) = P(2) = 6
```

7.
```
1 + i | 2      0       -3       4        0
      |     2 + 2i     4i    -7 + i   -4 - 2i
      |_____
        2   2 + 2i  -3 + 4i  -3 + i  -4 - 2i  ← P(c) = P(1 + i) = -4 - 2i
```

8.
```
2 | 1  -2   2   -4
  |     2   0    4
  |_____
    1   0   2    0   ← Since P(2) = 0, then x = c = 2 is a zero of P(x).
```

9.
```
-3 | 1   0  -7    6
   |    -3   9   -6
   |_____
     1  -3   2    0   ← Since P(-3) = 0, then x = c = -3 is a zero of P(x).
```

10.
```
4 | 3   0  -14     8
  |    12   48   136
  |_____
    3  12   34   144   ← Since P(4) = 144 ≠ 0, then x = c = 4 is not a zero of
                          P(x).
```

11.
```
2 | 2  -5   1   10
  |     4  -2   -2
  |_____
    2  -1  -1    8   ← Since P(2) = 8 ≠ 0, then x - 2 is not a factor of P(x).
```

12. $-i \mid \begin{array}{cccc} 2 & -4 & 2 & -4 \end{array}$

$\underline{\begin{array}{ccc} -2i & -2 + 4i & 4 \end{array}}$

$\begin{array}{cccc} 2 & -4 - 2i & 4i & 0 \end{array}$ ← Since $P(-i) = 0$, then $x + i$ is a factor of $P(x)$.

13. $\frac{1}{4} \mid \begin{array}{cccc} 4 & 7 & -62 & 15 \end{array}$

$\phantom{\frac{1}{4} \mid}\underline{\begin{array}{ccc} 1 & 2 & -15 \end{array}}$

$\phantom{\frac{1}{4} \mid}\underbrace{\begin{array}{ccc} 4 & 8 & -60 \end{array}}\ \ 0$

This gives
$$4x^2 + 8x - 60 = 0$$
$$4(x^2 + 2x - 15) = 0$$
$$4(x + 5)(x - 3) = 0$$
$$x + 5 = 0,\ x - 3 = 0$$
$$x = -5,\qquad x = 3$$
The other two zeros are -5 and 3.

14. Observe that $P(2) = -12$ and $P(3) = 20$ are of opposite signs so that one zero is between $x = 2$ and $x = 3$.

$2 \mid \begin{array}{ccccc} 1 & 1 & -9 & -7 & 14 \end{array} \qquad 3 \mid \begin{array}{ccccc} 1 & 1 & -9 & -7 & 14 \end{array}$

$\underline{\begin{array}{cccc} 2 & 6 & -6 & -26 \end{array}} \qquad\ \ \underline{\begin{array}{cccc} 3 & 12 & 9 & 6 \end{array}}$

$\begin{array}{ccccc} 1 & 3 & -3 & -13 & -12 \end{array}$ ← $P(2)$ $\qquad \begin{array}{ccccc} 1 & 4 & 3 & 2 & 20 \end{array}$ ← $P(3)$

15. Observe that $P(1) = -10$ and $P(2) = 10$ are of opposite signs so that one zero is between $x = 1$ and $x = 2$.

$1 \mid \begin{array}{ccccc} 1 & 2 & -1 & -6 & -6 \end{array} \qquad 2 \mid \begin{array}{ccccc} 1 & 2 & -1 & -6 & -6 \end{array}$

$\underline{\begin{array}{cccc} 1 & 3 & 2 & -4 \end{array}} \qquad\ \underline{\begin{array}{cccc} 2 & 8 & 14 & 16 \end{array}}$

$\begin{array}{ccccc} 1 & 3 & 2 & -4 & -10 \end{array} \qquad\ \begin{array}{ccccc} 1 & 4 & 7 & 8 & 10 \end{array}$

Also, observe that $P(-1) = -2$ and $P(-2) = 6$ are of opposite signs so that another zero is between $x = -1$ and $x = -2$.

$-1 \mid \begin{array}{ccccc} 1 & 2 & -1 & -6 & -6 \end{array} \qquad -2 \mid \begin{array}{ccccc} 1 & 2 & -1 & -6 & -6 \end{array}$

$\underline{\begin{array}{cccc} -1 & -1 & 2 & 4 \end{array}} \qquad\ \underline{\begin{array}{cccc} -2 & 0 & 2 & 12 \end{array}}$

$\begin{array}{ccccc} 1 & 1 & -2 & -4 & -2 \end{array} \qquad\ \begin{array}{ccccc} 1 & 0 & -1 & -6 & 6 \end{array}$

16. The factors are $(x - 2)^2$ and $[x - (-1)]^3 = (x + 1)^3$. The polynomial is $P(x) = (x - 2)^2 (x + 1)^3$.

17. If the zeros are 2, -1, and 3, then the factors are $(x - 2)$, $(x + 1)$, and $(x - 3)$ and the polynomial is $P(x) = a(x - 2)(x + 1)(x - 3)$.

Since $P(1) = -16$, this gives
$$-16 = a(1 - 2)(1 + 1)(1 - 3)$$
$$-16 = a(-1)(2)(-2)$$
$$-16 = a(4)$$
$$-\frac{16}{4} = a$$
$$-4 = a$$
Thus, $P(x) = -4(x - 2)(x + 1)(x - 3)$.

18. $P(x) = 5x^3 - x^2 + 4x + 9$ has two variations in signs. $P(x)$ has 2 or 0 positive real zeros.

Now, $P(-x) = 5(-x)^3 - (-x)^2 + 4(-x) + 9$

$= -5x^3 - x^2 - 4x + 9$ has 1 variation in sign. $P(x)$ has 1 negative real zero.

The possibilities are (1): 2 positive and 1 negative zeros
(2): 1 negative real and 2 nonreal complex zeros

19. $P(x) = 2x^4 + x^3 - 4x - 3$ has one variation in signs so that it has one positive zero

Now, $P(-x) = 2(-x)^4 + (-x)^3 - 4(-x) - 3$

$= 2x^4 - x^3 + 4x - 3$ has three variations in signs so that it has 1 or 3 negative zeros

The possibilities are (1): 1 positive, 3 negative zeros
(2): 1 positive, 1 negative, 2 nonreal complex zeros

20. $P(x) = 3x^4 - 5x^3 - 7x^2 - 4x + 6$ has 2 variations in signs. $P(x)$ has 2 or 0 positive real zeros

Now, $P(-x) = 3(-x)^4 - 5(-x)^3 - 7(-x)^2 - 4(-x) + 6$

$= 3x^4 + 5x^3 - 7x^2 + 4x + 6$ has 2 variations in sign $P(x)$ has 2 or 0 negative real zeros.

The possibilities are (1): 2 positive and 2 negative zeros
(2): 2 positive and 2 nonreal complex zeros
(3): 2 negative and 2 nonreal complex zeros
(4) 4 nonreal complex zeros

21. Since $1 - i$ is a zero, then its conjugate, $1 + i$, must also be a zero. For the 3rd degree polynomial, $P(x)$, the three zeros are 3, $1 - i$, and $1 + i$.

22. Since -2 is a zero, $x - (-2) = x + 2$ is one factor of $P(x)$.

$$
\begin{array}{r|rrrr}
-2 & 1 & -2 & -3 & 10 \\
   &   & -2 & 8 & -10 \\
\hline
   & 1 & -4 & 5 & 0 \leftarrow \text{Remainder}
\end{array}
$$

Quotient is $x^2 - 4x + 5$

The remaining two zeros come from $x^2 - 4x + 5 = 0$

Use the quadratic formula to get $x = \dfrac{-(-4) \pm \sqrt{(-4)^2 - 4(1)(5)}}{2(1)}$

$= \dfrac{4 \pm \sqrt{16 - 20}}{2} = \dfrac{4 \pm \sqrt{-4}}{2} = \dfrac{4 \pm 2i}{2} = 2 \pm i$

The other two zeros are $2 + i$ and $2 - i$.

23. Since $-i$ is a zero, then its conjugate, i, must also be a zero. For the 4th degree polynomial, two other zeros must be found. The first two factors are $x + i$ and $x - i$ and $(x + i)(x - i) = x^2 + 1$. Now,

$$
\begin{array}{r}
x^2 - x - 2 \\
x^2 + 1\overline{\smash{)}x^4 - x^3 - x^2 - x - 2} \\
\underline{x^4 \phantom{{}- x^3} + x^2} \\
-x^3 - 2x^2 - x \\
\underline{-x^3 \phantom{{}- 2x^2} - x} \\
-2x^2 \phantom{{}- x} - 2 \\
\underline{-2x^2 \phantom{{}- x} - 2} \\
0
\end{array}
$$

Thus, $P(x) = x^4 - x^3 - x^2 - x - 2$
$= (x^2 + 1)(x^2 - x - 2)$
$= (x^2 + 1)(x - 2)(x + 1)$

The other two zeros occur when $(x - 2)(x + 1) = 0$
$x - 2 = 0, \quad x + 1 = 0$
$x = 2, \qquad x = -1$
The four zeros are $-i,\ i,\ 2,\ -1$.

24. Since the zeros are 4 and $2i$, the factors are $x - 4$ and $x - 2i$ and the polynomial is $P(x) = (x - 4)(x - 2i)$
$= x^2 - 2ix - 4x + 8i$
$= x^2 - (4 + 2i)x + 8i$

25. Since $3i$ is a zero, then $-3i$ is also a zero. The four zeros are $3i$, $-3i$, -1, and -2. The factors are $x - 3i$, $x + 3i$, $x + 1$, and $x + 2$. The polynomial is:
$P(x) = (x - 3i)(x + 3i)(x + 1)(x + 2)$
$= (x^2 + 9)(x + 1)(x + 2)$
$= (x^3 + x^2 + 9x + 9)(x + 2)$
$= x^4 + 3x^3 + 11x^2 + 27x + 18$

26. a) Observe that
$$
\begin{array}{r}
2\,\vert\ \ 2 \quad 4 \quad -3 \quad -6 \\
\ \ 4 \quad 16 \quad 26 \\
\hline
2 \quad 8 \quad 13 \quad 20
\end{array}
\qquad \text{and} \qquad
\begin{array}{r}
1\,\vert\ \ 2 \quad 4 \quad -3 \quad -6 \\
\ \ 2 \quad 6 \quad 3 \\
\hline
2 \quad 6 \quad 3 \quad -3
\end{array}
$$
All positive values   Not all positive values

Here, $x = 2$ is an upper bound for the zeros.

b) Observe that
$$
\begin{array}{r}
-2\,\vert\ \ 2 \quad 4 \quad -3 \quad -6 \\
\ \ -4 \quad 0 \quad 6 \\
\hline
2 \quad 0 \quad -3 \quad 0
\end{array}
\qquad \text{and} \qquad
\begin{array}{r}
-3\,\vert\ \ 2 \quad 4 \quad -3 \quad -6 \\
\ \ -6 \quad 6 \quad -9 \\
\hline
2 \quad -2 \quad 3 \quad -15
\end{array}
$$
Alternating signs

Here, $x = -3$ is a lower bound for the zeros due to alternating signs upon using synthetic division.

27. The possibilities for rational zeros are ±6, ±3, ±2, ±1, $\pm\frac{1}{2}$, $\pm\frac{3}{2}$.

    Now -2 is a zero since $P(-2) = 0$.

    Also, $P(x) = 2x^3 + 4x^2 - 3x - 6 = (x + 2)(2x^2 - 3)$ ← No rational zeros from last factor.

    ```
    -2│2   4  -3  -6
      │   -4   0   6
       ─────────────
       2   0  -3   0  ← Remainder
    ```
    $Q(x) = 2x^2 - 3$

    Thus, -2 is the **only** rational zero for $P(x)$.

28. The list of possible rational zeros is ±1, ±2, ±3, ±6, $\pm\frac{1}{5}$, $\pm\frac{2}{5}$, $\pm\frac{3}{5}$, $\pm\frac{6}{5}$.

    Notice that none of these will produce a zero since $5x^6 > 0$ and $8x^4 > 0$ for all values (real numbers) of x. Hence, $P(x) = 5x^6 + 8x^4 + 6 > 0$ for all values of x. There are **no** rational zeros of $P(x)$.

29. The possibilities for rational zeros are ±2, ±1, $\pm\frac{1}{3}$, $\pm\frac{2}{3}$.

    Let's try 1 using synthetic division.

    ```
    1│3   2  -7   2
     │    3   5  -2
      ─────────────
      3   5  -2   0  ← Remainder
    ```
    $Q(x) = 3x^2 + 5x - 2$

    Since the remainder is 0, $x = 1$ is a zero and $x - 1$ is a factor.
    Thus, $3x^3 + 2x^2 - 7x + 2 = 0$ becomes $(x - 1)(3x^2 + 5x - 2) = 0$
    $$(x - 1)(3x - 1)(x + 2) = 0$$
    $$x - 1 = 0, \ 3x - 1 = 0, \ x + 2 = 0$$
    $$x = 1, \qquad x = \frac{1}{3}, \qquad x = -2$$

    Thus, the three zeros are 1, $\frac{1}{3}$, -2.

30. The list of possible rational zeros is ±1, ±2, ±5, ±10. Let's try 5 using synthetic division:

    ```
    5│1  -6  +3  +10
     │    5  -5  -10
      ──────────────
      1  -1  -2    0  ← Remainder
    ```
    $Q(x) = x^2 - x - 2$

    Since the remainder is 0, $x = 5$ is a zero and $x - 5$ is a factor.
    Thus, $x^3 - 6x^2 + 3x + 10 = 0$ becomes
    $$(x - 5)(x^2 - x - 2) = 0$$
    $$(x - 5)(x - 2)(x + 1) = 0$$
    $$x - 5 = 0, \ x - 2 = 0, \ x + 1 = 0$$
    $$x = 5, \qquad x = 2, \qquad x = -1$$
    The three zeros are 5, 2, and -1.

31.

$$x^3 - x^2 - x - 2 = 0$$
$$(x - 2)(x^2 + x + 1) = 0$$
$$x - 2 = 0, \quad x^2 + x + 1 = 0$$
$$x = 2, \qquad x = \frac{-1 \pm \sqrt{1^2 - 4(1)(1)}}{2(1)}$$
$$= \frac{-1 \pm \sqrt{-3}}{2}$$
$$= \frac{-1 \pm i\sqrt{3}}{2}$$

The three zeros are 2, $\dfrac{-1 \pm i\sqrt{3}}{2}$.

$$\begin{array}{r|rrrr} 2 & 1 & -1 & -1 & -2 \\ & & 2 & 2 & 2 \\ \hline & 1 & 1 & 1 & 0 \end{array}$$

$$Q(x) = x^2 + x + 1$$

32. The list of possible rational zeros is ± 1, ± 2, ± 4, ± 8, ± 16. Let's try 4 using synthetic division:

$$\begin{array}{r|rrrrr} 4 & 1 & -1 & -12 & -4 & 16 \\ & & 4 & 12 & 0 & -16 \\ \hline & 1 & 3 & 0 & -4 & 0 \leftarrow \text{Remainder} \end{array}$$

Quotient: $x^3 + 3x^2 - 4$.

Since the remainder is 0, $x = 4$ is a zero and $x - 4$ is a factor. Additional zeros are obtained from $x^3 + 3x^2 - 4 = 0$. Let's try 1 using synthetic division:

$$\begin{array}{r|rrrr} 1 & 1 & 3 & 0 & -4 \\ & & 1 & 4 & 4 \\ \hline & 1 & 4 & 4 & 0 \leftarrow \text{Remainder} \end{array}$$

Since the remainder is zero, $x = 1$ is a zero and $x - 1$ is a factor. Thus, the original equation $x^4 - x^3 - 12x^2 - 4x + 16 = 0$ becomes:

$$(x - 4)(x^3 + 3x^2 - 4) = 0$$
$$(x - 4)(x - 1)(x^2 + 4x + 4) = 0$$
$$(x - 4)(x - 1)(x + 2)(x + 2) = 0$$
$$x - 4 = 0, \quad x - 1 = 0, \quad x + 2 = 0, \quad x + 2 = 0$$
$$x = 4, \qquad x = 1, \qquad x = -2, \qquad x = -2$$

The four zeros are 4, 1, -2, -2.

33.

$$P(x) = 3x^3 - 7x^2 + 8x - 2$$
$$= \left(x - \frac{1}{3}\right)(3x^2 - 6x + 6)$$
$$= \left(x - \frac{1}{3}\right)(3)(x^2 - 2x + 2)$$
$$x - \frac{1}{3} = 0, \quad x^2 - 2x + 2 = 0$$
$$x = \frac{1}{3}, \qquad x = \frac{-(-2) \pm \sqrt{(-2)^2 - 4(1)(2)}}{2(1)}$$
$$= \frac{2 \pm \sqrt{4 - 8}}{2}$$
$$= \frac{2 \pm \sqrt{-4}}{2} = \frac{2 \pm 2i}{2} = 1 \pm i$$

$$\begin{array}{r|rrrr} \frac{1}{3} & 3 & -7 & 8 & -2 \\ & & 1 & -2 & 2 \\ \hline & 3 & -6 & 6 & 0 \end{array}$$

$$Q(x) = 3x^2 - 6x + 6$$

The three zeros are $\dfrac{1}{3}$, $1 \pm i$.

34. The list of possible rational zeros is ±1, ±2, ±4, ±5, ±10, ±20, $\pm\frac{1}{3}$, $\pm\frac{2}{3}$, $\pm\frac{4}{3}$, $\pm\frac{5}{3}$, $\pm\frac{10}{3}$, $\pm\frac{20}{3}$, $\pm\frac{1}{9}$, $\pm\frac{2}{9}$, $\pm\frac{4}{9}$, $\pm\frac{5}{9}$, $\pm\frac{10}{9}$, $\pm\frac{20}{9}$. Let's try -2 using synthetic division:

```
-2 | 9   27    8   -20
   |     -18  -18   20
   -------------------------
     9    9   -10    0  ← Remainder
```

Since the remainder is 0, $x = -2$ is a zero and $x + 2$ is a factor. Now, the original polynomial equation $P(x) = 0$ becomes:

$$9x^3 + 27x^2 + 8x - 20 = 0$$
$$(x + 2)(9x^2 + 9x - 10) = 0$$
$$(x + 2)(3x - 2)(3x + 5) = 0$$
$$x + 2 = 0, \quad 3x - 2 = 0, \quad 3x + 5 = 0$$
$$x = -2, \qquad x = \frac{2}{3}, \qquad x = -\frac{5}{3}$$

The three zeros are -2, $\frac{2}{3}$, and $-\frac{5}{3}$.

35. $P(x) = 3x^3 + 19x^2 + 30x + 8$
    $\quad = (x + 2)(3x^2 + 13x + 4)$
    $\quad = (x + 2)(3x + 1)(x + 4)$

```
-2 | 3   19   30    8
   |     -6  -26   -8
   --------------------
     3   13    4    0
```

$$Q(x) = 3x^2 + 13x + 4$$
$$x + 2 = 0, \quad 3x + 1 = 0, \quad x + 4 = 0$$
$$x = -2, \qquad x = -\frac{1}{3}, \qquad x = -4.$$

The zeros are -2, $-\frac{1}{3}$, and -4.

36. Observe that $P(1.1) = -0.7749 < 0$ and $P(1.2) = 0.7216 > 0$. Then $P(x)$ must have a zero in the interval $(1.1, 1.2)$. The length of this interval is 0.1.

```
1.1 | 1   1     3      -2       -5
    |     1.1   2.31   5.841    4.2251
    -----------------------------------
      1   2.1   5.31   3.841   -0.7749
```

```
1.2 | 1   1     3      -2       -5
    |     1.2   2.64   6.768    5.7216
    -----------------------------------
      1   2.2   5.64   4.768    0.7216
```

37. Observe that $P(-0.55) = -0.3145 < 0$ and $P(-0.65) = 0.3795 > 0$. The zero is between -0.55 and -0.65 which, when rounded off, is -0.6.

```
-0.55 | 2    0       1        -4        -3
      |      -1.10   0.605    -0.8828   2.6855
      -------------------------------------------
        2   -1.10    1.605    -4.8828  -0.3145
```

```
-0.65 | 2    0       1        -4        -3
      |      -1.30   0.845    -1.1993   3.3795
      -------------------------------------------
        2   -1.30    1.845    -5.1993   0.3795
```

38. $P(x) = 2x^3 + 5x + 2$ has **no** variation in sign. Thus, it has **no** positive real zero. Also, $P(-x) = 2(-x)^3 + 5(-x) = 2$
    $\qquad\qquad\qquad\qquad = -2x^3 - 5x + 2$ has one variation in sign. Thus, it has one negative real zero.

Observe that $P(-0.35) = 0.1642 > 0$ and $P(-0.45) = -0.4323 < 0$. The zero between -0.35 and -0.45 which, when rounded off, is -0.4.

```
-0.35 | 2    0      5       2
      |      -0.7   0.245  -1.8358
      --------------------------------
        2   -0.7    5.245   0.1642
```

```
-0.45 | 2    0      5       2
      |      -0.9   0.405  -2.4323
      --------------------------------
        2   -0.9    5.405  -0.4323
```

39. $P(x) = x(x + 2)(x - 3)$
   $P(x) = 0$ when $x(x + 2)(x - 3) = 0$
   $\quad\quad\quad x = 0, \quad x + 2 = 0, \quad x - 3 = 0$
   $\quad\quad\quad x = 0, \quad\quad x = -2, \quad\quad x = 3$

   The zeros are 0, -2, and 3.

x	y
-3	-18
-2	0
-1	4
0	0
1	-6
2	-8
3	0

   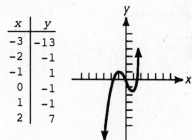

40. $P(x) = -x^2(x^2 + 1)$

   The only zero occurs when $x^2 = 0$
   $\quad\quad\quad\quad\quad\quad\quad$ or $x = 0$.

x	y
-2	-20
-1	-2
0	0
1	-2
2	-20

   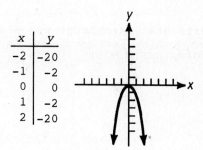

41. $P(x) = x^3 + x^2 - 2x - 1$
   The zeros are -1.8, -0.4, and 1.2.

x	y
-3	-13
-2	-1
-1	1
0	-1
1	-1
2	7

42. Vertical asymptote: $x - 1 = 0$
   $\quad\quad\quad\quad\quad$ or $\quad x = 1$
   Horizontal asymptote: $y = 0$ since (by definition) $n < m$, that is $0 < 1$

43. Horizontal asymptote: $y = \dfrac{a_n}{b_m} = \dfrac{1}{1} = 1$
   Vertical asymptote: $x + 2 = 0$
   $\quad\quad\quad\quad\quad\quad\quad x = -2$

44. Vertical asymptotes: $x - 1 = 0, \quad x + 2 = 0$
   $\quad\quad\quad\quad\quad\quad$ or $\quad x = 1, \quad\quad x = -2$
   Horizontal asymptote: $y = 0$ since (by definition) $n < m$, that is $1 < 2$

45. Express $f(x) = \dfrac{-x^2}{x - 3}$ as $-x - 3 - \dfrac{9}{x - 3}$:

$$
\begin{array}{r}
-x - 3 \\
x - 3 \overline{\smash{)}-x^2 + 0x + 0} \\
\underline{-x^2 + 3x} \\
-3x + 0 \\
\underline{-3x + 9} \\
-9
\end{array}
$$

   For the vertical asymptote: $x - 3 = 0$
   $\quad\quad\quad\quad\quad\quad\quad\quad\quad x = 3$
   For the oblique asymptote: $y = -x - 3$

46. Vertical asymptote: $x + 1 = 0$
$\qquad\qquad\qquad$ or $\quad x = -1$

Horizontal asymptote: $y = \dfrac{a_n}{b_m} = \dfrac{2}{1} = 2$

(Here, $n = m = 1$)

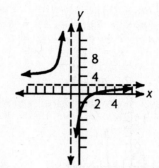

47. No horizontal asymptote.

Vertical asymptote: $x = 0$

$$f(x) = \frac{x^2 + 1}{x} = \frac{x^2}{x} + \frac{1}{x} = x + \frac{1}{x}$$

The oblique asymptote is $y = x$.

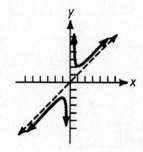

48. $f(x) = \dfrac{x^2 - 1}{x + 1} = \dfrac{(x + 1)(x - 1)}{x + 1} = x - 1$
for $x \neq -1$.

The graph is a straight line with a hole at $(-1, -2)$.

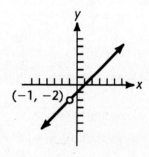

Exercises 6.1

1. $f(0) = \left(\dfrac{3}{2}\right)^0 = 1$

3. $f(2) = \left(\dfrac{3}{2}\right)^2 = \dfrac{9}{4}$

5. $f(-1) = \left(\dfrac{3}{2}\right)^{-1} = \dfrac{2}{3}$

7. $f\left(-\dfrac{1}{2}\right) = \left(\dfrac{3}{2}\right)^{-1/2} = \sqrt{\dfrac{2}{3}} = \dfrac{\sqrt{2}}{\sqrt{3}} \cdot \dfrac{\sqrt{3}}{\sqrt{3}} = \dfrac{\sqrt{6}}{3}$

9. $3^x = 27$
   $3^x = 3^3$ Now equate exponents
   $x = 3$

11. $2^x = \dfrac{1}{16}$
    $2^x = \dfrac{1}{2^4}$
    $2^x = 2^{-4}$ Now equate exponents
    $x = -4$

13. $2^x = 1$
    $2^x = 2^0$ Now equate exponents
    $x = 0$

15. $\left(\dfrac{1}{2}\right)^x = 8$
    $\left(\dfrac{1}{2}\right)^x = 2^3$
    $2^{-x} = 2^3$ Now equate exponents
    $-x = 3$
    $x = -3$

17. $\left(\dfrac{2}{3}\right)^x = \left(\dfrac{9}{4}\right)$
    $\left(\dfrac{2}{3}\right)^x = \left(\dfrac{3}{2}\right)^2$
    $\left(\dfrac{2}{3}\right)^x = \left(\dfrac{2}{3}\right)^{-2}$ Now equate exponents
    $x = -2$

19. $8^x = \dfrac{1}{16}$
    $2^{3x} = \dfrac{1}{2^4}$
    $2^{3x} = 2^{-4}$ Now equate exponents
    $3x = -4$
    $x = -\dfrac{4}{3}$

21. $5^{-x} = 625$
    $5^{-x} = 5^4$ Equate exponents
    $-x = 4$
    $x = -4$

23. $3^{2x-1} = \dfrac{1}{81}$
    $3^{2x-1} = 3^{-4}$ Equate exponents
    $2x - 1 = -4$
    $2x = -3$
    $x = -\dfrac{3}{2}$

25. $(81)^{2x+1} = \dfrac{1}{3}$
    $(81)^{2x+1} = 3^{-1}$
    $[3^4]^{2x+1} = 3^{-1}$
    $3^{8x+4} = 3^{-1}$ Equate exponents
    $8x + 4 = -1$
    $8x = -5$
    $x = -\dfrac{5}{8}$

27.

x	y
0	1
1	3
2	9
-1	$\frac{1}{3}$
-2	$\frac{1}{9}$

29.

x	y
0	1
1	$\frac{1}{3}$
2	$\frac{1}{9}$
-1	3
-2	9

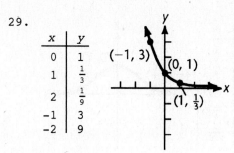

31.

x	y
0	-1
1	-2
2	-4
-1	$-\frac{1}{2}$
-2	$-\frac{1}{4}$

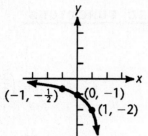

33.

x	y
0	-1
1	-3
2	-9
-1	$-\frac{1}{3}$
-2	$-\frac{1}{9}$

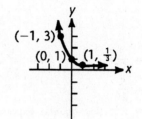

35.

x	y
0	-1
1	$-\frac{1}{3}$
2	$-\frac{1}{9}$
-1	-3
-2	-9

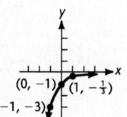

37.

x	y
0	1
1	$\frac{1}{3}$
2	$\frac{1}{9}$
-1	3
-2	9

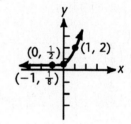

39.

x	y
0	1
1	$\frac{2}{3}$
2	$\frac{4}{9}$
-1	$\frac{3}{2}$
-2	$\frac{9}{4}$

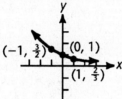

41.

x	y
0	$\frac{1}{2}$
1	2
2	8
-1	$\frac{1}{8}$
-2	$\frac{1}{32}$

43.

x	y
0	2
1	10
2	50
-1	$\frac{2}{5}$
-2	$\frac{2}{25}$

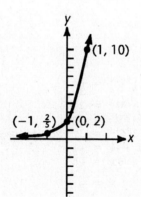

45.

x	y
0	2
1	$\frac{6}{5}$
2	$\frac{26}{25}$
-1	6
-2	26

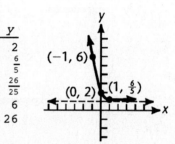

47.

x	y
0	1
1	$\frac{1}{4}$
2	$\frac{1}{256}$
-1	$\frac{1}{4}$
-2	$\frac{1}{256}$

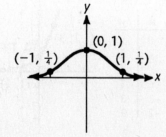

49. $P = 16,000 \cdot 2^{t/10}$ Let $t = 40$:
$$P = 16,000 \cdot 2^{40/10}$$
$$= 16,000 \cdot 2^4$$
$$= 16,000 \cdot 16$$
$$= 256,000$$

To double, let $P = 32,000$:
$$32,000 = 16,000 \cdot 2^{t/10}$$
$$\frac{32,000}{16,000} = 2^{t/10}$$

$2 = 2^{t/10}$ Equate exponents: $1 = \frac{t}{10}$
$$10 = t$$

The time is $t = 10$ yrs.

51. Use $A = P\left(1 + \dfrac{r}{n}\right)^{tn}$ where $P = 20,000$, $r = 12\% = .12$, $n = 4$, and $t = 8$:

$$A = 20,000\left(1 + \dfrac{.12}{4}\right)^{8\cdot4}$$
$$= 20,000(1 + .03)^{32}$$
$$= 20,000(1.03)^{32}$$
$$= 20,000(2.5750828)$$
$$= \$51,501.66$$

53. Use $A = P\left(1 + \dfrac{r}{n}\right)^{tn}$ where $P = 42,000$, $r = .09$, $n = 12$, and $t = 5$:

$$A = 42,000\left(1 + \dfrac{.09}{12}\right)^{5\cdot12}$$
$$= 42,000(1 + .0075)^{60}$$
$$= 42,000(1.0075)^{60}$$
$$= 42,000(1.5656810)$$
$$= \$65,758.60$$

Now, the interest =
$\$65,758.60 - \$42,000 = \$23,758.60$

55. $A = C(1 - r)^t$ where $C = 5000$, $r = 10\% = .10$, and $t = 4$:
$$A = 5000(1 - .10)^4$$
$$= 5000(.90)^4$$
$$= 5000(.6561)$$
$$= \$3280.50$$

57. $N = 2000 \cdot 2^{t/2}$ where $t = 24$:
$$N = 2000 \cdot 2^{24/2}$$
$$= 2000 \cdot 2^{12}$$
$$= 2000 \cdot 4096$$
$$= 8,192,000$$

59. $A(t) = 1000\left(\dfrac{1}{2}\right)^t$

61.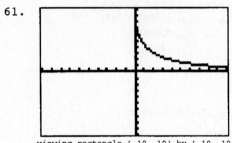

viewing rectangle [-10, 10] by [-10, 10]

63.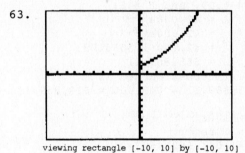

viewing rectangle [-10, 10] by [-10, 10]

65. -0.8 67. $(2.81, \infty)$ 69. $(0.31, 4.00)$

71. The graph of $y = 3^x$ is given in figure 6.1. Notice that the graph is increasing throughout its domain. The graph of $y = b$ (for $b > 0$) is a horizontal line which intersects the graph in only one point. Thus, equation y values produces the equation $3^x = b$ which has only one solution.

Exercises 6.2

1.

x	y
-5	0.6
0	1
5	1.6
10	2.7

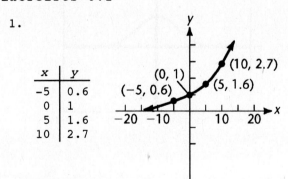

3.

x	y
-5	2.7
-2	1.5
0	1
5	0.4

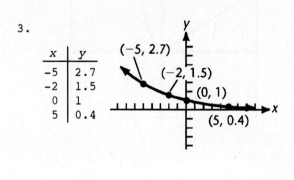

5.

x	y
-1	1.4
0	2
1	3.7
2	8.4

(2, 8.4)
(1, 3.7)
(−1, 1.4)
(0, 2)

7.

x	y
0	0.4
1	1
2	2.7
3	7.4

(3, 7.4)
(2, 2.7)
(0, 0.4)
(1, 1)

9.

x	y
-2	3.8
-1	1.5
0	1
1	1.5
2	3.8

(−2, 3.8)
(2, 3.8)
(−1, 1.5)
(1, 1.5)
(0, 1)

11. Use $A = Pe^{rt}$ where $P = 20{,}000$,
$r = 0.12$ and $t = 8$.

$A = 20{,}000e^{0.12(8)}$
$= 20{,}000e^{.96}$
$= 20{,}000(2.611696473)$
$= \$52{,}233.93$

13. Use $A = Pe^{rt}$ where $P = 42{,}000$,
$r = 0.09$, and $t = 5$:
$A = 42{,}000e^{0.09(5)}$
$= 42{,}000e^{.45}$
$= 42{,}000(1.5683129)$
$= \$65{,}869.11$
The interest earned is
$\$65{,}869.11 - \$42{,}000 = \$23{,}869.11$

15. Use $A = A_0e^{-0.000124t}$ where
$t = 1988 - 1390 = 598$:
$A = A_0e^{-0.000124(598)}$
$= A_0e^{-0.074152}$
$= A_0(.9285)$
Thus, 92.85% of ^{14}C remains.

17. $A = 100e^{-0.000411(1000)}$
$= 100e^{-0.411}$
$= 100(.66298693)$
$= 66.30$ milligrams rounded to
    two decimal places

19. $A = A_0e^{-0.000124(3000)}$
$= A_0e^{-0.372}$
$= A_0(.68935424)$
Thus, 68.94% remains

21. $N = 60e^{-0.163(2)}$
$= 60e^{-0.326}$
$= 60(.72180519)$
$= 43$ rounded to the nearest integer

23.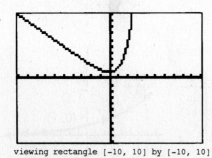

viewing rectangle [−10, 10] by [−10, 10]

25.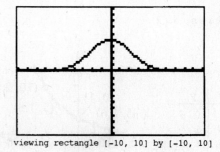

viewing rectangle [−10, 10] by [−10, 10]

27. 1.9

29. Comparing both graphs on the graphics calculator, the graphs are virtually identical.

31. a) For $f(x) = e^x$, the line $y = 0$ is a horizontal asymptote which is a feature not found for any polynomial function.

b) The graph of the polynomial function of odd degree, $y = P(x)$, will always possess an x intercept, whereas the graph of $f(x) = e^x$ does not.

Exercises 6.3

1. $4^2 = 16$ is expressed as $\log_4 16 = 2$ 3. $3^4 = 81$ is expressed as $\log_3 81 = 4$

5. $3^{-2} = \dfrac{1}{9}$ is expressed as $\log_3 \dfrac{1}{9} = -2$ 7. $10^{-2} = \dfrac{1}{100}$ is expressed as $\log_{10} \dfrac{1}{100} = -2$

9. $5^0 = 1$ is expressed as $\log_5 1 = 0$ (11.) $4^{-1/2} = \dfrac{1}{2}$ is expressed as $\log_4 \dfrac{1}{2} = -\dfrac{1}{2}$

13. $\log_2 64 = 6$ is expressed as $2^6 = 64$ 15. $\log_3 \dfrac{1}{27} = -3$ is expressed as $3^{-3} = \dfrac{1}{27}$

(17.) $\log_{3/4} \dfrac{9}{16} = 2$ is expressed as $\left(\dfrac{3}{4}\right)^2 = \dfrac{9}{16}$

19. $\log_{10}(0.01) = -2$ is expressed as $(10)^{-2} = 0.01$

21.

x	y
1	0
3	1
9	2
$\frac{1}{3}$	-1
$\frac{1}{9}$	-2

23.

x	y
1	0
$\frac{1}{3}$	1
$\frac{1}{9}$	2
3	-1
9	-2

25.

x	y
1	0
8	1
64	2
$\frac{1}{8}$	-1
$\frac{1}{64}$	-2

27.

x	y
-3	0
-1	1
5	2
$-\frac{11}{3}$	-1
$-\frac{35}{9}$	-2

29. $y = \log_2 1$ is expressed as $2^y = 1$
    or $y = 0$

31. $y = \log_2 8$ is expressed as $2^y = 8$
    $2^y = 2^3$
    $y = 3$

33. $y = \log_{1/3} \dfrac{1}{9}$ is expressed as $\left(\dfrac{1}{3}\right)^y = \dfrac{1}{9}$
    $\left(\dfrac{1}{3}\right)^y = \left(\dfrac{1}{3}\right)^2$
    $y = 2$

(35.) $2 = \log_3 x$ is expressed as $3^2 = x$
    $9 = x$

37. $\log_3 x = -4$ is expressed as $3^{-4} = x$ 39. $y = \log_{1/10} 100$ is expressed as $\left(\dfrac{1}{10}\right)^y = 100$

$$\dfrac{1}{81} = x$$

$$10^{-y} = 10^2$$
$$-y = 2$$
$$y = -2$$

41. $y = \log_{1/3} 9$ is expressed as $\left(\dfrac{1}{3}\right)^y = 9$ 43. $3 = \log_a 8$ is expressed as $a^3 = 8$

$$3^{-y} = 3^2$$
$$-y = 2$$
$$y = -2$$

$$a^3 = 2^3$$
$$a = 2$$

45. $\log_a 125 = -3$ is expressed as $a^{-3} = 125$

$$a^{-3} = 5^3$$

$$\text{Express } 5 = \left(\dfrac{1}{5}\right)^{-1}: \quad a^{-3} = \left(\dfrac{1}{5}\right)^{-3}$$

$$a = \dfrac{1}{5}$$

47. $y = \log_5 \dfrac{1}{125}$ is expressed as $5^y = \dfrac{1}{125}$ 49. $y = \log_2 4$ is expressed as $2^y = 4$

$$5^y = 5^{-3}$$
$$y = -3$$

$$2^y = 2^2$$
$$y = 2$$

51. $\log_{10} 15 = \log_{10}(3 \cdot 5) = \log_{10} 3 + \log_{10} 5$ 53. $\log_{10} 27 = \log_{10} 3^3 = 3 \log_{10} 3$

55. $\log_{10}(5)^{1/2} = \dfrac{1}{2} \log_{10} 5$ 57. $\log_{10} \dfrac{3}{5} = \log_{10} 3 - \log_{10} 5$

59. $\log_{10} \dfrac{9}{25} = \log_{10} 9 - \log_{10} 25 = \log_{10} 3^2 - \log_{10} 5^2 = 2 \log_{10} 3 - 2 \log_{10} 5$

61. $\log_a xy = \log_a x + \log_a y$ 63. $\log_a \dfrac{x^2 y^3}{z^4} = \log_a x^2 y^3 - \log_a z^4$

$$= \log_a x^2 + \log_a y^3 - \log_a z^4$$
$$= 2 \log_a x + 3 \log_a y - 4 \log_a z$$

65. $\log_a \dfrac{x^3}{y^5 z^2} = \log_a x^3 - \log_a y^5 z^2$ 67. $\log_a \dfrac{\sqrt[5]{xz^2}}{y} = \log_a (xz^2)^{1/5} - \log_a y$

$$= \log_a x^3 - \log_a y^5 - \log_a z^2$$

$$= \dfrac{1}{5} \log_a xz^2 - \log_a y$$

$$= 3 \log_a x - 5 \log_a y - 2 \log_a z$$

$$= \dfrac{1}{5} \log_a x + \dfrac{2}{5} \log_a z - \log_a y$$

69. $\log_a \sqrt{\dfrac{xy^4}{z^3}} = \log_a \left(\dfrac{xy^4}{z^3}\right)^{1/2} = \dfrac{1}{2} \log_a \dfrac{xy^4}{z^3}$

$$= \dfrac{1}{2} [\log_a xy^4 - \log_a z^3]$$

$$= \dfrac{1}{2} [\log_a x + \log_a y^4 - \log_a z^3]$$

$$= \dfrac{1}{2} [\log_a x + 4 \log_a y - 3 \log_a z]$$

$$= \dfrac{1}{2} \log_a x + 2 \log_a y - \dfrac{3}{2} \log_a z$$

71. $4 \log_{10} 5 - 2 \log_{10} 5 + 2 = \log_{10} 5^4 - \log_{10} 5^2 + \log_{10} 10^2$

use $2 = \log_{10} 10^2$

$= \log_{10} \dfrac{(5^4)(10^2)}{5^2} = \log_{10} (5^2)(10^2) = \log_{10} 2500$

73. $4 \log_{10} 2 + \log_{10} 2 - 1 = \log_{10} 2^4 + \log_{10} 2 - \log_{10} 10^1$

use $1 = \log_{10} 10^1$

$= \log_{10} \left(\dfrac{2^4 \cdot 2}{10}\right) = \log_{10} \dfrac{2^5}{10} = \log_{10} \dfrac{2^4}{5} = \log_{10} \dfrac{16}{5} = \log_{10} 3.2$

75. $3 \log_{10} 4 - \log_{10} 4 - 2 = \log_{10} 4^3 - \log_{10} 4 - \log_{10} 10^2$

use $2 = \log_{10} 10^2$

$= \log_{10} \dfrac{4^3}{4 \cdot 10^2} = \log_{10} \dfrac{4^2}{10^2} = \log_{10} 0.16$

77. $\log_a x - 2 \log_a y + \dfrac{1}{2} \log_a z = \log_a x + \log_a z^{1/2} - \log_a y^2 = \log_a \dfrac{x\sqrt{z}}{y^2}$

79. $\log_a x^2 y + 2 \log_a 5xy^3 - \log_a 10x^2 y^2$

$= \log_a x^2 y + \log_a (5xy^3)^2 - \log_a 10x^2 y^2 = \log_a \dfrac{x^2 y (5xy^3)^2}{10x^2 y^2}$

$= \log_a \dfrac{x^2 y (25x^2 y^6)}{10x^2 y^2} = \log_a \dfrac{25x^4 y^7}{10x^2 y^2} = \log_a \dfrac{5x^2 y^5}{2}$

81. $\dfrac{2}{3} \log_a 8x^2 z^3 + \log_a 3 + \dfrac{1}{3} \log_a 27x^4 y^6 z^9$

$= \log_a (8x^2 z^3)^{2/3} + \log_a 3 + \log_a (27x^4 y^6 z^9)^{1/3}$

$= \log_a [3(8x^2 z^3)^{2/3} (27x^4 y^6 z^9)^{1/3}]$

$= \log_a [3(4x^{4/3} z^2)(3x^{4/3} y^2 z^3)]$

$= \log_a [36x^{8/3} y^2 z^5]$

83. Let $x = \log_a u$ and $y = \log_a v$

Then, $u = a^x$ and $v = a^y$.

By the quotient rule in the Law of Exponents

$\dfrac{u}{v} = \dfrac{a^x}{a^y} = a^{x-y}$

and, therefore, $\log_a \left(\dfrac{u}{v}\right) = x - y = \log_a u - \log_a v$

85. a) By definition, $\log_a x$ is the power to which a must be raised to obtain x.

Suppose $a < 0$ and $\log_a x = \dfrac{1}{2}$. Then,

$a^{1/2} = x$

or $\quad \sqrt{a} = x$

Since $a < 0$, then $\sqrt{a}$ is **not** a real number. Thus, a must be positive.

b) Suppose $a = 1$, $x \neq 1$, and $\log_a x = y$. Then,

$$a^y = x$$

or $\quad 1^y = x$

or $\quad 1 = x$ which is a contradiction. Thus, $a \neq 1$.

Also, the line $1 = x$ is a vertical line which is **not** a one-to-one function.

87. Use $A(t) = Pe^{rt}$ with $e = b^{\log_b e}$ to obtain $A(t) = P(b^{\log_b e})^{rt} = Pb^{(\log_b e)rt} = Pb^{(r\log_b e)t} = Pb^{kt}$. The advantage of choosing e as the base is the fact that scientific and graphing calculators have the function $\log_e$ built into them but do not have $\log_b$ built in (for $b \neq 10$).

Exercises 6.4

1. $\log(2x + 1) - \log 5 = \log x$

$$\log \frac{2x + 1}{5} = \log x \longrightarrow \frac{2x + 1}{5} = x$$

$$2x + 1 = 5x$$
$$1 = 3x$$
$$\frac{1}{3} = x \text{ and it checks}$$

3. $\log x^2 - \log 9 = \log x$

$$\log \frac{x^2}{9} = \log x \longrightarrow \frac{x^2}{9} = x$$

$$x^2 = 9x$$
$$x^2 - 9x = 0$$
$$x(x - 9) = 0$$
$$x = 0, \quad x - 9 = 0$$
$$x = 0, \qquad x = 9 \text{ Discard } x = 0 \text{ since it is } \textbf{not} \text{ possible to compute } \log 0.$$

5. $\log_5(4x - 1) - \log_5 3 = \log_5(x + 2)$

$$\log_5\left[\frac{4x - 1}{3}\right] = \log_5(x + 2) \longrightarrow \frac{4x - 1}{3} = x + 2$$

$$4x - 1 = 3(x + 2)$$
$$4x - 1 = 3x + 6$$
$$x - 1 = 6$$
$$x = 7 \text{ and it checks}$$

7. $\log_2 x - \log_2(x - 1) = 3$

$$\log_2\left[\frac{x}{x - 1}\right] = 3 \xrightarrow[\substack{\text{Exponential} \\ \text{Form}}]{} 2^3 = \frac{x}{x - 1}$$

$$8 = \frac{x}{x - 1}$$
$$8(x - 1) = x$$
$$8x - 8 = x$$
$$7x = 8$$
$$x = \frac{8}{7} \text{ and it checks}$$

9. $\log x + \log(3x - 7) = 1$

$$\log[x(3x - 7)] = 1 \xrightarrow[\substack{\text{Exponential} \\ \text{Form}}]{} 10^1 = x(3x - 7)$$

$$10 = 3x^2 - 7x$$
$$0 = 3x^2 - 7x - 10$$
$$0 = (3x - 10)(x + 1)$$

$$0 = 3x - 10, \quad 0 = x + 1$$

$x = \dfrac{10}{3}, \qquad x = -1$ Here $\dfrac{10}{3}$ checks whereas -1

does not since $\log x$ and $\log(3x - 7)$ are undefined when $x = -1$.

11. $\log 8x + \log(x + 2) = 1$

$\log[8x(x + 2)] = 1 \xrightarrow[\text{Form}]{\text{Exponential}} 10^1 = 8x(x + 2)$

$$10 = 8x^2 + 16x$$
$$0 = 8x^2 + 16x - 10 \quad \text{Now} \div \text{by 2:}$$
$$0 = 4x^2 + 8x - 5$$
$$0 = (2x - 1)(2x + 5)$$
$$0 = 2x - 1, \quad 0 = 2x + 5$$
$$1 = 2x, \qquad -5 = 2x$$
$$\frac{1}{2} = x, \qquad -\frac{5}{2} = x \quad \text{Here only } \frac{1}{2} \text{ checks.}$$

13. $\log 2x + \log(13x - 1) = 2$

$\log[2x(13x - 1)] = 2 \xrightarrow[\text{Form}]{\text{Exponential}} 10^2 = 2x(13x - 1)$

$$100 = 26x^2 - 2x$$
$$0 = 26x^2 - 2x - 100 \quad \text{Now} \div \text{by 2:}$$
$$0 = 13x^2 - x - 50$$
$$0 = (x - 2)(13x + 25)$$
$$0 = x - 2, \quad 0 = 13x + 25$$
$$2 = x, \qquad -\frac{25}{13} = x \qquad \text{Here only 2 checks.}$$

15. $\log_4 x + \log_4 (x + 6) = 2$

$\log_4 [x(x + 6)] = 2 \xrightarrow[\text{Form}]{\text{Exponential}} 4^2 = x(x + 6)$

$$16 = x^2 + 6x$$
$$0 = x^2 + 6x - 16$$
$$0 = (x + 8)(x - 2)$$
$$0 = x + 8, \quad 0 = x - 2$$
$$-8 = x, \qquad 2 = x \qquad \text{Here only 2 checks.}$$

17. $e^{\ln x^2} = 4$ becomes

$x^2 = 4$ since $e^{\ln x^2} = x^2$

$x = \pm\sqrt{4} = \pm 2$ and both values check

19. $\ln e^{4x} = 8$ becomes

$4x = 8$ since $\ln e^{4x} = 4x$

$x = \dfrac{8}{4} = 2$ and it checks

21. $3^x = 17$. Take the log of each side: $\log 3^x = \log 17$

$$x \log 3 = \log 17$$
$$x = \frac{\log 17}{\log 3} \approx 2.58$$

23. $4^{1-x} = 19$. Take the log of each side: $\log 4^{1-x} = \log 19$

$$(1 - x)\log 4 = \log 19$$
$$1 - x = \frac{\log 19}{\log 4}$$
$$-x = -1 + \frac{\log 19}{\log 4}$$
$$x = 1 - \frac{\log 19}{\log 4} \approx -1.12$$

25. $9^{2x^2-1} = 11$. Take the log of each side: $\log 9^{2x^2-1} = \log 11$

$$(2x^2 - 1)\log 9 = \log 11$$

$$2x^2 - 1 = \frac{\log 11}{\log 9}$$

$$2x^2 = 1 + \frac{\log 11}{\log 9}$$

$$x^2 = \frac{1}{2}\left(1 + \frac{\log 11}{\log 9}\right)$$

$$x = \pm\sqrt{\frac{1}{2}\left(1 + \frac{\log 11}{\log 9}\right)} \approx \pm 1.02$$

27. $3^x = 2^{4x+3}$. Take the log of each side: $\log 3^x = \log 2^{4x+3}$

$$x \log 3 = (4x + 3)\log 2$$

$$x \log 3 = 4x \log 2 + 3 \log 2$$

$$x \log 3 - 4x \log 2 = 3 \log 2$$

$$x(\log 3 - 4 \log 2) = 3 \log 2$$

$$x = \frac{3 \log 2}{\log 3 - 4 \log 2} \approx -1.24$$

29. $5^{x-3} = 9^{x+4}$. Take the log of each side: $\log 5^{x-3} = \log 9^{x+4}$

$$(x - 3)\log 5 = (x + 4)\log 9$$

$$x \log 5 - 3 \log 5 = x \log 9 + 4 \log 9$$

$$x \log 5 - x \log 9 - 3 \log 5 = 4 \log 9$$

$$x \log 5 - x \log 9 = 4 \log 9 + 3 \log 5$$

$$x(\log 5 - \log 9) = 4 \log 9 + 3 \log 5$$

$$x = \frac{4 \log 9 + 3 \log 5}{\log 5 - \log 9} \approx -23.2$$

31. $9^{x+2} = 7^{3x-1}$. Take the log of each side: $\log 9^{x+2} = \log 7^{3x-1}$

$$(x + 2)\log 9 = (3x - 1)\log 7$$

$$x \log 9 + 2 \log 9 = 3x \log 7 - \log 7$$

$$x \log 9 - 3x \log 7 + 2 \log 9 = -\log 7$$

$$x \log 9 - 3x \log 7 = -\log 7 - 2 \log 9$$

$$x(\log 9 - 3 \log 7) = -\log 7 - 2 \log 9$$

$$x = \frac{-\log 7 - 2 \log 9}{\log 9 - 3 \log 7} \approx 1.74$$

33. $(1 + x)^3 = 3.12$. Take the cube root: $1 + x = \sqrt[3]{3.12}$

$$x = \sqrt[3]{3.12} - 1 \approx 0.461$$

35. $(1 + 0.06)^x = 3.17$ can be written as

$(1.06)^x = 3.17$. Now, take the log of each side: $\log(1.06)^x = \log 3.17$

$$x \log 1.06 = \log 3.17$$

$$x = \frac{\log 3.17}{\log 1.06} \approx 19.8$$

37. $\log_3 7 = \dfrac{\log 7}{\log 3} \approx 1.77$

39. $\log_6 43 = \dfrac{\log 43}{\log 6} \approx 2.10$

41. $\log_2 28.6 = \dfrac{\log 28.6}{\log 2} \approx 4.84$

43. $\log_7 132 = \dfrac{\log 132}{\log 7} \approx 2.51$

45. $e^x = 4.2$. Take ln of each side: $\ln e^x = \ln 4.2$
$$x \ln e = \ln 4.2$$
$$x = \ln 4.2 \approx 1.44$$

47. $e^{-x} = 5.3$. Take ln of each side: $\ln e^{-x} = \ln 5.3$
$$-x \ln e = \ln 5.3$$
$$-x = \ln 5.3$$
$$x = -\ln 5.3 \approx -1.67$$

49. $e^{x+1} = 67.1$. Take ln of each side: $\ln(e^{x+1}) = \ln 67.1$
$$(x + 1)\ln e = \ln 67.1$$
$$x + 1 = \ln 67.1$$
$$x = \ln 67.1 - 1 \approx 3.21$$

51. $e^{2x+1} = 114$. Take ln of each side: $\ln(e^{2x+1}) = \ln 114$
$$(2x + 1)\ln e = \ln 114$$
$$2x + 1 = \ln 114$$
$$2x = \ln 114 - 1$$
$$x = \frac{1}{2}(\ln 114 - 1) \approx 1.87$$

53. $A = 1000\left(1 + \dfrac{0.12}{4}\right)^{5 \cdot 4} = 1000(1.03)^{20} \approx 1000(1.80611) = \$1806.11$

55. Use $A = P\left(1 + \dfrac{r}{n}\right)^{tn}$ where $A = 2P$, $r = 0.14$, and $n = 2$ to get
$$2P = P(1 + 0.07)^{2t} = P(1.07)^{2t}$$
Now $\div$ by P to get: $\quad 2 = (1.07)^{2t}$
Take log of each side $\quad \log 2 = \log(1.07)^{2t}$
$$\log 2 = 2t \log(1.07)$$
$$\frac{\log 2}{2 \log 1.07} = t \quad \text{or} \quad t \approx 5.12 \text{ years}$$

57. Let A be the amount to be invested now.

$$\$10,000 = A\left(1 + \frac{0.10}{4}\right)^{6 \cdot 4}$$
$$10,000 = A(1.025)^{24} \qquad \text{Take log of each side}$$
$$\log 10,000 = \log[A(1.025)^{24}]$$
$$4 = \log A + 24 \log 1.025$$
$$4 - 24 \log 1.025 = \log A$$
$$3.74263 = \log A$$
$$10^{3.74263} = A \quad \text{or} \quad A \approx \$5528.75$$

Another method for solving this problem is to start with
$$10,000 = A(1.025)^{24}$$
$$10,000 = A(1.808726)$$
$$\frac{10,000}{1.808726} = A \quad \text{or} \quad A \approx \$5528.75$$

59. Use $A = Pe^{0.10t}$ with $A = 2P$ to get
$$2P = Pe^{0.10t}$$
$2 = e^{0.10t}$. Take ln of each side: $\quad \ln 2 = \ln e^{0.10t}$
$$\ln 2 = 0.10t(\ln e)$$
$$\ln 2 = 0.10t$$
$$\frac{\ln 2}{0.10} = t \quad \text{or} \quad t \approx 6.93 \text{ years}$$

61. Use $P = 2000e^{0.14t}$ with $P = 8000$ to get

$$8000 = 2000e^{0.14t}$$
$$4 = e^{0.14t}. \text{ Take ln of each side: } \ln 4 = \ln e^{0.14t}$$
$$\ln 4 = 0.14t(\ln e)$$
$$\ln 4 = 0.14t$$
$$\frac{\ln 4}{0.14} = t \text{ or } t \approx 9.90 \text{ days}$$

63. Use $P = 3000e^{0.18t}$ with $P = 9000$ to get

$$9000 = 3000e^{0.18t}$$
$$3 = e^{0.18t}. \text{ Take ln of each side: } \ln 3 = \ln e^{0.18t}$$
$$\ln 3 = 0.18t(\ln e)$$
$$\ln 3 = 0.18t$$
$$\frac{\ln 3}{0.18} = t \text{ or } t \approx 6.10 \text{ days}$$

65. Use $A = 40e^{-0.000124x}$ with $A = 10$ to get

$$10 = 40e^{-0.000124x}$$
$$0.25 = e^{-0.000124x}. \text{ Take ln of each side: } \ln 0.25 = \ln e^{-0.000124x}$$
$$\ln(0.25) = -0.000124x(\ln e)$$
$$\ln(0.25) = -0.000124x$$
$$\frac{\ln(0.25)}{-0.000124} = x \text{ or } x \approx 11,179 \text{ years}$$

(or 11,200 years when rounded)

67. Use $A = 100e^{-0.000411x}$ with $A = 50$ to get

$$50 = 100e^{-0.000411x}$$
$$0.5 = e^{-0.000411x}. \text{ Take ln of each side: } \ln(0.5) = \ln e^{-0.000411x}$$
$$\ln(0.5) = -0.000411x(\ln e)$$
$$\ln(0.5) = -0.000411x$$
$$\frac{\ln(0.5)}{-0.000411} = x \text{ or } x \approx 1686 \text{ years}$$

(or 1700 years when rounded)

69. $L = 10 \log(I \times 10^{12}) = 10 \log I + 10 \log 10^{12}$
$$= 10 \log I + 10 \cdot 12 \log 10$$
$$= 10 \log I + 120(1)$$
$$= 10 \log I + 120$$

Now, if I is changed to $3I$, then the new value for L, called L^* is
$$L^* = 10 \log (3I) + 120$$
$$= 10 \log 3 + 10 \log I + 120$$

The difference $L^* - L$ is $(10 \log 3 + 10 \log I + 120) - (10 \log I + 120)$
$$= 10 \log 3$$
$$\approx 10(0.4771)$$
$$= 4.771$$
That is, the increase is slightly less than 5 decibels.

71.

73.

75.

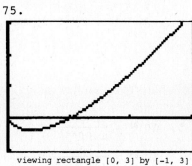

viewing rectangle [-1, 10] by [-2, 2]   viewing retangle [-1, 10] by [-2, 2]   viewing rectangle [0, 3] by [-1, 3]

77.

79.

81.

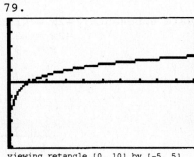

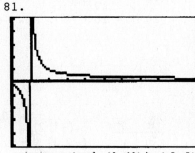

viewing retangle [-10, 10] by [-10, 10]   viewing rectangle [0, 10] by [-5, 5]   viewing rectangle [0, 10] by [-5, 5]

83. 0.57

85. 1.36

87. u and v must be positive for $\log_a u$ and $\log_b v$ to be defined.

CHAPTER REVIEW

1. a) $f(-2) = \left(\frac{2}{3}\right)^{-2} = \frac{1}{\left(\frac{2}{3}\right)^2} = \frac{1}{\frac{4}{9}} = \frac{9}{4}$

    b) $f\left(\frac{1}{2}\right) = \left(\frac{2}{3}\right)^{1/2} = \sqrt{\frac{2}{3}} = \frac{\sqrt{2}}{\sqrt{3}} \cdot \frac{\sqrt{3}}{\sqrt{3}} = \frac{\sqrt{6}}{\sqrt{9}} = \frac{\sqrt{6}}{3}$

2. $(81)^x = \frac{1}{27}$

    $(3^4)^x = \frac{1}{3^3}$

    $3^{4x} = 3^{-3}$. Now, equate exponents: $4x = -3$

    $x = -\frac{3}{4}$

3. $8^x = \frac{1}{32}$

    $(2^3)^x = \frac{1}{2^5}$

    $2^{3x} = 2^{-5}$ Now, equate exponents: $3x = -5$

    $x = -\frac{5}{3}$

4. $\quad 4^{x+3} = 32^{2x-5}$
$(2^2)^{x+3} = (2^5)^{2x-5}$
$2^{2(x+3)} = 2^{5(2x-5)}$. Now, equate exponents:

$$2(x + 3) = 5(2x - 5)$$
$$2x + 6 = 10x - 25$$
$$2x + 31 = 10x$$
$$31 = 8x$$
$$\frac{31}{8} = x$$

5. $10^{-2x} = \dfrac{1}{1000^{x+3}}$

$10^{-2x} = \dfrac{1}{(10^3)^{x+3}}$

$10^{-2x} = \dfrac{1}{10^{3x+9}}$

$10^{-2x} = 10^{-(3x+9)}$. Now, equate exponents:

$$-2x = -(3x + 9)$$
$$-2x = -3x - 9$$
$$x = -9$$

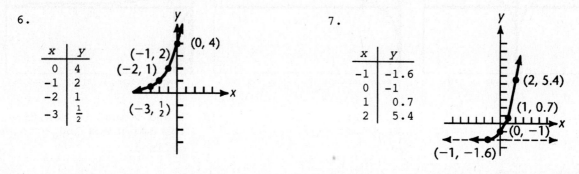

6.

x	y
0	4
-1	2
-2	1
-3	$\frac{1}{2}$

(−1, 2) (0, 4)
(−2, 1)
(−3, $\frac{1}{2}$)

7.

x	y
-1	-1.6
0	-1
1	0.7
2	5.4

(2, 5.4)
(1, 0.7)
(0, −1)
(−1, −1.6)

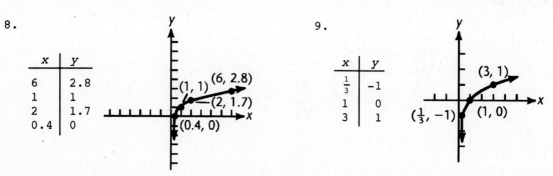

8.

x	y
6	2.8
1	1
2	1.7
0.4	0

(1, 1) (6, 2.8)
(2, 1.7)
(0.4, 0)

9.

x	y
$\frac{1}{3}$	-1
1	0
3	1

(3, 1)
($\frac{1}{3}$, −1) (1, 0)

10. $\log_a 64 = \dfrac{3}{2}$ becomes
$a^{3/2} = 64$
$(a^{3/2})^{2/3} = (64)^{2/3}$

$a = 64^{2/3} = (\sqrt[3]{64})^2 = 4^2 = 16$

11. $y = \log_{16} \dfrac{1}{32}$ can be written as

$16^y = \dfrac{1}{32}$

$(2^4)^y = \dfrac{1}{2^5}$

$2^{4y} = 2^{-5}$. Equate exponents: $4y = -5$

$$y = \frac{-5}{4} = -\frac{5}{4}$$

12. $\log_4 x = -\dfrac{5}{2}$ becomes

$4^{-5/2} = x$

$\dfrac{1}{4^{5/2}} = x$

$\dfrac{1}{(\sqrt{4}\,)^5} = x$

$\dfrac{1}{2^5} = x$

$\dfrac{1}{32} = x$

13. $\log_3 3^{4x} = 18$

$4x \log_3 3 = 18$

$4x(1) = 18$

$4x = 18$

$x = \dfrac{18}{4} = \dfrac{9}{2}$

14. $e^{\ln e^3} = y$ becomes

$e^3 = y$

15. $\log_{10} 10^x = -3$

$x \log_{10} 10 = -3$

$x(1) = -3$

$x = -3$

16. $\log_a \left(\dfrac{x^2 z}{y^3}\right) = \log_a(x^2 z) - \log_a y^3$

$= \log_a x^2 + \log_a z - \log_a y^3$

$= 2\log_a x + \log_a z - 3\log_a y$

17. $\log_a \dfrac{\sqrt[3]{x^2 y}}{z^4} = \log_a \sqrt[3]{x^2 y} - \log_a z^4$

$= \log_a x^{2/3} y^{1/3} - \log_a z^4$

$= \log_a x^{2/3} + \log_a y^{1/3} - \log_a z^4$

$= \dfrac{2}{3}\log_a x + \dfrac{1}{3}\log_a y - 4\log_a z$

18. $2\log_a x^2 z + \log_a 9xy^3 - 2\log_a 4yz^3$

$= \log_a (x^2 z)^2 + \log_a 9xy^3 - \log_a (4yz^3)^2 = \log_a \dfrac{(x^2 z)^2 (9xy^3)}{(4yz^3)^2}$

$= \log_a \dfrac{x^4 z^2 \cdot 9xy^3}{16y^2 z^6} = \log_a \dfrac{9x^5 y}{16z^4}$

19. $2\log_a x^2 yz + 3\log_a 2xyz^2 - \log_a 4x^3 y^2 z^4$

$= \log_a (x^2 yz)^2 + \log_a (2xyz^2)^3 - \log_a 4x^3 y^2 z^4$

$= \log_a \dfrac{(x^2 yz)^2 (2xyz^2)^3}{4x^3 y^2 z^4} = \log_a \dfrac{x^4 y^2 z^2 \cdot 8x^3 y^3 z^6}{4x^3 y^2 z^4}$

$= \log_a \dfrac{8x^7 y^5 z^8}{4x^3 y^2 z^4} = \log_a 2x^4 y^3 z^4$

20. $x = \log_3 17 = \dfrac{\log 17}{\log 3}$

$\approx \dfrac{1.23045}{0.47712} \approx 2.58$

21. $e^x = 4.7$. Take ln of each side: $\ln e^x = \ln 4.7$

$x \ln e = \ln 4.7$

$x(1) = \ln 4.7$

$x = \ln 4.7 \approx 1.55$

22. $5^{x+1} = 192$. Take the log of each side: $\log 5^{x+1} = \log 192$

$$(x + 1)\log 5 = \log 192$$

$$x + 1 = \frac{\log 192}{\log 5}$$

$$x = \frac{\log 192}{\log 5} - 1$$

$$\approx \frac{2.28330}{0.69897} - 1 \approx 2.27$$

23. $5^{2x-1} = 8^{x+1}$. Take log of each side: $\log 5^{2x-1} = \log 8^{x+1}$

$$(2x - 1)\log 5 = (x + 1)\log 8$$

$$2x \log 5 - \log 5 = x \log 8 + \log 8$$

$$2x \log 5 - x \log 8 = \log 8 + \log 5$$

$$x(2 \log 5 - \log 8) = \log 8 + \log 5$$

$$x = \frac{\log 8 + \log 5}{2 \log 5 - \log 8} \approx \frac{1.60206}{0.49485} \approx 3.24$$

24. $\log x + \log(3x + 1) = 1$

$\log[x(3x + 1)] = 1$. Convert to the exponential form: $x(3x + 1) = 10^1$

$$3x^2 + x = 10$$

$$3x^2 + x - 10 = 0$$

$$(3x - 5)(x + 2) = 0$$

$$3x - 5 = 0, \quad x + 2 = 0$$

$$x = \frac{5}{3}, \qquad x = -2$$

Only $x = \frac{5}{3}$ checks and $x = -2$ does **not** check since $\log x$ becomes $\log(-2)$ which is **not** defined.

25. $\log_8 (x + 5) + \log_8 (3x - 1) = 2$

$$\log_8 [(x + 5)(3x - 1)] = 2$$

$$(x + 5)(3x - 1) = 8^2$$

$$(x + 5)(3x - 1) = 64$$

$$3x^2 + 14x - 5 = 64$$

$$3x^2 + 14x - 69 = 0$$

$$(x - 3)(3x + 23) = 0$$

$$x - 3 = 0, \quad 3x + 23 = 0$$

$$x = 3, \qquad x = -\frac{23}{3}$$ Here only $x = 3$ checks. When $x = -\frac{23}{3}$, the $\log_8 (3x - 1)$ and $\log_8 (x + 5)$ are **not** defined.

26. $W(t) = 50 - 45(2)^{-0.361t}$. Let $t = 4$ to obtain:

$$W(4) = 50 - 45(2)^{-0.361(4)}$$

$$= 50 - 45(2)^{-1.444}$$

$$\approx 50 - 45(0.36755)$$

$$= 50 - 16.53975$$

$$= 33.46025 \text{ which rounds to 33 words per minute}$$

27. $F(t) = 35 + 177(2)^{-0.0256t}$. Let $t = 1$ hour $= 60$ minutes to get:

$$F(60) = 35 + 177(2)^{-0.0256(60)}$$

$$= 35 + 177(2)^{-1.536}$$

$$\approx 35 + 177(0.34484)$$

$$= 35 + 61.0367$$

$$\approx 96^\circ \text{ F}$$

28. The formula is $A = P\left(1 + \dfrac{0.10}{12}\right)^{12t}$ where $12t$ represents the number of months after t years. Now, let $A = P + 50\% \ P = P + 0.5P = 1.5P$:

$$1.5P = P\left(1 + \frac{0.10}{12}\right)^{12t} \quad \text{First} \div \text{ by } P:$$

$$1.5 = \left(1 + \frac{0.10}{12}\right)^{12t}$$

$$\log 1.5 = \log\left(1 + \frac{0.10}{12}\right)^{12t}$$

$$\log 1.5 = 12t \log\left(1 + \frac{0.10}{12}\right)$$

$$\frac{\log 1.5}{\log\left(1 + \frac{0.10}{12}\right)} = 12t$$

or $\quad \dfrac{0.17609}{0.003604} \approx 12t$

or $\quad\quad\quad 48.8 \approx 12t \quad$ That is, it takes approximately 49 months for the principal to increase by 50%.

29. $P = 130{,}000e^{-0.26t}$. Let $P = 10{,}000$ to obtain: $10{,}000 = 130{,}000e^{-0.26t}$

$$\frac{10{,}000}{130{,}000} = e^{-0.26t}$$

$$\frac{1}{13} \approx e^{-0.26t}$$

$$\ln\left(\frac{1}{13}\right) = -0.26t$$

$$\frac{\ln(1/13)}{-0.26} = t$$

or $\quad 9.87$ hours $\approx t$

30. $W(t) = 50 - 45(2)^{-0.361t}$. Let $W = 42$ to obtain: $42 = 50 - 45(2)^{-0.361t}$

$$-8 = -45(2)^{-0.361t}$$

$$\frac{-8}{-45} = 2^{-0.361t}$$

$$\frac{8}{45} = 2^{-0.361t}$$

$$\log\left(\frac{8}{45}\right) = \log 2^{-0.361t}$$

$$\log\left(\frac{8}{45}\right) = -0.361t(\log 2)$$

$$\frac{\log(8/45)}{-0.361(\log 2)} = t$$

or $\quad \dfrac{-0.75012}{(-0.361)(0.30103)} \approx t \quad$ or $\quad t \approx 6.9$ months

31. $N = 60e^{-0.163x}$. Let $N = 60 - 40 = 20$ to obtain:

$$20 = 60e^{-0.163x}$$
$$\frac{20}{60} = e^{-0.163x}$$
$$\frac{1}{3} = e^{-0.163x}$$
$$\ln\left(\frac{1}{3}\right) = -0.163x$$
$$\frac{\ln(1/3)}{-0.163} = x$$

or 6.74 years $\approx x$

32. $A(x) = A_0 e^{-0.000495x}$. Let $A(x) = 0.37A_0$ to obtain:

$$0.37A_0 = A_0 e^{-0.000495x}$$
$$0.37 = e^{-0.000495x}$$
$$\ln(0.37) = \ln(e^{-0.000495x}) = -0.000495x(\ln e) = -0.000495x$$

Now, $\dfrac{\ln(0.37)}{-0.000495} = x$

or $\dfrac{-0.99425}{-0.000495} \approx x$ or $x \approx 2008.5$ days which rounds to 2009 days

Exercises 7.1

1. $\{(-2, 3)\}$

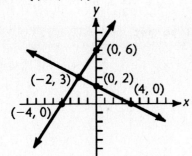

3. $\varnothing$

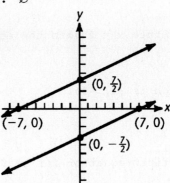

5. $2x - 5y = 11$
    $5x + y = 14$
    Solve the second equation for y: $5x + y = 14$
    $$y = -5x + 14$$
    Substitute this result into the first equation: $2x - 5y = 11$
    $$2x - 5(-5x + 14) = 11$$
    $$2x + 25x - 70 = 11$$
    $$27x - 70 = 11$$
    $$27x = 81$$
    $$x = 3$$

    Now, substitute $x = 3$ into $y = -5x + 14$ to get
    $$y = -5(3) + 14$$
    $$y = -15 + 14$$
    $$y = -1$$
    The solution is $(3, -1)$.

7. $x + 2y = -5$
    $2x + y = 2$
    Solve the first equation for x: $x + 2y = -5$
    $$x = -2y - 5$$
    Substitute this result into the second equation: $2x + y = 2$
    $$2(-2y - 5) + y = 2$$
    $$-4y - 10 + y = 2$$
    $$-3y - 10 = 2$$
    $$-3y = 12$$
    $$y = \frac{12}{-3} = -4$$

    Now, substitute $y = -4$ into the equation $x = -2y - 5$ to get
    $$x = -2(-4) - 5$$
    $$x = 8 - 5$$
    $$x = 3$$
    The solution is $(3, -4)$.

9. $7x + 3y = 5$
    $5x - y = 13$
    Solve the second equation for y: $5x - y = 13$
    $$-y = -5x + 13$$
    $$y = 5x - 13$$

Substitute this result into the first equation: $7x + 3y = 5$

$$7x + 3(5x - 13) = 5$$
$$7x + 15x - 39 = 5$$
$$22x = 44$$
$$x = \frac{44}{22} = 2$$

Now, substitute $x = 2$ into the equation $y = 5x - 13$ to get

$$y = 5 \cdot 2 - 13$$
$$y = 10 - 13$$
$$y = -3$$

The solution is $(2, -3)$.

11. $9x + 2y = 3$
$2x - 3y = 11$
Solve the first equation for y: $9x + 2y = 3$

$$2y = -9x + 3$$
$$y = -\frac{9}{2}x + \frac{3}{2}$$

Substitute this result into the second equation: $2x - 3y = 11$

$$2x - 3\left[-\frac{9}{2}x + \frac{3}{2}\right] = 11$$
$$2x + \frac{27}{2}x - \frac{9}{2} = 11$$

Multiply by 2:
$$4x + 27x - 9 = 22$$
$$31x - 9 = 22$$
$$31x = 31$$
$$x = \frac{31}{31} = 1$$

Now, substitute $x = 1$ into the equation $y = -\frac{9}{2}x + \frac{3}{2}$ to get

$$y = -\frac{9}{2}(1) + \frac{3}{2}$$
$$y = -\frac{9}{2} + \frac{3}{2}$$
$$y = -\frac{6}{2}$$
$$y = -3$$

The solution is $(1, -3)$.

13. $2x - 5y = -13$ $2x - 5y = -13$

$4x - y = 1 \xrightarrow{\text{Mult by -5}}$ $-20x + 5y = -5$

Add: $-18x = -18$

$$x = \frac{-18}{-18} = 1$$

Now, substitute $x = 1$ into the equation $4x - y = 1$ to get

$$4 \cdot 1 - y = 1$$
$$4 - y = 1$$
$$-y = -3$$
$$y = 3$$

The solution is $(1, 3)$.

15. $2x + 5y = 5$ $\xrightarrow{\text{Mult by 3}}$ $6x + 15y = 15$

$3x + 4y = 18$ $\xrightarrow{\text{Mult by -2}}$ $\underline{-6x - 8y = -36}$

Add: $\quad 7y = -21$

$$y = -\frac{21}{7} = -3$$

Now, substitute $y = -3$ into the equation $2x + 5y = 5$

$$2x + 5(-3) = 5$$
$$2x - 15 = 5$$
$$2x = 20$$
$$x = \frac{20}{2} = 10$$

The solution is $(10, -3)$.

17. $2x - 3y = -2$ $\xrightarrow{\text{Mult by 3}}$ $6x - 9y = -6$

$-3x + 4y = 0$ $\xrightarrow{\text{Mult by 2}}$ $\underline{-6x + 8y = 0}$

Add: $\quad -y = -6$

$$y = 6$$

Now, substitute $y = 6$ into the equation $-3x + 4y = 0$ to get

$$-3x + 4 \cdot 6 = 0$$
$$-3x + 24 = 0$$
$$-3x = -24$$
$$x = \frac{-24}{-3} = 8$$

The solution is $(8, 6)$.

19. $7x + 5y = 2$ $\xrightarrow{\text{Mult by 3}}$ $21x + 15y = -6$

$4x + 3y = -8$ $\xrightarrow{\text{Mult by -5}}$ $\underline{-20x - 15y = 40}$

Add: $\quad x \quad\quad = 46$

Now, substitute $x = 46$ into the equation $4x + 3y = -8$ to get

$$4 \cdot 46 + 3y = -8$$
$$184 + 3y = -8$$
$$3y = -192$$
$$y = \frac{-192}{3} = -64$$

The solution is $(46, -64)$.

21. $3x + 4y = 1$ $\qquad\qquad 3x + 4y = 1$

$5x - 2y = 19$ $\xrightarrow{\text{Mult by 2}}$ $\underline{10x - 4y = 38}$

Add: $\quad 13x \quad\quad = 39$

$$x = \frac{39}{13} = 3$$

Now, substitute $x = 3$ into the equation $3x + 4y = 1$ to get

$$3 \cdot 3 + 4y = 1$$
$$9 + 4y = 1$$
$$4y = -8$$
$$y = \frac{-8}{4} = -2$$

The solution is $(3, -2)$.

23. $3x - 2y = 7 \xrightarrow{\text{Mult by -2}} -6x + 4y = -14$

      $6x - 4y = 4$ $\underline{6x - 4y = 4}$

                     Add: $0 = -10$ which is impossible. Thus, there is **no** solution.

25. $2x - 3y = -12$ $2x - 3y = -12$

       $x + 2y = 8 \xrightarrow{\text{Mult by -2}} \underline{-2x - 4y = -16}$

                   Add: $-7y = -28$

$$y = \frac{-28}{-7} = 4$$

Now, substitute $y = 4$ into the equation $x + 2y = 8$ to get
$$x + 2 \cdot 4 = 8$$
$$x + 8 = 8$$
$$x = 0$$

The solution is $(0, 4)$.

27. $2x = 3y - 1$

      $6y = 4x + 5$

Solve the first equation for x: $2x = 3y - 1$
$$x = \frac{3}{2}y - \frac{1}{2}$$

Substitute this result into the second equation to get $6y = 4\left(\frac{3}{2}y - \frac{1}{2}\right) + 5$
$$6y = 6y - 2 + 5$$
$$6y = 6y + 3$$
$$0 = 3$$

which is impossible. Thus there is **no** solution.

29. $2x - 4 = 3y$

      $6x = 9y + 12$

Solve the first equation for y: $2x - 4 = 3y$
$$\frac{2}{3}x - \frac{4}{3} = y$$

Now, substitute this value into the equation $6x = 9y + 12$ to get
$$6x = 9\left(\frac{2}{3}x - \frac{4}{3}\right) + 12$$
$$6x = 6x - 12 + 12$$
$$6x = 6x$$
$$0 = 0$$

Since this is an identity, there are infinitely many solutions. This is expressed as $\{(x, y) \mid 2x - 3y = 4\}$.

31. $3x = 12 - 6y$

      $4y = 8 - 2x$

Solve the first equation for x: $3x = 12 - 6y$
$$x = 4 - 2y$$

Substitute this result into the second equation to get $4y = 8 - 2(4 - 2y)$
$$4y = 8 - 8 + 4y$$
$$4y = 0 + 4y$$
$$0 = 0$$

Since this is an identity, there are infinitely many solutions. This is expressed as $\{(x, y) \mid x + 2y = 4\}$.

33. $\dfrac{1}{x} + \dfrac{1}{y} = 5$

    becomes $\quad u + v = 5 \quad$ <u>Mult by -2</u> → $\quad -2u - 2v = -10$

    $\dfrac{2}{x} + \dfrac{3}{y} = 14 \qquad\qquad 2u + 3v = 14 \qquad\qquad\qquad \underline{2u + 3v = \;\;\;14}$

    $\qquad\qquad\qquad\qquad\qquad\qquad\qquad\qquad$ Add: $\qquad\qquad v = \;\;\;\;\;4$

Now, substitute $v = 4$ into the equation $u + v = 5$

$$u + 4 = 5$$
$$u = 1$$

At this point, $u = 1$ and $v = 4$. Since $u = \dfrac{1}{x}$ and $v = \dfrac{1}{y}$, we have

$$\dfrac{1}{x} = 1 \text{ and } \dfrac{1}{y} = 4$$

$$\text{or} \quad x = 1 \text{ and } y = \dfrac{1}{4}$$

The solution is $\left(1, \dfrac{1}{4}\right)$.

35. $\dfrac{2}{x} + \dfrac{3}{y} = 66$

    becomes $\quad 2u + 3v = 66 \quad$ <u>Mult by -3</u> → $\quad -6u - 9v = -198$

    $\dfrac{3}{x} + \dfrac{4}{y} = 91 \qquad\qquad 3u + 4v = 91 \quad$ <u>Mult by 2</u> → $\quad \underline{6u + 8v = \;\;\;182}$

    $\qquad\qquad\qquad\qquad\qquad\qquad\qquad\qquad$ Add: $\qquad -v = -16$

    $\qquad\qquad\qquad\qquad\qquad\qquad\qquad\qquad\qquad\qquad\qquad\;\; v = \;\;\;\;16$

Now, substitute $v = 16$ into the equation $2u + 3v = 66$

$$2u + 3 \cdot 16 = 66$$
$$2u + 48 = 66$$
$$2u = 18$$
$$u = 9$$

At this point, $u = 9$ and $v = 16$. Since $u = \dfrac{1}{x}$ and $v = \dfrac{1}{y}$, we have

$$\dfrac{1}{x} = 9 \text{ and } \dfrac{1}{y} = 16$$

$$\text{or} \qquad x = \dfrac{1}{9} \text{ and } y = \dfrac{1}{16}$$

The solution is $\left(\dfrac{1}{9}, \dfrac{1}{16}\right)$.

37. Let x be the amount invested at 9% and y the amount invested at 5%. The system is:

    $x + \quad y = 15,000 \qquad\qquad x + y = \;\;\;15,000$

    $.09x + .05y = 1130 \quad$ <u>Mult by -20</u> → $\quad -1.8x - y = -22,600$

    $\qquad\qquad\qquad\qquad$ Add: $\quad -.8x \qquad = -7,600$

    $\qquad\qquad\qquad\qquad\qquad\qquad\qquad x = \dfrac{-7600}{-.8} = \$9,500$

Substitute $x = 9,500$ into the equation $x + y = 15,000$ to get

$$9,500 + y = 15,000$$
$$y = \$5,500$$

Thus, $9,500 was invested at 9% and $5,500 was invested at 5%.

39. Let c be the rate of the current and r be the rate in still water.
The system is $\quad 30(r - c) = 8$

    $\qquad\qquad\qquad\; 20(r + c) = 8$

using minutes for the time. It may be a good idea to convert minutes to

hours. That is 30 min $= 30\left(\dfrac{1}{60}\right)$hr $= \dfrac{1}{2}$ hr and 20 min $= 20\left(\dfrac{1}{60}\right)$hr $= \dfrac{1}{3}$ hr.

The system is:

$$\frac{1}{2}(r - c) = 8 \xrightarrow{\text{Mult by 2}} r - c = 16$$

$$\frac{1}{3}(r + c) = 8 \xrightarrow{\text{Mult by 3}} \underline{r + c = 24}$$

$$\text{Add:} \quad 2r \quad\quad = 40$$
$$r \quad\quad = 20 \text{ miles/hour}$$

To find c, use $r + c = 24$ with $r = 20$.

$$20 + c = 24$$
$$c = 4 \text{ miles/ hour}$$

41. Let x be the amount of 50% protein (in grams) and y be the amount of 25% protein (in grams) needed. The system is

$$x + \quad y = 28 \quad\quad\quad\quad x + y = 28$$

$$.50x + .25y = .30(28) \xrightarrow{\text{Mult by -4}} \underline{-2x - y = -33.6}$$
$$\text{Add:} \quad -x \quad\quad = -5.6$$
$$x = 5.6 \text{ grams}$$

Now, substitute $x = 5.6$ into the equation $x + y = 28$ to obtain

$$5.6 + y = 28$$
$$y = 22.4 \text{ grams}$$

Thus, we need 5.6 grams of 50% protein and 22.4 grams of 25% protein.

43. Let x be the amount of peanuts and y be the amount of cashews. The system is

$$x + \quad y = 100 \xrightarrow{\text{Mult by -2.50}} -2.50x - 2.50y = -250.00$$

$$2.50x + 3.60y = 316.00 \quad\quad\quad \underline{2.50x + 3.60y = \quad 316.00}$$
$$\text{Add:} \quad\quad\quad 1.10y = \quad 66.00$$
$$y = \frac{66.00}{1.10} = 60 \text{ lbs.}$$

Now, substitute $y = 60$ into the equation $x + y = 100$ to obtain

$$x + 60 = 100$$
$$x = 40 \text{ lbs.}$$

Thus, we need 40 lbs of peanuts and 60 lbs of cashews.

45. Let r be the cost of reserved tickets and a the cost of general admission tickets. The system is

$$r = 2a$$
$$1000r + 11,000a = 91,000$$

To solve, substitute $r = 2a$ into the second equation:

$$1,000r + 11,000a = 91,000$$
$$1,000(2a) + 11,000a = 91,000$$
$$2,000a + 11,000a = 91,000$$
$$13,000a = 91,000$$
$$a = \frac{91,000}{13,000} = \$7$$

Substitute $a = 7$ into the equation $r = 2a$ to get

$$r = 2 \cdot 7$$
$$r = \$14$$

Thus, reserved seat tickets are \$14 and general admission tickets are \$7.

47. $D = 100 - \dfrac{5}{2}p$ and $S = \dfrac{5}{4}p - \dfrac{25}{2}$

Equate D with S to get $100 - \dfrac{5}{2}p = \dfrac{5}{4}p - \dfrac{25}{2}$

Multiply by 4:

$$4\left(100 - \dfrac{5}{2}p\right) = 4\left(\dfrac{5}{4}p - \dfrac{25}{2}\right)$$

$$400 - 10p = 5p - 50$$
$$400 = 15p - 50$$
$$450 = 15p$$
$$\dfrac{450}{15} = p$$
$$30 = p$$

49. Let x be the number of 1st type desks and y be the number of 2nd type desks needed. The system is

$$10x + 6y = 1000 \qquad\qquad 10x + 6y = 1000$$

$$5x + 4y = 600 \xrightarrow{\text{Mult by -2}} \quad \underline{-10x - 8y = -1200}$$

$$\text{Add:} \qquad -2y = -200$$

$$y = \dfrac{-200}{-2} = 100 \text{ desks}$$

Now, let $y = 100$ in the equation $5x + 4y = 600$ to get

$$5x + 4 \cdot 100 = 600$$
$$5x + 400 = 600$$
$$5x = 200$$
$$x = \dfrac{200}{5} = 40 \text{ desks}$$

Thus, 40 desks of the 1st type and 100 desks of the 2nd type can be manufactured.

51. Let A be the time for machine a and B be the time for machine b to do a job. The system is

$$\dfrac{1}{A} + \dfrac{1}{B} = \dfrac{1}{2} \xrightarrow{\text{Mult by -4}} \quad -\dfrac{4}{A} - \dfrac{4}{B} = -2$$

$$\dfrac{4}{A} + \dfrac{1}{B} = 1 \qquad\qquad\quad \dfrac{4}{A} + \dfrac{1}{B} = 1$$

$$\text{Add:} \qquad -\dfrac{3}{B} = -1$$

Now, solve for B: $-\dfrac{3}{B} = -1$

$$-3 = -B$$
$$3 = B$$

Thus, machine B can do the job in 3 hours.

53. (3.01, -1.95)   55. (1.15, -1.39)

57. Write the system as $a_1x + b_1y = c_1$
$\qquad\qquad\qquad\quad a_2x + b_2y = c_2$

If a_2 and b_2 are the **same** multiple of a_1 and b_1 and c_2 is that same multiple of c_1, then the system is dependent with an infinite number of solutions (graph is the same straight line).

If a_2 and b_2 are the **same** multiple of a_1 and b_1 but c_2 is **not** that same multiple of c_1, then the system is inconsistent and no solution exists (graph consists of parallel lines).

For all other possibilities, the system is independent and a unique solution exists. Here, we do not want a_2 and b_2 to be the same multiple of a_1 and b_1.

Exercises 7.2

1. $x + y \quad\;\; = 1$
   $\quad\;\; y - z = 1$
   $x + y + z = 2$

   To start the solution, work with the first and third equations:

   $x + y \quad\;\; = 1 \xrightarrow{\text{Mult by } -1} -x - y \quad\;\; = -1$

   $x + y + z = 2 \qquad\qquad\quad\underline{x + y + z = \;\; 2}$

   $\qquad\qquad\qquad\text{Add:} \qquad\qquad\quad z = \;\; 1$

   Substitute $z = 1$ into $y - z = 1$ to get
   $\qquad\qquad\qquad y - 1 = 1$
   $\qquad\qquad\qquad\quad\;\; y = 2$

   Finally, substitute $y = 2$ into $x + y = 1$ to have
   $\qquad\qquad\qquad\qquad x + 2 = 1$
   $\qquad\qquad\qquad\qquad\qquad x = -1$

   The solution is $(-1, 2, 1)$.

3. $x + \;\; y - 2z = -3$
   $\qquad 2y + 4z = 4$
   $x \qquad\;\; + \;\; z = 5$

   To start the solution, work with the first and third equations:

   $x + y - 2z = -3 \xrightarrow{\text{Mult by } -1} -x - y + 2z = \;\; 3$

   $x \qquad\; + \;\; z = 5 \qquad\qquad\quad\underline{x \qquad\;\; + \;\; z = 5}$

   $\qquad\qquad\quad\text{Add:} \qquad\quad -y + 3z = 8$

   $2y + 4z = 4 \xrightarrow{\text{Mult by } \frac{1}{2}} y + 2z = \;\; 2$

   $-y + 3z = 8 \qquad\qquad\quad \underline{-y + 3z = \;\; 8}$

   $\qquad\qquad\text{Add:} \qquad\qquad 5z = 10$

   $\qquad\qquad\qquad\qquad\qquad\;\; z = 2$

   Substitute $z = 2$ into $x + z = 5$ to get
   $\qquad\qquad\qquad x + 2 = 5$
   $\qquad\qquad\qquad\qquad x = 3$

   Finally, substitute $z = 2$ into $2y + 4z = 4$ to have
   $\qquad\qquad\qquad 2y + 4 \cdot 2 = 4$
   $\qquad\qquad\qquad\;\; 2y + 8 = 4$
   $\qquad\qquad\qquad\qquad\;\; 2y = -4$
   $\qquad\qquad\qquad\qquad\quad\; y = -2$

   The solution is $(3, -2, 2)$.

5. $x - 2y + 3z = 3$
    $x - y + z = 1$
    $x - y + 3z = 5$

$x - 2y + 3z = 3$ $\xrightarrow{\text{Mult by -1}}$ $-x + 2y - 3z = -3$

$x - y + z = 1$ $\phantom{\xrightarrow{\text{Mult by -1}}}$ $\underline{x - y + z = 1}$

 Add: $y - 2z = -2$

$x - y + z = 1$ $\xrightarrow{\text{Mult by -1}}$ $-x + y - z = -1$

$x - y + 3z = 5$ $\phantom{\xrightarrow{\text{Mult by -1}}}$ $\underline{x - y + 3z = 5}$

 Add: $2z = 4$

 $z = 2$

Substitute $z = 2$ into $y - 2z = -2$ to get
$$y - 2 \cdot 2 = -2$$
$$y - 4 = -2$$
$$y = 2$$

Finally, substitute $y = 2$ and $z = 2$ into $x - y + z = 1$ to have
$$x - 2 + 2 = 1$$
$$x = 1$$

The solution is $(1, 2, 2)$.

7. $7x - 8y - z = 3$
    $x - y = -1$
    $3x - 3y - z = 4$

$x - y = -1$ $\xrightarrow{\text{Mult by -3}}$ $-3x + 3y = 3$

$3x - 3y - z = 4$ $\phantom{\xrightarrow{\text{Mult by -3}}}$ $\underline{3x - 3y - z = 4}$

 Add: $-z = 7$

 $z = -7$

Now, use the first equation with $z = -7$ and the second equation to get:

$7x - 8y + 7 = 3$ $7x - 8y = -4$ $\phantom{\xrightarrow{\text{Mult by -7}}}$ $7x - 8y = -4$

$x - y = -1$ or $x - y = -1$ $\xrightarrow{\text{Mult by -7}}$ $\underline{-7x + 7y = 7}$

 Add: $-y = 3$

 $y = -3$

Finally, substitute $y = -3$ into $x - y = -1$ to have
$$x - (-3) = -1$$
$$x + 3 = -1$$
$$x = -4$$

The solution is $(-4, -3, -7)$.

9. $x - 2y + z = 1$
    $3y - z = -5$
    $x + y + 2z = 0$

$x - 2y + z = 1$ $\xrightarrow{\text{Mult by -1}}$ $-x + 2y - z = -1$

$x + y + 2z = 0$ $\phantom{\xrightarrow{\text{Mult by -1}}}$ $\underline{x + y + 2z = 0}$

 Add: $3y + z = -1$

 $3y - z = -5$

 $\underline{3y + z = -1}$

Add: $6y = -6 \rightarrow y = -1$

Substitute $y = -1$ into $3y - z = -5$ to get

$$3(-1) - z = -5$$
$$-3 - z = -5$$
$$-z = -2$$
$$z = 2$$

Finally, substitute $y = -1$ and $z = 2$ into $x + y + 2z = 0$ to have

$$x + (-1) + 2 \cdot 2 = 0$$
$$x + 3 = 0$$
$$x = -3$$

The solution is $(-3, -1, 2)$.

11. $x - 2y - z = 11$
     $2x - y - z = -5$
     $x \quad\quad - z = -7$

$x - 2y - z = 11$ $x - 2y - z = 11$

$x \quad\quad - z = -7$ $\xrightarrow{\text{Mult by } -1}$ $\underline{-x \quad\quad + z = 7}$

           Add: $\quad -2y \quad\quad = 18 \rightarrow y = -9$

Use the 2nd and 3rd equations with $y = -9$:

$2x + 9 - z = -5$ or $2x - z = -14$ $\xrightarrow{\text{Mult by } -1}$ $2x - z = -14$

$x \quad\quad - z = -7$ $x - z = -7$ $\underline{-x + z = 7}$

                                    Add: $x \quad\quad = -7$

Finally, substitute $x = -7$ into $x - z = -7$ to have

$$-7 - z = -7$$
$$-z = 0$$
$$z = 0$$

The solution is $(-7, -9, 0)$.

13. $x + y - z = 6$
     $-2x \quad\quad + 3z = -14$
     $x - y - z = 6$

$x + y - z = 6$ $\xrightarrow{\text{Mult by } -1}$ $-x - y + z = -6$

$x - y - z = 6$ $\underline{x - y - z = 6}$

           Add: $\quad -2y \quad\quad = 0$

                      $y = 0$

Use the 1st and 2nd equations with $y = 0$:

$x + 0 - z = 6$ or $x - z = 6$ $\xrightarrow{\text{Mult by } 2}$ $2x - 2z = 12$

$-2x \quad\quad + 3z = -14$ $-2x + 3z = -14$ $\underline{-2x + 3z = -14}$

                            Add: $\quad\quad\quad z = -2$

Finally, substitute $y = 0$ and $z = -2$ into $x + y - z = 6$ to have

$$x + 0 - (-2) = 6$$
$$x + 2 = 6$$
$$x = 4$$

The solution is $(4, 0, -2)$.

15.
$$x - y + z = 0$$
$$4x + z = 5$$
$$7x + y + z = 11$$

$$x - y + z = 0$$
$$7x + y + z = 0$$

Add: $8x + 2z = 0$ or upon $\div$ by 2: $4x + z = 0$

Now, the two equations with x and z are:
$$4x + z = 5 \qquad\qquad 4x + z = 5$$
$$4x + z = 0 \xrightarrow{\text{Mult by -1}} -4x - z = 0$$

Add: $0 = 5$ which is impossible. This means the system is inconsistent and **no** solution occurs.

17.
$$4x + y + 2z = -2$$
$$x - y - z = -9$$
$$-5x - 5y - 7z = 100$$

$$4x + y + 2z = -2$$
$$x - y - z = -9$$

Add: $5x + z = -11$

$$x - y - z = -9 \xrightarrow{\text{Mult by -5}} -5x + 5y + 5z = 45$$
$$-5x - 5y - 7z = 100 \qquad\qquad -5x - 5y - 7z = 100$$

Add: $-10x - 2z = 145$

Now, the two equations with x and z are:
$$5x + z = -11 \xrightarrow{\text{Mult by 2}} 10x + 2z = -22$$
$$-10x - 2z = 145 \qquad\qquad -10x - 2z = 145$$

Add: $0 = 123$ which is impossible. This means the system is inconsistent and **no** solution occurs.

19.
$$2x - 3y + z = 1$$
$$x - y + 2z = 0$$
$$x + y + 3z = 3$$

$$2x - 3y + z = 1 \qquad\qquad 2x - 3y + z = 1$$
$$x - y + 2z = 0 \xrightarrow{\text{Mult by -3}} -3x + 3y - 6z = 0$$

Add: $-x - 5z = 1$

$$x - y + 2z = 0$$
$$x + y + 3z = 3$$

Add: $2x + 5z = 3$

Now, $-x - 5z = 1$
$$2x + 5z = 3$$

Add: $x = 4$ Substitute $x = 4$ into $2x + 5z = 3$ to get
$$2 \cdot 4 + 5z = 3$$
$$8 + 5z = 3$$
$$5z = -5$$
$$z = -1$$

Finally, substitute $x = 4$ and $z = -1$ into $x + y + 3z = 3$ to have

$$4 + y + 3(-1) = 3$$
$$4 + y - 3 = 3$$
$$y + 1 = 3$$
$$y = 2$$

The solution is $(4, 2, -1)$.

21. For the system, substitute u for $\frac{1}{x}$, v for $\frac{1}{y}$, and w for $\frac{1}{z}$ to get

$$3u + v + w = 5$$
$$u + 2v - w = -3$$
$$u + v + 3w = 3$$

$$\begin{array}{l} 3u + v + w = 5 \\ \underline{u + 2v - w = -3} \\ \text{Add: } 4u + 3v = 2 \end{array}$$

$$u + 2v - w = -3 \xrightarrow{\text{Mult by 3}} 3u + 6v - 3w = -9$$
$$u + v + 3w = 3 \qquad\qquad \underline{u + v + 3w = 3}$$
$$\text{Add: } 4u + 7v = -6$$

Now, the two equations with u and v are

$$4u + 3v = 2 \xrightarrow{\text{Mult by -1}} -4u - 3v = -2$$
$$4u + 7v = -6 \qquad\qquad \underline{4u + 7v = -6}$$
$$\text{Add: } 4v = -8$$
$$v = -2$$

Substitute $v = -2$ into $4u + 3v = 2$ to get

$$4u + 3(-2) = 2$$
$$4u - 6 = 2$$
$$4u = 8$$
$$u = 2$$

Finally, substitute $u = 2$ and $v = -2$ into $3u + v + w = 5$ to have

$$3 \cdot 2 + (-2) + w = 5$$
$$6 - 2 + w = 5$$
$$4 + w = 5$$
$$w = 1$$

At this point, $u = 2$, $v = -2$, and $w = 1$. But, $u = 2$ becomes $\frac{1}{x} = 2$ or $x = \frac{1}{2}$

$$v = -2 \text{ becomes } \frac{1}{y} = -2 \text{ or } y = -\frac{1}{2}$$

$$w = 1 \text{ becomes } \frac{1}{z} = 1 \text{ or } z = 1$$

The solution is $\left(\frac{1}{2}, -\frac{1}{2}, 1\right)$.

23. Again, substitute u for $\dfrac{1}{x}$, v for $\dfrac{1}{y}$, and w for $\dfrac{1}{z}$.

$$u - 2v + 3w = -5$$
$$2u - v + w = 2$$
$$8u + 3v + 2w = 7$$

$$u - 2v + 3w = -5 \qquad\qquad\qquad u - 2v + 3w = -5$$
$$2u - v + w = 2 \xrightarrow{\text{Mult by } -3} -6u + 3v - 3w = -6$$
$$\text{Add:} \quad -5u + v = -11$$

$$2u - v + w = 2 \xrightarrow{\text{Mult by } -2} -4u + 2v - 2w = -4$$
$$8u + 3v + 2w = 7 \qquad\qquad 8u + 3v + 2w = 7$$
$$\text{Add:} \quad 4u + 5v = 3$$

Now, the two equations with u and v are

$$-5u + v = -11 \xrightarrow{\text{Mult by } -5} 25u - 5v = 55$$
$$4u + 5v = 3 \qquad\qquad 4u + 5v = 3$$
$$\text{Add:} \quad 29u = 58$$
$$u = 2$$

Substitute $u = 2$ into $4u + 5v = 3$ to get
$$4 \cdot 2 + 5v = 3$$
$$8 + 5v = 3$$
$$5v = -5$$
$$v = -1$$

Finally, substitute $u = 2$ and $v = -1$ into $2u - v + w = 2$ to have
$$2 \cdot 2 - (-1) + w = 2$$
$$4 + 1 + w = 2$$
$$5 + w = 2$$
$$w = -3$$

At this point, $u = 2$, $v = -1$, and $w = -3$. But, $u = 2$ becomes $\dfrac{1}{x} = 2$ or $x = \dfrac{1}{2}$

$$v = -1 \text{ becomes } \dfrac{1}{y} = -1 \text{ or } y = -1$$

$$w = -3 \text{ becomes } \dfrac{1}{z} = -3 \text{ or } z = -\dfrac{1}{3}$$

The solution is $\left(\dfrac{1}{2}, -1, -\dfrac{1}{3} \right)$

25. For the point $(-1, 1)$, substitute -1 for x and 1 for y. The equation is:
$$1 = a(-1)^2 + b(-1) + c$$
or $\quad 1 = a - b + c$

For the point $(0, 3)$, substitute 0 for x and 3 for y. The equation is:
$$3 = a(0)^2 + b(0) + c$$
or $\quad 3 = c$

For the point $(1, 9)$, substitute 1 for x and 9 for y. The equation is:
$$9 = a(1)^2 + b(1) + c$$
or $\quad 9 = a + b + c$

The system is
$$a - b + c = 1$$
$$c = 3$$
$$a + b + c = 9$$

Since $c = 3$ (from the 2nd equation), substitute this into the 1st and 3rd equations:

$$a - b + 3 = 1 \qquad a - b = -2$$
$$a + b + 3 = 9 \quad \text{or} \quad \underline{a + b = 6}$$
$$\text{Add: } 2a \qquad = 4$$
$$a = 2$$

Now, substitute $a = 2$ into the equation $a + b = 6$ to have
$$2 + b = 6$$
$$b = 4$$

The values are $a = 2$, $b = 4$, and $c = 3$ and the equation of the parabola becomes $y = 2x^2 + 4x + 3$.

27. For the point $(-1, -1)$, substitute -1 for x and -1 for y. The equation is:
$$(-1)^2 + (-1)^2 + a(-1) + b(-1) + c = 0$$
$$1 + 1 - a - b + c = 0$$
$$-a - b + c = -2$$

For the point $(2, 2)$, substitute 2 for x and 2 for y. The equation is:
$$2^2 + 2^2 + a(2) + b(2) + c = 0$$
$$4 + 4 + 2a + 2b + c = 0$$
$$2a + 2b + c = -8$$

For the point $(5, -1)$, substitute 5 for x and -1 for y. The equation is:
$$5^2 + (-1)^2 + a(5) + b(-1) + c = 0$$
$$25 + 1 + 5a - b + c = 0$$
$$5a - b + c = -26$$

The system is
$$-a - b + c = -2$$
$$2a + 2b + c = -8$$
$$5a - b + c = -26$$

To find c, use $-a - b + c = -2$ $\xrightarrow{\text{Mult by 2}}$ $-2a - 2b + 2c = -4$

$\qquad\qquad\qquad 2a + 2b + c = -8 \qquad\qquad\qquad \underline{2a + 2b + c = -8}$

$\qquad\qquad\qquad\qquad\qquad\qquad\qquad \text{Add:} \qquad\qquad 3c = -12$

$\qquad\qquad\qquad\qquad\qquad\qquad\qquad\qquad\qquad\qquad\qquad c = -4$

To find a, use $-a - b + c = -2$ $\xrightarrow{\text{Mult by -1}}$ $a + b - c = 2$

$\qquad\qquad\qquad 5a - b + c = -26 \qquad\qquad\qquad \underline{5a - b + c = -26}$

$\qquad\qquad\qquad\qquad\qquad\qquad\qquad \text{Add:} \quad 6a \qquad\qquad = -24$

$\qquad\qquad\qquad\qquad\qquad\qquad\qquad\qquad\qquad\qquad\qquad a = -4$

Finally, to find b, use $-a - b + c = -2$ with $a = -4$ and $c = -4$:
$$-(-4) - b + (-4) = -2$$
$$-b = -2$$
$$b = 2$$

The equation is $x^2 + y^2 - 4x + 2y - 4 = 0$.

29. Let x, y, and z be the number of girls on the third grade, fourth grade, and fifth grade teams respectively. The system of equations is

$$x + y + z = 30 \qquad\qquad\qquad x + y + z = 30$$
$$x = y + z - 8 \quad\longrightarrow\quad x - y - z = -8$$
$$z = \frac{1}{2}(x + y) \qquad\qquad\quad -x - y + 2z = 0$$

We need to find the value for x (number of girls on the third grade team). To do this, simply add together the first and second equations:

$$x + y + z = 30$$
$$\underline{x - y - z = -8}$$

Add: $2x \qquad\quad = 22$
$$x = 11$$

There are 11 girls on the third grade team.

31. Let x be the time of the smallest pump, y be the time of the middle pump, and z be the time of the largest pump. The system of equations is

$$\frac{1}{x} + \frac{1}{y} + \frac{1}{z} = \frac{1}{2}$$
$$\frac{1}{x} \qquad + \frac{1}{z} = \frac{1}{3}$$
$$\frac{1}{x} + \frac{1}{y} \qquad = \frac{1}{4}$$

We need to find the value for x. First, add together the 2nd and 3rd equations:

$$\frac{1}{x} \qquad\quad + \frac{1}{z} = \frac{1}{3}$$
$$\underline{\frac{1}{x} + \frac{1}{y} \qquad\quad = \frac{1}{4}}$$

Add: $\dfrac{2}{x} + \dfrac{1}{y} + \dfrac{1}{z} = \dfrac{7}{12}$

This equation and the 1st equation produce

$$\frac{1}{x} + \frac{1}{y} + \frac{1}{z} = \frac{1}{2} \xrightarrow{\text{Mult by -1}} -\frac{1}{x} - \frac{1}{y} - \frac{1}{z} = -\frac{1}{2}$$
$$\frac{2}{x} + \frac{1}{y} + \frac{1}{z} = \frac{7}{12} \qquad\qquad\qquad \underline{\frac{2}{x} + \frac{1}{y} + \frac{1}{z} = \frac{7}{12}}$$

Add: $\qquad \dfrac{1}{x} \qquad\qquad = \dfrac{7}{12} - \dfrac{1}{2}$

$$\frac{1}{x} = \frac{7}{12} - \frac{6}{12} = \frac{1}{12}$$

Then, $x = 12$ days. Working alone, the smaller pump requires 12 days to fill the tanker.

33. Let x be the measure of the smallest angle, y be the measure of the middle angle, and z be the measure of the largest angle. The system of equations is

$$x + y + z = 180 \qquad\qquad\qquad x + y + z = 180$$
$$z = x + y - 20 \longrightarrow -x - y + z = -20$$
$$z = 2x - 10 \qquad\qquad\qquad -2x \qquad + z = -10$$

To find z, add together the first and second equations:

$$x + y + z = 180$$
$$\underline{-x - y + z = -20}$$

Add: $\qquad\quad 2z = 160$
$$z = 80°$$

To find x, use $z = 2x - 10$ with $z = 80$:
$$80 = 2x - 10$$
$$90 = 2x$$
$$45° = x$$

Finally, to find y, use $x + y + z = 180$ with $x = 45$ and $z = 80$:
$$45 + y + 80 = 180$$
$$y + 125 = 180$$
$$y = 55°$$

The three angles are $45°$, $55°$, and $80°$.

35. Let x be the amount of 8-4-8 fertilizer, y be the amount of 15-30-15 fertilizer, and z be the amount of 12-6-12 fertilizer needed to produce the final mixture. One equation is $x + y + z = 200$. Looking at the nitrogen, phosphoric acid, and potash from the three fertilizers produces these 3 equations:

Nitrogen: $8x + 15y + 12z = 13(200)$
Phosphoric Acid: $4x + 30y + 6z = 20(200)$
Potash: $8x + 15y + 12z = 13(200)$

Since the 3rd equation is identical to the 1st equation, it is not needed for the system. The resulting system is
$$x + y + z = 200$$
$$8x + 15y + 12z = 2600$$
$$4x + 30y + 6z = 4000$$

To solve, start with

$8x + 15y + 12z = 2600$ $8x + 15y + 12z = 2600$

$4x + 30y + 6z = 4000$ $\xrightarrow{\text{Mult by } -2}$ $-8x - 60y - 12z = -8000$

Add: $\qquad -45y = -5400$

$$y = \frac{-5400}{-45} = 120 \text{ lb}$$

Now, let $y = 120$ in the 1st and 2nd equations:

$x + 120 + z = 200$ or $x + z = 80$ $\xrightarrow{\text{Mult by } -8}$ $-8x - 8z = -640$

$8x + 15(120) + 12z = 2600$ $8x + 12z = 800$ $8x + 12z = 800$

Add: $\qquad 4z = 160$

$$z = 40 \text{ lb}$$

Finally, substitute $z = 40$ into $x + z = 80$ to obtain
$$x + 40 = 80$$
$$x = 40 \text{ lb}$$

Thus, the mixture requires 40 pounds of 8-4-8 fertilizer, 120 pounds of 15-30-15 fertilizer, and 40 pounds of 12-6-12 fertilizer.

37. a) Select one of the equations and solve for one of the variables in terms of the other two variables.

b) Substitute the expression found in step (a) into the other two equations. This produces a system of two equations in two variables.

c) Solve the system found in step (b) by the substitution method.

d) Substitute the values found in step (c) into the expression obtained in step (a) to find the value of the remaining variable.

Exercises 7-3

1. $y = 2x + 6$
    $x^2 = 2y$

    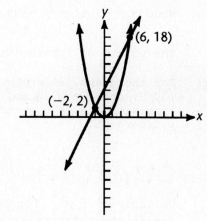

    Substitute $y = 2x + 6$ into $x^2 = 2y$ to obtain
    $$x^2 = 2(2x + 6)$$
    $$x^2 = 4x + 12$$
    $$x^2 - 4x - 12 = 0$$
    $$(x - 6)(x + 2) = 0$$
    $$x - 6 = 0, \quad x + 2 = 0$$
    $$x = 6, \qquad x = -2$$

    When $x = 6$, then $y = 2x + 6 = 2 \cdot 6 + 6$
    $$= 12 + 6 = 18.$$

    When $x = -2$, then $y = 2x + 6 = 2(-2) + 6$
    $$= -4 + 6 = 2.$$

    The two points of intersection are (6, 18) and (-2, 2). For the graph, $y = 2x + 6$ is a straight line and $x^2 = 2y$ is a parabola.

3. $y = 3x^2 + 12x$
    $2x - y = 16$

    Substitute $y = 3x^2 + 12x$ into $2x - y = 16$ to get
    $$2x - (3x^2 + 12x) = 16$$
    $$2x - 3x^2 - 12x = 16$$
    $$-3x^2 - 10x - 16 = 0$$
    $$3x^2 + 10x + 16 = 0$$

    By the quadratic formula $x = \dfrac{-10 \pm \sqrt{(10)^2 - 4(3)(16)}}{2 \cdot 3} = \dfrac{-10 \pm \sqrt{100 - 192}}{6}$

    $= \dfrac{-10 \pm \sqrt{-92}}{6}$ which does not produce real values

    Thus, there is **no** solution and no points of intersection.

    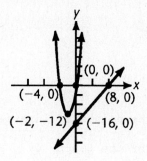

    For the graph $y = 3x^2 + 12x$ is a parabola with vertex at (-2, -12) and $2x - y = 16$ is a straight line with 8 as the x intercept and -16 for the y intercept.

5. $x^2 + y^2 = 4$
    $2x - y = 2$

    Solve $2x - y = 2$ for y: $2x - y = 2$
    $$-y = -2x + 2$$
    $$y = 2x - 2$$
    Substitute $y = 2x - 2$ into $x^2 + y^2 = 4$ to get
    $$x^2 + (2x - 2)^2 = 4$$
    $$x^2 + 4x^2 - 8x + 4 = 4$$
    $$5x^2 - 8x = 0$$
    $$x(5x - 8) = 0$$
    $$x = 0, \quad 5x - 8 = 0$$
    $$x = 0 \quad \text{or} \quad x = \frac{8}{5}$$

When $x = 0$, then $y = 2x - 2 = 2 \cdot 0 - 2 = 0 - 2 = -2$.

When $x = \frac{8}{5}$, then $y = 2x - 2 = 2\left(\frac{8}{5}\right) - 2 = \frac{16}{5} - 2 = \frac{16}{5} - \frac{10}{5} = \frac{6}{5}$.

The points of intersection are $(0, -2)$ and $\left(\frac{8}{5}, \frac{6}{5}\right)$.

For the graph, $x^2 + y^2 = 4$ is a circle with center at $(0, 0)$ and $r = 2$. Also, $2x - y = 2$ is a straight line with 1 as the x intercept and -2 as the y intercept.

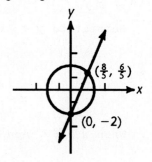

7. $x^2 + y^2 = 10x$
   $4y = 3x - 8$

Multiply $x^2 + y^2 = 10x$ by 16: $16x^2 + 16y^2 = 160x$

Square the equation $4y = 3x - 8$: $(4y)^2 = (3x - 8)^2$ or $16y^2 = 9x^2 - 48x + 64$

Substitute $16y^2 = 9x^2 - 48x + 64$ into $16x^2 + 16y^2 = 160x$

$$16x^2 + (9x^2 - 48x + 64) = 160x$$
$$25x^2 - 48x + 64 = 160x$$
$$25x^2 - 208x + 64 = 0$$
$$(25x - 8)(x - 8) = 0$$
$$25x - 8 = 0, \quad x - 8 = 0$$
$$x = \frac{8}{25}, \qquad x = 8$$

When $x = 8$, then $4y = 3x - 8$ becomes $4y = 3 \cdot 8 - 8$
$$4y = 24 - 8$$
$$4y = 16$$
$$y = 4$$

When $x = \frac{8}{25}$, then $4y = 3x - 8$ becomes $4y = 3\left(\frac{8}{25}\right) - 8$

$$4y = \frac{24}{25} - 8$$
$$4y = -\frac{176}{25}$$
$$y = -\frac{44}{25}$$

The points of intersection are $(8, 4)$ and $\left(\frac{8}{25}, -\frac{44}{25}\right)$.

For the graph, $4y = 3x - 8$ is a straight line with -2 as the y intercept and $\frac{8}{3}$ for the x intercept. Also, the equation $x^2 + y^2 = 10x$ becomes
$$x^2 - 10x + y^2 = 0$$
$$x^2 - 10x + \underline{25} + y^2 = 0 + \underline{25}$$
$$(x - 5)^2 + y^2 = 25$$
Which is a circle with center at $(5, 0)$ and $r = 5$.

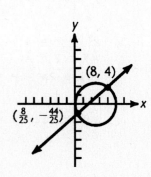

9. $3x^2 + 2y^2 = 5$
   $3x + 2y = -1$ or $2y = -3x - 1$

   Multiply $3x^2 + 2y^2 = 5$ by 2: $6x^2 + 4y^2 = 10$
   Square the equation $2y = -3x - 1$ to get $(2y)^2 = (-3x - 1)^2$
   $\qquad\qquad\qquad\qquad\qquad\qquad$ or $\quad 4y^2 = 9x^2 + 6x + 1$

   Now, substitute $4y^2 = 9x^2 + 6x + 1$ into $6x^2 + 4y^2 = 10$ to get
   $$6x^2 + (9x^2 + 6x + 1) = 10$$
   $$15x^2 + 6x + 1 = 10$$
   $$15x^2 + 6x - 9 = 0$$
   Divide by 3: $\qquad\qquad 5x^2 + 2x - 3 = 0$
   $$(5x - 3)(x + 1) = 0$$
   $$5x - 3 = 0, \ x + 1 = 0$$
   $$x = \frac{3}{5}, \qquad x = -1$$

   When $x = -1$, then $2y = -3x - 1$ becomes $2y = -3(-1) - 1$
   $$2y = 3 - 1$$
   $$2y = 2$$
   $$y = 1$$

   When $x = \frac{3}{5}$, then $2y = -3x - 1$ becomes $2y = -3\left(\frac{3}{5}\right) - 1$
   $$2y = -\frac{9}{5} - 1$$
   $$2y = -\frac{14}{5}$$
   $$y = -\frac{7}{5}$$

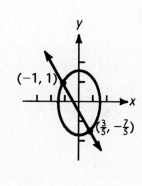

   The points of intersection are $(-1, 1)$ and $\left(\frac{3}{5}, -\frac{7}{5}\right)$.

   For the graphs, $3x^2 + 2y^2 = 5$ is an ellipse with $\pm\sqrt{\dfrac{5}{3}}$ for the x intercepts

   and $\pm\sqrt{\dfrac{5}{2}}$ for the y intercepts. Also, the equation $3x + 2y = -1$ is a
   straight line.

11. $36x^2 + 16y^2 = 25$
    $x - 2y = 6$

    Solve $x - 2y = 6$ for x: $x - 2y = 6$
    $$x = 2y + 6$$

    Substitute $x = 2y + 6$ into $36x^2 + 16y^2 = 25$ to obtain
    $$36(2y + 6)^2 + 16y^2 = 25$$
    $$36(4y^2 + 24y + 36) + 16y^2 = 25$$
    $$144y^2 + 864y + 1296 + 16y^2 = 25$$
    $$160y^2 + 864y + 1271 = 0$$

    By the quadratic formula, $y = \dfrac{-864 \pm \sqrt{(864)^2 - 4(160)(1271)}}{2(160)}$

    $\qquad\qquad\qquad\qquad\qquad = \dfrac{-864 \pm \sqrt{746,496 - 813,440}}{320}$

225

$$= \frac{-864 \pm \sqrt{-66,944}}{320}$$ which has **no** real solutions.

Thus, there is **no** solution and **no** points of intersection.

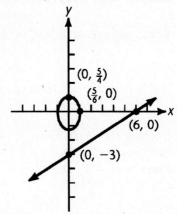

For the graph, $36x^2 + 16y^2 = 25$ is an ellipse with $\pm\frac{5}{6}$ for the x intercepts and $\pm\frac{5}{4}$ for the y intercepts. Also, $x - 2y = 6$ is a straight line.

13. $x^2 + y^2 = 25$
$(3x - 4y)(3x + 4y) = 0$ or $9x^2 - 16y^2 = 0$

$$x^2 + y^2 = 25 \xrightarrow{\text{Mult by -9}} \quad -9x^2 - 9y^2 = -225$$
$$9x^2 - 16y^2 = 0 \qquad\qquad\quad \underline{9x^2 - 16y^2 = \quad\ 0}$$
$$\text{Add:} \qquad -25y^2 = -225$$
$$y^2 = 9$$
$$y = \pm 3$$

Now, substitute $y^2 = 9$ into
$$x^2 + y^2 = 25$$
$$x^2 + 9 = 25$$
$$x^2 = 16$$
$$x = \pm 4$$

The four points of intersection are
$(4, 3)$, $(4, -3)$, $(-4, 3)$, and $(-4, -3)$.

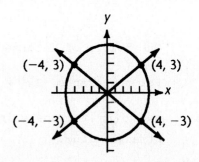

For the graph, $x^2 + y^2 = 25$ is a circle with center at $(0, 0)$ and $r = 5$.
Also, $(3x - 4y)(3x + 4y) = 0$ can be written as $3x - 4y = 0$, $3x + 4y = 0$ and each produces a straight line passing through the origin.

15. $x^2 + y^2 = 9$
$y^2 + 2x = 10$

Solve $y^2 + 2x = 10$ for y^2: $\quad y^2 + 2x = 10$
$$y^2 = -2x + 10$$

Substitute $y^2 = -2x + 10$ into $x^2 + y^2 = 9$ to obtain
$$x^2 + (-2x + 10) = 9$$
$$x^2 - 2x + 10 = 9$$
$$x^2 - 2x + 1 = 0$$
$$(x - 1)^2 = 0$$
$$x - 1 = 0$$
$$x = 1$$

Substitute $x = 1$ into $x^2 + y^2 = 9$ to have
$$1^2 + y^2 = 9$$
$$1 + y^2 = 9$$
$$y^2 = 8$$
$$y = \pm\sqrt{8} = \pm2\sqrt{2}$$

The points are $(1, 2\sqrt{2})$ and $(1, -2\sqrt{2})$.

For the graph, $x^2 + y^2 = 9$ is a circle with center at $(0, 0)$ and $r = 3$. Also, $y^2 + 2x = 10$ is $y^2 = -2x + 10$ which is a parabola opening to the left and with vertex at $(5, 0)$.

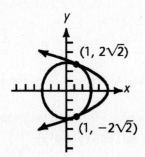

17.
$$x^2 + y^2 = 1$$
$$x^2 - y^2 = 1$$
Add: $\overline{2x^2 = 2}$
$$x^2 = 1$$
$$x = \pm\sqrt{1} = \pm1$$

Substitute $x^2 = 1$ into $x^2 + y^2 = 1$ to get
$$1 + y^2 = 1$$
$$y^2 = 0$$
$$y = 0$$
The solution is $(1, 0)$, $(-1, 0)$.

19. $\dfrac{x^2}{9} + \dfrac{y^2}{16} = 1$ $\xrightarrow{\text{Mult by -9}}$ $-x^2 - \dfrac{9}{16}y^2 = -9$

$x^2 + y^2 = 1$ $\underline{x^2 + y^2 = 1}$

 Add: $\dfrac{7}{16}y^2 = -8$ which has **no** solution since the left side is always positive while the right side is negative.

There is **no** solution to this system.

21. $2x + 7y = 15$

 $xy = 1 \rightarrow y = \dfrac{1}{x}$

Substitute $y = \dfrac{1}{x}$ into $2x + 7y = 15$ to get

$$2x + 7\left(\frac{1}{x}\right) = 15$$

Multiply by x: $2x^2 + 7 = 15x$
$$2x^2 - 15x + 7 = 0$$
$$(2x - 1)(x - 7) = 0$$
$$2x - 1 = 0, \quad x - 7 = 0$$
$$x = \frac{1}{2}, \quad\quad x = 7$$

Since $y = \dfrac{1}{x}$, when $x = \dfrac{1}{2}$, then $y = \dfrac{1}{1/2} = 2$ and when $x = 7$, then $y = \dfrac{1}{7}$.

The solution is $\left(\dfrac{1}{2}, 2\right)$, $\left(7, \dfrac{1}{7}\right)$.

23. $3x + 2y = -5$

$$xy = -1 \rightarrow y = -\frac{1}{x}$$

Substitute $y = -\frac{1}{x}$ into $3x + 2y = -5$ to get

$$3x + 2\left(-\frac{1}{x}\right) = -5$$

Multiply by x: $\qquad 3x^2 - 2 = -5x$

$$3x^2 + 5x - 2 = 0$$
$$(3x - 1)(x + 2) = 0$$
$$3x - 1 = 0, \quad x + 2 = 0$$
$$x = \frac{1}{3}, \qquad x = -2$$

Since $y = -\frac{1}{x}$, when $x = \frac{1}{3}$, then $y = -\frac{1}{1/3} = -3$ and when $x = -2$,

then $y = -\frac{1}{-2} = \frac{1}{2}$.

The solution is $\left(\frac{1}{3}, -3\right)$, $\left(-2, \frac{1}{2}\right)$.

25. $9y^2 - 4x^2 = 7$

$$xy = -2 \rightarrow y = -\frac{2}{x}$$

Substitute $y = -\frac{2}{x}$ into $9y^2 - 4x^2 = 7$ to get

$$9\left(-\frac{2}{x}\right)^2 - 4x^2 = 7$$

$$9\left(\frac{4}{x^2}\right) - 4x^2 = 7$$

Multiply by x^2: $\qquad 36 - 4x^4 = 7x^2$

$$-4x^4 - 7x^2 + 36 = 0$$
$$4x^4 + 7x^2 - 36 = 0$$
$$(4x^2 - 9)(x^2 + 4) = 0$$
$$4x^2 - 9 = 0, \quad \underbrace{x^2 + 4 = 0}$$

$$4x^2 = 9 \qquad \text{No values here}$$

$$x^2 = \frac{9}{4}$$

$$x = \pm\sqrt{\frac{9}{4}} = \pm\frac{3}{2}$$

Since $y = -\frac{2}{x}$, when $x = \frac{3}{2}$, then $y = -\frac{2}{3/2} = -\frac{4}{3}$ and when $x = -\frac{3}{2}$, then

$y = -\frac{2}{-3/2} = \frac{4}{3}$.

The solution is $\left(\frac{3}{2}, -\frac{4}{3}\right)$, $\left(-\frac{3}{2}, \frac{4}{3}\right)$.

27. $2x^2 + 3xy + y^2 = 12$
     $2x^2 - xy + y^2 = 4$

Subtract the 2nd equation from the first: $4xy = 8$
Then solve for y: $xy = 2$
$$y = \frac{2}{x}$$

Substitute $y = \frac{2}{x}$ into the 1st equation: $2x^2 + 3xy + y^2 = 12$

$$2x^2 + 3x\left(\frac{2}{x}\right) + \left(\frac{2}{x}\right)^2 = 12$$

$$2x^2 + 6 + \frac{4}{x^2} = 12$$

$$2x^2 + \frac{4}{x^2} = 6$$

Now, multiply by x^2: $2x^4 + 4 = 6x^2$
$$2x^4 - 6x^2 + 4 = 0$$
$$x^4 - 3x^2 + 2 = 0$$
$$(x^2 - 2)(x^2 - 1) = 0$$
$$x^2 - 2 = 0, \; x^2 - 1 = 0$$
$$x^2 = 2, \qquad x^2 = 1$$
$$x = \pm\sqrt{2}, \qquad x = \pm\sqrt{1} = \pm 1$$

Since $y = \frac{2}{x}$ when $x = \sqrt{2}$, then $y = \frac{2}{\sqrt{2}} = \sqrt{2}$

when $x = -\sqrt{2}$, then $y = \frac{2}{-\sqrt{2}} = -\sqrt{2}$

when $x = 1$, then $y = \frac{2}{1} = 2$

and when $x = -1$, then $y = \frac{2}{-1} = -2$

The solution is $(1, 2)$, $(-1, -2)$, $(\sqrt{2}, \sqrt{2})$, $(-\sqrt{2}, -\sqrt{2})$.

29. $5x^2 - 4xy - 3y^2 = -8$
     $5x^2 + 5xy - 3y^2 = 28$

Subtract the 1st equation from the 2nd: $9xy = 36$
Then, solve for y: $xy = 4$
$$y = \frac{4}{x}$$

Substitute $y = \frac{4}{x}$ into the 1st equation: $5x^2 - 4xy - 3y^2 = -8$

$$5x^2 - 4x\left(\frac{4}{x}\right) - 3\left(\frac{4}{x}\right)^2 = -8$$

$$5x^2 - 16 - \frac{48}{x^2} = -8$$

$$5x^2 - \frac{48}{x^2} = 8$$

Now, multiply by x^2:
$$5x^4 - 48 = 8x^2$$
$$5x^4 - 8x^2 - 48 = 0$$
$$(5x^2 + 12)(x^2 - 4) = 0$$
$$x^2 - 4 = 0, \quad \underbrace{5x^2 + 12 = 0}$$
$$x^2 = 4 \qquad \text{No values here}$$
$$x = \pm\sqrt{4} = \pm 2$$

Since $y = \dfrac{4}{x}$ when $x = 2$, then $y = \dfrac{4}{2} = 2$

and when $x = -2$, then $y = \dfrac{4}{-2} = -2$

The solution is $(2, 2)$, $(-2, -2)$.

31. $x^2 - 3y^2 = -2$
$xy + 2y^2 = 3$

Solve $xy + 2y^2 = 3$ for x:
$$xy + 2y^2 = 3$$
$$xy = -2y^2 + 3$$
$$x = -2y + \frac{3}{y}$$

Substitute $x = -2y + \dfrac{3}{y}$ into $x^2 - 3y^2 = -2$ to get

$$\left(-2y + \frac{3}{y}\right)^2 - 3y^2 = -2$$

$$4y^2 - 12 + \frac{9}{y^2} - 3y^2 = -2$$

$$y^2 + \frac{9}{y^2} = 10$$

Multiply by y^2:
$$y^4 + 9 = 10y^2$$
$$y^4 - 10y^2 + 9 = 0$$
$$(y^2 - 1)(y^2 - 9) = 0$$
$$y^2 - 1 = 0, \quad y^2 - 9 = 0$$
$$y^2 = 1, \qquad y^2 = 9$$
$$y = \pm 1, \qquad y = \pm 3$$

Since $x = -2y + \dfrac{3}{y}$ when $y = 1$, then $x = -2(1) + \dfrac{3}{1} = -2 + 3 = 1$

when $y = -1$, then $x = -2(-1) + \dfrac{3}{-1} = 2 - 3 = -1$

when $y = 3$, then $x = -2(3) + \dfrac{3}{3} = -6 + 1 = -5$

and when $y = -3$, then $x = -2(-3) + \dfrac{3}{-3} = 6 - 1 = 5$

The solution is $(1, 1)$, $(-1, -1)$, $(-5, 3)$, $(5, -3)$.

33. $x^2 - y = 2$
$x = |y| \rightarrow x^2 = y^2$

Substitute $x^2 = y^2$ into $x^2 - y = 2$ to get
$$y^2 - y = 2$$
$$y^2 - y - 2 = 0$$
$$(y - 2)(y + 1) = 0$$
$$y - 2 = 0, \quad y + 1 = 0$$
$$y = 2, \qquad y = -1$$

Since $x = |y|$ when $y = 2$, then $x = |2| = 2$

and when $y = -1$, then $x = |-1| = 1$

The solution is $(1, -1)$, $(2, 2)$.

35. $x^2 + y^2 = 25$

    $y - |x| = 1 \rightarrow -|x| = 1 - y$

                or $|x| = -1 + y$

                or $|x|^2 = (-1 + y)^2$

                or $x^2 = 1 - 2y + y^2$

Substitute $x^2 = 1 - 2y + y^2$ into $x^2 + y^2 = 25$ to get

$$(1 - 2y + y^2) + y^2 = 25$$
$$2y^2 - 2y + 1 = 25$$
$$2y^2 - 2y - 24 = 0$$

Divide by 2: $y^2 - y - 12 = 0$
$$(y - 4)(y + 3) = 0$$
$$y - 4 = 0, \quad y + 3 = 0$$
$$y = 4, \qquad y = -3$$

Since $y - |x| = 1$ when $y = 4$, then $4 - |x| = 1$ or $-|x| = -3$ or $|x| = 3$ which produces $x = 3$ or $x = -3$.

When $y = -3$, then $-3 - |x| = 1$ or $-|x| = 4$ or $|x| = -4$ which has **no** values.

The solution is $(3, 4)$, $(-3, 4)$.

37. $C(x) = 2000 - 50x$ and $R(x) = 50x$

Set $C(x) = R(x)$ to get
$$2000 - 50x = 50x$$
$$2000 = 100x$$
$$\frac{2000}{100} = x$$
$$20 = x$$

When $x = 20$, then $C(20)$ and $R(20)$ produce the same value of 1000. The solution is $(20, 1000)$.

39. Let x and y be the numbers. The system is $x - y = 15$

                                           $xy = 76$

Solve $x - y = 15$ for x: $x - y = 15$
$$x = y + 15 \text{ and substitute into } xy = 76$$
$$(y + 15)y = 76$$
$$y^2 + 15y = 76$$
$$y^2 + 15y - 76 = 0$$
$$(y + 19)(y - 4) = 0$$
$$y + 19 = 0, \quad y - 4 = 0$$
$$y = -19, \quad y = 4$$

Since $x = y + 15$ when $y = -19$, then $x = -19 + 15 = -4$

        and when $y = 4$,   then $x = 4 + 15 = 19$

The solution is 4 and 19 or -4 and -19.

41. Let x and y be the numbers. The system is:
$$x^2 + y^2 = 65$$
$$x^2 - y^2 = 33$$

Add: $2x^2 \qquad = 98$
$$x^2 = 49$$
$$x = \pm\sqrt{49} = \pm 7$$

Substitute $x^2 = 49$ into $x^2 + y^2 = 65$ to get
$$49 + y^2 = 65$$
$$y^2 = 16$$
$$y = \pm\sqrt{16} = \pm 4$$

The solution is 4 and 7, -4 and 7, 4 and -7, or -4 and -7.

43. Let (a, b) be the point(s) on the circle. The slope of the line joining $(0, 0)$ to (a, b) is $\dfrac{b}{a}$. Since the line $y = mx - 5$ is tangent to the circle at (a, b), this line is $\perp$ to the line joining $(0, 0)$ to (a, b). Thus, $m = -1(b/a) = -\dfrac{a}{b}$. The equation of the tangent line is $y = -\dfrac{a}{b}x - 5$. Now let $y = b$ and $x = a$:

$$b = -\dfrac{a}{b}(a) - 5$$
$$b^2 = -a^2 - 5b$$
$$a^2 + b^2 = -5b$$

Also, substitute $x = a$ and $y = b$ into $x^2 + y^2 = 9$

$$a^2 + b^2 = 9$$

The system is $a^2 + b^2 = -5b$
$\qquad\qquad\quad\; a^2 + b^2 = 9$

Subtract the first equation from the second equation:
$$0 = 9 + 5b$$
$$-9 = 5b$$
$$-\dfrac{9}{5} = b$$

Now, substitute $b = -\dfrac{9}{5}$ into $a^2 + b^2 = 9$ to get

$$a^2 + \left(-\dfrac{9}{5}\right)^2 = 9$$

$$a^2 + \dfrac{81}{25} = 9$$

$$a^2 = 9 - \dfrac{81}{25} = \dfrac{225}{25} - \dfrac{81}{25} = \dfrac{144}{25} \text{ so that } a = \pm\dfrac{12}{5}$$

Now, $m = -\dfrac{a}{b} = \pm\dfrac{12/5}{9/5} = \pm\dfrac{12}{9} = \pm\dfrac{4}{3}$.

45. $C(x) = 24 + 20x$ and $R(x) = x(10 + x)$
Set $C(x) = R(x)$ to get $24 + 20x = x(10 + x)$
$$24 + 20x = 10x + x^2$$
$$-x^2 + 10x + 24 = 0$$
$$x^2 - 10x - 24 = 0$$
$$(x - 12)(x + 2) = 0$$
$$x - 12 = 0, \; x + 2 = 0$$
$$x = 12, \qquad x = -2 \leftarrow \text{Discard since } x \geq 0.$$

When $x = 12$, then $C(12) = R(12) = 264$. That is, $x = 12$ becomes 1200 items and the cost or revenue becomes \$264,000.

47. Let s be the length of one side of the end of the carrier and let ℓ be the length. There are two equations containing s and ℓ.

One equation is the volume is 24 (cubic feet). But, the volume of the carrier is area of triangle · length.

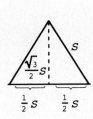

That is, V = area of triangle · length

$$= \frac{1}{2} (\text{base})(\text{height})(\text{length})$$

$$= \frac{1}{2} (s)\left(\frac{\sqrt{3}}{2} s\right)(\ell)$$

$$= \frac{\sqrt{3}}{4} s^2 \ell$$

This equation is $\frac{\sqrt{3}}{4} s^2 \ell = 24$.

For the other equation, it states that 3 times more wood is used for the three sides than for the two ends.
That is   3 sides = 3 · 2 ends

$$3(s\ell) = 3 \cdot 2\left(\frac{\sqrt{3}}{4} s^2\right)$$

or upon dividing by 3: $s\ell = 2\left(\frac{\sqrt{3}}{4} s^2\right)$

or $\ell = \frac{\sqrt{3}}{2} s$

The system is $\frac{\sqrt{3}}{4} s^2 \ell = 24$

$$\ell = \frac{\sqrt{3}}{2} s$$

Use the substitution method to let $\ell = \frac{\sqrt{3}}{2} s$ into the equation $\frac{\sqrt{3}}{4} s^2 \ell = 24$ to get:

$$\frac{\sqrt{3}}{4} s^2 \left(\frac{\sqrt{3}}{2} s\right) = 24$$

$$\frac{3}{8} s^3 = 24$$

$$s^3 = 64$$

$$s = \sqrt[3]{64} = 4 \text{ ft}$$

Then, $\ell = \frac{\sqrt{3}}{2} s = \frac{\sqrt{3}}{2} (4) = 2\sqrt{3}$ ft. The dimensions are $s = 4$ ft and $\ell = 2\sqrt{3}$ ft.

49. $(1, 0)$, $(0.23, -1.46)$ 51. $(1.44, 1.39)$

53. $x^2 + y^2 = 1$ $x^2 + y^2 = 1$

   $x^2 - y = 4$ $\xrightarrow{\text{Mult by } -1}$ $-x^2 + y = -4$

   $\phantom{x^2 - y = 4 \xrightarrow{\text{Mult by}}}$ Add: $\overline{ y^2 + y = -3}$

   $\phantom{x^2 - y = 4 \xrightarrow{\text{Mult by}}}$ or $y^2 + y + 3 = 0$

To solve, use the quadratic formula to get

$$y = \frac{-1 \pm \sqrt{1^2 - 4(1)(3)}}{2(1)} = \frac{-1 \pm \sqrt{1 - 12}}{2} = \frac{-1 \pm \sqrt{-11}}{2} = \frac{-1 \pm \sqrt{11}\, i}{2}$$

Now, when $y = \dfrac{-1 + \sqrt{11}\,i}{2} = -\dfrac{1}{2} + \dfrac{\sqrt{11}\,i}{2}$, then the equation $x^2 - y = 4$

becomes $x^2 - \left(-\dfrac{1}{2} + \dfrac{\sqrt{11}\,i}{2}\right) = 4$

$$x^2 = 4 + \left(-\dfrac{1}{2} + \dfrac{\sqrt{11}\,i}{2}\right) = \dfrac{7}{2} + \dfrac{\sqrt{11}\,i}{2}$$

and $\qquad x = \pm\sqrt{\dfrac{7}{2} + \dfrac{\sqrt{11}\,i}{2}}$

Also, when $y = \dfrac{-1 - \sqrt{11}\,i}{2} = -\dfrac{1}{2} - \dfrac{\sqrt{11}\,i}{2}$, then the equation $x^2 - y = 4$

becomes $x^2 - \left(-\dfrac{1}{2} - \dfrac{\sqrt{11}\,i}{2}\right) = 4$

$$x^2 = 4 + \left(-\dfrac{1}{2} - \dfrac{\sqrt{11}\,i}{2}\right) = \dfrac{7}{2} - \dfrac{\sqrt{11}\,i}{2}$$

and $\qquad x = \pm\sqrt{\dfrac{7}{2} - \dfrac{\sqrt{11}\,i}{2}}$

The solution occurs from the system of complex numbers.

Exercises 7.4

1. $3x + 2y \le 12$
   First, $3x + 2y = 12$ is a straight line with 6 as
   the y intercept and 4 as the x intercept. For the
   test point use $(0, 0)$.
   $$3x + 2y \le 12$$
   $$3(0) + 2(0) \le 12$$
   $$0 \le 12$$
   Thus, the region containing the origin, $(0, 0)$,
   is shaded. The boundary is included (solid line).

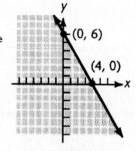

3. $y > (x - 1)^2$
   First, $y = (x - 1)^2$ is a parabola opening upward
   with vertex at $(1, 0)$. For the test point use
   $(0, 0)$.
   $$y > (x - 1)^2$$
   $$0 > (0 - 1)^2$$
   $$0 \not> 1$$
   Thus, the region **not** containing $(0, 0)$ is shaded.
   The boundary is not included (dashed curve).

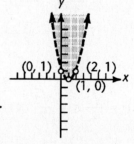

5. $y \le 4 - x^2$
   First, $y = 4 - x^2$ is a parabola opening downward
   with vertex at $(0, 4)$. For the test point use
   $(0, 0)$.
   $$y \le 4 - x^2$$
   $$0 \le 4 - 0^2$$
   $$0 \le 4$$
   Thus, the region containing $(0, 0)$ is shaded.
   The boundary is included (solid curve).

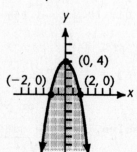

7. $x^2 + y^2 \leq 16$

First, $x^2 + y^2 = 16$ is a circle with center at
(0, 0) and $r = 4$. For the test point, use (0, 0).
$$x^2 + y^2 \leq 16$$
$$0^2 + 0^2 \leq 16$$
$$0 \leq 16$$
Thus, the region containing (0, 0) is shaded.
The boundary is included (solid curve/circle).

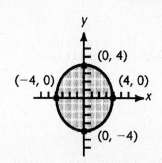

For Problems 9—23, the curves were sketched and the points of intersection were identified in the previous exercise set. The inequalities are graphed here.

9.   11. No solution   13.   15.

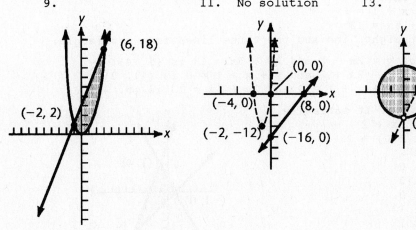

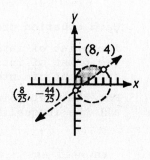

17.   19. ∅   21.   23.

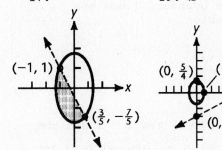

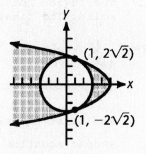

25. $x - y > 3$
     $x + 2y > 4$
The boundary lines are $x - y = 3$
$\qquad\qquad\qquad\qquad x + 2y = 4$
Each equation produces a straight line. For the point of intersection,
subtract the first equation from the second: $3y = 1$
$$y = \frac{1}{3}$$

Substitute $y = \frac{1}{3}$ into the second equation: $x + 2\left(\frac{1}{3}\right) = 4$

$$x = 4 - \frac{2}{3} = \frac{12}{3} - \frac{2}{3} = \frac{10}{3}$$

The point is $\left(\frac{10}{3}, \frac{1}{3}\right)$.

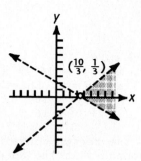

For the test point, use (6, 0). It satisfies
both inequalities:

$$x - y > 3$$
$$6 - 0 > 3$$
$$6 > 3$$

as well as
$$x + 2y > 4$$
$$6 + 2(0) > 4$$
$$6 > 4$$

The region containing (6, 0) is shaded and both boundary lines are dashed.

27. $y \leq 2x + 2$
$y + x + 1 \geq 0$
$2y + 5x \leq 13$

The boundary lines are $y = 2x + 2$
$$y + x + 1 = 0$$
$$2y + 5x = 13$$

Each equation produces a straight line and the three lines form a triangle.

The point of intersection of $y = 2x + 2$ and $2y + 5x = 13$ is (1, 4).
The point of intersection of $y = 2x + 2$ and $y + x + 1 = 0$ is (-1, 0).
The point of intersection of $y + x + 1 = 0$ and $2y + 5x = 13$ is (5, -6).

For the test point, use (0, 0). It satisfies
all the inequalities: $y \leq 2x + 2$
$$0 \leq 2(0) + 2$$
$$0 \leq 2$$

as well as
$$y + x + 1 \geq 0$$
$$0 + 0 + 1 \geq 0$$
$$1 \geq 0$$

as well as
$$2y + 5x \leq 13$$
$$2(0) + 5(0) \leq 13$$
$$0 \leq 13$$

The region containing (0, 0) is shaded and all three boundary lines are
solid. The final region is the triangle and its interior.

29. $x^2 + y^2 > 4$
$4x^2 + 9y^2 \leq 36$

The boundary curves are $x^2 + y^2 = 4$
$$4x^2 + 9y^2 = 36$$

The first equation produces a circle with center at (0, 0) and $r = 2$ and the
second equation can be written as $\dfrac{x^2}{9} + \dfrac{y^2}{4} = 1$ which is an ellipse with ±3 for
the x intercepts and ±2 for the y intercepts.

For the points of intersection, solve the system:

$$x^2 + y^2 = 4 \xrightarrow{\text{Mult by } -4} -4x^2 - 4y^2 = -16$$
$$4x^2 + 9y^2 = 36 \qquad\qquad \underline{4x^2 + 9y^2 = 36}$$
$$\text{Add:} \qquad 5y^2 = 20$$
$$y^2 = 4$$
$$y = \pm\sqrt{4} = \pm 2$$

Use $y^2 = 4$ in $x^2 + y^2 = 4$ to get
$$x^2 + 4 = 4$$
$$x^2 = 0$$
$$x = 0$$

The points are (0, 2) and (0, -2).

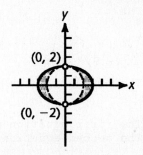

At this time, the graph of the circle is inside the ellipse. For test points, use (2.5, 0) or (-2.5, 0) and note that each one satisfies the inequalities:

$$x^2 + y^2 > 4$$
$$(\pm 2.5)^2 + 0^2 > 4$$
$$6.25 > 4$$

as well as
$$4x^2 + 9y^2 \leq 36$$
$$4(\pm 2.5)^2 + 9(0)^2 \leq 36$$
$$25 \leq 36$$

The regions containing (2.5, 0) and (-2.5, 0) are shaded with the circle being a dashed curve and the ellipse being a solid curve. *Note:* Any other test point not in the shaded region would not satisfy at least one of the inequalities.

31. $x^2 + 2y^2 \geq 18$
$2x^2 + y^2 \leq 33$

The boundary curves are $x^2 + 2y^2 = 18$
$$2x^2 + y^2 = 33$$
Each equation produces an ellipse: $x^2 + 2y^2 = 18$
$$\frac{x^2}{18} + \frac{y^2}{9} = 1$$

x intercepts $\pm\sqrt{18} \approx 4.2$, y intercepts ± 3

$$2x^2 + y^2 = 33$$
$$\frac{x^2}{\frac{33}{2}} + \frac{y^2}{33} = 1$$

x intercepts $\pm\sqrt{\frac{33}{2}} \approx \pm 4.1$, y intercepts $\pm\sqrt{33} \approx 5.7$

For the points of intersection solve the system:

$x^2 + 2y^2 = 18$ $\xrightarrow{\text{Mult by -2}}$ $-2x^2 - 4y^2 = -36$
$2x^2 + y^2 = 33$ $\underline{\quad 2x^2 + y^2 = \quad 33}$
Add: $\quad -3y^2 = \quad -3$
$$y^2 = 1$$
$$y = \pm\sqrt{1} = \pm 1$$

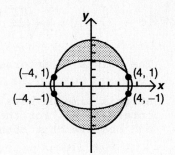

Substitute $y^2 = 1$ into $x^2 + 2y^2 = 18$ to get
$$x^2 + 2(1) = 18$$
$$x^2 = 16$$
$$x = \pm\sqrt{16} = \pm 4$$

The points of intersection are (-4, -1), (-4, 1), (4, 1), and (4, -1). At this time we have two intersecting ellipses. For the test points, use (0, 4) and (0, -4) and observe that both inequalities are satisfied since
$$x^2 + 2y^2 \geq 18$$
$$0^2 + 2(\pm 4)^2 \geq 18$$
$$32 \geq 18$$
as well as $2x^2 + y^2 \leq 33$
$$2(0)^2 + (\pm 4)^2 \leq 33$$
$$16 \leq 33$$

The regions containing (0, 4) and (0, -4) are shaded with both curves being solid. *Note:* Any test point not in the shaded region would not satisfy at least one of the inequalities.

33. $x^2 - y^2 > 1$
    $4x^2 + 9y^2 < 36$

The boundary curves are $x^2 - y^2 = 1$
$$4x^2 + 9y^2 = 36$$

The first equation produces a hyperbola with vertices at (1, 0) and (-1, 0).
The second equation is $4x^2 + 9y^2 = 36$

$$\frac{x^2}{9} + \frac{y^2}{4} = 1 \quad \text{which is an ellipse with } x \text{ intercepts } \pm 3 \text{ and } y \text{ intercepts } \pm 2$$

For the point of intersection, solve the system:

$$x^2 - y^2 = 1 \xrightarrow{\text{Mult by 9}} 9x^2 - 9y^2 = 9$$
$$4x^2 + 9y^2 = 36 \qquad\qquad \underline{4x^2 + 9y^2 = 36}$$
$$\text{Add:} \quad 13x^2 \qquad = 45$$
$$x^2 = \frac{45}{13}$$

$$x = \pm\sqrt{\frac{45}{13}} = \pm\sqrt{\frac{45}{13} \cdot \frac{13}{13}} = \pm\sqrt{\frac{9 \cdot 5 \cdot 13}{13 \cdot 13}}$$
$$= \pm\frac{3\sqrt{65}}{13}$$

To find y, substitute $x^2 = \frac{45}{13}$ into $x^2 - y^2 = 1$ to get

$$x^2 - y^2 = 1$$
$$\frac{45}{13} - y^2 = 1$$
$$-y^2 = 1 - \frac{45}{13}$$
$$-y^2 = -\frac{32}{13}$$
$$y^2 = \frac{32}{13}$$

$$y = \pm\sqrt{\frac{32}{13}} = \pm\sqrt{\frac{32 \cdot 13}{13 \cdot 13}} = \pm\sqrt{\frac{16 \cdot 2 \cdot 13}{13 \cdot 13}} = \pm\frac{4\sqrt{26}}{13}$$

The points of intersection are $\left(\pm\frac{3\sqrt{65}}{13}, \pm\frac{4\sqrt{26}}{13} \right)$.

At this time we have an ellipse and a hyperbola
intersecting. For the test points, use (2, 0) and (-2, 0)
and observe that both inequalities are satisfied:

$$x^2 - y^2 > 1 \qquad \text{as well as} \qquad 4x^2 + 9y^2 < 36$$
$$(\pm 2)^2 - (0)^2 > 1 \qquad\qquad 4(\pm 2)^2 + 9(0)^2 < 36$$
$$4 - 0 > 1 \qquad\qquad 8 + 0 < 36$$
$$4 > 1 \qquad\qquad 8 < 36$$

The regions containing (2, 0) and (-2, 0) are shaded with both curves being
dashed lines.

35. $y \geq 2^x$
    $y \geq 2^{-x}$

The boundary curves are $y = 2^x$
$\qquad\qquad\qquad\qquad y = 2^{-x}$

Each produces an exponential curve. For the point of intersection equate $y = 2^x$
and $y = 2^{-x}$ to get $\qquad 2^x = 2^{-x}$
$\qquad\qquad\qquad\qquad\qquad x = -x$
$\qquad\qquad\qquad\qquad\quad 2x = 0$
$\qquad\qquad\qquad\qquad\quad\; x = 0$

When $x = 0$, the corresonding y value is 1. For
the test point use (0, 5) observing that both
inequalities are satisfied:

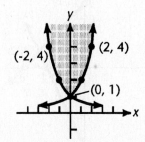

$\qquad\qquad\qquad\qquad y \geq 2^x$
$\qquad\qquad\qquad\qquad 5 \geq 2^0$
$\qquad\qquad\qquad\qquad 5 \geq 1$
as well as $\qquad\quad y \geq 2^{-x}$
$\qquad\qquad\qquad\qquad 5 \geq 2^{-0}$
$\qquad\qquad\qquad\qquad 5 \geq 1$

The region containing (0, 5) is shaded with both
curves being solid.

Note: Any test point not in the shaded region
would not satisfy at least one of the
inequalities.

37. The region 'above' the line $x + 2y = 4$ is given as $x + 2y \geq 4$.
    The region 'above' the line $x - y = 1$ is given as $x - y \leq 1$.
    Solution is $x + 2y \geq 4$, $x - y \leq 1$.

39. The region 'below' the parabola $y = -x^2 - 2x + 3$ is given as $y \leq -x^2 - 2x + 3$.
    The region to the 'left' of the line $x = 0$ is given as $x \leq 0$.
    Solution is $y \leq -x^2 - 2x + 3$, $x \leq 0$.

41. The region 'above' the parabola $y = x^2$ is given as $y > x^2$.
    The region 'below' the line $y - x = 2$ is given as $y - x < 2$.
    Solution is $y > x^2$, $y - x < 2$.

43. The region 'above' the exponential curve $y = 2^x$ is given as $y \geq 2^x$.
    The region to the 'right' of the line $x = -2$ is given as $x \geq -2$.
    The region 'below' the line $x + y = 1$ is given as $x + y \leq 1$.
    Solution is $y \geq 2^x$, $x + y \leq 1$, $x \geq -2$.

45.

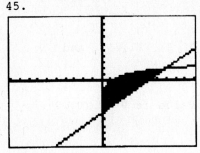

viewing rectangle [-10, 10] by [-10, 10]

47.

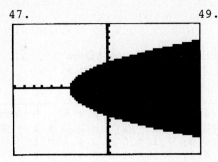

viewing rectangle [-10, 10] by [-10, 10]

49.

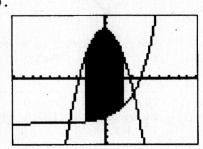

viewing rectangle [-10, 10] by [-10, 10]

51. a) For $y - mx = b$, the points are on the line.

   b) For $y - mx > b$, the points are 'above' the line.

   c) For $y - mx < b$, the points are 'below' the line.

Exercises 7.5

1.

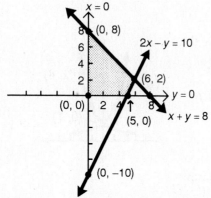

The convex set is a polygon with vertices $(0, 0)$, $(0, 8)$, $(6, 2)$, and $(5, 0)$. Construct the following table.

Vertex	$f = 2x + y$
$(0, 0)$	$2(0) + 0 = 0 + 0 = 0$
$(0, 8)$	$2(0) + 8 = 0 + 8 = 8$
$(6, 2)$	$2(6) + 2 = 12 + 2 = 14$ ← Max value is 14 at $(6, 2)$
$(5, 0)$	$2(5) + 0 = 10 + 0 = 10$

3.

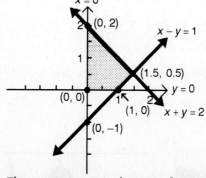

The convex set is a polygon with vertices $(0, 0)$, $(0, 2)$, $(1, 0)$, and $(1.5, 0.5)$. Construct the following table.

Vertex	$f = x + 4y + 1$
$(0, 2)$	$0 + 4(2) + 1 = 8 + 1 = 9$ ← Max value is 9 at $(0, 2)$
$(1.5, 0.5)$	$1.5 + 4(0.5) + 1 = 1.5 + 2.0 + 1 = 4.5$
$(0, 0)$	$0 + 4(0) + 1 = 0 + 0 + 1 = 1$
$(1, 0)$	$1 + 4(0) + 1 = 1 + 0 + 1 = 2$

240

5.

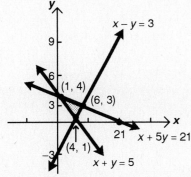

The convex set is the triangle with vertices (1, 4), (4, 1), and (6, 3).
Construct the following table.

Vertex	$f = 3x + y + 4$
(1, 4)	$3(1) + 4 + 4 = 11$ ← Min value is 11 at (1, 4)
(4, 1)	$3(4) + 1 + 4 = 17$
(6, 3)	$3(6) + 3 + 4 = 25$

7.

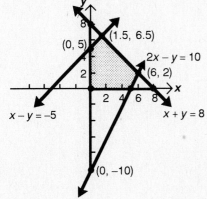

The convex set is a polygon with vertices (0, 0), (0, 5), (1.5, 6.5), (5, 0), and (6, 2). Construct the following table.

Vertex	$f = 2x + y$
(0, 0)	$2(0) + (0) = 0$
(0, 5)	$2(0) + 5 = 5$
(1.5, 6.5)	$2(1.5) + 6.5 = 9.5$
(6, 2)	$2(6) + 2 = 14$ ← Max value is 14 at (6, 2)
(5, 0)	$2(5) + 0 = 10$

9.

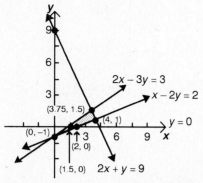

The convex set is a polygon with vertices (1.5, 0), (2, 0), (4, 1), and (3.75, 1.5). Construct the following table.

Vertex	$f = y - x + 3$
(1.5, 0)	0 - 1.5 + 3 = 1.5
(2, 0)	0 - 2 + 3 = 1
(4, 1)	1 - 4 + 3 = 0 ← Min value is 0 at (4, 1)
(3.75, 1.5)	1.5 - 3.75 + 3 = 0.75

11.

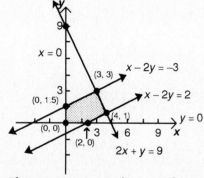

The convex set is a polygon with vertices (0, 0), (2, 0), (0, 1.5), (3, 3), and (4, 1). Construct the following table.

Vertex	$f = x + y - 2$
(0, 0)	0 + 0 - 2 = -2
(2, 0)	2 + 0 - 2 = 0
(0, 1.5)	0 + 1.5 - 2 = -0.5
(3, 3)	3 + 3 - 2 = 4 ← Max value is 4 at (3, 3)
(4, 1)	4 + 1 - 2 = 3

13. Let x be the number of cans of the first
mixture and y be the number of cans of the
second mixture. The system of inequalities is

$$6x + 5y \leq 240$$
$$2x + 3y \leq 96$$
$$x \geq 0$$
$$y \geq 0$$

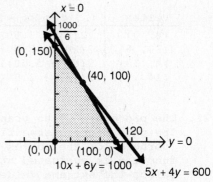

The convex set is a polygon with vertices
(0, 0), (0, 32), (40, 0), and (30, 12).
The function to be maximized is $f = 4x + 5y$.
The following table is constructed.

Vertex	$f = 4x + 5y$
(0, 0)	$4(0) + 5(0) = \$0$
(0, 32)	$4(0) + 5(32) = \$160$
(40, 0)	$4(40) + 5(0) = \$160$
(30, 12)	$4(30) + 5(12) = \$180$ ← Max profit is \$180 when 30 cans of the first mixture and 12 cans of the second mixture are formed.

15. Let x be the number of desks of the first type
and y be the number of desks of the second type.

$$10x + 6y \leq 1000$$
$$5x + 4y \leq 600$$
$$x \geq 0$$
$$y \geq 0$$

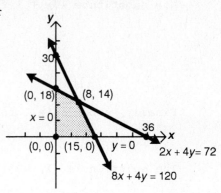

The convex set is a polygon with vertices (0, 0),
(0, 150), (100, 0), and (40, 100). The function
to be maximized is $f = 45x + 30y$. The following
table is constructed.

Vertex	$f = 45x + 30y$
(0, 0)	$45(0) + 30(0) = \$0$
(0, 150)	$45(0) + 30(150) = \$4500$
(100, 0)	$45(100) + 30(0) = \$4500$
(40, 100)	$45(40) + 30(100) = \$4800$ ← Max profit is \$4800 when 40 desks of the first type and 100 desks of the second type are manufactured.

17. Let x be the number of chairs and y be the number
of stools manufactured.

$$1\frac{3}{5}x + \frac{4}{5}y \leq 24 \xrightarrow{\text{Mult by 5}} 8x + 4y \leq 120$$
$$\frac{2}{3}x + \frac{4}{3}y \leq 24 \xrightarrow{\text{Mult by 3}} 2x + 4y \leq 72$$
$$x \geq 0 \qquad\qquad x \geq 0$$
$$y \geq 0 \qquad\qquad y \geq 0$$

The convex set is the polygon with vertices
(0, 0), (0, 18), (15, 0), and (8, 14). The
function to be maximized is $f = 30x + 20y$.
The following is constructed.

Vertex	$f = 30x + 20y$
(0, 0)	$30(0) + 20(0) = \$0$
(0, 18)	$30(0) + 20(18) = \$360$
(15, 0)	$30(15) + 20(0) = \$450$
(8, 14)	$30(8) + 20(14) = \$520$ ← Max profit is \$520 when 8 chairs and 14 stools are manufactured.

19. Let x be the number of cold cut trays and y be the number of seafood trays.

$$10x + 25y \leq 365 \quad \textit{Note: } 6 \text{ hrs and 5 min is } 6(60) + 5 = 365 \text{ min}$$
$$15x + 10y \leq 300 \quad 5 \text{ hrs is } 5(60) = 300 \text{ min}$$
$$x \geq 4$$
$$y \geq 0$$

The convex set is the polygon with vertices (4, 0), (20, 0), (4, 13), and (14, 9). The function to be maximized is $f = 2x + 3.50y$. The following table is constructed.

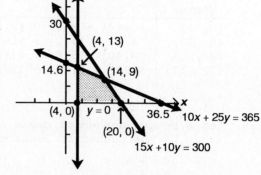

Vertex	$f = 2x + 3.50y$
(4, 0)	$2(4) + 3.50(0) = \$8$
(20, 0)	$2(20) + 3.50(0) = \$40$
(4, 13)	$2(4) + 3.50(13) = \$53.50$
(14, 9)	$2(14) + 3.50(9) = \$59.50$ ← Max profit is \$59.50 when 14 cold cut trays and 9 seafood trays are produced.

21. The procedure is to graph the inequalities to form the convex set. Identify the points of intersection (vertices) where pairs of lines form the 'corners' of the convex set. Set up a table of values where the linear function is evaluated at each vertex. From the list of functional values, select the extreme values (maximum value or minimum value).

CHAPTER REVIEW

1. $3x + y = 1$
   $x + 3y = -5$

Solve $3x + y = 1$ for y: $3x + y = 1$
$$y = -3x + 1$$

Now substitute $-3x + 1$ for y into $x + 3y = -5$ to obtain
$$x + 3(-3x + 1) = -5$$
$$x - 9x + 3 = -5$$
$$-8x + 3 = -5$$
$$-8x = -8$$
$$x = \frac{-8}{-8} = 1$$

Finally, substitute 1 for x into $y = -3x + 1$
$$= -3(1) + 1$$
$$= -3 + 1$$
$$= -2$$

The solution is (1, -2).

2. $4x - 5y = 13$
   $x + 2y = 0$

   Solve $x + 2y = 0$ for x: $x + 2y = 0$
   $$x = -2y$$

   Now, substitute $-2y$ for x into the first equation: $4x - 5y = 13$
   $$4(-2y) - 5y = 13$$
   $$-8y - 5y = 13$$
   $$-13y = 13$$
   $$y = \frac{13}{-13} = -1$$

   Substitute -1 for y into the equation $x = -2y$
   $$x = -2(-1)$$
   $$x = 2$$

   The solution is $(2, -1)$.

3. $3x + 2y = 1$ $\xrightarrow{\text{Mult by } -2}$ $-6x - 4y = -2$

   $2x + 3y = 9$ $\xrightarrow{\text{Mult by } 3}$ $\underline{6x + 9y = 27}$

   Add: $\qquad\qquad\qquad\qquad 5y = 25$
   $$y = \frac{25}{5} = 5$$

   Now, substitute 5 for y into $3x + 2y = 1$
   $$3x + 2(5) = 1$$
   $$3x + 10 = 1$$
   $$3x = -9$$
   $$x = \frac{-9}{3} = -3$$

   The solution is $(-3, 5)$.

4. $9x + 2y = 3$ $\xrightarrow{\text{Mult by } -5}$ $-45x - 10y = -15$

   $7x + 5y = -8$ $\xrightarrow{\text{Mult by } 2}$ $\underline{14x + 10y = -16}$

   Add: $\qquad\qquad -31x \qquad\;\; = -31$
   $$x = \frac{-31}{-31} = 1$$

   Substitute 1 for x into the equation $9x + 2y = 3$
   $$9(1) + 2y = 3$$
   $$9 + 2y = 3$$
   $$2y = -6$$
   $$y = \frac{-6}{2} = -3$$

   The solution is $(1, -3)$.

5. $2x - 5y = 3$ $\qquad\qquad 2x - 5y = 3$

   $3x - y = -2$ $\xrightarrow{\text{Mult by } -5}$ $\underline{-15x + 5y = 10}$

   Add: $\qquad\qquad -13x \qquad\;\; = 13$
   $$x = \frac{13}{-13} = -1$$

   Now, substitute -1 for x into $2x - 5y = 3$
   $$2(-1) - 5y = 3$$
   $$-2 - 5y = 3$$
   $$-5y = 5$$
   $$y = \frac{5}{-5} = -1$$

   The solution is $(-1, -1)$.

6. $4x - 3y = 5$ $\qquad\qquad$ $4x - 3y = 5$

$3x + y = 7$ $\xrightarrow{\text{Mult by 3}}$ $\dfrac{9x + 3y = 21}{}$

$\qquad\qquad\qquad$ Add: $\quad 13x \qquad = 26$

$$x = \frac{26}{13} = 2$$

Substitute 2 for x into the equation $3x + y = 7$ to get
$$3(2) + y = 7$$
$$6 + y = 7$$
$$y = 1$$

The solution is $(2, 1)$.

7. $3x + 5y = 7$ $\xrightarrow{\text{Mult by 3}}$ $9x + 15y = 21$

$2x - 3y = 11$ $\xrightarrow{\text{Mult by 5}}$ $\dfrac{10x - 15y = 55}{}$

$\qquad\qquad\qquad$ Add: $\quad 19x \qquad = 76$

$$x = \frac{76}{19} = 4$$

Now, substitute 4 for x into $3x + 5y = 7$
$$3(4) + 5y = 7$$
$$12 + 5y = 7$$
$$5y = -5$$
$$y = \frac{-5}{5} = -1$$

The solution is $(4, -1)$.

8. $2x - 3y = 5$ $\xrightarrow{\text{Mult by -2}}$ $-4x + 6y = -10$

$4x - 6y = -10$ $\qquad\qquad$ $\dfrac{4x - 6y = -10}{}$

$\qquad\qquad\qquad$ Add: $\qquad\qquad 0 = -20$ Impossible

There is **no** solution and the system is inconsistent.

9. $2x + 4y = 4$ $\xrightarrow{\text{Mult by -1.5}}$ $-3x - 6y = -6$

$3x + 6y = 6$ $\qquad\qquad$ $\dfrac{3x + 6y = 6}{}$

$\qquad\qquad\qquad$ Add: $\qquad\qquad 0 = 0$ an identity

Thus, the solution is all values satisfying the equation $2x + 4y = 4$ which reduces (simplifies) to $x + 2y = 2$
$$\text{or} \qquad x = 2 - 2y$$
The answer is $\{(x, y) \mid x = 2 - 2y\}$.

10. Let b be the rate of the boat and c the rate of the current. The system is:

upstream: $3(b - c) = 36$ $\quad$ or $\quad$ $b - c = 12$

downstream: $2(b + c) = 36$ $\qquad\qquad$ $\dfrac{b + c = 18}{}$

$\qquad\qquad\qquad$ Add: $2b \qquad = 30$

$$b = \frac{30}{2} = 15 \text{ miles per hour}$$

Substitute 15 for b into the equation $b + c = 18$ to get
$$15 + c = 18$$
$$c = 3 \text{ miles per hour}$$

The rate of the current is 3 mph and the rate of the boat is 15 mph.

11. Let x be the speed of the plane in still air and y the speed of the wind. The two equations are:

Against wind: $9(x - y) = 900$ $\xrightarrow{\text{Divide by 9}}$ $x - y = 100$

with wind: $6(x + y) = 900$ $\xrightarrow{\text{Divide by 6}}$ $\underline{x + y = 150}$

$$\text{Add:} \quad 2x \quad\quad = 250$$

$$x = \frac{250}{2} = 125 \text{ mph}$$

Now, substitute 125 for x into $x + y = 150$

$$125 + y = 150$$

$$y = 25 \text{ mph}$$

Thus, the speed of the plane in still air is 125 miles per hour and the speed of the air is 25 miles per hour.

12. Let x be the amount of the first alloy and y the amount of the second alloy. The system is

$$\begin{array}{l} x + y = 300 \\ 0.7x + 0.2y = (300)(0.4) \end{array} \text{ or } \begin{array}{l} x + y = 300 \\ 7x + 2y = 1200 \end{array} \xrightarrow{\text{Mult by -2}} \begin{array}{l} -2x - 2y = -600 \\ \underline{7x + 2y = 1200} \end{array}$$

$$\text{Add:} \quad 5x \quad\quad = 600$$

$$x = \frac{600}{5} = 120 \text{ lbs.}$$

Now, substitute 120 for x into the equation $x + y = 300$ to get

$$120 + y = 300$$

$$y = 180 \text{ pounds}$$

Thus, 120 pounds of the first alloy and 180 pounds of the second alloy are needed.

13. $\dfrac{x^2}{a^2} - \dfrac{y^2}{b^2} = 1$ Let $x = 1$ and $y = 1$: $\dfrac{1^2}{a^2} - \dfrac{1^2}{b^2} = 1$

$$\text{or} \quad \frac{1}{a^2} - \frac{1}{b^2} = 1$$

Let $x = \dfrac{5}{7}$ and $y = \dfrac{1}{7}$ $\dfrac{(5/7)^2}{a^2} - \dfrac{(1/7)^2}{b^2} = 1$

$$\text{or} \quad \frac{25/49}{a^2} - \frac{1/49}{b^2} = 1$$

Multiply by 49: $\dfrac{25}{a^2} - \dfrac{1}{b^2} = 49$

The system is $\dfrac{1}{a^2} - \dfrac{1}{b^2} = 1$ $\xrightarrow{\text{Mult by -1}}$ $-\dfrac{1}{a^2} + \dfrac{1}{b^2} = -1$

$$\frac{25}{a^2} - \frac{1}{b^2} = 49 \quad\quad\quad\quad \frac{25}{a^2} - \frac{1}{b^2} = 49$$

$$\text{Add:} \quad \frac{24}{a^2} \quad\quad = 48$$

$$24 = 48a^2$$

$$\frac{24}{48} = a^2$$

$$\frac{1}{2} = a^2$$

$$\sqrt{\frac{1}{2}} = a \quad \text{(Use only positive value here.)}$$

$$\frac{\sqrt{2}}{2} = a$$

Now, substitute $a^2 = \dfrac{1}{2}$ into $\dfrac{1}{a^2} - \dfrac{1}{b^2} = 1$

$$\dfrac{1}{\dfrac{1}{2}} - \dfrac{1}{b^2} = 1$$

$$2 - \dfrac{1}{b^2} = 1$$

$$-\dfrac{1}{b^2} = -1$$

$$\dfrac{1}{b^2} = 1$$

$$1 = b^2$$

$$\sqrt{1} = b \text{ (Again, use only the positive value here.)}$$
$$1 = b$$

Thus, $a = \dfrac{\sqrt{2}}{2}$ and $b = 1$.

14. $x + y - z = -5$
   $3y + 5z = 3$
   $2x \quad - 3z = -5$

First eliminate x between the 1st and 3rd equations:

$x + y - z = -5$ $\xrightarrow{\text{Mult by -2}}$ $-2x - 2y + 2z = 10$

$2x \quad - 3z = -5$ $\underline{\qquad 2x \qquad - 3z = -5}$

Add: $\qquad -2y - z = 5$

Combine this equation with the 2nd equation to get a system in 2 unknowns:

$-2y - z = 5$ $\xrightarrow{\text{Mult by 5}}$ $-10y - 5z = 25$

$3y + 5z = 3$ $\underline{\qquad 3y + 5z = 3}$

Add: $-7y \qquad = 28$ or $y = \dfrac{28}{-7} = -4$

To find z, substitute -4 for y into the equation $3y + 5z = 3$

$$3(-4) + 5z = 3$$
$$-12 + 5z = 3$$
$$5z = 15$$
$$z = \dfrac{15}{5} = 3$$

To find x, substitute 3 for z into the equation $2x - 3z = -5$

$$2x - 3(3) = -5$$
$$2x - 9 = -5$$
$$2x = 4$$
$$x = \dfrac{4}{2} = 2$$

The solution is $(2, -4, 3)$.

15. $x - 3y - 4z = -10$
   $2x + y + 3z = 0$
   $x \quad - 3z = 6$

First, eliminate y between the 1st and 2nd equations:

$x - 3y - 4z = -10$ $\qquad\qquad x - 3y - 4z = -10$

$2x + y + 3z = 0$ $\xrightarrow{\text{Mult by 3}}$ $\underline{6x + 3y + 9z = 0}$

Add: $7x \qquad + 5z = -10$

Combine this equation with the 3rd equation to get a system in 2 unknowns:

$7x + 5z = -10$ $7x + 5z = -10$

$x - 3z = 6$ $\xrightarrow{\text{Mult by -7}}$ $\underline{-7x + 21z = -42}$

Add: $\qquad\qquad\quad 26z = -52$

$$z = -\frac{52}{26} = -2$$

Now, substitute -2 for z into $x - 3z = 6$

$$x - 3(-2) = 6$$
$$x + 6 = 6$$
$$x = 0$$

Finally, substitute $x = 0$ and $z = -2$ into $2x + y + 3z = 0$

$$2(0) + y + 3(-2) = 0$$
$$y - 6 = 0$$
$$y = 6$$

The solution is $(0, 6, -2)$.

16. $2x - y - 3z = 5$
$x - 2y - z = -2$
$3x + 8y + 2z = 7$

First eliminate x between the 1st and 2nd equations:

$2x - y - 3z = 5$ $2x - y - 3z = 5$

$x - 2y - z = -2$ $\xrightarrow{\text{Mult by -2}}$ $\underline{-2x + 4y + 2z = 4}$

Add: $\qquad\qquad\qquad\quad 3y - z = 9$

Now, eliminate x between the 2nd and 3rd equations:

$x - 2y - z = -2$ $\xrightarrow{\text{Mult by -3}}$ $-3x + 6y + 3z = 6$

$3x + 8y + 2z = 7$ $\qquad\qquad\quad$ $\underline{3x + 8y + 2z = 7}$

Add: $\qquad\qquad\qquad\qquad 14y + 5z = 13$

The system of 2 equations in 2 unknowns is

$3y - z = 9$ $\xrightarrow{\text{Mult by 5}}$ $15y - 5z = 45$

$14y + 5z = 13$ $\qquad\qquad$ $\underline{14y + 5z = 13}$

Add: $29y = 58$

$$y = \frac{58}{29} = 2$$

To find z, substitute 2 for y into the equation $3y - z = 9$

$$3(2) - z = 9$$
$$6 - z = 9$$
$$-z = 3$$
$$z = -3$$

To find x, substitute -3 for z and 2 for y into the equation $x - 2y - z = -2$ to get

$$x - 2(2) - (-3) = -2$$
$$x - 4 + 3 = -2$$
$$x - 1 = -2$$
$$x = -1$$

The solution is $(-1, 2, -3)$.

17. $x - 2y - 2z = 9$
    $2x + y + 3z = 0$
    $5x + 5y + 11z = 9$

First, eliminate y between the 1st and 2nd equations:

$x - 2y - 2z = 9$ $\qquad\qquad\qquad$ $x - 2y - 2z = 9$

$2x + y + 3z = 0$ $\xrightarrow{\text{Mult by 2}}$ $4x + 2y + 6z = 0$

$\qquad\qquad\qquad\qquad$ Add: $\quad 5x \qquad + 4z = 9$

Now, eliminate y between the 2nd and 3rd equations:

$2x + y + 3z = 0$ $\xrightarrow{\text{Mult by -5}}$ $-10x - 5y - 15z = 0$

$5x + 5y + 11z = 9$ $\qquad\qquad\qquad$ $5x + 5y + 11z = 9$

$\qquad\qquad\qquad\qquad$ Add: $\quad -5x \qquad - 4z = 9$

The system of two equations in two unknowns is
$\qquad 5x + 4z = 9$

$\qquad \underline{-5x - 4z = 9}$

Add: $\qquad\quad 0 = 18$ which is **not** possible.

Thus, there is no solution and the system is inconsistent.

18. $2x - 3y - 9z = -2$
    $x + 2y + 6z = 6$
    $4x - 2y - 6z = -4$

First eliminate x between the 1st and 2nd equations:
$\qquad 2x - 3y - 9z = -2$ $\qquad\qquad\qquad$ $2x - 3y - 9z = \quad -2$

$\qquad x + 2y + 6z = 6$ $\xrightarrow{\text{Mult by -2}}$ $-2x - 4y - 12z = -12$

$\qquad\qquad\qquad$ Add: $\qquad\qquad -7y - 21z = -14$

$\qquad\qquad\qquad\qquad$ or $\quad y + 3z = 2$

Now, eliminate x between the 2nd and 3rd equations:

$\qquad x + 2y + 6z = 6$ $\xrightarrow{\text{Mult by -4}}$ $-4x - 8y - 24z = -24$

$\qquad 4x - 2y - 6z = 4$ $\qquad\qquad\qquad$ $4x - 2y - 6z = \quad -4$

$\qquad\qquad\qquad$ Add: $\qquad\qquad -10y - 30z = -28$

$\qquad\qquad\qquad\qquad$ or $\quad 5y + 15z = 14$

The system of 2 equations in 2 unknowns is:

$\qquad y + 3z = 2$ $\xrightarrow{\text{Mult by -5}}$ $-5y - 15z = -10$

$\qquad 5y + 15z = 14$ $\qquad\qquad\qquad$ $5y + 15z = \quad 14$

$\qquad\qquad\qquad$ Add: $\qquad\qquad 0 = 4$ Impossible

Thus, there is **no** solution and the system is inconsistent.

19. $ax + by + cz = 3$. Let $x = 1$, $y = 3$, $z = 2$: $\quad a + 3b + 2c = 3$
    $\qquad\qquad\qquad\qquad$ $x = 3$, $y = -2$, $z = 1$: $3a - 2b + c = 3$
    $\qquad\qquad\qquad\qquad$ $x = 2$, $y = -4$, $z = -3$: $2a - 4b - 3c = 3$
    $\qquad\qquad\qquad\qquad\qquad\qquad$ This is the system to be solved.

$\qquad a + 3b + 2c = 3$ $\qquad\qquad\qquad$ $a + 3b + 2c = \quad 3$

$\qquad 3a - 2b + c = 3$ $\xrightarrow{\text{Mult by -2}}$ $-6a + 4b - 2c = -6$

$\qquad\qquad\qquad$ Add: $\qquad -5a + 7b \qquad = -3$

$$3a - 2b + c = 3 \xrightarrow{\text{Mult by 3}} 9a - 6b + 3c = 9$$

$$2a - 4b - 3c = 3 \qquad\qquad\qquad \underline{2a - 4b - 3c = 3}$$

$$\text{Add:} \quad 11a - 10b \quad\;\; = 12$$

The system of two equations is

$$-5a + 7b = -3 \xrightarrow{\text{Mult by 11}} -55a + 77b = -33$$

$$11a - 10b = 12 \xrightarrow{\text{Mult by 5}} \underline{55a - 50b = 60}$$

$$\text{Add:} \qquad\qquad 27b = 27$$

$$b = \frac{27}{27} = 1$$

Substitute 1 for b into $11a - 10b = 12$

$$11a - 10(1) = 12$$
$$11a - 10 = 12$$
$$11a = 22$$
$$a = \frac{22}{11} = 2$$

Finally, substitute $a = 2$ and $b = 1$ into $3a - 2b + c = 3$

$$3(2) - 2(1) + c = 3$$
$$6 - 2 + c = 3$$
$$4 + c = 3$$
$$c = -1$$

The solution is $a = 2$, $b = 1$, and $c = -1$.

20. Let n be the number of nickels, d the number of dimes, and q the number of quarters. The system is:

$$n + d + q = 41$$
$$d = 2q$$
$$5n + 10d + 25q = 505$$

Substitute $2q$ for d into the 1st and 3rd equations:

$$n + d + q = 41$$
$$5n + 10d + 25q = 505$$ becomes
$$n + 2q + q = 41$$
$$5n + 10(2q) + 25q = 505$$ or

$$n + 3q = 41 \xrightarrow{\text{Mult by -15}} -15n - 45q = -615$$

$$5n + 45q = 505 \qquad\qquad\qquad \underline{5n + 45q = 505}$$

$$\text{Add:} \quad -10n \qquad = -110$$

$$\text{or} \quad n = \frac{-110}{-10} = 11$$

Substitute 11 for n into the equation $n + 3q = 41$ to get

$$11 + 3q = 41$$
$$3q = 30$$
$$q = 10$$

Finally, substitute 10 for q into the equation $d = 2q$

$$d = 2(10) = 20$$

Thus, there were 11 nickels, 10 quarters, and 20 dimes.

21. Let x be the time for backhoe A, y the time for backhoe B, and z the time for backhoe C. Then $\frac{1}{x}$, $\frac{1}{y}$, $\frac{1}{z}$ represent the part of the project completed each hour by the three backhoes. The system is

$$\frac{2}{x} + \frac{2}{y} + \frac{2}{z} = 1$$

$$\frac{2}{x} \qquad + \frac{6}{z} = 1$$

$$\frac{4}{y} + \frac{4}{z} = 1$$

$$\frac{2}{x} + \frac{2}{y} + \frac{2}{z} = 1 \qquad\qquad \frac{2}{x} + \frac{2}{y} + \frac{2}{z} = 1$$

$$\frac{2}{x} \qquad + \frac{6}{z} = 1 \quad\xrightarrow{\text{Mult by -1}}\quad -\frac{2}{x} \qquad - \frac{6}{z} = -1$$

$$\text{Add:} \qquad\qquad \frac{2}{y} - \frac{4}{z} = 0$$

The system of two equations is:

$$\frac{2}{y} - \frac{4}{z} = 0$$

$$\frac{4}{y} + \frac{4}{z} = 1$$

Add: $\dfrac{6}{y} \qquad = 1 \quad$ or $\quad 6 = y$

Substitute 6 for y into $\dfrac{2}{y} - \dfrac{4}{z} = 0$

$$\frac{2}{6} - \frac{4}{z} = 0$$

$$\frac{1}{3} - \frac{4}{z} = 0$$

$$\frac{1}{3} = \frac{4}{z} \quad \text{Cross multiply: } z = 12.$$

Finally, substitute 12 for z into $\dfrac{2}{x} + \dfrac{6}{z} = 1$

$$\frac{2}{x} + \frac{6}{12} = 1$$

$$\frac{2}{x} + \frac{1}{2} = 1$$

$$\frac{2}{x} = \frac{1}{2} \quad \text{Cross multiply: } x = 4.$$

The solution is $x = 4$ hours for backhoes A, $y = 6$ hours for backhoe B, and $z = 12$ hours for backhoe C.

22. $x = y^2 + 1$
    $x - y = 3$

Substitute $y^2 + 1$ for x into the second equation $\quad x - y = 3$

$$y^2 + 1 - y = 3$$

$$y^2 - y - 2 = 0$$

$$(y - 2)(y + 1) = 0$$

$$y - 2 = 0, \quad y + 1 = 0$$

$$y = 2, \qquad\quad y = -1$$

When $y = 2$, the equation $x - y = 3$ becomes
$$x - 2 = 3$$
$$x = 5$$

When $y = -1$, the equation $x - y = 3$ becomes
$$x - (-1) = 3$$
$$x + 1 = 3$$
$$x = 2$$

The solution is $(5, 2)$ and $(2, -1)$.

23. $x = y^2 - 4y + 3$
    $x - y = 3$

Solve $x - y = 3$ for x: $x - y = 3$
$$x = y + 3$$

Now, substitute $y + 3$ for x into the first equation $x = y^2 - 4y + 3$
$$y + 3 = y^2 - 4y + 3$$
$$0 = y^2 - 5y$$
$$0 = y(y - 5)$$
$$y = 0, \quad y - 5 = 0$$
$$y = 0, \quad\quad y = 5$$

When $y = 0$, the equation $x = y + 3$ becomes $x = 0 + 3 = 3$.
When $y = 5$, the equation $x = y + 3$ becomes $x = 5 + 3 = 8$.

The solution is $(3, 0)$, $(8, 5)$.

24. $x^2 - 3y = 1 \quad\quad\quad x^2 - 3y = 1$

$2x - y = 3 \xrightarrow{\text{Mult by -3}} -6x + 3y = -9$
$$\text{Add:} \quad x^2 - 6x = -8$$
$$x^2 - 6x + 8 = 0$$
$$(x - 4)(x - 2) = 0$$
$$x - 4 = 0, \ x - 2 = 0$$
$$x = 4, \quad\quad x = 2$$

When $x = 4$, the equation $2x - y = 3$ becomes
$$2(4) - y = 3$$
$$8 - y = 3$$
$$-y = -5$$
$$y = 5$$

When $x = 2$, the equation $2x - y = 3$ becomes
$$2(2) - y = 3$$
$$4 - y = 3$$
$$-y = -1$$
$$y = 1$$

The solution is $(4, 5)$ and $(2, 1)$.

25. $x^2 + 3xy - 6y^2 = 8$
    $x^2 - xy - 6y^2 = 4$

Subtract the second equation from the first: $4xy = 4$
$$xy = 1$$
$$y = \frac{1}{x}$$

Now, substitute $\frac{1}{x}$ for y into the first equation: $x^2 + 3xy - 6y^2 = 8$

$$x^2 + 3x\left(\frac{1}{x}\right) - 6\left(\frac{1}{x}\right)^2 = 8$$

$$x^2 + 3 - \frac{6}{x^2} = 8$$

$$x^2 - 5 - \frac{6}{x^2} = 0$$

Multiply by x^2: $\quad x^4 - 5x^2 - 6 = 0$

$$(x^2 - 6)(x^2 + 1) = 0$$

$$x^2 - 6 = 0, \quad \underbrace{x^2 + 1 = 0}$$

$$x^2 = 6 \quad \text{No values here}$$

$$x = \pm\sqrt{6}$$

When $x = \sqrt{6}$, then $y = \frac{1}{x}$ becomes $y = \frac{1}{\sqrt{6}} = \frac{\sqrt{6}}{6}$.

When $x = -\sqrt{6}$, then $y = \frac{1}{x}$ becomes $y = \frac{1}{-\sqrt{6}} = -\frac{\sqrt{6}}{6}$.

The solution is $\left(\sqrt{6}, \frac{\sqrt{6}}{6}\right)$, $\left(-\sqrt{6}, -\frac{\sqrt{6}}{6}\right)$.

26. $\begin{array}{l} 4y^2 - x = 0 \\ x + 2y = 6 \end{array}$ or $\begin{array}{l} -x + 4y^2 = 0 \\ \underline{x + 2y = 6} \end{array}$

Add: $4y^2 + 2y = 6$

$$4y^2 + 2y - 6 = 0$$

$$2y^2 + y - 3 = 0$$

$$(2y + 3)(y - 1) = 0$$

$$2y + 3 = 0, \quad y - 1 = 0$$

$$y = -\frac{3}{2}, \quad y = 1$$

When $y = -\frac{3}{2}$, then $x + 2y = 6$ becomes

$$x + 2\left(-\frac{3}{2}\right) = 6$$

$$x - 3 = 6$$

$$x = 9$$

When $y = 1$, then $x + 2y = 6$ becomes

$$x + 2(1) = 6$$

$$x + 2 = 6$$

$$x = 4$$

The solution is $\left(9, -\frac{3}{2}\right)$, $(4, 1)$.

27. $x^2 - 2x + 3y - 3 = 0$
$x + 2y = 4$

Solve $x + 2y = 4$ for x: $x + 2y = 4$

$$x = -2y + 4$$

Substitute $-2y + 4$ for x into the first equation: $x^2 - 2x + 3y - 3 = 0$

$$(-2y + 4)^2 - 2(-2y + 4) + 3y - 3 = 0$$
$$4y^2 - 16y + 16 + 4y - 8 + 3y - 3 = 0$$
$$4y^2 - 9y + 5 = 0$$
$$(4y - 5)(y - 1) = 0$$
$$4y - 5 = 0, \quad y - 1 = 0$$
$$y = \frac{5}{4}, \qquad y = 1$$

When $y = 1$, then $x = -2y + 4$ becomes $x = -2(1) + 4 = 2$.

When $y = \frac{5}{4}$, then $x = -2y + 4$ becomes $x = -2\left(\frac{5}{4}\right) + 4 = -\frac{5}{2} + 4 = -\frac{5}{2} + \frac{8}{2} = \frac{3}{2}$.

The solution is $(2, 1)$, $\left(\frac{3}{2}, \frac{5}{4}\right)$.

28. $3x^2 + y^2 = 15 \xrightarrow{\text{Mult by } -2} -6x^2 - 2y^2 = -30$

$x^2 + 2y^2 = 10$

$$\underline{x^2 + 2y^2 = 10}$$
$$\text{Add:} \quad -5x^2 = -20$$
$$x^2 = 4$$
$$x = \pm\sqrt{4} = \pm 2$$

When $x = 2$, then $x^2 + 2y^2 = 10$ becomes

$$(2)^2 + 2y^2 = 10$$
$$4 + 2y^2 = 10$$
$$2y^2 = 6$$
$$y^2 = 3$$
$$y = \pm\sqrt{3}$$

When $x = -2$, then $x^2 + 2y^2 = 10$ becomes

$$(-2)^2 + 2y^2 = 10$$
$$4 + 2y^2 = 10$$
$$2y^2 = 6$$
$$y^2 = 3$$
$$y = \pm\sqrt{3}$$

The solution is $(2, \sqrt{3})$, $(2, -\sqrt{3})$, $(-2, \sqrt{3})$, $(-2, -\sqrt{3})$.

29. $3x + y > 7$ The boundary lines are $3x + y = 7$

$x - y \le 1$ $x - y = 1$

$ y < 4$ $y = 4$

which produce three straight lines with

$3x + y = 7$ and $x - y = 1$ intersecting at $(2, 1)$

$x - y = 1$ and $y = 4$ intersecting at $(5, 4)$

and $3x + y = 7$ and $y = 4$ intersecting at $(1, 4)$.

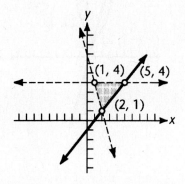

This produces a triangular region and using the point $(2, 3)$ as a test point satisfies all 3 inequalities.

$3x + y > 7$	$x - y \le 1$	$y < 4$
$3(2) + 3 > 7$	$2 - 3 \le 1$	$3 < 4$
$6 + 3 > 7$	$-1 \le 1$	
$9 > 7$		

The solution is the region inside the triangle along with part of the boundary.

30. $2x + y < -2$
     $4x - 3y \geq -24$
     $y \geq -4$

The boundary lines are $2x + y = -2$
$4x - 3y = -24$
$y = -4$
which produce three straight lines with
$2x + y = -2$ and $y = -4$ intersecting at $(1, -4)$
$4x - 3y = -24$ and $y = -4$ intersecting at $(-9, -4)$
and $2x + y = -2$ and $4x - 3y = -24$ intersecting at $(-3, 4)$.

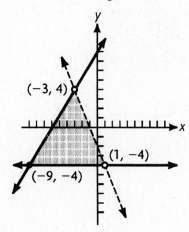

This produces a triangular region and using the point $(-3, 0)$ as a test point satisfies all 3 inequalities:

$2x + y < -2$	$4x - 3y \geq -24$	$y \geq -4$
$2(-3) + 0 < -2$	$4(-3) - 3(0) \geq -24$	$0 \geq -4$
$-6 + 0 < -2$	$-12 - 0 \geq -24$	
$-6 < -2$	$-12 \geq -24$	

The solution is the region inside the triangle along with part of the boundary.

31. $x^2 + 2x + 3 \leq y$
     $3x + y + 1 \leq 0$

The boundary curves are $x^2 + 2x + 3 = y$
$3x + y + 1 = 0$
To obtain the points of intersection, solve
$3x + y + 1 = 0$ for y: $3x + y + 1 = 0$
$y = -3x - 1$
Now, substitute $-3x - 1$ for y into the first equation:
$x^2 + 2x + 3 = y$
$x^2 + 2x + 3 = -3x - 1$
$x^2 + 5x + 4 = 0$
$(x + 4)(x + 1) = 0$
$x + 4 = 0, \; x + 1 = 0$
$x = -4, \qquad x = -1$

When $x = -4$, then $y = -3x - 1$ becomes $y = -3(-4) - 1 = 12 - 1 = 11$.
When $x = -1$, then $y = -3x - 1$ becomes $y = -3(-1) - 1 = 3 - 1 = 2$.
The points of intersection are $(-4, 11)$ and $(-1, 2)$. The graph consists of a line $(3x + y + 1 = 0)$ intersecting a parabola $(x^2 + 2x + 3 = y)$. For the test point, use $(-2, 4)$ which satisfies both inequalities.

$x^2 + 2x - 3 \leq y$	$3x + y + 1 \leq 0$
$(-2)^2 + 2(-2) + 3 \leq 4$	$3(-2) + 4 + 1 \leq 0$
$4 - 4 + 3 \leq 4$	$-6 + 4 + 1 \leq 0$
$3 \leq 4$	$-1 \leq 0$

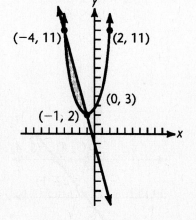

The solution is the region between the line and the parabola including the entire boundary.

32. $x^2 + y^2 \geq 9$ The boundary curves are $x^2 + y^2 = 9$
    $x^2 + 2y \leq 10$ $x^2 + 2y = 10$
                    To obtain the points of intersection, subtract the second
                    equation from the first to obtain $y^2 - 2y = 9 - 10$
                                    $y^2 - 2y + 1 = 0$
                                    $(y - 1)^2 = 0$
                                    $y - 1 = 0$
                                    $y = 1$

When $y = 1$, then $x^2 + y^2 = 9$ becomes $x^2 + (1)^2 = 9$
                                    $x^2 + 1 = 9$
                                    $x^2 = 8$
                                    $x = \pm\sqrt{8} = \pm 2\sqrt{2}$

The points of intersection are $(2\sqrt{2}, 1)$ and
$(-2\sqrt{2}, 1)$. The graph consists of a parabola
intersecting a circle. For a test point, use
$(0, 4)$ which satisfies both inequalities:

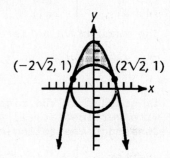

$x^2 + y^2 \geq 9$	$x^2 + 2y \leq 10$
$0^2 + 4^2 \geq 9$	$0^2 + 2(4) \leq 10$
$0 + 16 \geq 9$	$0 + 8 \leq 10$
$16 \geq 9$	$8 \leq 10$

The solution is the region above the circle and
below the parabola including the entire boundary.

33. $x^2 + y^2 \leq 4$
    $3y + x^2 \geq 0$

The boundary curves are $x^2 + y^2 = 4$
                        $3y + x^2 = 0$
To obtain the points of intersection, solve $3y + x^2 = 0$ for x^2: $3y + x^2 = 0$
                                                            $x^2 = -3y$

Now, substitute $-3y$ for x^2 into the first equation: $x^2 + y^2 = 4$
                                                $-3y + y^2 = 4$
                                                $y^2 - 3y - 4 = 0$
                                                $(y - 4)(y + 1) = 0$
                                                $y - 4 = 0, \; y + 1 = 0$
                                                $y = 4, \qquad y = -1$

When $y = 4$, then $x^2 = -3y$ becomes $x^2 = -3(4) = -12$ which has no real values.
When $y = -1$, then $x^2 = -3y$ becomes $x^2 = -3(-1) = 3$ so that $x = \pm\sqrt{3}$.

The points of intersection are $(\sqrt{3}, -1)$ and
$(-\sqrt{3}, -1)$. The graph consists of a parabola
($3y + x^2 = 0$) intersecting a circle ($x^2 + y^2 = 4$).

For the test point, use $(0, 1)$ which satisfies
both inequalities:

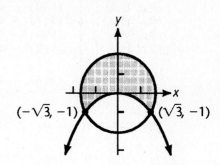

$x^2 + y^2 \leq 4$	$3y + x^2 \geq 0$
$0^2 + 1^2 \leq 4$	$3(1) + 0^2 \geq 0$
$0 + 1 \leq 4$	$3 + 0 \geq 0$
$1 \leq 4$	$3 \geq 0$

The solution is the region inside the circle that
is above the parabola.

34. The region given by $3x + 4y \leq 24$
$$x - 2y \geq -2$$
$$x \geq 0$$
$$y \geq 0$$
is graphed at the right.

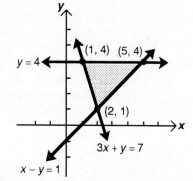

The convex set is a polygon with vertices at $(0, 0)$, $(0, 1)$, $(8, 0)$, and $(4, 3)$. Construct the following table:

Vertex	$f = 2x + 7y$
$(0, 0)$	$2(0) + 7(0) = 0 + 0 = 0$
$(0, 1)$	$2(0) + 7(1) = 0 + 7 = 7$
$(8, 0)$	$2(8) + 7(0) = 16 + 0 = 16$
$(4, 3)$	$2(4) + 7(3) = 8 + 21 = 29$

The maximum value is 29 which occurs at $(4, 3)$.

35. The region given by $3x + y \geq 7$
$$x - y \leq 1$$
$$y \leq 4$$
is graphed at the right. It is a triangular region with vertices at $(1, 4)$, $(5, 4)$, and $(2, 1)$. Construct the following table:

Vertex	$f = 4x + 5y$
$(1, 4)$	$4(1) + 5(4) = 4 + 20 = 24$
$(5, 4)$	$4(5) + 5(4) = 20 + 20 = 40$
$(2, 1)$	$4(2) + 5(1) = 8 + 5 = 13$

The minimum value is 13 which occurs at $(2, 1)$.

36. The region given by $3x + 2y \leq 12$
$$3x - 2y \geq 6$$
$$x - 2y \geq -4$$
$$x \geq 0$$
$$y \geq 0$$
is graphed at the right.

The convex set is a polygon with vertices at

$(0, 0)$, $(0, 2)$, $(2, 0)$, $(2, 3)$ and $\left(3, \frac{3}{2}\right)$.

Construct the following table:

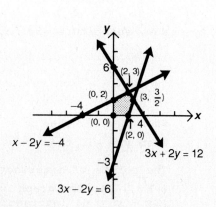

Vertex	$f = 15x - 4y$
$(0, 0)$	$15(0) - 4(0) = 0 - 0 = 0$
$(0, 2)$	$15(0) - 4(2) = 0 - 8 = -8$
$(2, 0)$	$15(2) - 4(0) = 30 - 0 = 30$
$(2, 3)$	$15(2) - 4(3) = 30 - 12 = 18$
$\left(3, \frac{3}{2}\right)$	$15(3) - 4\left(\frac{3}{2}\right) = 45 - 6 = 39$

The maximum value is 39 which occurs at $\left(3, \frac{3}{2}\right)$.

37. Let x be the amount of $6 per pound mix and y be the amount of $5 per pound mix required to produce revenue. The system of inequalities is:

$$x \geq 0$$
$$y \geq 0$$

peanuts: $\frac{1}{2}x + \frac{3}{4}y \leq 400$

cashews: $\frac{1}{2}x + \frac{1}{4}y \leq 200$

The region appears as follows:

The vertices are $(0, 0)$, $(400, 0)$, $(200, 400)$,

$\left(0, \frac{1600}{3}\right)$. The function to be maximized is

$f = 6x + 5y$ and the table is:

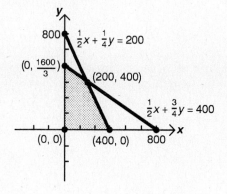

Vertex	$f = 6x + 5y$
$(0, 0)$	$6(0) + 5(0) = \$0$
$(400, 0)$	$6(400) + 5(0) = \$2400$
$\left(0, \frac{1600}{3}\right)$	$6(0) + 5\left(\frac{1600}{3}\right) \approx \$2667$
$(200, 400)$	$6(200) + 5(400) = 1200 + 2000 = \$3200$

The maximum revenue occurs when 200 pounds of the $6 mixture and 400 pounds of the $5 mixture are used.

38. Let S be the number of scientific calculators and b be the number of business calculators required.

The system of inequalities is

$$S \geq 0$$
$$b \geq 0$$
$$b \leq 400$$
$$S + b \leq 500$$

The region appears as follows.

The vertices are $(0, 0)$, $(400, 0)$, $(400, 100)$, and $(0, 500)$. The function to be maximized is $f = 4b + 3S$.

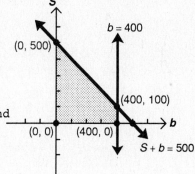

Vertex	$f = 4b + 3S$
$(0, 0)$	$4(0) + 3(0) = 0$
$(400, 0)$	$4(400) + 3(0) = \$1600$
$(400, 100)$	$4(400) + 3(100) = \$1900$ ← Max profit
$(0, 500)$	$4(0) + 3(500) = \$1500$

The maximum profit occurs when 400 business calculators and 100 scientific calculators are produced.

CHAPTER 8   MATRICES AND DETERMINANTS

Exercises 8.1

1. 2 × 3 with 2 rows and 3 columns

3. 3 × 1 with 3 rows and 1 column. It is a column matrix.

5. 2 × 4 with 2 rows and 4 columns

7. 1 × 1 with 1 row and 1 column. It is a column matrix, a row matrix, a square matrix, and a diagonal matrix.

9. 1 × 2 with 1 row and 2 columns. It is a row matrix.

11. 3 × 3 with 3 rows and 3 columns. It is a square matrix and a diagonal matrix.

13. $a_{11} = 2(1) + 1 = 2 + 1 = 3$
    $a_{21} = 2(2) + 1 = 4 + 1 = 5$
    $a_{12} = 2(1) + 2 = 2 + 2 = 4$
    $a_{22} = 2(2) + 2 = 4 + 2 = 6$
    $a_{13} = 2(1) + 3 = 2 + 3 = 5$
    $a_{23} = 2(2) + 3 = 4 + 3 = 7$
    $a_{14} = 2(1) + 4 = 2 + 4 = 6$
    $a_{24} = 2(2) + 4 = 4 + 4 = 8$

    The matrix appears as $\begin{bmatrix} 3 & 4 & 5 & 6 \\ 5 & 6 & 7 & 8 \end{bmatrix}$.

15. $a_{11} = (-1)^1 \cdot 1 = -1 \cdot 1 = -1$
    $a_{12} = (-1)^1 \cdot 2 = -1 \cdot 2 = -2$
    $a_{21} = (-1)^2 \cdot 1 = 1 \cdot 1 = 1$
    $a_{22} = (-1)^2 \cdot 2 = 1 \cdot 2 = 2$
    $a_{31} = (-1)^3 \cdot 1 = -1 \cdot 1 = -1$
    $a_{32} = (-1)^3 \cdot 2 = -1 \cdot 2 = -2$
    $a_{41} = (-1)^4 \cdot 1 = 1 \cdot 1 = 1$
    $a_{42} = (-1)^4 \cdot 2 = 1 \cdot 2 = 2$

    The matrix appears as $\begin{bmatrix} -1 & -2 \\ 1 & 2 \\ -1 & -2 \\ 1 & 2 \end{bmatrix}$.

17. Here $a_{11} = a_{21} = a_{22} = 0$ and
    $a_{12} = a_{13} = a_{14} = a_{23} = a_{24} = 1$ and the matrix appears as

    $$\begin{bmatrix} 0 & 1 & 1 & 1 \\ 0 & 0 & 1 & 1 \end{bmatrix}$$

19. Here $a_{11} = a_{21} = a_{22} = a_{31} = a_{32} = a_{33} = a_{41} = a_{42} = a_{43} = 1$
    and $a_{12} = 2 \cdot 1 + 3 \cdot 2 = 2 + 6 = 8$
    $a_{13} = 2 \cdot 1 + 3 \cdot 3 = 2 + 9 = 11$
    $a_{23} = 2 \cdot 2 + 3 \cdot 3 = 4 + 9 = 13$

    The matrix appears as $\begin{bmatrix} 1 & 8 & 11 \\ 1 & 1 & 13 \\ 1 & 1 & 1 \\ 1 & 1 & 1 \end{bmatrix}$.

21. Not equal since corresponding entries are not equal.

23. The second matrix can be simplified to $\begin{bmatrix} 1 & -7 \\ 5 & 3.2 \end{bmatrix}$ and we see that it is equal to the first matrix.

25. Not equal since the dimensions are different for the two matrices.

27. Not equal since the dimensions are different for the two matrices.

29. Equal only if $x = 2$ and $y = 3$.

31. Equal only if $2x - 5 = -1$
$$2x = 4$$
$$x = 2$$

33. $\begin{bmatrix} 3 & -1 & 1 \\ 2 & 7 & -4 \end{bmatrix} + \begin{bmatrix} 2 & 1 & 0 \\ 1 & 3 & -1 \end{bmatrix} = \begin{bmatrix} 3+2 & -1+1 & 1+0 \\ 2+1 & 7+3 & -4+(-1) \end{bmatrix} = \begin{bmatrix} 5 & 0 & 1 \\ 3 & 10 & -5 \end{bmatrix}$

35. Not possible to add since dimensions are different.

37. $\begin{bmatrix} 4 & -2 \\ 1 & 7 \end{bmatrix} - \begin{bmatrix} 2 & 6 \\ -3 & 0 \end{bmatrix} = \begin{bmatrix} 4-2 & -2-6 \\ 1-(-3) & 7-0 \end{bmatrix} = \begin{bmatrix} 2 & -8 \\ 4 & 7 \end{bmatrix}$

39. $3\begin{bmatrix} -1 & 3 \\ 2 & 0 \\ -1 & 1 \end{bmatrix} - 2\begin{bmatrix} 4 & -2 \\ -5 & 4 \\ 0 & -3 \end{bmatrix} = \begin{bmatrix} 3(-1)-2(4) & 3(3)-2(-2) \\ 3(2)-2(-5) & 3(0)-2(4) \\ 3(-1)-2(0) & 3(1)-2(-3) \end{bmatrix} = \begin{bmatrix} -11 & 13 \\ 16 & -8 \\ -3 & 9 \end{bmatrix}$

41. Not possible to add since the dimensions are different.

43. $-3\begin{bmatrix} -1 \\ 6 \end{bmatrix} + 2\begin{bmatrix} -2 \\ 5 \end{bmatrix} = \begin{bmatrix} -3(-1)+2(-2) \\ -3(6)+2(5) \end{bmatrix} = \begin{bmatrix} -1 \\ -8 \end{bmatrix}$

45. $7\begin{bmatrix} 2 & 1 \\ 3 & 5 \\ 0 & 2 \end{bmatrix} - \begin{bmatrix} 6 & 1 \\ 2 & 3 \\ -2 & 1 \end{bmatrix} = \begin{bmatrix} 7(2)-1(6) & 7(1)-1(1) \\ 7(3)-1(2) & 7(5)-1(3) \\ 7(0)-1(-2) & 7(2)-1(1) \end{bmatrix} = \begin{bmatrix} 8 & 6 \\ 19 & 32 \\ 2 & 13 \end{bmatrix}$

47. $2\begin{bmatrix} 4 & -2 & 5 & 1 \\ 7 & 1 & 8 & 2 \end{bmatrix} + 3\begin{bmatrix} 6 & 2 & 1 & -3 \\ 8 & -1 & 4 & -2 \end{bmatrix} - 4\begin{bmatrix} -6 & -1 & 0 & 4 \\ 1 & 5 & 1 & 3 \end{bmatrix}$

$= \begin{bmatrix} 2(4)+3(6)-4(-6) & 2(-2)+3(2)-4(-1) & 2(5)+3(1)-4(0) & 2(1)+3(-3)-4(4) \\ 2(7)+3(8)-4(1) & 2(1)+3(-1)-4(5) & 2(8)+3(4)-4(1) & 2(2)+3(-2)-4(3) \end{bmatrix}$

$= \begin{bmatrix} 50 & 6 & 13 & -23 \\ 34 & -21 & 24 & -14 \end{bmatrix}$

49. Let $A = [a_{ij}]$, $B = [b_{ij}]$ and $C = [c_{ij}]$
    Then $A + (B + C) = [a_{ij}] + ([b_{ij}] + [c_{ij}])$
    $$= [a_{ij}] + [b_{ij} + c_{ij}]$$
    $$= [a_{ij} + b_{ij} + c_{ij}] \quad (*)$$

    Also, $(A + B) + C = ([a_{ij}] + [b_{ij}]) + [c_{ij}]$
    $$= [a_{ij} + b_{ij}] + [c_{ij}]$$
    $$= [a_{ij} + b_{ij} + c_{ij}] \quad (**)$$

    Since (*) and (**) are the same, this means that $A + (B + C) = (A + B) + C$.

51. a) Let $A = [a_{ij}]$
       Then $1 \cdot A = 1 \cdot [a_{ij}] = [1 \cdot a_{ij}] = [a_{ij}] = A$

    b) If $CA = \mathbf{0}$, then either $C = 0$ or $A = \mathbf{0}$.

Exercises 8.2

1. $A_{(2,\ 3)} \cdot B_{(3,\ 7)}$: conformable with order being 2 × 7

      Equal

3. $B_{(3,\ 2)} \cdot C_{(3,\ 3)}$: not conformable

      Not equal

5. $A_{(1,\ 3)} \cdot C_{(3,\ 1)}$: conformable with order 1 × 1

      Equal

7. $I_2 \cdot A_{(2,\ 7)} = I_{(2,\ 2)} \cdot A_{(2,\ 7)}$: conformable with order 2 × 7

      Equal

9. $X_{(8,\ 2)} \cdot Y_{(2,\ 1)}$: conformable with order 8 × 1

      Equal

11. $A_{(3,\ 3)} \cdot A_{(3,\ 3)}$: conformable with order 3 × 3

      Equal

13. $X_{(2,\ 3)} \cdot Z_{(3,\ 3)}$: not conformable

      Not equal

15. $D_{(1,\ 1)} \cdot C_{(1,\ 12)}$: conformable with order 1 × 12

      Equal

17. $\begin{bmatrix} 1 & 3 & 0 \\ -2 & 4 & 1 \end{bmatrix} \begin{bmatrix} 1 & 2 \\ 2 & -1 \\ 1 & 3 \end{bmatrix} = \begin{bmatrix} 1(1) + 3(2) + 0(1) & 1(2) + 3(-1) + 0(3) \\ -2(1) + 4(2) + 1(1) & -2(2) + 4(-1) + 1(3) \end{bmatrix} = \begin{bmatrix} 7 & -1 \\ 7 & -5 \end{bmatrix}$

19. $\begin{bmatrix} -2 & 1 & 3 \\ 4 & 7 & 0 \\ -1 & 4 & -2 \end{bmatrix} \begin{bmatrix} 3 & 5 \\ -1 & 0 \\ 2 & 4 \end{bmatrix} = \begin{bmatrix} -2(3) + 1(-1) + 3(2) & -2(5) + 1(0) + 3(4) \\ 4(3) + 7(-1) + 0(2) & 4(5) + 7(0) + 0(4) \\ -1(3) + 4(-1) - 2(2) & -1(5) + 4(0) - 2(4) \end{bmatrix} = \begin{bmatrix} -1 & 2 \\ 5 & 20 \\ -11 & -13 \end{bmatrix}$

21. $\begin{bmatrix} 1 & 0 \\ 5 & -1 \end{bmatrix} I_3$ is **not** possible since the dimension of $\begin{bmatrix} 1 & 0 \\ 5 & -1 \end{bmatrix}$ is 2 × 2 and the dimension of I_3 is 3 × 3.

      Not equal

23. $[-3 \quad -2 \quad 0 \quad 1][1 \quad 0 \quad 3 \quad -1]$ is **not** possible since the dimension of first matrix is 1 × 4 and dimension of second is 1 × 4.

      Not equal

25. $[0 \quad -3 \quad 1] \begin{bmatrix} 10 \\ -4 \\ 10 \end{bmatrix} = [0(10) - 3(-4) + 1(10)] = [22]$

27. $\begin{bmatrix} -2 \\ 1 \\ 1 \end{bmatrix} \begin{bmatrix} 4 & 11 & -2 \end{bmatrix} = \begin{bmatrix} -2(4) & -2(11) & -2(-2) \\ 1(4) & 1(11) & 1(-2) \\ 1(4) & 1(11) & 1(-2) \end{bmatrix} = \begin{bmatrix} -8 & -22 & 4 \\ 4 & 11 & -2 \\ 4 & 11 & -2 \end{bmatrix}$

29. $\begin{bmatrix} 1 & 0 \\ 0 & 1 \\ 0 & 0 \end{bmatrix} \begin{bmatrix} 1 & 3 \\ 2 & -2 \\ 1 & 0 \end{bmatrix}$ is **not** possible since the dimension of the first matrix is 3×2 and dimension of second is 3×2.

Not equal

31. $\begin{bmatrix} 0 & 0 & 0 & 0 \\ 0 & 0 & 0 & 0 \\ 0 & 0 & 0 & 0 \end{bmatrix} \begin{bmatrix} 1 & 0 \\ 5 & 1 \\ -3 & 4 \\ 2 & -1 \end{bmatrix} = \begin{bmatrix} 0(1) + 0(5) + 0(-3) + 0(2) & 0(0) + 0(1) + 0(4) + 0(-1) \\ 0(1) + 0(5) + 0(-3) + 0(2) & 0(0) + 0(1) + 0(4) + 0(-1) \\ 0(1) + 0(5) + 0(-3) + 0(2) & 0(0) + 0(1) + 0(4) + 0(-1) \end{bmatrix}$

$= \begin{bmatrix} 0 & 0 \\ 0 & 0 \\ 0 & 0 \end{bmatrix}$

33. $\begin{bmatrix} 3 & -1 \\ 8 & 1 \end{bmatrix} \begin{bmatrix} 0 & 0 \end{bmatrix}$ is **not** possible since the dimension of the first matrix is 2×2 and the dimension of the second is 1×2

Not equal

35. $\begin{bmatrix} 11 & 3 \\ 4 & 1 \end{bmatrix} \begin{bmatrix} -1 & 3 \\ 4 & -11 \end{bmatrix} = \begin{bmatrix} 11(-1) + 3(4) & 11(3) + 3(-11) \\ 4(-1) + 1(4) & 4(3) + 1(-11) \end{bmatrix} = \begin{bmatrix} 1 & 0 \\ 0 & 1 \end{bmatrix}$

37. $\begin{bmatrix} 10 & 1 & -4 \\ -8 & 1 & 5 \\ -1 & -1 & 4 \end{bmatrix} \begin{bmatrix} \frac{1}{9} & 0 & \frac{1}{9} \\ \frac{3}{9} & \frac{4}{9} & -\frac{2}{9} \\ \frac{1}{9} & \frac{1}{9} & \frac{2}{9} \end{bmatrix}$

$= \begin{bmatrix} 10(\frac{1}{9}) + 1(\frac{3}{9}) - 4(\frac{1}{9}) & 10(0) + 1(\frac{4}{9}) - 4(\frac{1}{9}) & 10(\frac{1}{9}) + 1(-\frac{2}{9}) - 4(\frac{2}{9}) \\ -8(\frac{1}{9}) + 1(\frac{3}{9}) + 5(\frac{1}{9}) & -8(0) + 1(\frac{4}{9}) + 5(\frac{1}{9}) & -8(\frac{1}{9}) + 1(-\frac{2}{9}) + 5(\frac{2}{9}) \\ -1(\frac{1}{9}) - 1(\frac{3}{9}) + 4(\frac{1}{9}) & -1(0) - 1(\frac{4}{9}) + 4(\frac{1}{9}) & -1(\frac{1}{9}) - 1(-\frac{2}{9}) + 4(\frac{2}{9}) \end{bmatrix} = \begin{bmatrix} 1 & 0 & 0 \\ 0 & 1 & 0 \\ 0 & 0 & 1 \end{bmatrix}$

39. $BA = \begin{bmatrix} -5 & -2 \\ 1 & 0 \\ 4 & -2 \end{bmatrix} \begin{bmatrix} -1 & 3 \\ -2 & 1 \end{bmatrix} = \begin{bmatrix} -5(-1) - 2(-2) & -5(3) - 2(1) \\ 1(-1) + 0(-2) & 1(3) + 0(1) \\ 4(-1) - 2(-2) & 4(3) - 2(1) \end{bmatrix} = \begin{bmatrix} 9 & -17 \\ -1 & 3 \\ 0 & 10 \end{bmatrix}$

and

$CB = \begin{bmatrix} 1 & 1 & 0 \\ 0 & 1 & 1 \\ -1 & 0 & -1 \end{bmatrix} \begin{bmatrix} -5 & -2 \\ 1 & 0 \\ 4 & -2 \end{bmatrix} = \begin{bmatrix} 1(-5) + 1(1) + 0(4) & 1(-2) + 1(0) + 0(-2) \\ 0(-5) + 1(1) + 1(4) & 0(-2) + 1(0) + 1(-2) \\ -1(-5) + 0(1) - 1(4) & -1(-2) + 0(0) - 1(-2) \end{bmatrix}$

$= \begin{bmatrix} -4 & -2 \\ 5 & -2 \\ 1 & 4 \end{bmatrix}$

41. $AC = \begin{bmatrix} 1 & -1 & 0 & 2 \\ 1 & 1 & 0 & 4 \\ 0 & 0 & 1 & 5 \\ 0 & 1 & -1 & 0 \end{bmatrix} \begin{bmatrix} 1 & 4 \\ 4 & 0 \\ 1 & 2 \\ 0 & -1 \end{bmatrix}$

$= \begin{bmatrix} 1(1) - 1(4) + 0(1) + 2(0) & 1(4) - 1(0) + 0(2) + 2(-1) \\ 1(1) + 1(4) + 0(1) + 4(0) & 1(4) + 1(0) + 0(2) + 4(-1) \\ 0(1) + 0(4) + 1(1) + 5(0) & 0(4) + 0(0) + 1(2) + 5(-1) \\ 0(1) + 1(4) - 1(1) + 0(0) & 0(4) + 1(0) - 1(2) + 0(-1) \end{bmatrix} = \begin{bmatrix} -3 & 2 \\ 5 & 0 \\ 1 & -3 \\ 3 & -2 \end{bmatrix}$

$$BC = \begin{bmatrix} 1 & -1 & 2 & 0 \\ 0 & 1 & 5 & 3 \end{bmatrix} \begin{bmatrix} 1 & 4 \\ 4 & 0 \\ 1 & 2 \\ 0 & -1 \end{bmatrix}$$

$$= \begin{bmatrix} 1(1) - 1(4) + 2(1) + 0(0) & 1(4) - 1(0) + 2(2) + 0(-1) \\ 0(1) + 1(4) + 5(1) + 3(0) & 0(4) + 1(0) + 5(2) + 3(-1) \end{bmatrix} = \begin{bmatrix} -1 & 8 \\ 9 & 7 \end{bmatrix}$$

$$BA = \begin{bmatrix} 1 & -1 & 2 & 0 \\ 0 & 1 & 5 & 3 \end{bmatrix} \begin{bmatrix} 1 & -1 & 0 & 2 \\ 1 & 1 & 0 & 4 \\ 0 & 0 & 1 & 5 \\ 0 & 1 & -1 & 0 \end{bmatrix}$$

$$= \begin{bmatrix} 1(1) - 1(1) + 2(0) + 0(0) & 1(-1) - 1(1) + 2(0) + 0(1) & 1(0) - 1(0) + 2(1) + 0(-1) & 1(2) - 1(4) + 2(5) + 0(0) \\ 0(1) + 1(1) + 5(0) + 3(0) & 0(-1) + 1(1) + 5(0) + 3(1) & 0(0) + 1(0) + 5(1) + 3(-1) & 0(2) + 1(4) + 5(5) + 3(0) \end{bmatrix}$$

$$= \begin{bmatrix} 0 & -2 & 2 & 8 \\ 1 & 4 & 2 & 29 \end{bmatrix}$$

$$CB = \begin{bmatrix} 1 & 4 \\ 4 & 0 \\ 1 & 2 \\ 0 & -1 \end{bmatrix} \begin{bmatrix} 1 & -1 & 2 & 0 \\ 0 & 1 & 5 & 3 \end{bmatrix}$$

$$= \begin{bmatrix} 1(1) + 4(0) & 1(-1) + 4(1) & 1(2) + 4(5) & 1(0) + 4(3) \\ 4(1) + 0(0) & 4(-1) + 0(1) & 4(2) + 0(5) & 4(0) + 0(3) \\ 1(1) + 2(0) & 1(-1) + 2(1) & 1(2) + 2(5) & 1(0) + 2(3) \\ 0(1) - 1(0) & 0(-1) - 1(1) & 0(2) - 1(5) & 0(0) - 1(3) \end{bmatrix}$$

$$= \begin{bmatrix} 1 & 3 & 22 & 12 \\ 4 & -4 & 8 & 0 \\ 1 & 1 & 12 & 6 \\ 0 & -1 & -5 & -3 \end{bmatrix}$$

43. Let A be any 2×3 matrix and B be any 3×4 matrix.
Then AB is conformable since $A_{(2,\,3)} \cdot B_{(3,\,4)}$ forms a 2×4 matrix

Equal

whereas BA is **not** conformable since $B_{(3,\,4)} \cdot A_{(2,\,3)}$.

Not equal

45. Several answers are possible. One such combination is

$$A = \begin{bmatrix} 1 & 2 & -1 \\ 0 & 1 & 2 \\ 1 & 0 & 0 \end{bmatrix} \text{ and } B = \begin{bmatrix} 1 & 0 & 0 \\ 0 & -1 & 2 \\ 1 & 1 & 1 \end{bmatrix}$$

where $AB = \begin{bmatrix} 0 & -3 & 3 \\ 2 & 1 & 4 \\ 1 & 0 & 0 \end{bmatrix}$ and $BA = \begin{bmatrix} 1 & 2 & -1 \\ 2 & -1 & -2 \\ 2 & 3 & 1 \end{bmatrix}$

47. Several answers are possible. One such combination is

$$A = \begin{bmatrix} 1 & 0 \\ 1 & 2 \end{bmatrix} \text{ and } B = \begin{bmatrix} 0 & 0 \\ 2 & 2 \end{bmatrix} \text{ where } AB = BA = \begin{bmatrix} 0 & 0 \\ 4 & 4 \end{bmatrix}$$

49. $AB = \begin{bmatrix} -1 & 3 \\ -2 & 0 \\ -1 & 1 \end{bmatrix} \begin{bmatrix} 2 & 1 \\ -3 & 0 \end{bmatrix} = \begin{bmatrix} -11 & -1 \\ -4 & -2 \\ -5 & -1 \end{bmatrix}$, $AC = \begin{bmatrix} -1 & 3 \\ -2 & 0 \\ -1 & 1 \end{bmatrix} \begin{bmatrix} 5 & 3 \\ 0 & -1 \end{bmatrix} = \begin{bmatrix} -5 & -6 \\ -10 & -6 \\ -5 & -4 \end{bmatrix}$

and the sum $AB + AC$ is $\begin{bmatrix} -11 & -1 \\ -4 & -2 \\ -5 & -1 \end{bmatrix} + \begin{bmatrix} -5 & -6 \\ -10 & -6 \\ -5 & -4 \end{bmatrix} = \begin{bmatrix} -16 & -7 \\ -14 & -8 \\ -10 & -5 \end{bmatrix}$.

Now $B + C = \begin{bmatrix} 2 & 1 \\ -3 & 0 \end{bmatrix} + \begin{bmatrix} 5 & 3 \\ 0 & -1 \end{bmatrix} = \begin{bmatrix} 7 & 4 \\ -3 & -1 \end{bmatrix}$

and then $A(B + C) = \begin{bmatrix} -1 & 3 \\ -2 & 0 \\ -1 & 1 \end{bmatrix} \begin{bmatrix} 7 & 4 \\ -3 & -1 \end{bmatrix} = \begin{bmatrix} -16 & -7 \\ -14 & -8 \\ -10 & -5 \end{bmatrix}$

and we see that $A(B + C) = AB + AC$.

51. $A^2 = \begin{bmatrix} -6 & 4 \\ 1 & 3 \end{bmatrix} \begin{bmatrix} -6 & 4 \\ 1 & 3 \end{bmatrix} = \begin{bmatrix} 40 & -12 \\ -3 & 13 \end{bmatrix}$, $B^2 = \begin{bmatrix} 0 & 1 \\ 1 & 2 \end{bmatrix} \begin{bmatrix} 0 & 1 \\ 1 & 2 \end{bmatrix} = \begin{bmatrix} 1 & 2 \\ 2 & 5 \end{bmatrix}$

and then $A^2 - B^2 = \begin{bmatrix} 40 & -12 \\ -3 & 13 \end{bmatrix} - \begin{bmatrix} 1 & 2 \\ 2 & 5 \end{bmatrix} = \begin{bmatrix} 39 & -14 \\ -5 & 8 \end{bmatrix}$.

Now, $A - B = \begin{bmatrix} -6 & 4 \\ 1 & 3 \end{bmatrix} - \begin{bmatrix} 0 & 1 \\ 1 & 2 \end{bmatrix} = \begin{bmatrix} -6 & 3 \\ 0 & 1 \end{bmatrix}$,

$A + B = \begin{bmatrix} -6 & 4 \\ 1 & 3 \end{bmatrix} + \begin{bmatrix} 0 & 1 \\ 1 & 2 \end{bmatrix} = \begin{bmatrix} -6 & 5 \\ 2 & 5 \end{bmatrix}$

and then, $(A - B)(A + B) = \begin{bmatrix} -6 & 3 \\ 0 & 1 \end{bmatrix} \begin{bmatrix} -6 & 5 \\ 2 & 5 \end{bmatrix} = \begin{bmatrix} 42 & -15 \\ 2 & 5 \end{bmatrix}$

53. Store A: $\begin{bmatrix} 2 & 1 & \frac{1}{2} & 1 \end{bmatrix} \begin{bmatrix} 65¢ \\ 69¢ \\ 84¢ \\ 103¢ \end{bmatrix} = [2(65¢) + 1(69¢) + \frac{1}{2}(84¢) + 1(103¢)] = [344¢]$
   or [\$3.44]

   Store B: $\begin{bmatrix} 2 & 1 & \frac{1}{2} & 1 \end{bmatrix} \begin{bmatrix} 63¢ \\ 89¢ \\ 62¢ \\ 78¢ \end{bmatrix} = [2(63¢) + 1(89¢) + \frac{1}{2}(62¢) + 1(78¢)] = [324¢]$
   or [\$3.24]

   Store C: $\begin{bmatrix} 2 & 1 & \frac{1}{2} & 1 \end{bmatrix} \begin{bmatrix} 72¢ \\ 90¢ \\ 78¢ \\ 82¢ \end{bmatrix} = [2(72¢) + 1(90¢) + \frac{1}{2}(78¢) + 1(82¢)] = [355¢]$
   or [\$3.55]

55. To solve this problem, let $A = \begin{bmatrix} a_1 & a_2 \\ a_3 & a_4 \end{bmatrix}$, $B = \begin{bmatrix} b_1 & b_2 \\ b_3 & b_4 \end{bmatrix}$, and $C = \begin{bmatrix} c_1 & c_2 \\ c_3 & c_4 \end{bmatrix}$.
Note that we are using single subscripts which will aid in the ease of computations.

If necessary, depending upon the instructions of your instructor, you can use double subscripts by changing a_1 to a_{11}, a_2 to a_{12}, a_3 to a_{21}, etc.

We want to show $A(BC) = (AB)C$.

The left side becomes

$A(BC) = \begin{bmatrix} a_1 & a_2 \\ a_3 & a_4 \end{bmatrix} \left\{ \begin{bmatrix} b_1 & b_2 \\ b_3 & b_4 \end{bmatrix} \begin{bmatrix} c_1 & c_2 \\ c_3 & c_4 \end{bmatrix} \right\}$

$= \begin{bmatrix} a_1 & a_2 \\ a_3 & a_4 \end{bmatrix} \begin{bmatrix} b_1c_1 + b_2c_3 & b_1c_2 + b_2c_4 \\ b_3c_1 + b_4c_3 & b_3c_2 + b_4c_4 \end{bmatrix}$

$$= \begin{bmatrix} a_1(b_1c_1 + b_2c_3) + a_2(b_3c_1 + b_4c_3) & a_1(b_1c_2 + b_2c_4) + a_2(b_3c_2 + b_4c_4) \\ a_3(b_1c_1 + b_2c_3) + a_4(b_3c_1 + b_4c_3) & a_3(b_1c_2 + b_2c_4) + a_4(b_3c_2 + b_4c_4) \end{bmatrix}$$

$$= \begin{bmatrix} a_1b_1c_1 + a_1b_2c_3 + a_2b_3c_1 + a_2b_4c_3 & a_1b_1c_2 + a_1b_2c_4 + a_2b_3c_2 + a_2b_4c_4 \\ a_3b_1c_1 + a_3b_2c_3 + a_4b_3c_1 + a_4b_4c_3 & a_3b_1c_2 + a_3b_2c_4 + a_4b_3c_2 + a_4b_4c_4 \end{bmatrix}$$

The right side becomes

$$(AB)C = \left\{ \begin{bmatrix} a_1 & a_2 \\ a_3 & a_4 \end{bmatrix} \begin{bmatrix} b_1 & b_2 \\ b_3 & b_4 \end{bmatrix} \right\} \begin{bmatrix} c_1 & c_2 \\ c_3 & c_4 \end{bmatrix}$$

$$= \begin{bmatrix} a_1b_1 + a_2b_3 & a_1b_2 + a_2b_4 \\ a_3b_1 + a_4b_3 & a_3b_2 + a_4b_4 \end{bmatrix} \begin{bmatrix} c_1 & c_2 \\ c_3 & c_4 \end{bmatrix}$$

$$= \begin{bmatrix} (a_1b_1 + a_2b_3)c_1 + (a_1b_2 + a_2b_4)c_3 & (a_1b_1 + a_2b_3)c_2 + (a_1b_2 + a_2b_4)c_4 \\ (a_3b_1 + a_4b_3)c_1 + (a_3b_2 + a_4b_4)c_3 & (a_3b_1 + a_4b_3)c_2 + (a_3b_2 + a_4b_4)c_4 \end{bmatrix}$$

$$= \begin{bmatrix} a_1b_1c_1 + a_2b_3c_1 + a_1b_2c_3 + a_2b_4c_3 & a_1b_1c_2 + a_2b_3c_2 + a_1b_2c_4 + a_2b_4c_4 \\ a_3b_1c_1 + a_4b_3c_1 + a_3b_2c_3 + a_4b_4c_3 & a_3b_1c_2 + a_4b_3c_2 + a_3b_2c_4 + a_4b_4c_4 \end{bmatrix}$$

Comparing both results yields $A(BC) = (AB)C$.

57. Let $B = \begin{bmatrix} b_1 & b_2 \\ b_3 & b_4 \end{bmatrix}$, $C = \begin{bmatrix} c_1 & c_2 \\ c_3 & c_4 \end{bmatrix}$, and $D = \begin{bmatrix} d_1 & d_2 \\ d_3 & d_4 \end{bmatrix}$

Again, we are using single instead of double subscripts. We want to show that $(B + C)D = BD + CD$.

The left side becomes

$$(B + C)D = \left\{ \begin{bmatrix} b_1 & b_2 \\ b_3 & b_4 \end{bmatrix} + \begin{bmatrix} c_1 & c_2 \\ c_3 & c_4 \end{bmatrix} \right\} \begin{bmatrix} d_1 & d_2 \\ d_3 & d_4 \end{bmatrix}$$

$$= \begin{bmatrix} b_1 + c_1 & b_2 + c_2 \\ b_3 + c_3 & b_4 + c_4 \end{bmatrix} \begin{bmatrix} d_1 & d_2 \\ d_3 & d_4 \end{bmatrix}$$

$$= \begin{bmatrix} (b_1 + c_1)d_1 + (b_2 + c_2)d_3 & (b_1 + c_1)d_2 + (b_2 + c_2)d_4 \\ (b_3 + c_3)d_1 + (b_4 + c_4)d_3 & (b_3 + c_3)d_2 + (b_4 + c_4)d_4 \end{bmatrix}$$

$$= \begin{bmatrix} b_1d_1 + c_1d_1 + b_2d_3 + c_2d_3 & b_1d_2 + c_1d_2 + b_2d_4 + c_2d_4 \\ b_3d_1 + c_3d_1 + b_4d_3 + c_4d_3 & b_3d_2 + c_3d_2 + b_4d_4 + c_4d_4 \end{bmatrix}$$

The right side becomes

$$BD + CD = \begin{bmatrix} b_1 & b_2 \\ b_3 & b_4 \end{bmatrix} \begin{bmatrix} d_1 & d_2 \\ d_3 & d_4 \end{bmatrix} + \begin{bmatrix} c_1 & c_2 \\ c_3 & c_4 \end{bmatrix} \begin{bmatrix} d_1 & d_2 \\ d_3 & d_4 \end{bmatrix}$$

$$= \begin{bmatrix} b_1d_1 + b_2d_3 & b_1d_2 + b_2d_4 \\ b_3d_1 + b_4d_3 & b_3d_2 + b_4d_4 \end{bmatrix} + \begin{bmatrix} c_1d_1 + c_2d_3 & c_1d_2 + c_2d_4 \\ c_3d_1 + c_4d_3 & c_3d_2 + c_4d_4 \end{bmatrix}$$

$$= \begin{bmatrix} b_1d_1 + b_2d_3 + c_1d_1 + c_2d_3 & b_1d_2 + b_2d_4 + c_1d_2 + c_2d_4 \\ b_3d_1 + b_4d_3 + c_3d_1 + c_4d_3 & b_3d_2 + b_4d_4 + c_3d_2 + c_4d_4 \end{bmatrix}$$

Comparing both results yields $(B + C)D = BD + CD$.

59. Let $A = [a_{ij}]$

Then $(a + b)A = (a + b)[a_{ij}] = [(a + b)a_{ij}] = [aa_{ij} + ba_{ij}]$

$$= [aa_{ij}] + [ba_{ij}]$$
$$= a[a_{ij}] + b[a_{ij}]$$
$$= aA + bA$$

61. In explaining your answer, use that fact that in general, AB is **not** equal to BA.

Exercises 8.3

1. $\begin{bmatrix} 3 & -1 & \vdots & 0 \\ -1 & 1 & \vdots & 1 \end{bmatrix}$

3. $\begin{bmatrix} 3 & -2 & 5 & \vdots & 0 \\ 4 & 7 & -1 & \vdots & 0 \\ 1 & 0 & 1 & \vdots & 2 \end{bmatrix}$

5. $\begin{bmatrix} 1 & -1 & 0 & 0 & \vdots & 0 \\ 0 & 0 & 1 & 1 & \vdots & 0 \\ 3 & 0 & 0 & 2 & \vdots & 0 \\ 0 & 5 & -1 & 0 & \vdots & 0 \end{bmatrix}$

7. The system is
$$x + 2y = 5$$
$$3x + 4y = 6$$

9. The system is
$$x = a$$
$$z = b$$
$$y = c$$

11. The system is
$$x \qquad + z \qquad = 0$$
$$2y + z + 3w = 7$$
$$3x + y + z + w = 1$$
$$-3x + y - z + 2w = 5$$

13. $\begin{bmatrix} 1 & 1 & \vdots & 1 \\ 2 & 3 & \vdots & -2 \end{bmatrix} \xrightarrow{-2R_1 + R_3} \begin{bmatrix} 1 & 1 & \vdots & 1 \\ 0 & 1 & \vdots & -4 \end{bmatrix}$ This gives the system $x + y = 1$
$$y = -4$$

Now, $x + y = 1$
$$x + (-4) = 1$$
$$x = 5$$
The solution is $x = 5$, $y = -4$.

15. $\begin{bmatrix} 4 & 5 & \vdots & -22 \\ 3 & -4 & \vdots & -1 \end{bmatrix} \xrightarrow{\frac{1}{4}R_1} \begin{bmatrix} 1 & \frac{5}{4} & \vdots & -\frac{22}{4} \\ 3 & -4 & \vdots & -1 \end{bmatrix} \xrightarrow{-3R_1 + R_2} \begin{bmatrix} 1 & \frac{5}{4} & \vdots & -\frac{22}{4} \\ 0 & -\frac{31}{4} & \vdots & \frac{62}{4} \end{bmatrix} \xrightarrow{-\frac{4}{31}R_2} \begin{bmatrix} 1 & \frac{5}{4} & \vdots & -\frac{22}{4} \\ 0 & 1 & \vdots & -2 \end{bmatrix}$

This gives the system $a + \frac{5}{4}b = -\frac{22}{4}$
$$b = -2$$

The second equation is $b = -2$.

The first equation gives $a + \frac{5}{4}b = -\frac{22}{4}$
$$a + \frac{5}{4}(-2) = -\frac{22}{4}$$
$$a - \frac{10}{4} = -\frac{22}{4}$$
$$a = -\frac{22}{4} + \frac{10}{4} = -\frac{12}{4} = -3$$
The solution is $a = -3$, $b = -2$.

17. $\begin{bmatrix} 1 & -2 & \vdots & 0 \\ 2 & -4 & \vdots & 1 \end{bmatrix} \xrightarrow{-2R_1 + R_2} \begin{bmatrix} 1 & -2 & \vdots & 0 \\ 0 & 0 & \vdots & 1 \end{bmatrix}$ which provides **no** solution since the last row becomes $0 = 1$

19. $\begin{bmatrix} 2 & -3 & \vdots & -1 \\ -5 & 8 & \vdots & 0 \end{bmatrix} \xrightarrow{\frac{1}{2}R_1} \begin{bmatrix} 1 & -\frac{3}{2} & \vdots & -\frac{1}{2} \\ -5 & 8 & \vdots & 0 \end{bmatrix}$

$\xrightarrow{5R_1 + R_2} \begin{bmatrix} 1 & -\frac{3}{2} & \vdots & -\frac{1}{2} \\ 0 & \frac{1}{2} & \vdots & -\frac{5}{2} \end{bmatrix}$

$\xrightarrow{2R_2} \begin{bmatrix} 1 & -\frac{3}{2} & \vdots & -\frac{1}{2} \\ 0 & 1 & \vdots & -5 \end{bmatrix}$ The system is $x - \frac{3}{2} = -\frac{1}{2}$
$$y = -5$$

The second equation is $y = -5$.

The first equation gives $x - \dfrac{3}{2}y = -\dfrac{1}{2}$

$$x - \frac{3}{2}(-5) = -\frac{1}{2}$$

$$x + \frac{15}{2} = -\frac{1}{2}$$

$$x = -\frac{1}{2} - \frac{15}{2} = -\frac{16}{2} = -8$$

The solution is $x = -8$, $y = -5$.

21. $\begin{bmatrix} 2 & 7 & \vdots & 0 \\ 1 & -2 & \vdots & 0 \end{bmatrix} \xrightarrow{R_1 \leftrightarrow R_2} \begin{bmatrix} 1 & -2 & \vdots & 0 \\ 2 & 7 & \vdots & 0 \end{bmatrix}$

$\xrightarrow{-2R_1 + R_2} \begin{bmatrix} 1 & -2 & \vdots & 0 \\ 0 & 11 & \vdots & 0 \end{bmatrix}$
The system is $x - 2y = 0$
$\qquad\qquad\qquad 11y = 0$
which gives $\quad y = 0$
$\qquad\qquad\qquad\quad x = 0.$

23. $\begin{bmatrix} 1 & -2 & 0 & \vdots & 1 \\ 0 & 1 & 1 & \vdots & 0 \\ 2 & 0 & 3 & \vdots & 3 \end{bmatrix} \xrightarrow{-2R_1 + R_3} \begin{bmatrix} 1 & -2 & 0 & \vdots & 1 \\ 0 & 1 & 1 & \vdots & 0 \\ 0 & 4 & 3 & \vdots & 1 \end{bmatrix} \xrightarrow{-4R_2 + R_3} \begin{bmatrix} 1 & -2 & 0 & \vdots & 1 \\ 0 & 1 & 1 & \vdots & 0 \\ 0 & 0 & -1 & \vdots & 1 \end{bmatrix}$

$\xrightarrow{-R_3} \begin{bmatrix} 1 & -2 & 0 & \vdots & 1 \\ 0 & 1 & 1 & \vdots & 0 \\ 0 & 0 & 1 & \vdots & -1 \end{bmatrix}$
The system is $x - 2y = 1$
$\qquad\qquad\qquad\quad y + z = 0$
$\qquad\qquad\qquad\qquad\quad z = -1$

The third equation is $z = -1$
The second equation gives $y + z = 0$
$$y + (-1) = 0$$
$$y = 1$$

The first equation gives $x - 2y = 1$
$$x - 2(1) = 1$$
$$x - 2 = 1$$
$$x = 3$$
The solution is $x = 3$, $y = 1$, $z = -1$.

25. $\begin{bmatrix} 1 & -1 & -4 & \vdots & -4 \\ -3 & 4 & 2 & \vdots & 6 \\ -1 & 3 & 2 & \vdots & 10 \end{bmatrix} \xrightarrow{3R_1 + R_2} \begin{bmatrix} 1 & -1 & -4 & \vdots & -4 \\ 0 & 1 & -10 & \vdots & -6 \\ -1 & 3 & 2 & \vdots & 10 \end{bmatrix} \xrightarrow{R_1 + R_3} \begin{bmatrix} 1 & -1 & -4 & \vdots & -4 \\ 0 & 1 & -10 & \vdots & -6 \\ 0 & 2 & -2 & \vdots & 6 \end{bmatrix}$

$\xrightarrow{-2R_2 + R_3} \begin{bmatrix} 1 & -1 & -4 & \vdots & -4 \\ 0 & 1 & -10 & \vdots & -6 \\ 0 & 0 & 18 & \vdots & 18 \end{bmatrix}$

$\xrightarrow{\frac{1}{18}R_3} \begin{bmatrix} 1 & -1 & -4 & \vdots & -4 \\ 0 & 1 & -10 & \vdots & -6 \\ 0 & 0 & 1 & \vdots & 1 \end{bmatrix}$
The system is $x - y - 4z = -4$
$\qquad\qquad\qquad\quad y - 10z = -6$
$\qquad\qquad\qquad\qquad\quad z = 1$

The third equation is $z = 1$.

The second equation gives $y - 10z = -6$
$$y - 10(1) = -6$$
$$y = 4$$

The first equation gives $x - y - 4z = -4$
$$x - 4 - 4(1) = -4$$
$$x - 8 = -4$$
$$x = 4$$
The solution is $x = 4$, $y = 4$, $z = 1$.

27. $\begin{bmatrix} 0 & -2 & -2 & \vdots & -2 \\ 2 & -1 & 1 & \vdots & -3 \\ 1 & 1 & 3 & \vdots & -2 \end{bmatrix} \xrightarrow{R_1 \leftrightarrow R_3} \begin{bmatrix} 1 & 1 & 3 & \vdots & -2 \\ 2 & -1 & 1 & \vdots & -3 \\ 0 & -2 & -2 & \vdots & -2 \end{bmatrix} \xrightarrow{-2R_1 + R_2} \begin{bmatrix} 1 & 1 & 3 & \vdots & -2 \\ 0 & -3 & -5 & \vdots & 1 \\ 0 & -2 & -2 & \vdots & -2 \end{bmatrix}$

$\xrightarrow{-\frac{1}{2}R_3} \begin{bmatrix} 1 & 1 & 3 & \vdots & -2 \\ 0 & -3 & -5 & \vdots & 1 \\ 0 & 1 & 1 & \vdots & 1 \end{bmatrix} \xrightarrow{R_2 \leftrightarrow R_3} \begin{bmatrix} 1 & 1 & 3 & \vdots & -2 \\ 0 & 1 & 1 & \vdots & 1 \\ 0 & -3 & -5 & \vdots & 1 \end{bmatrix}$

$\xrightarrow{3R_2 + R_3} \begin{bmatrix} 1 & 1 & 3 & \vdots & -2 \\ 0 & 1 & 1 & \vdots & 1 \\ 0 & 0 & -2 & \vdots & 4 \end{bmatrix} \xrightarrow{-\frac{1}{2}R_3} \begin{bmatrix} 1 & 1 & 3 & \vdots & -2 \\ 0 & 1 & 1 & \vdots & 1 \\ 0 & 0 & 1 & \vdots & -2 \end{bmatrix}$ The system is $\begin{aligned} x + y + 3z &= -2 \\ y + z &= 1 \\ z &= -2 \end{aligned}$

The third equation is $z = -2$.

The second equation gives $y + z = 1$
$$y - 2 = 1$$
$$y = 3.$$

The first equation gives $x + y + 3z = -2$
$$x + 3 + 3(-2) = -2$$
$$x - 3 = -2$$
$$x = 1$$
The solution is $x = 1$, $y = 3$, $z = -2$.

29. $\begin{bmatrix} 1 & 0 & -1 & \vdots & 1 \\ 3 & 1 & 0 & \vdots & 6 \\ 0 & 5 & 6 & \vdots & -12 \end{bmatrix} \xrightarrow{-3R_1 + R_2} \begin{bmatrix} 1 & 0 & -1 & \vdots & 1 \\ 0 & 1 & 3 & \vdots & 3 \\ 0 & 5 & 6 & \vdots & -12 \end{bmatrix} \xrightarrow{-5R_2 + R_3} \begin{bmatrix} 1 & 0 & -1 & \vdots & 1 \\ 0 & 1 & 3 & \vdots & 3 \\ 0 & 0 & -9 & \vdots & -27 \end{bmatrix}$

$\xrightarrow{-\frac{1}{9}R_3} \begin{bmatrix} 1 & 0 & -1 & \vdots & 1 \\ 0 & 1 & 3 & \vdots & 3 \\ 0 & 0 & 1 & \vdots & 3 \end{bmatrix}$ The system is $\begin{aligned} r - t &= 1 \\ s + 3t &= 3 \\ t &= 3 \end{aligned}$

The third equation is $t = 3$.

The second equation gives $s + 3t = 3$
$$s + 3(3) = 3$$
$$s = -6$$

The first equation gives $r - t = 1$
$$r - 3 = 1$$
$$r = 4$$
The solution is $r = 4$, $s = -6$, $t = 3$.

31. $\begin{bmatrix} 2 & -2 & -2 & \vdots & -2 \\ 1 & -4 & 1 & \vdots & -2 \\ 5 & -8 & -3 & \vdots & -2 \end{bmatrix} \xrightarrow{\frac{1}{2}R_1} \begin{bmatrix} 1 & -1 & -1 & \vdots & -1 \\ 1 & -4 & 1 & \vdots & -2 \\ 5 & -8 & -3 & \vdots & -2 \end{bmatrix} \xrightarrow{-R_1 + R_2} \begin{bmatrix} 1 & -1 & -1 & \vdots & -1 \\ 0 & -3 & 2 & \vdots & -1 \\ 5 & -8 & -3 & \vdots & -2 \end{bmatrix}$

$\xrightarrow{-5R_1 + R_3} \begin{bmatrix} 1 & -1 & -1 & \vdots & -1 \\ 0 & -3 & 2 & \vdots & -1 \\ 0 & -3 & 2 & \vdots & 3 \end{bmatrix} \xrightarrow{-R_2 + R_3} \begin{bmatrix} 1 & -1 & -1 & \vdots & -1 \\ 0 & -3 & 2 & \vdots & -1 \\ 0 & 0 & 0 & \vdots & 4 \end{bmatrix}$ which results in **no** solution since the third row gives $0 = 4$.

33. $\begin{bmatrix} 1 & 1 & 0 & -1 & \vdots & 2 \\ 0 & 1 & 1 & 2 & \vdots & -3 \\ -2 & 0 & 0 & 1 & \vdots & -4 \\ 1 & 1 & 1 & 1 & \vdots & 0 \end{bmatrix} \xrightarrow{2R_1 + R_3} \begin{bmatrix} 1 & 1 & 0 & -1 & \vdots & 2 \\ 0 & 1 & 1 & 2 & \vdots & -3 \\ 0 & 2 & 0 & -1 & \vdots & 0 \\ 1 & 1 & 1 & 1 & \vdots & 0 \end{bmatrix}$

$\xrightarrow{-R_1 + R_4} \begin{bmatrix} 1 & 1 & 0 & -1 & \vdots & 2 \\ 0 & 1 & 1 & 2 & \vdots & -3 \\ 0 & 2 & 0 & -1 & \vdots & 0 \\ 0 & 0 & 1 & 2 & \vdots & -2 \end{bmatrix} \xrightarrow{-2R_2 + R_3} \begin{bmatrix} 1 & 1 & 0 & -1 & \vdots & 2 \\ 0 & 1 & 1 & 2 & \vdots & -3 \\ 0 & 0 & -2 & -5 & \vdots & 6 \\ 0 & 0 & 1 & 2 & \vdots & -2 \end{bmatrix}$

$\xrightarrow{R_3 \leftrightarrow R_4} \begin{bmatrix} 1 & 1 & 0 & -1 & \vdots & 2 \\ 0 & 1 & 1 & 2 & \vdots & -3 \\ 0 & 0 & 1 & 2 & \vdots & -2 \\ 0 & 0 & -2 & -5 & \vdots & 6 \end{bmatrix} \xrightarrow{2R_3 + R_4} \begin{bmatrix} 1 & 1 & 0 & -1 & \vdots & 2 \\ 0 & 1 & 1 & 2 & \vdots & -3 \\ 0 & 0 & 1 & 2 & \vdots & -2 \\ 0 & 0 & 0 & -1 & \vdots & 2 \end{bmatrix}$

$\xrightarrow{-R_4} \begin{bmatrix} 1 & 1 & 0 & -1 & \vdots & 2 \\ 0 & 1 & 1 & 2 & \vdots & -3 \\ 0 & 0 & 1 & 2 & \vdots & -2 \\ 0 & 0 & 0 & 1 & \vdots & -2 \end{bmatrix}$ The system is $\begin{aligned} x + y \quad - \ t &= 2 \\ y + z + 2t &= -3 \\ z + 2t &= -2 \\ t &= -2 \end{aligned}$

The fourth equation is $t = -2$.

The third equation gives $\begin{aligned} z + 2t &= -2 \\ z + 2(-2) &= -2 \\ z - 4 &= -2 \\ z &= 2 \end{aligned}$

The second equation gives $\begin{aligned} y + z + 2t &= -3 \\ y + 2 + 2(-2) &= -3 \\ y - 2 &= -3 \\ y &= -1 \end{aligned}$

The first equation gives $\begin{aligned} x + y - t &= 2 \\ x + (-1) - (-2) &= 2 \\ x + 1 &= 2 \\ x &= 1 \end{aligned}$

The solution is $x = 1$, $y = -1$, $z = 2$, $t = -2$.

35. $\begin{bmatrix} -1 & 1 & -1 & 0 & \vdots & 5 \\ 0 & 1 & -1 & 3 & \vdots & 10 \\ 1 & 0 & -1 & -1 & \vdots & 5 \\ 1 & 0 & 0 & 3 & \vdots & 5 \end{bmatrix} \xrightarrow[R_1 + R_4]{R_1 + R_3} \begin{bmatrix} -1 & 1 & -1 & 0 & \vdots & 5 \\ 0 & 1 & -1 & 3 & \vdots & 10 \\ 0 & 1 & -2 & -1 & \vdots & 10 \\ 0 & 1 & -1 & 3 & \vdots & 10 \end{bmatrix}$

$\xrightarrow{-R_1} \begin{bmatrix} 1 & -1 & 1 & 0 & \vdots & -5 \\ 0 & 1 & -1 & 3 & \vdots & 10 \\ 0 & 1 & -2 & -1 & \vdots & 10 \\ 0 & 1 & -1 & 3 & \vdots & 10 \end{bmatrix} \xrightarrow{R_2 + R_1} \begin{bmatrix} 1 & 0 & 0 & 3 & \vdots & 5 \\ 0 & 1 & -1 & 3 & \vdots & 10 \\ 0 & 1 & -2 & -1 & \vdots & 10 \\ 0 & 1 & -1 & 3 & \vdots & 10 \end{bmatrix}$

$\xrightarrow[-R_2 + R_4]{-R_2 + R_3} \begin{bmatrix} 1 & 0 & 0 & 3 & \vdots & 5 \\ 0 & 1 & -1 & 3 & \vdots & 10 \\ 0 & 0 & -1 & -4 & \vdots & 0 \\ 0 & 0 & 0 & 0 & \vdots & 0 \end{bmatrix}$ Since the last row produces $0 = 0$, there are infinitely many solutions. Let $d = r$ where r is any real number.

The third row now gives $\begin{aligned} -c - 4d &= 0 \\ -c - 4r &= 0 \\ -c &= 4r \\ c &= -4r \end{aligned}$ The second row gives $\begin{aligned} b - c + 3d &= 10 \\ b - (-4r) + 3r &= 10 \\ b + 7r &= 10 \\ b &= 10 - 7r \end{aligned}$

The first row gives $\begin{aligned} a + 3d &= 5 \\ a + 3r &= 5 \\ a &= 5 - 3r \end{aligned}$

The solution is represented by $a = 5 - 3r$, $b = 10 - 7r$, $c = -4r$, $d = r$.

37. $\begin{bmatrix} 1 & 2 & \vdots & 1 \\ 2 & c^2 & \vdots & c \end{bmatrix} \xrightarrow{-2R_1 + R_2} \begin{bmatrix} 1 & 2 & \vdots & 1 \\ 0 & c^2 - 4 & \vdots & c - 2 \end{bmatrix}$

There are several cases to consider. If $c = -2$, then the last row becomes $[0 \quad 0 \vdots -4]$ which translates into $0 = -4$ and **no** solution exists.

If $c = 2$, then the last row becomes $[0 \quad 0 \vdots 0]$ which translates into $0 = 0$ and infinitely many solutions exist.

Finally, if $c \neq 2$, $c \neq -2$, then exactly one solution exists since

$\begin{bmatrix} 1 & 2 & \vdots & 1 \\ 0 & c^2 - 4 & \vdots & c - 2 \end{bmatrix} \xrightarrow{\frac{1}{c^2 - 4} R_2} \begin{bmatrix} 1 & 2 & \vdots & 1 \\ 0 & 1 & \vdots & \frac{c - 2}{c^2 - 4} \end{bmatrix} = \begin{bmatrix} 1 & 2 & \vdots & 1 \\ 0 & 1 & \vdots & \frac{1}{c + 2} \end{bmatrix}$

39. $\begin{bmatrix} 1 & 2 & -1 & \vdots & 1 \\ 1 & 1 & 0 & \vdots & 0 \\ 1 & 2 & c^2 - 1 & \vdots & c + 1 \end{bmatrix} \xrightarrow{-R_1 + R_3} \begin{bmatrix} 1 & 2 & -1 & \vdots & 1 \\ 1 & 1 & 0 & \vdots & 0 \\ 0 & 0 & c^2 & \vdots & c \end{bmatrix}$

Here, solutions exist for all values of c. However, if $c = 0$, then the last row becomes $[0 \quad 0 \quad 0 \vdots 0]$ which translates into $0 = 0$ and infinitely many solutions exist.

If $c \neq 0$, then exactly one solution exists and the values for x, y, z can be obtained.

41. On the left side of the dashed line, 1's are required on the diagonal and 0's are required below the diagonal for both methods. However, for the Gauss-Jordan Elimination Method, 0's are required above the diagonal also. For the Gauss Elimination Method, 0's are not required above the diagonal. It is more convenient to solve a system of linear equations by Gaussian elimination. The corresponding system of equations can be solved by back substitution. Both methods are well suited for computer computation because they are systematic.

Exercises 8.4

1. Let $A = \begin{bmatrix} 0 & 3 \\ -2 & 4 \end{bmatrix}$. Then, $\delta(A) = ad - bc = (0)(4) - (3)(-2) = 0 + 6 = 6$

and $A^{-1} = \frac{1}{6}\begin{bmatrix} 4 & -3 \\ 2 & 0 \end{bmatrix} = \begin{bmatrix} \frac{4}{6} & -\frac{3}{6} \\ \frac{2}{6} & \frac{0}{6} \end{bmatrix} = \begin{bmatrix} \frac{2}{3} & -\frac{1}{2} \\ \frac{1}{3} & 0 \end{bmatrix}$

3. Let $A = \begin{bmatrix} -1 & 3 \\ 2 & 2 \end{bmatrix}$. Then, $\delta(A) = ad - bc = (-1)(2) - (3)(2) = -2 - 6 = -8$

and $A^{-1} = \frac{1}{-8}\begin{bmatrix} 2 & -3 \\ -2 & -1 \end{bmatrix} = \begin{bmatrix} \frac{2}{-8} & \frac{-3}{-8} \\ \frac{-2}{-8} & \frac{-1}{-8} \end{bmatrix} = \begin{bmatrix} -\frac{1}{4} & \frac{3}{8} \\ \frac{1}{4} & \frac{1}{8} \end{bmatrix}$

5. Let $A = \begin{bmatrix} -3 & 5 \\ 12 & -20 \end{bmatrix}$ Then $\delta(A) = (-3)(-20) - (5)(12) = 60 - 60 = 0$

and, consequently, the inverse does **not** exist.

7. Let $A = \begin{bmatrix} -2 & 3 \\ 5 & 7 \end{bmatrix}$. Then $\delta(A) = (-2)(7) - (3)(5) = -14 - 15 = -29$

and $A^{-1} = \dfrac{1}{-29}\begin{bmatrix} 7 & -3 \\ -5 & -2 \end{bmatrix} = \begin{bmatrix} \frac{7}{-29} & \frac{-3}{-29} \\ \frac{-5}{-29} & \frac{-2}{-29} \end{bmatrix} = \begin{bmatrix} -\frac{7}{29} & \frac{3}{29} \\ \frac{5}{29} & \frac{2}{29} \end{bmatrix}$

9. $\begin{bmatrix} 0 & 1 & 0 & \vdots & 1 & 0 & 0 \\ 1 & 4 & 1 & \vdots & 0 & 1 & 0 \\ 0 & 3 & -1 & \vdots & 0 & 0 & 1 \end{bmatrix} \xrightarrow{R_1 \leftrightarrow R_2} \begin{bmatrix} 1 & 4 & 1 & \vdots & 0 & 1 & 0 \\ 0 & 1 & 0 & \vdots & 1 & 0 & 0 \\ 0 & 3 & -1 & \vdots & 0 & 0 & 1 \end{bmatrix}$

$\xrightarrow{-4R_2 + R_1} \begin{bmatrix} 1 & 0 & 1 & \vdots & -4 & 1 & 0 \\ 0 & 1 & 0 & \vdots & 1 & 0 & 0 \\ 0 & 3 & -1 & \vdots & 0 & 0 & 1 \end{bmatrix} \xrightarrow{-3R_2 + R_3} \begin{bmatrix} 1 & 0 & 1 & \vdots & -4 & 1 & 0 \\ 0 & 1 & 0 & \vdots & 1 & 0 & 0 \\ 0 & 0 & -1 & \vdots & -3 & 0 & 1 \end{bmatrix}$

$\xrightarrow{-R_3} \begin{bmatrix} 1 & 0 & 1 & \vdots & -4 & 1 & 0 \\ 0 & 1 & 0 & \vdots & 1 & 0 & 0 \\ 0 & 0 & 1 & \vdots & 3 & 0 & -1 \end{bmatrix} \xrightarrow{-R_3 + R_1} \begin{bmatrix} 1 & 0 & 0 & \vdots & -7 & 1 & 1 \\ 0 & 1 & 0 & \vdots & 1 & 0 & 0 \\ 0 & 0 & 1 & \vdots & 3 & 0 & -1 \end{bmatrix}$

The inverse is $\begin{bmatrix} -7 & 1 & 1 \\ 1 & 0 & 0 \\ 3 & 0 & -1 \end{bmatrix}$.

11. $\begin{bmatrix} 1 & -4 & 2 & \vdots & 1 & 0 & 0 \\ 2 & -9 & 5 & \vdots & 0 & 1 & 0 \\ 1 & -5 & 4 & \vdots & 0 & 0 & 1 \end{bmatrix} \xrightarrow{-2R_1 + R_2} \begin{bmatrix} 1 & -4 & 2 & \vdots & 1 & 0 & 0 \\ 0 & -1 & 1 & \vdots & -2 & 1 & 0 \\ 1 & -5 & 4 & \vdots & 0 & 0 & 1 \end{bmatrix}$

$\xrightarrow{-R_1 + R_3} \begin{bmatrix} 1 & -4 & 2 & \vdots & 1 & 0 & 0 \\ 0 & -1 & 1 & \vdots & -2 & 1 & 0 \\ 0 & -1 & 2 & \vdots & -1 & 0 & 1 \end{bmatrix} \xrightarrow{-R_2} \begin{bmatrix} 1 & -4 & 2 & \vdots & 1 & 0 & 0 \\ 0 & 1 & -1 & \vdots & 2 & -1 & 0 \\ 0 & -1 & 2 & \vdots & -1 & 0 & 1 \end{bmatrix}$

$\xrightarrow{4R_2 + R_1} \begin{bmatrix} 1 & 0 & -2 & \vdots & 9 & -4 & 0 \\ 0 & 1 & -1 & \vdots & 2 & -1 & 0 \\ 0 & -1 & 2 & \vdots & -1 & 0 & 1 \end{bmatrix} \xrightarrow{R_2 + R_3} \begin{bmatrix} 1 & 0 & -2 & \vdots & 9 & -4 & 0 \\ 0 & 1 & -1 & \vdots & 2 & -1 & 0 \\ 0 & 0 & 1 & \vdots & 1 & -1 & 1 \end{bmatrix}$

$\xrightarrow{2R_3 + R_1} \begin{bmatrix} 1 & 0 & 0 & \vdots & 11 & -6 & 2 \\ 0 & 1 & -1 & \vdots & 2 & -1 & 0 \\ 0 & 0 & 1 & \vdots & 1 & -1 & 1 \end{bmatrix} \xrightarrow{R_3 + R_2} \begin{bmatrix} 1 & 0 & 0 & \vdots & 11 & -6 & 2 \\ 0 & 1 & 0 & \vdots & 3 & -2 & 1 \\ 0 & 0 & 1 & \vdots & 1 & -1 & 1 \end{bmatrix}$

The inverse is $\begin{bmatrix} 11 & -6 & 2 \\ 3 & -2 & 1 \\ 1 & -1 & 1 \end{bmatrix}$.

13. $\begin{bmatrix} 1 & 0 & 1 & \vdots & 1 & 0 & 0 \\ -1 & 2 & 1 & \vdots & 0 & 1 & 0 \\ 0 & 1 & 3 & \vdots & 0 & 0 & 1 \end{bmatrix} \xrightarrow{R_1 + R_2} \begin{bmatrix} 1 & 0 & 1 & \vdots & 1 & 0 & 0 \\ 0 & 2 & 2 & \vdots & 1 & 1 & 0 \\ 0 & 1 & 3 & \vdots & 0 & 0 & 1 \end{bmatrix}$

$\xrightarrow{R_2 \leftrightarrow R_3} \begin{bmatrix} 1 & 0 & 1 & \vdots & 1 & 0 & 0 \\ 0 & 1 & 3 & \vdots & 0 & 0 & 1 \\ 0 & 2 & 2 & \vdots & 1 & 1 & 0 \end{bmatrix} \xrightarrow{-2R_2 + R_3} \begin{bmatrix} 1 & 0 & 1 & \vdots & 1 & 0 & 0 \\ 0 & 1 & 3 & \vdots & 0 & 0 & 1 \\ 0 & 0 & -4 & \vdots & 1 & 1 & -2 \end{bmatrix}$

$\xrightarrow{-\frac{1}{4}R_3} \begin{bmatrix} 1 & 0 & 1 & \vdots & 1 & 0 & 0 \\ 0 & 1 & 3 & \vdots & 0 & 0 & 1 \\ 0 & 0 & 1 & \vdots & -\frac{1}{4} & -\frac{1}{4} & \frac{1}{2} \end{bmatrix} \xrightarrow{-R_3 + R_1} \begin{bmatrix} 1 & 0 & 0 & \vdots & \frac{5}{4} & \frac{1}{4} & -\frac{1}{2} \\ 0 & 1 & 3 & \vdots & 0 & 0 & 1 \\ 0 & 0 & 1 & \vdots & -\frac{1}{4} & -\frac{1}{4} & \frac{1}{2} \end{bmatrix}$

$\xrightarrow{-3R_3 + R_2} \begin{bmatrix} 1 & 0 & 0 & \vdots & \frac{5}{4} & \frac{1}{4} & -\frac{1}{2} \\ 0 & 1 & 0 & \vdots & \frac{3}{4} & \frac{3}{4} & -\frac{1}{2} \\ 0 & 0 & 1 & \vdots & -\frac{1}{4} & -\frac{1}{4} & \frac{1}{2} \end{bmatrix}$ The inverse is $\begin{bmatrix} \frac{5}{4} & \frac{1}{4} & -\frac{1}{2} \\ \frac{3}{4} & \frac{3}{4} & -\frac{1}{2} \\ -\frac{1}{4} & -\frac{1}{4} & \frac{1}{2} \end{bmatrix}$.

15. $\begin{bmatrix} 5 & 3 & -2 & \vdots & 1 & 0 & 0 \\ -1 & 2 & 5 & \vdots & 0 & 1 & 0 \\ 11 & 4 & -9 & \vdots & 0 & 0 & 1 \end{bmatrix} \xrightarrow{R_2 \leftrightarrow R_1} \begin{bmatrix} -1 & 2 & 5 & \vdots & 0 & 1 & 0 \\ 5 & 3 & -2 & \vdots & 1 & 0 & 0 \\ 11 & 4 & -9 & \vdots & 0 & 0 & 1 \end{bmatrix}$

$\xrightarrow{-R_1} \begin{bmatrix} 1 & -2 & -5 & \vdots & 0 & -1 & 0 \\ 5 & 3 & -2 & \vdots & 1 & 0 & 0 \\ 11 & 4 & -9 & \vdots & 0 & 0 & 1 \end{bmatrix} \xrightarrow[-11R_1 + R_3]{-5R_1 + R_2} \begin{bmatrix} 1 & -2 & -5 & \vdots & 0 & -1 & 0 \\ 0 & 13 & 23 & \vdots & 1 & 5 & 0 \\ 0 & 26 & 46 & \vdots & 0 & 11 & 1 \end{bmatrix}$

$\xrightarrow{-2R_2 + R_3} \begin{bmatrix} 1 & -2 & -5 & \vdots & 0 & -1 & 0 \\ 0 & 13 & 23 & \vdots & 1 & 5 & 0 \\ 0 & 0 & 0 & \vdots & -2 & 1 & 1 \end{bmatrix}$

The inverse does not exist since three 0's appear on the left of the dashed line in the third row.

17. $\begin{bmatrix} 1 & 0 & -1 & 2 & \vdots & 1 & 0 & 0 & 0 \\ 3 & -1 & -1 & 6 & \vdots & 0 & 1 & 0 & 0 \\ 2 & 0 & -3 & 8 & \vdots & 0 & 0 & 1 & 0 \\ 1 & 2 & -2 & -9 & \vdots & 0 & 0 & 0 & 1 \end{bmatrix} \xrightarrow[\substack{-2R_1 + R_3 \\ -R_1 + R_4}]{-3R_1 + R_2} \begin{bmatrix} 1 & 0 & -1 & 2 & \vdots & 1 & 0 & 0 & 0 \\ 0 & -1 & 2 & 0 & \vdots & -3 & 1 & 0 & 0 \\ 0 & 0 & -1 & 4 & \vdots & -2 & 0 & 1 & 0 \\ 0 & 2 & -1 & -11 & \vdots & -1 & 0 & 0 & 1 \end{bmatrix}$

$\xrightarrow{-R_2} \begin{bmatrix} 1 & 0 & -1 & 2 & \vdots & 1 & 0 & 0 & 0 \\ 0 & 1 & -2 & 0 & \vdots & 3 & -1 & 0 & 0 \\ 0 & 0 & -1 & 4 & \vdots & -2 & 0 & 1 & 0 \\ 0 & 2 & -1 & -11 & \vdots & -1 & 0 & 0 & 1 \end{bmatrix} \xrightarrow{-2R_2 + R_4} \begin{bmatrix} 1 & 0 & -1 & 2 & \vdots & 1 & 0 & 0 & 0 \\ 0 & 1 & -2 & 0 & \vdots & 3 & -1 & 0 & 0 \\ 0 & 0 & -1 & 4 & \vdots & -2 & 0 & 1 & 0 \\ 0 & 0 & 3 & -11 & \vdots & -7 & 2 & 0 & 1 \end{bmatrix}$

$\xrightarrow{-R_3} \begin{bmatrix} 1 & 0 & -1 & 2 & \vdots & 1 & 0 & 0 & 0 \\ 0 & 1 & -2 & 0 & \vdots & 3 & -1 & 0 & 0 \\ 0 & 0 & 1 & -4 & \vdots & 2 & 0 & -1 & 0 \\ 0 & 0 & 3 & -11 & \vdots & -7 & 2 & 0 & 1 \end{bmatrix} \xrightarrow[\substack{2R_3 + R_2 \\ -3R_3 + R_4}]{R_3 + R_1} \begin{bmatrix} 1 & 0 & 0 & -2 & \vdots & 3 & 0 & -1 & 0 \\ 0 & 1 & 0 & -8 & \vdots & 7 & -1 & -2 & 0 \\ 0 & 0 & 1 & -4 & \vdots & 2 & 0 & -1 & 0 \\ 0 & 0 & 0 & 1 & \vdots & -13 & 2 & 3 & 1 \end{bmatrix}$

$\xrightarrow[\substack{8R_4 + R_2 \\ 4R_4 + R_3}]{2R_4 + R_1} \begin{bmatrix} 1 & 0 & 0 & 0 & \vdots & -23 & 4 & 5 & 2 \\ 0 & 1 & 0 & 0 & \vdots & -97 & 15 & 22 & 8 \\ 0 & 0 & 1 & 0 & \vdots & -50 & 8 & 11 & 4 \\ 0 & 0 & 0 & 1 & \vdots & -13 & 2 & 3 & 1 \end{bmatrix}$

The inverse is $\begin{bmatrix} -23 & 4 & 5 & 2 \\ -97 & 15 & 22 & 8 \\ -50 & 8 & 11 & 4 \\ -13 & 2 & 3 & 1 \end{bmatrix}$

19. $2x + 3y = 1$
$2x + y = 7$

Let $A = \begin{bmatrix} 2 & 3 \\ 2 & 1 \end{bmatrix}$, $X = \begin{bmatrix} x \\ y \end{bmatrix}$, and $B = \begin{bmatrix} 1 \\ 7 \end{bmatrix}$.

Now, $\delta(A) = (2)(1) - (2)(3) = 2 - 6 = -4$ and $A^{-1} = \dfrac{1}{-4}\begin{bmatrix} 1 & -3 \\ -2 & 2 \end{bmatrix}$.

Finally, $X = A^{-1}B$ or $\begin{bmatrix} x \\ y \end{bmatrix} = -\dfrac{1}{4}\begin{bmatrix} 1 & -3 \\ -2 & 2 \end{bmatrix}\begin{bmatrix} 1 \\ 7 \end{bmatrix} = -\dfrac{1}{4}\begin{bmatrix} -20 \\ 12 \end{bmatrix} = \begin{bmatrix} 5 \\ -3 \end{bmatrix}$ which means that
$x = 5$ and $y = -3$.

21. $x_1 - 3x_2 = -6$
$6x_1 - 3x_2 = 9$

Let $A = \begin{bmatrix} 1 & -3 \\ 6 & -3 \end{bmatrix}$, $X = \begin{bmatrix} x_1 \\ x_2 \end{bmatrix}$, and $B = \begin{bmatrix} -6 \\ 9 \end{bmatrix}$.

Now, $\delta(A) = (1)(-3) - (-3)(6) = -3 + 18 = 15$ and $A^{-1} = \dfrac{1}{15}\begin{bmatrix} -3 & 3 \\ -6 & 1 \end{bmatrix}$.

Finally, $X = A^{-1}B = \dfrac{1}{15}\begin{bmatrix} -3 & 3 \\ -6 & 1 \end{bmatrix}\begin{bmatrix} -6 \\ 9 \end{bmatrix} = \dfrac{1}{15}\begin{bmatrix} 45 \\ 45 \end{bmatrix} = \begin{bmatrix} 3 \\ 3 \end{bmatrix}$ which means that

$x_1 = 3$ and $x_2 = 3$.

23. $3x + 2y = 22$
   $x + 2y = 10$

   Let $A = \begin{bmatrix} 3 & 2 \\ 1 & 2 \end{bmatrix}$, $X = \begin{bmatrix} x \\ y \end{bmatrix}$, and $B = \begin{bmatrix} 22 \\ 10 \end{bmatrix}$. Now, $\delta(A) = 3(2) - (2)(1) = 6 - 2 = 4$

   and $A^{-1} = \dfrac{1}{4}\begin{bmatrix} 2 & -2 \\ -1 & 3 \end{bmatrix}$. Finally, $X = A^{-1}B = \dfrac{1}{4}\begin{bmatrix} 2 & -2 \\ -1 & 3 \end{bmatrix}\begin{bmatrix} 22 \\ 10 \end{bmatrix} = \dfrac{1}{4}\begin{bmatrix} 24 \\ 8 \end{bmatrix} = \begin{bmatrix} 6 \\ 2 \end{bmatrix}$ which

   means that $x = 6$ and $y = 2$.

25. $x + y = 1$
   $7x + 3y = 0$

   Let $A = \begin{bmatrix} 1 & 1 \\ 7 & 3 \end{bmatrix}$, $X = \begin{bmatrix} x \\ y \end{bmatrix}$, and $B = \begin{bmatrix} 1 \\ 0 \end{bmatrix}$. Now, $\delta(A) = (1)(3) - (1)(7) = 3 - 7 = -4$

   and $A^{-1} = \dfrac{1}{-4}\begin{bmatrix} 3 & -1 \\ -7 & 1 \end{bmatrix}$. Finally, $X = A^{-1}B = -\dfrac{1}{4}\begin{bmatrix} 3 & -1 \\ -7 & 1 \end{bmatrix}\begin{bmatrix} 1 \\ 0 \end{bmatrix} = -\dfrac{1}{4}\begin{bmatrix} 3 \\ -7 \end{bmatrix} = \begin{bmatrix} -\frac{3}{4} \\ \frac{7}{4} \end{bmatrix}$

   which means that $x = -\dfrac{3}{4}$ and $y = \dfrac{7}{4}$.

27. $x_1 + 5x_2 = -4$ Let $A = \begin{bmatrix} 1 & 5 & 0 \\ 2 & 0 & -2 \\ 0 & 4 & -1 \end{bmatrix}$, $X = \begin{bmatrix} x_1 \\ x_2 \\ x_3 \end{bmatrix}$, and $B = \begin{bmatrix} -4 \\ 0 \\ 4 \end{bmatrix}$
   $2x_1 - 2x_3 = 0$
   $4x_2 - x_3 = 4$

   For the inverse of A, we have

   $\begin{bmatrix} 1 & 5 & 0 & \vdots & 1 & 0 & 0 \\ 2 & 0 & -2 & \vdots & 0 & 1 & 0 \\ 0 & 4 & -1 & \vdots & 0 & 0 & 1 \end{bmatrix} \xrightarrow{-2R_1 + R_2} \begin{bmatrix} 1 & 5 & 0 & \vdots & 1 & 0 & 0 \\ 0 & -10 & -2 & \vdots & -2 & 1 & 0 \\ 0 & 4 & -1 & \vdots & 0 & 0 & 1 \end{bmatrix}$

   $\xrightarrow{R_2 \leftrightarrow R_3} \begin{bmatrix} 1 & 5 & 0 & \vdots & 1 & 0 & 0 \\ 0 & 4 & -1 & \vdots & 0 & 0 & 1 \\ 0 & -10 & -2 & \vdots & -2 & 1 & 0 \end{bmatrix} \xrightarrow{\frac{1}{4}R_2} \begin{bmatrix} 1 & 5 & 0 & \vdots & 1 & 0 & 0 \\ 0 & 1 & -\frac{1}{4} & \vdots & 0 & 0 & \frac{1}{4} \\ 0 & -10 & -2 & \vdots & -2 & 1 & 0 \end{bmatrix}$

   $\xrightarrow[10R_2 + R_3]{-5R_2 + R_1} \begin{bmatrix} 1 & 0 & \frac{5}{4} & \vdots & 1 & 0 & -\frac{5}{4} \\ 0 & 1 & -\frac{1}{4} & \vdots & 0 & 0 & \frac{1}{4} \\ 0 & 0 & -\frac{9}{2} & \vdots & -2 & 1 & \frac{5}{2} \end{bmatrix} \xrightarrow{-\frac{2}{9}R_3} \begin{bmatrix} 1 & 0 & \frac{5}{4} & \vdots & 1 & 0 & -\frac{5}{4} \\ 0 & 1 & -\frac{1}{4} & \vdots & 0 & 0 & \frac{1}{4} \\ 0 & 0 & 1 & \vdots & \frac{4}{9} & -\frac{2}{9} & -\frac{5}{9} \end{bmatrix}$

   $\xrightarrow[\frac{1}{4}R_3 + R_2]{-\frac{5}{4}R_3 + R_1} \begin{bmatrix} 1 & 0 & 0 & \vdots & \frac{4}{9} & \frac{5}{18} & -\frac{5}{9} \\ 0 & 1 & 0 & \vdots & \frac{1}{9} & -\frac{1}{18} & \frac{1}{9} \\ 0 & 0 & 1 & \vdots & \frac{4}{9} & -\frac{2}{9} & -\frac{5}{9} \end{bmatrix}$

   Now, $A^{-1} = \begin{bmatrix} \frac{4}{9} & \frac{5}{18} & -\frac{5}{9} \\ \frac{1}{9} & -\frac{1}{18} & \frac{1}{9} \\ \frac{4}{9} & -\frac{2}{9} & -\frac{5}{9} \end{bmatrix}$ and $X = A^{-1}B = \begin{bmatrix} \frac{4}{9} & \frac{5}{18} & -\frac{5}{9} \\ \frac{1}{9} & -\frac{1}{18} & \frac{1}{9} \\ \frac{4}{9} & -\frac{2}{9} & -\frac{5}{9} \end{bmatrix}\begin{bmatrix} -4 \\ 0 \\ 4 \end{bmatrix} = \begin{bmatrix} -4 \\ 0 \\ -4 \end{bmatrix}$ so that $x_1 = -4$,
   $x_2 = 0$,
   $x_3 = -4$

29. $\begin{aligned} x - y + z &= -2 \\ y - 3z &= 10 \\ 3x - 3y + 2z &= -3 \end{aligned}$ Let $A = \begin{bmatrix} 1 & -1 & 1 \\ 0 & 1 & -3 \\ 3 & -3 & 2 \end{bmatrix}$, $X = \begin{bmatrix} x \\ y \\ z \end{bmatrix}$, and $B = \begin{bmatrix} -2 \\ 10 \\ -3 \end{bmatrix}$

For the inverse of A, we have

$\begin{bmatrix} 1 & -1 & 1 & \vdots & 1 & 0 & 0 \\ 0 & 1 & -3 & \vdots & 0 & 1 & 0 \\ 3 & -3 & 2 & \vdots & 0 & 0 & 1 \end{bmatrix} \xrightarrow{-3R_1 + R_3} \begin{bmatrix} 1 & -1 & 1 & \vdots & 1 & 0 & 0 \\ 0 & 1 & -3 & \vdots & 0 & 1 & 0 \\ 0 & 0 & -1 & \vdots & -3 & 0 & 1 \end{bmatrix}$

$\xrightarrow{R_2 + R_1} \begin{bmatrix} 1 & 0 & -2 & \vdots & 1 & 1 & 0 \\ 0 & 1 & -3 & \vdots & 0 & 1 & 0 \\ 0 & 0 & -1 & \vdots & -3 & 0 & 1 \end{bmatrix} \xrightarrow{-R_3} \begin{bmatrix} 1 & 0 & -2 & \vdots & 1 & 1 & 0 \\ 0 & 1 & -3 & \vdots & 0 & 1 & 0 \\ 0 & 0 & 1 & \vdots & 3 & 0 & -1 \end{bmatrix}$

$\xrightarrow[3R_3 + R_1]{2R_3 + R_1} \begin{bmatrix} 1 & 0 & 0 & \vdots & 7 & 1 & -2 \\ 0 & 1 & 0 & \vdots & 9 & 1 & -3 \\ 0 & 0 & 1 & \vdots & 3 & 0 & -1 \end{bmatrix}$

Now, $A^{-1} = \begin{bmatrix} 7 & 1 & -2 \\ 9 & 1 & -3 \\ 3 & 0 & -1 \end{bmatrix}$ and $X = A^{-1}B = \begin{bmatrix} 7 & 1 & -2 \\ 9 & 1 & -3 \\ 3 & 0 & -1 \end{bmatrix} \begin{bmatrix} -2 \\ 10 \\ -3 \end{bmatrix} = \begin{bmatrix} 2 \\ 1 \\ -3 \end{bmatrix}$ so that $\begin{aligned} x &= 2, \\ y &= 1, \\ z &= -3 \end{aligned}$

31. $\begin{aligned} -y + 3z &= 0 \\ -x + 5z &= 0 \\ x - y - z &= 1 \end{aligned}$ Let $A = \begin{bmatrix} 0 & -1 & 3 \\ -1 & 0 & 5 \\ 1 & -1 & -1 \end{bmatrix}$, $X = \begin{bmatrix} x \\ y \\ z \end{bmatrix}$, and $B = \begin{bmatrix} 0 \\ 0 \\ 1 \end{bmatrix}$.

For the inverse of A, we have

$\begin{bmatrix} 0 & -1 & 3 & \vdots & 1 & 0 & 0 \\ -1 & 0 & 5 & \vdots & 0 & 1 & 0 \\ 1 & -1 & -1 & \vdots & 0 & 0 & 1 \end{bmatrix} \xrightarrow{R_1 \leftrightarrow R_3} \begin{bmatrix} 1 & -1 & -1 & \vdots & 0 & 0 & 1 \\ -1 & 0 & 5 & \vdots & 0 & 1 & 0 \\ 0 & -1 & 3 & \vdots & 1 & 0 & 0 \end{bmatrix}$

$\xrightarrow{R_1 + R_2} \begin{bmatrix} 1 & -1 & -1 & \vdots & 0 & 0 & 1 \\ 0 & -1 & 4 & \vdots & 0 & 1 & 1 \\ 0 & -1 & 3 & \vdots & 1 & 0 & 0 \end{bmatrix} \xrightarrow{-R_2} \begin{bmatrix} 1 & -1 & -1 & \vdots & 0 & 0 & 1 \\ 0 & 1 & -4 & \vdots & 0 & -1 & -1 \\ 0 & -1 & 3 & \vdots & 1 & 0 & 0 \end{bmatrix}$

$\xrightarrow[R_2 + R_3]{R_2 + R_1} \begin{bmatrix} 1 & 0 & -5 & \vdots & 0 & -1 & 0 \\ 0 & 1 & -4 & \vdots & 0 & -1 & -1 \\ 0 & 0 & -1 & \vdots & 1 & -1 & -1 \end{bmatrix} \xrightarrow{-R_3} \begin{bmatrix} 1 & 0 & -5 & \vdots & 0 & -1 & 0 \\ 0 & 1 & -4 & \vdots & 0 & -1 & -1 \\ 0 & 0 & 1 & \vdots & -1 & 1 & 1 \end{bmatrix}$

$\xrightarrow[4R_3 + R_2]{5R_3 + R_1} \begin{bmatrix} 1 & 0 & 0 & \vdots & -5 & 4 & 5 \\ 0 & 1 & 0 & \vdots & -4 & 3 & 3 \\ 0 & 0 & 1 & \vdots & -1 & 1 & 1 \end{bmatrix}$

Now, $A^{-1} = \begin{bmatrix} -5 & 4 & 5 \\ -4 & 3 & 3 \\ -1 & 1 & 1 \end{bmatrix}$ and $X = A^{-1}B = \begin{bmatrix} -5 & 4 & 5 \\ -4 & 3 & 3 \\ -1 & 1 & 1 \end{bmatrix} \begin{bmatrix} 0 \\ 0 \\ 1 \end{bmatrix} = \begin{bmatrix} 5 \\ 3 \\ 1 \end{bmatrix}$ so that $\begin{aligned} x &= 5, \\ y &= 3, \\ z &= 1 \end{aligned}$

33. $\begin{aligned} x + 2y + z &= 4 \\ y + z &= 0 \\ x + y + z &= 3 \end{aligned}$ Let $A = \begin{bmatrix} 1 & 2 & 1 \\ 0 & 1 & 1 \\ 1 & 1 & 1 \end{bmatrix}$, $X = \begin{bmatrix} x \\ y \\ z \end{bmatrix}$, and $B = \begin{bmatrix} 4 \\ 0 \\ 3 \end{bmatrix}$

For the inverse of A, we have

$\begin{bmatrix} 1 & 2 & 1 & \vdots & 1 & 0 & 0 \\ 0 & 1 & 1 & \vdots & 0 & 1 & 0 \\ 1 & 1 & 1 & \vdots & 0 & 0 & 1 \end{bmatrix} \xrightarrow{-R_1 + R_3} \begin{bmatrix} 1 & 2 & 1 & \vdots & 1 & 0 & 0 \\ 0 & 1 & 1 & \vdots & 0 & 1 & 0 \\ 0 & -1 & 0 & \vdots & -1 & 0 & 1 \end{bmatrix}$

$$\xrightarrow[\substack{R_2 + R_3}]{-2R_2 + R_1} \begin{bmatrix} 1 & 0 & -1 & \vdots & 1 & -2 & 0 \\ 0 & 1 & 1 & \vdots & 0 & 1 & 0 \\ 0 & 0 & 1 & \vdots & -1 & 1 & 1 \end{bmatrix} \xrightarrow[\substack{-R_3 + R_2}]{R_3 + R_1} \begin{bmatrix} 1 & 0 & 0 & \vdots & 0 & -1 & 1 \\ 0 & 1 & 0 & \vdots & 1 & 0 & -1 \\ 0 & 0 & 1 & \vdots & -1 & 1 & 1 \end{bmatrix}$$

Now, $A^{-1} = \begin{bmatrix} 0 & -1 & 1 \\ 1 & 0 & -1 \\ -1 & 1 & 1 \end{bmatrix}$ and $X = A^{-1}B = \begin{bmatrix} 0 & -1 & 1 \\ 1 & 0 & -1 \\ -1 & 0 & 1 \end{bmatrix}\begin{bmatrix} 4 \\ 0 \\ 3 \end{bmatrix} = \begin{bmatrix} 3 \\ 1 \\ -1 \end{bmatrix}$ so that $x = 3,$ $y = 1,$ $z = -1$

35.
$$\begin{aligned} x + y + 2z - w &= 0 \\ -2x - y - 2z + 2w &= -1 \\ 4x - 2y + z \phantom{{}- w} &= -4 \\ y + z - w &= 1 \end{aligned}$$
Let $A = \begin{bmatrix} 1 & 1 & 2 & -1 \\ -2 & -1 & -2 & 2 \\ 4 & -2 & 1 & 0 \\ 0 & 1 & 1 & -1 \end{bmatrix}$, $X = \begin{bmatrix} x \\ y \\ z \\ w \end{bmatrix}$, $B = \begin{bmatrix} 0 \\ -1 \\ -4 \\ 1 \end{bmatrix}$

For the inverse of A, we have

$$\begin{bmatrix} 1 & 1 & 2 & -1 & \vdots & 1 & 0 & 0 & 0 \\ -2 & -1 & -2 & 2 & \vdots & 0 & 1 & 0 & 0 \\ 4 & -2 & 1 & 0 & \vdots & 0 & 0 & 1 & 0 \\ 0 & 1 & 1 & -1 & \vdots & 0 & 0 & 0 & 1 \end{bmatrix} \xrightarrow[\substack{-4R_1 + R_3}]{2R_1 + R_2} \begin{bmatrix} 1 & 1 & 2 & -1 & \vdots & 1 & 0 & 0 & 0 \\ 0 & 1 & 2 & 0 & \vdots & 2 & 1 & 0 & 0 \\ 0 & -6 & -7 & 4 & \vdots & -4 & 0 & 1 & 0 \\ 0 & 1 & 1 & -1 & \vdots & 0 & 0 & 0 & 1 \end{bmatrix}$$

$$\xrightarrow[\substack{6R_2 + R_3 \\ -R_2 + R_4}]{-R_2 + R_1} \begin{bmatrix} 1 & 0 & 0 & -1 & \vdots & -1 & -1 & 0 & 0 \\ 0 & 1 & 2 & 0 & \vdots & 2 & 1 & 0 & 0 \\ 0 & 0 & 5 & 4 & \vdots & 8 & 6 & 1 & 0 \\ 0 & 0 & -1 & -1 & \vdots & -2 & -1 & 0 & 1 \end{bmatrix}$$

$$\xrightarrow{R_3 \leftrightarrow R_4} \begin{bmatrix} 1 & 0 & 0 & -1 & \vdots & -1 & -1 & 0 & 0 \\ 0 & 1 & 2 & 0 & \vdots & 2 & 1 & 0 & 0 \\ 0 & 0 & -1 & -1 & \vdots & -2 & -1 & 0 & 1 \\ 0 & 0 & 5 & 4 & \vdots & 8 & 6 & 1 & 0 \end{bmatrix}$$

$$\xrightarrow{-R_3} \begin{bmatrix} 1 & 0 & 0 & -1 & \vdots & -1 & -1 & 0 & 0 \\ 0 & 1 & 2 & 0 & \vdots & 2 & 1 & 0 & 0 \\ 0 & 0 & 1 & 1 & \vdots & 2 & 1 & 0 & -1 \\ 0 & 0 & 5 & 4 & \vdots & 8 & 6 & 1 & 0 \end{bmatrix} \xrightarrow[\substack{-5R_3 + R_4}]{-2R_3 + R_2} \begin{bmatrix} 1 & 0 & 0 & -1 & \vdots & -1 & -1 & 0 & 0 \\ 0 & 1 & 0 & -2 & \vdots & -2 & -1 & 0 & 2 \\ 0 & 0 & 1 & 1 & \vdots & 2 & 1 & 0 & -1 \\ 0 & 0 & 0 & -1 & \vdots & -2 & 1 & 1 & 5 \end{bmatrix}$$

$$\xrightarrow{-R_4} \begin{bmatrix} 1 & 0 & 0 & -1 & \vdots & -1 & -1 & 0 & 0 \\ 0 & 1 & 0 & -2 & \vdots & -2 & -1 & 0 & 2 \\ 0 & 0 & 1 & 1 & \vdots & 2 & 1 & 0 & -1 \\ 0 & 0 & 0 & 1 & \vdots & 2 & -1 & -1 & -5 \end{bmatrix} \xrightarrow[\substack{2R_4 + R_2 \\ -R_4 + R_3}]{R_4 + R_1} \begin{bmatrix} 1 & 0 & 0 & 0 & \vdots & 1 & -2 & -1 & -5 \\ 0 & 1 & 0 & 0 & \vdots & 2 & -3 & -2 & -8 \\ 0 & 0 & 1 & 0 & \vdots & 0 & 2 & 1 & 4 \\ 0 & 0 & 0 & 1 & \vdots & 2 & -1 & -1 & -5 \end{bmatrix}$$

Now, $A^{-1} = \begin{bmatrix} 1 & -2 & -1 & -5 \\ 2 & -3 & -2 & -8 \\ 0 & 2 & 1 & 4 \\ 2 & -1 & -1 & -5 \end{bmatrix}$ and $X = A^{-1}B = \begin{bmatrix} 1 & -2 & -1 & -5 \\ 2 & -3 & -2 & -8 \\ 0 & 2 & 1 & 4 \\ 2 & -1 & -1 & -5 \end{bmatrix}\begin{bmatrix} 0 \\ -1 \\ -4 \\ 1 \end{bmatrix} = \begin{bmatrix} 1 \\ 3 \\ -2 \\ 0 \end{bmatrix}$

so that $x = 1,\ y = 3,\ z = -2,\ w = 0.$

37.
$$\begin{bmatrix} a & 0 & 0 & \vdots & 1 & 0 & 0 \\ 0 & b & 0 & \vdots & 0 & 1 & 0 \\ 0 & 0 & c & \vdots & 0 & 0 & 1 \end{bmatrix} \xrightarrow{\frac{1}{a}R_1} \begin{bmatrix} 1 & 0 & 0 & \vdots & \frac{1}{a} & 0 & 0 \\ 0 & b & 0 & \vdots & 0 & 1 & 0 \\ 0 & 0 & c & \vdots & 0 & 0 & 1 \end{bmatrix}$$

$$\xrightarrow{\frac{1}{b}R_2} \begin{bmatrix} 1 & 0 & 0 & \vdots & \frac{1}{a} & 0 & 0 \\ 0 & 1 & 0 & \vdots & 0 & \frac{1}{b} & 0 \\ 0 & 0 & c & \vdots & 0 & 0 & 1 \end{bmatrix} \xrightarrow{\frac{1}{c}R_3} \begin{bmatrix} 1 & 0 & 0 & \vdots & \frac{1}{a} & 0 & 0 \\ 0 & 1 & 0 & \vdots & 0 & \frac{1}{b} & 0 \\ 0 & 0 & 1 & \vdots & 0 & 0 & \frac{1}{c} \end{bmatrix}$$

The inverse is $\begin{bmatrix} \frac{1}{a} & 0 & 0 \\ 0 & \frac{1}{b} & 0 \\ 0 & 0 & \frac{1}{c} \end{bmatrix}$.

39. Start with $AXC = B$. Multiply by A^{-1} on the left of each side:

$$A^{-1}(AXC) = A^{-1}B$$
$$(A^{-1}A)XC = A^{-1}B$$
$$I \cdot XC = A^{-1}B$$
$$XC = A^{-1}B. \text{ Now multiply by } C^{-1} \text{ on the right of each side.}$$
$$(XC)C^{-1} = A^{-1}BC^{-1}$$
$$X(CC^{-1}) = A^{-1}BC^{-1}$$
$$X \cdot I = A^{-1}BC^{-1}$$
$$X = A^{-1}BC^{-1}$$

41.
$$\begin{bmatrix} a & b & \vdots & 1 & 0 \\ c & d & \vdots & 0 & 1 \end{bmatrix} \xrightarrow{\frac{1}{a}R_1} \begin{bmatrix} 1 & \frac{b}{a} & \vdots & \frac{1}{a} & 0 \\ c & d & \vdots & 0 & 1 \end{bmatrix} \xrightarrow{-cR_1 + R_2} \begin{bmatrix} 1 & \frac{b}{a} & \vdots & \frac{1}{a} & 0 \\ 0 & -\frac{cb}{a} + d & \vdots & -\frac{c}{a} & 1 \end{bmatrix}$$

or $\begin{bmatrix} 1 & \frac{b}{a} & \vdots & \frac{1}{a} & 0 \\ 0 & \frac{ad-bc}{a} & \vdots & -\frac{c}{a} & 1 \end{bmatrix} \xrightarrow[\text{(providing } ad-bc \neq 0)]{\frac{a}{ad-bc}R_2} \begin{bmatrix} 1 & \frac{b}{a} & \vdots & \frac{1}{a} & 0 \\ 0 & 1 & \vdots & \frac{-c}{ad-bc} & \frac{a}{ad-bc} \end{bmatrix}$

$\xrightarrow{-\frac{b}{a}R_2 + R_1} \begin{bmatrix} 1 & 0 & \vdots & \frac{1}{a} + \frac{bc}{a(ad-bc)} & \frac{-b}{ad-bc} \\ 0 & 1 & \vdots & \frac{-c}{ad-bc} & \frac{a}{ad-bc} \end{bmatrix}$ or $\begin{bmatrix} 1 & 0 & \vdots & \frac{1}{a}\left(1 + \frac{bc}{ad-bc}\right) & \frac{-b}{ad-bc} \\ 0 & 1 & \vdots & \frac{-c}{ad-bc} & \frac{a}{ad-bc} \end{bmatrix}$

or $\begin{bmatrix} 1 & 0 & \vdots & \frac{d}{ad-bc} & \frac{-b}{ad-bc} \\ 0 & 1 & \vdots & \frac{-c}{ad-bc} & \frac{a}{ad-bc} \end{bmatrix}$. The inverse is $\begin{bmatrix} \frac{d}{ad-bc} & \frac{-b}{ad-bc} \\ \frac{-c}{ad-bc} & \frac{a}{ad-bc} \end{bmatrix}$ or $\frac{1}{ad-bc}\begin{bmatrix} d & -b \\ -c & a \end{bmatrix}$.

43. Confusion may arise because we may believe that these notations

$$(B)\left(\frac{1}{A}\right), \qquad \left(\frac{1}{A}\right)(B), \qquad \text{and} \qquad \frac{B}{A}$$

mean to divide matrix B by matrix A rather than to multiply matrix B by the inverse of matrix A. Recall that division of matrices is **not** defined.

Exercises 8.5

1. $\begin{vmatrix} -3 & 2 \\ 1 & 1 \end{vmatrix} = (-3)(1) - (2)(1)$
   $= -3 - 2 = -5$

3. $\begin{vmatrix} 5 & 8 \\ 2 & 3 \end{vmatrix} = (5)(3) - (8)(2)$
   $= 15 - 16 = -1$

5. $\begin{vmatrix} 7 & 0 \\ 0 & -3 \end{vmatrix} = (7)(-3) - (0)(0)$
   $= -21 - 0 = -21$

7. $\begin{vmatrix} -1 & -1 \\ 1 & 1 \end{vmatrix} = (-1)(1) - (-1)(1)$
   $= -1 + 1 = 0$

9. $\begin{vmatrix} -\frac{1}{2} & \frac{1}{3} \\ \frac{1}{6} & \frac{1}{9} \end{vmatrix} = \left(-\frac{1}{2}\right)\left(\frac{1}{9}\right) - \left(\frac{1}{3}\right)\left(\frac{1}{6}\right) = -\frac{1}{18} - \frac{1}{18} = -\frac{2}{18} = -\frac{1}{9}$

11. $\begin{vmatrix} 0 & 3 \\ -2 & 1 \end{vmatrix} = (0)(1) - (3)(-2) = 0 - (-6) = 0 + 6 = 6$

13. $\begin{vmatrix} 1 & 11 \\ -2 & 4 \end{vmatrix} = (1)(4) - (11)(-2) = 4 - (-22) = 4 + 22 = 26$

15. $\begin{vmatrix} x & -x \\ 2 & 3 \end{vmatrix} = (x)(3) - (-x)(2) = 3x - (-2x) = 3x + 2x = 5x$

17. $\begin{vmatrix} 2 & -1 & 2 \\ 0 & 2 & 0 \\ -3 & 1 & 0 \end{vmatrix} = a_{13}C_{13} + a_{23}C_{23} + a_{33}C_{33}$

    $= 2C_{13} + 0C_{23} + 0C_{33} = 2C_{13} = 2(-1)^4 \begin{vmatrix} 0 & 2 \\ -3 & 1 \end{vmatrix}$
    $= 2[0 - (-6)] = 2[6] = 12$

19. $\begin{vmatrix} 2 & 0 & -1 \\ -1 & 2 & 0 \\ 0 & -1 & 2 \end{vmatrix} = a_{11}C_{11} + a_{12}C_{12} + a_{13}C_{13}$

    $= 2C_{11} + 0C_{12} + (-1)C_{13}$
    $= 2C_{11} + (-1)C_{13}$
    $= 2(-1)^2 \begin{vmatrix} 2 & 0 \\ -1 & 2 \end{vmatrix} + (-1)(-1)^4 \begin{vmatrix} -1 & 2 \\ 0 & -1 \end{vmatrix}$
    $= 2(4 - 0) - (1 - 0) = 2(4) - (1) = 8 - 1 = 7$

21. $\begin{vmatrix} 1 & 4 & 0 \\ -1 & 2 & -1 \\ 0 & 3 & -1 \end{vmatrix} = a_{11}C_{11} + a_{12}C_{12} + a_{13}C_{13}$

    $= 1C_{11} + 4C_{12} + 0C_{13}$
    $= C_{11} + 4C_{12}$
    $= (-1)^2 \begin{vmatrix} 2 & -1 \\ 3 & -1 \end{vmatrix} + 4(-1)^3 \begin{vmatrix} -1 & -1 \\ 0 & -1 \end{vmatrix}$
    $= 1[-2 - (-3)] - 4[1 - 0]$
    $= (-2 + 3) - 4(1)$
    $= 1 - 4 = -3$

23. $\begin{vmatrix} 1 & 3 & -2 \\ 1 & 4 & 0 \\ 2 & 5 & -7 \end{vmatrix} = a_{21}C_{21} + a_{22}C_{22} + a_{23}C_{23}$

$= 1C_{21} + 4C_{22} + 0C_{23}$

$= C_{21} + 4C_{22}$

$= (-1)^3 \begin{vmatrix} 3 & -2 \\ 5 & -7 \end{vmatrix} + 4(-1)^4 \begin{vmatrix} 1 & -2 \\ 2 & -7 \end{vmatrix}$

$= -[-21 - (-10)] + 4[-7 - (-4)]$

$= -(-21 + 10) + 4(-7 + 4)$

$= -(-11) + 4(-3) = 11 - 12 = -1$

25. $\begin{vmatrix} -1 & 1 & 1 \\ 1 & -1 & 1 \\ 1 & 1 & 1 \end{vmatrix} = a_{31}C_{31} + a_{32}C_{32} + a_{33}C_{33}$

$= 1C_{31} + 1C_{32} + 1C_{33}$

$= C_{31} + C_{32} + C_{33}$

$= (-1)^4 \begin{vmatrix} 1 & 1 \\ -1 & 1 \end{vmatrix} + (-1)^5 \begin{vmatrix} -1 & 1 \\ 1 & 1 \end{vmatrix} + (-1)^6 \begin{vmatrix} -1 & 1 \\ 1 & -1 \end{vmatrix}$

$= 1[1 - (-1)] - (-1 - 1) + (1 - 1)$

$= (1 + 1) - (-2) + 0 = 2 + 2 + 0 = 4$

27. $\begin{vmatrix} 1 & -2 & 3 \\ 2 & 1 & 2 \\ 3 & -3 & 6 \end{vmatrix} = a_{31}C_{31} + a_{32}C_{32} + a_{33}C_{33}$

$= 3C_{31} - 3C_{32} + 6C_{33}$

$= 3(-1)^4 \begin{vmatrix} -2 & 3 \\ 1 & 2 \end{vmatrix} - 3(-1)^5 \begin{vmatrix} 1 & 3 \\ 2 & 2 \end{vmatrix} + 6(-1)^6 \begin{vmatrix} 1 & -2 \\ 2 & 1 \end{vmatrix}$

$= 3(-4 - 3) + 3(2 - 6) + 6[1 - (-4)]$

$= 3(-7) + 3(-4) + 6(5) = -21 - 12 + 30 = -3$

29. $\begin{vmatrix} 4 & -5 & 1 \\ 5 & -9 & 2 \\ 2 & -4 & 1 \end{vmatrix} = a_{12}C_{12} + a_{22}C_{22} + a_{32}C_{32}$

$= -5C_{12} - 9C_{22} - 4C_{32}$

$= -5(-1)^3 \begin{vmatrix} 5 & 2 \\ 2 & 1 \end{vmatrix} - 9(-1)^4 \begin{vmatrix} 4 & 1 \\ 2 & 1 \end{vmatrix} - 4(-1)^5 \begin{vmatrix} 4 & 1 \\ 5 & 2 \end{vmatrix}$

$= 5(5 - 4) - 9(4 - 2) + 4(8 - 5)$

$= 5(1) - 9(2) + 4(3) = 5 - 18 + 12 = -1$

31. $\begin{vmatrix} 2 & 4 & 3 \\ 5 & -9 & -2 \\ -1 & 11 & 5 \end{vmatrix} = a_{12}C_{12} + a_{22}C_{22} + a_{32}C_{32}$

$= 4C_{12} - 9C_{22} + 11C_{32}$

$= 4(-1)^3 \begin{vmatrix} 5 & -2 \\ -1 & 5 \end{vmatrix} - 9(-1)^4 \begin{vmatrix} 2 & 3 \\ -1 & 5 \end{vmatrix} + 11(-1)^5 \begin{vmatrix} 2 & 3 \\ 5 & -2 \end{vmatrix}$

$= -4(25 - 2) - 9[10 - (-3)] - 11(-4 - 15)$

$= -4(23) - 9(13) - 11(-19)$

$= -92 - 117 + 209 = -209 + 209 = 0$

33. $\begin{vmatrix} 1 & 0 & 1 & 2 \\ -2 & 1 & -1 & -4 \\ 1 & 0 & 0 & 1 \\ 0 & -2 & 1 & 5 \end{vmatrix} = a_{31}C_{31} + a_{32}C_{32} + a_{33}C_{33} + a_{34}C_{34}$

$\qquad\qquad\qquad = 1C_{31} + 0C_{32} + 0C_{33} + 1C_{34}$

$\qquad\qquad\qquad = C_{31} + C_{34}$

$= (-1)^4 \begin{vmatrix} 0 & 1 & 2 \\ 1 & -1 & -4 \\ -2 & 1 & 5 \end{vmatrix} + (-1)^7 \begin{vmatrix} 1 & 0 & 1 \\ -2 & 1 & -1 \\ 0 & -2 & 1 \end{vmatrix} = \begin{vmatrix} 0 & 1 & 2 \\ 1 & -1 & -4 \\ -2 & 1 & 5 \end{vmatrix} - \begin{vmatrix} 1 & 0 & 1 \\ -2 & 1 & -1 \\ 0 & -2 & 1 \end{vmatrix}$

$= b_{11}B_{11} + b_{12}B_{12} + b_{13}B_{13} - (d_{11}D_{11} + d_{12}D_{12} + d_{13}D_{13})$

$= 0B_{11} + 1B_{12} + 2B_{13} - (1D_{11} + 0D_{12} + 1D_{13})$

$= B_{12} + 2B_{13} - (D_{11} + D_{13})$

$= (-1)^3 \begin{vmatrix} 1 & -4 \\ -2 & 5 \end{vmatrix} + 2(-1)^4 \begin{vmatrix} 1 & -1 \\ -2 & 1 \end{vmatrix} - \left[(-1)^2 \begin{vmatrix} 1 & -1 \\ -2 & 1 \end{vmatrix} + (-1)^4 \begin{vmatrix} -2 & 1 \\ 0 & -2 \end{vmatrix} \right]$

$= -[5 - 8] + 2[1 - 2] - [(1 - 2) + (4 - 0)]$

$= -(-3) + 2(-1) - [-1 + 4] = 3 - 2 - 3 = -2$

35. $\begin{vmatrix} 0 & 4 & -2 & 1 \\ -1 & -2 & -1 & 1 \\ -2 & -1 & -3 & 3 \\ 1 & 0 & 2 & -1 \end{vmatrix} = a_{11}C_{11} + a_{12}C_{12} + a_{13}C_{13} + a_{14}C_{14}$

$\qquad\qquad\qquad\qquad = 0C_{11} + 4C_{12} - 2C_{13} + 1C_{14}$

$\qquad\qquad\qquad\qquad = 4C_{12} - 2C_{13} + C_{14}$

$= 4(-1)^3 \begin{vmatrix} -1 & -1 & 1 \\ -2 & -3 & 3 \\ 1 & 2 & -1 \end{vmatrix} - 2(-1)^4 \begin{vmatrix} -1 & -2 & 1 \\ -2 & -1 & 3 \\ 1 & 0 & -1 \end{vmatrix} + (-1)^5 \begin{vmatrix} -1 & -2 & -1 \\ -2 & -1 & -3 \\ 1 & 0 & 2 \end{vmatrix}$

$= -4 \begin{vmatrix} -1 & -1 & 1 \\ -2 & -3 & 3 \\ 1 & 2 & -1 \end{vmatrix} - 2 \begin{vmatrix} -1 & -2 & 1 \\ -2 & -1 & 3 \\ 1 & 0 & -1 \end{vmatrix} - \begin{vmatrix} -1 & -2 & -1 \\ -2 & -1 & -3 \\ 1 & 0 & 2 \end{vmatrix}$

Expand about last row

$= -4 \left[1(-1)^4 \begin{vmatrix} -1 & 1 \\ -3 & 3 \end{vmatrix} + 2(-1)^5 \begin{vmatrix} -1 & 1 \\ -2 & 3 \end{vmatrix} + (-1)(-1)^6 \begin{vmatrix} -1 & -1 \\ -2 & -3 \end{vmatrix} \right]$

$\quad - 2 \left[1(-1)^4 \begin{vmatrix} -2 & 1 \\ -1 & 3 \end{vmatrix} + 0(-1)^5 \begin{vmatrix} -1 & 1 \\ -2 & 3 \end{vmatrix} + (-1)(-1)^6 \begin{vmatrix} -1 & -2 \\ -2 & -1 \end{vmatrix} \right]$

$\quad - \left[1(-1)^4 \begin{vmatrix} -2 & -1 \\ -1 & -3 \end{vmatrix} + 0(-1)^5 \begin{vmatrix} -1 & -1 \\ -2 & -3 \end{vmatrix} + 2(-1)^6 \begin{vmatrix} -1 & -2 \\ -2 & -1 \end{vmatrix} \right]$

$= -4[0 + 2 - 1] - 2[-5 + 0 + 3] - [5 + 0 - 6]$

$= -4(1) - 2(-2) - (-1) = -4 + 4 + 1 = 1$

37. $\begin{vmatrix} 3 & x \\ -2 & 1 \end{vmatrix} = 1$

$\qquad 3 - (-2x) = 1$

$\qquad\quad 3 + 2x = 1$

$\qquad\qquad 2x = -2$

$\qquad\qquad\quad x = -1$

39. $\begin{vmatrix} 2x - 1 & 2 \\ 1 & 4 \end{vmatrix} = 10$

$\qquad 4(2x - 1) - 2 = 10$

$\qquad\quad 8x - 4 - 2 = 10$

$\qquad\qquad 8x - 6 = 10$

$\qquad\qquad\qquad 8x = 16$

$\qquad\qquad\qquad\quad x = 2$

41. $\begin{vmatrix} 2 & 0 & 1 \\ 4 & x & 5 \\ 1 & 4 & -3 \end{vmatrix} = -10$ or $a_{12}C_{12} + a_{22}C_{22} + a_{32}C_{32} = -10$

$$\text{or} \qquad 0C_{12} + xC_{22} + 4C_{32} = -10$$
$$\text{or} \qquad xC_{22} + 4C_{32} = -10$$

$$\text{or} \qquad x(-1)^4 \begin{vmatrix} 2 & 1 \\ 1 & -3 \end{vmatrix} + 4(-1)^5 \begin{vmatrix} 2 & 1 \\ 4 & 5 \end{vmatrix} = -10$$

$$x(-6 - 1) - 4(10 - 4) = -10$$
$$x(-7) - 4(6) = -10$$
$$-7x - 24 = -10$$
$$-7x = 14$$
$$x = -2$$

43. $\begin{vmatrix} 3 & x & x \\ -2 & 1 & 0 \\ 4 & 1 & -1 \end{vmatrix} = 21$ or $a_{21}C_{21} + a_{22}C_{22} + a_{23}C_{23} = 21$

$$\text{or} \qquad -2C_{21} + 1C_{22} + 0C_{23} = 21$$
$$\text{or} \qquad -2C_{21} + C_{22} = 21$$

$$\text{or } -2(-1)^3 \begin{vmatrix} x & x \\ 1 & -1 \end{vmatrix} + (-1)^4 \begin{vmatrix} 3 & x \\ 4 & -1 \end{vmatrix} = 21$$

$$2(-x - x) + (-3 - 4x) = 21$$
$$2(-2x) - 3 - 4x = 21$$
$$-4x - 3 - 4x = 21$$
$$-8x - 3 = 21$$
$$-8x = 24$$
$$x = -3$$

45. $A - xI = \begin{bmatrix} 2 & 0 \\ 0 & -3 \end{bmatrix} - x \begin{bmatrix} 1 & 0 \\ 0 & 1 \end{bmatrix} = \begin{bmatrix} 2 - x & 0 \\ 0 & -3 - x \end{bmatrix}$

Now $|A - xI| = 0$ becomes $\begin{vmatrix} 2 - x & 0 \\ 0 & -3 - x \end{vmatrix} = 0$

$$\text{or} \quad (2 - x)(-3 - x) - 0 = 0$$
$$(2 - x)(-3 - x) = 0$$

Set each factor equal to 0: $2 - x = 0, \quad -3 - x = 0$
$$x = 2, \qquad x = -3$$

47. $A - xI = \begin{bmatrix} 1 & -2 \\ 3 & -4 \end{bmatrix} - x \begin{bmatrix} 1 & 0 \\ 0 & 1 \end{bmatrix} = \begin{bmatrix} 1 - x & -2 \\ 3 & -4 - x \end{bmatrix}$

Now $|A - xI| = 0$ becomes $\begin{vmatrix} 1 - x & -2 \\ 3 & -4 - x \end{vmatrix} = 0$

$$\text{or } (1 - x)(-4 - x) - (-6) = 0$$
$$x^2 + 3x - 4 + 6 = 0$$
$$x^2 + 3x + 2 = 0$$
$$(x + 2)(x + 1) = 0$$

Set each factor equal to 0: $x + 2 = 0, \; x + 1 = 0$
$$x = -2, \qquad x = -1$$

49. $A - xI = \begin{bmatrix} 5 & 1 & -1 \\ 0 & -3 & 2 \\ 0 & 0 & 2 \end{bmatrix} - x \begin{bmatrix} 1 & 0 & 0 \\ 0 & 1 & 0 \\ 0 & 0 & 1 \end{bmatrix} = \begin{bmatrix} 5-x & 1 & -1 \\ 0 & -3-x & 2 \\ 0 & 0 & 2-x \end{bmatrix}$

Now $|A - xI| = 0$ becomes $\begin{vmatrix} 5-x & 1 & -1 \\ 0 & -3-x & 2 \\ 0 & 0 & 2-x \end{vmatrix} = 0$

or $\qquad a_{11}C_{11} + a_{21}C_{21} + a_{31}C_{31} = 0$

$$(5 - x)C_{11} + 0C_{21} + 0C_{31} = 0$$

$$(5 - x)C_{11} = 0$$

$$(5 - x)(-1)^2 \begin{vmatrix} -3-x & 2 \\ 0 & 2-x \end{vmatrix} = 0$$

or $(5 - x)[(-3 - x)(2 - x) - 0] = 0$

$$(5 - x)(-3 - x)(2 - x) = 0$$

$$5 - x = 0, \quad -3 - x = 0, \quad 2 - x = 0$$

$$x = 5, \qquad x = -3, \qquad x = 2$$

51. $A - xI = \begin{bmatrix} 1 & -2 & 0 \\ 0 & 0 & -3 \\ 2 & -4 & 0 \end{bmatrix} - x \begin{bmatrix} 1 & 0 & 0 \\ 0 & 1 & 0 \\ 0 & 0 & 1 \end{bmatrix} = \begin{bmatrix} 1-x & -2 & 0 \\ 0 & -x & -3 \\ 2 & -4 & -x \end{bmatrix}$

Now $|A - xI| = 0$ becomes $\begin{vmatrix} 1-x & -2 & 0 \\ 0 & -x & -3 \\ 2 & -4 & -x \end{vmatrix} = 0$

or $\qquad a_{11}C_{11} + a_{12}C_{12} + a_{13}C_{13} = 0$

or $\quad (1 - x)C_{11} + (-2)C_{12} + 0C_{13} = 0$

or $\qquad\qquad (1 - x)C_{11} - 2C_{12} = 0$

or $\quad (1 - x)(-1)^2 \begin{bmatrix} -x & -3 \\ -4 & -x \end{bmatrix} - 2(-1)^3 \begin{bmatrix} 0 & -3 \\ 2 & -x \end{bmatrix} = 0$

or $\qquad (1 - x)[x^2 - 12] + 2[0 - (-6)] = 0$

$$(1 - x)(x^2 - 12) + 12 = 0$$

$$-x^3 + x^2 + 12x - 12 + 12 = 0$$

$$-x^3 + x^2 + 12x = 0$$

$$-x(x^2 - x - 12) = 0$$

$$-x(x - 4)(x + 3) = 0$$

$$-x = 0, \quad x - 4 = 0, \quad x + 3 = 0$$

$$x = 0, \qquad x = 4, \qquad x = -3$$

53. a) $|A| = a_{11}a_{22}a_{33} + a_{12}a_{23}a_{31} + a_{13}a_{21}a_{32} - a_{11}a_{23}a_{32} - a_{12}a_{21}a_{33} - a_{13}a_{22}a_{31}$

$\qquad = a_{11}(a_{22}a_{33} - a_{23}a_{32}) - a_{12}(a_{21}a_{33} - a_{23}a_{31}) + a_{13}(a_{21}a_{32} - a_{22}a_{31})$

$\qquad = a_{11} \begin{vmatrix} a_{22} & a_{23} \\ a_{32} & a_{33} \end{vmatrix} - a_{12} \begin{vmatrix} a_{21} & a_{23} \\ a_{31} & a_{33} \end{vmatrix} + a_{13} \begin{vmatrix} a_{21} & a_{22} \\ a_{31} & a_{32} \end{vmatrix}$

$\qquad = a_{11}C_{11} + a_{12}C_{12} + a_{13}C_{13}$

b) $|A| = a_{11}a_{22}a_{33} + a_{12}a_{23}a_{31} + a_{13}a_{21}a_{32} - a_{11}a_{23}a_{32} - a_{12}a_{21}a_{33} - a_{13}a_{22}a_{31}$

$\qquad = -a_{21}(a_{12}a_{33} - a_{13}a_{32}) + a_{22}(a_{11}a_{33} - a_{13}a_{31}) - a_{23}(a_{11}a_{32} - a_{12}a_{31})$

$\qquad = -a_{21} \begin{vmatrix} a_{12} & a_{13} \\ a_{32} & a_{33} \end{vmatrix} + a_{22} \begin{vmatrix} a_{11} & a_{13} \\ a_{31} & a_{33} \end{vmatrix} - a_{23} \begin{vmatrix} a_{11} & a_{12} \\ a_{31} & a_{32} \end{vmatrix}$

$\qquad = a_{21}C_{21} + a_{22}C_{22} + a_{23}C_{23}$

55. The domain is the square matrix A and the range is any real number.

Exercises 8.6

1. $\begin{vmatrix} 3 & -4 \\ 1 & 5 \end{vmatrix} = -\begin{vmatrix} 1 & 5 \\ 3 & x \end{vmatrix}$

   $\downarrow$ use interchange property: Interchange two rows

   $-\begin{vmatrix} 1 & 5 \\ 3 & -4 \end{vmatrix} = -\begin{vmatrix} 1 & 5 \\ 3 & x \end{vmatrix}$

   and by equality, $x = -4$

3. $\begin{vmatrix} 5 & -1 \\ 2 & -3 \end{vmatrix} = a\begin{vmatrix} 5 & 1 \\ 2 & 3 \end{vmatrix}$

   $\downarrow$ use multiplication property: Second column is multiplied by -1

   $-1\begin{vmatrix} 5 & 1 \\ 2 & 3 \end{vmatrix} = a\begin{vmatrix} 5 & 1 \\ 2 & 3 \end{vmatrix}$

   and by equality, $a = -1$

5. $\begin{vmatrix} 1 & 0 & 2 \\ -2 & 1 & 4 \\ 0 & 1 & 5 \end{vmatrix} = x\begin{vmatrix} 0 & 2 & 1 \\ 1 & 4 & -2 \\ 1 & 5 & 0 \end{vmatrix}$

   $\downarrow$ use interchange property: Interchange 1st and 2nd columns

   $-\begin{vmatrix} 0 & 1 & 2 \\ 1 & -2 & 4 \\ 1 & 0 & 5 \end{vmatrix} = x\begin{vmatrix} 0 & 2 & 1 \\ 1 & 4 & -2 \\ 1 & 5 & 0 \end{vmatrix}$

   $\downarrow$ use interchange property again: Interchange 2nd and 3rd columns

   $\begin{vmatrix} 0 & 2 & 1 \\ 1 & 4 & -2 \\ 1 & 5 & 0 \end{vmatrix} = x\begin{vmatrix} 0 & 2 & 1 \\ 1 & 4 & -2 \\ 1 & 5 & 0 \end{vmatrix}$

   and by equality, $x = 1$

7. $\begin{vmatrix} -3 & 4 \\ 9 & -2 \end{vmatrix} = x\begin{vmatrix} -1 & -2 \\ 3 & 1 \end{vmatrix}$

   use multiplication property: 1st column is multiplied by 3
   $\downarrow$ 2nd column is multiplied by -2

   $3(-2)\begin{vmatrix} -1 & -2 \\ 3 & 1 \end{vmatrix} = x\begin{vmatrix} -1 & -2 \\ 3 & 1 \end{vmatrix}$

   and by equality, $x = 3(-2) = -6$

9. $\begin{vmatrix} 3 & 2 & 1 \\ 5 & -4 & -3 \\ 1 & 1 & 2 \end{vmatrix} = \begin{vmatrix} 3 & 2 & 1 \\ 14 & x & 0 \\ 1 & 1 & 2 \end{vmatrix}$

   $\downarrow$ use $3R_1 + R_2$

   $\begin{vmatrix} 3 & 2 & 1 \\ 14 & 2 & 0 \\ 1 & 1 & 2 \end{vmatrix} = \begin{vmatrix} 3 & 2 & 1 \\ 14 & x & 0 \\ 1 & 1 & 2 \end{vmatrix}$

   and by equality, $x = 2$

11.
$$\begin{vmatrix} -1 & 11 & -1 & 4 \\ 0 & 3 & 0 & 1 \\ 1 & 1 & -2 & 1 \\ 4 & 1 & 5 & -2 \end{vmatrix} = \begin{vmatrix} -1 & -1 & -1 & 4 \\ 0 & 0 & 0 & 1 \\ 1 & x & -2 & 1 \\ 4 & 7 & 5 & -2 \end{vmatrix}$$

$\downarrow$ use $-3C_4 + C_2$

$$\begin{vmatrix} -1 & -1 & -1 & 4 \\ 0 & 0 & 0 & 1 \\ 1 & -2 & -2 & 1 \\ 4 & 7 & 5 & -2 \end{vmatrix} = \begin{vmatrix} -1 & -1 & -1 & 4 \\ 0 & 0 & 0 & 1 \\ 1 & x & -2 & 1 \\ 4 & 7 & 5 & -2 \end{vmatrix}$$

and by equality, $x = -2$

13.
$$\begin{vmatrix} 2 & -1 & 3 \\ 1 & 2 & -1 \\ 0 & 2 & 1 \end{vmatrix} \xrightarrow[=]{-2R_2 + R_1} \begin{vmatrix} 0 & -5 & 5 \\ 1 & 2 & -1 \\ 0 & 2 & 1 \end{vmatrix} = 5\begin{vmatrix} 0 & -1 & 1 \\ 1 & 2 & -1 \\ 0 & 2 & 1 \end{vmatrix}$$

$= 5(a_{11}C_{11} + a_{21}C_{21} + a_{31}C_{31})$

$= 5(0C_{11} + 1C_{21} + 0C_{31}) = 5C_{21} = 5(-1)^3\begin{vmatrix} -1 & 1 \\ 2 & 1 \end{vmatrix} = -5(-1-2) = -5(-3) = 15$

15.
$$\begin{vmatrix} 7 & 4 & -2 \\ 1 & 2 & -1 \\ 4 & 2 & 0 \end{vmatrix} \xrightarrow[=]{-2R_2 + R_1} \begin{vmatrix} 5 & 0 & 0 \\ 1 & 2 & -1 \\ 4 & 2 & 0 \end{vmatrix} = a_{11}C_{11} + a_{12}C_{12} + a_{13}C_{13}$$

$= 5C_{11} + 0C_{12} + 0C_{13} = 5C_{11}$

$= 5(-1)^2\begin{vmatrix} 2 & -1 \\ 2 & 0 \end{vmatrix} = 5[0 - (-2)] = 5(2) = 10$

17.
$$\begin{vmatrix} 1 & 3 & -1 \\ 3 & -2 & 4 \\ 2 & 1 & 3 \end{vmatrix} \xrightarrow[=]{-3R_1 + R_2} \begin{vmatrix} 1 & 3 & -1 \\ 0 & -11 & 7 \\ 2 & 1 & 3 \end{vmatrix} \xrightarrow[=]{-2R_1 + R_3} \begin{vmatrix} 1 & 3 & -1 \\ 0 & -11 & 7 \\ 0 & -5 & 5 \end{vmatrix} = 5\begin{vmatrix} 1 & 3 & -1 \\ 0 & -11 & 7 \\ 0 & -1 & 1 \end{vmatrix}$$

$= 5(a_{11}C_{11} + a_{21}C_{21} + a_{31}C_{31})$

$= 5(1C_{11} + 0C_{21} + 0C_{31})$

$= 5C_{11} = 5(-1)^2\begin{vmatrix} -11 & 7 \\ -1 & 1 \end{vmatrix}$

$= 5[-11 - (-7)] = 5(-11 + 7) = 5(-4) = -20$

19.
$$\begin{vmatrix} 1 & 2 & 3 \\ 1 & 1 & 2 \\ 1 & 1 & 3 \end{vmatrix} \xrightarrow[=]{-R_2 + R_3} \begin{vmatrix} 1 & 2 & 3 \\ 1 & 1 & 2 \\ 0 & 0 & 1 \end{vmatrix} = a_{31}C_{31} + a_{32}C_{32} + a_{33}C_{33}$$

$= 0C_{31} + 0C_{32} + 1C_{33}$

$= C_{33} = (-1)^6\begin{vmatrix} 1 & 2 \\ 1 & 1 \end{vmatrix}$

$= 1(1 - 2) = -1$

21.
$$\begin{vmatrix} 2 & -1 & 2 \\ 3 & -5 & 1 \\ 2 & 1 & 3 \end{vmatrix} \xrightarrow[=]{2C_2 + C_1} \begin{vmatrix} 0 & -1 & 2 \\ -7 & -5 & 1 \\ 4 & 1 & 3 \end{vmatrix} \xrightarrow[=]{2C_2 + C_3} \begin{vmatrix} 0 & -1 & 0 \\ -7 & -5 & -9 \\ 4 & 1 & 5 \end{vmatrix}$$

$= a_{11}C_{11} + a_{12}C_{12} + a_{13}C_{13}$

$= 0C_{11} + (-1)C_{12} + 0C_{13} = -C_{12}$

$= (-1)(-1)^3\begin{vmatrix} -7 & -9 \\ 4 & 5 \end{vmatrix}$

$= 1[-35 - (-36)]$

$= (-35 + 36) = 1$

23. $\begin{vmatrix} -2 & 1 & -2 \\ 2 & -1 & 3 \\ -4 & 2 & 1 \end{vmatrix} \xrightarrow[=]{2C_2 + C_1} \begin{vmatrix} 0 & 1 & -2 \\ 0 & -1 & 3 \\ 0 & 2 & 1 \end{vmatrix} = a_{11}C_{11} + a_{21}C_{21} + a_{31}C_{31}$

$\qquad\qquad\qquad\qquad\qquad\qquad\qquad = 0C_{11} + 0C_{21} + 0C_{31}$

$\qquad\qquad\qquad\qquad\qquad\qquad\qquad = 0$

25. $\begin{vmatrix} 5 & 0 & 0 & 3 \\ 2 & 4 & -3 & 1 \\ 0 & 1 & 0 & -1 \\ 1 & -1 & 2 & -1 \end{vmatrix} \xrightarrow[=]{C_2 + C_4} \begin{vmatrix} 5 & 0 & 0 & 3 \\ 2 & 4 & -3 & 5 \\ 0 & 1 & 0 & 0 \\ 1 & -1 & 2 & -2 \end{vmatrix}$

$\qquad = a_{31}C_{31} + a_{32}C_{32} + a_{33}C_{33} + a_{34}C_{34}$

$\qquad = 0C_{31} + 1C_{32} + 0C_{33} + 0C_{34} = C_{32}$

$\qquad = (-1)^5 \begin{vmatrix} 5 & 0 & 3 \\ 2 & -3 & 5 \\ 1 & 2 & -2 \end{vmatrix} = -\begin{vmatrix} 5 & 0 & 3 \\ 2 & -3 & 5 \\ 1 & 2 & -2 \end{vmatrix}$

$\qquad \xrightarrow[=]{-2R_3 + R_2} -\begin{vmatrix} 5 & 0 & 3 \\ 0 & -7 & 9 \\ 1 & 2 & -2 \end{vmatrix} \xrightarrow[=]{-5R_3 + R_1} -\begin{vmatrix} 0 & -10 & 13 \\ 0 & -7 & 9 \\ 1 & 2 & -2 \end{vmatrix}$

$\qquad = -a_{31}C_{31} = -(1)C_{31} = -C_{31} = -(-1)^6 \begin{vmatrix} -10 & 13 \\ -7 & 9 \end{vmatrix}$

$\qquad = -[-90 - (-91)] = -(-90 + 91) = -(1) = -1$

27. $\begin{vmatrix} 1 & 0 & 1 & 0 \\ 0 & 1 & 2 & 1 \\ 4 & 0 & 0 & -2 \\ 1 & 2 & 3 & 0 \end{vmatrix} \xrightarrow[=]{-C_1 + C_3} \begin{vmatrix} 1 & 0 & 0 & 0 \\ 0 & 1 & 2 & 1 \\ 4 & 0 & -4 & -2 \\ 1 & 2 & 2 & 0 \end{vmatrix} = a_{11}C_{11} = 1C_{11}$

$= C_{11} = \begin{vmatrix} 1 & 2 & 1 \\ 0 & -4 & -2 \\ 2 & 2 & 0 \end{vmatrix} \xrightarrow[=]{2R_1 + R_2} \begin{vmatrix} 1 & 2 & 1 \\ 2 & 0 & 0 \\ 2 & 2 & 0 \end{vmatrix} = a_{21}C_{21} = 2C_{21}$

$\qquad\qquad\qquad\qquad\qquad\qquad\qquad = 2(-1)^3 \begin{vmatrix} 2 & 1 \\ 2 & 0 \end{vmatrix}$

$\qquad\qquad\qquad\qquad\qquad\qquad\qquad = -2(0 - 2)$

$\qquad\qquad\qquad\qquad\qquad\qquad\qquad = 4$

29. $\begin{vmatrix} 5 & -2 & 2 & 3 \\ 0 & 1 & 1 & 3 \\ 1 & 0 & 1 & 1 \\ -2 & -1 & 0 & 6 \end{vmatrix} \xrightarrow[=]{-C_3 + C_1} \begin{vmatrix} 3 & -2 & 2 & 3 \\ -1 & 1 & 1 & 3 \\ 0 & 0 & 1 & 1 \\ -2 & -1 & 0 & 6 \end{vmatrix} \xrightarrow[=]{-C_3 + C_4} \begin{vmatrix} 3 & -2 & 2 & 1 \\ -1 & 1 & 1 & 2 \\ 0 & 0 & 1 & 0 \\ -2 & -1 & 0 & 6 \end{vmatrix}$

$= a_{33}C_{33} = 1C_{33} = C_{33} = \begin{vmatrix} 3 & -2 & 1 \\ -1 & 1 & 2 \\ -2 & -1 & 6 \end{vmatrix} \xrightarrow[=]{C_1 + C_2} \begin{vmatrix} 3 & 1 & 1 \\ -1 & 0 & 2 \\ -2 & -3 & 6 \end{vmatrix}$

$\xrightarrow[=]{2C_1 + C_3} \begin{vmatrix} 3 & 1 & 7 \\ -1 & 0 & 0 \\ -2 & -3 & 2 \end{vmatrix} = a_{21}C_{21} = -1C_{21} = -C_{21}$

$\qquad\qquad\qquad\qquad\qquad = -(-1)^3 \begin{vmatrix} 1 & 7 \\ -3 & 2 \end{vmatrix}$

$\qquad\qquad\qquad\qquad\qquad = 1[2 - (-21)]$

$\qquad\qquad\qquad\qquad\qquad = (2 + 21)$

$\qquad\qquad\qquad\qquad\qquad = 23$

31. $\begin{vmatrix} 2 & 2 & -1 & 3 \\ 1 & 1 & -1 & 1 \\ 1 & -1 & 1 & 1 \\ 4 & 2 & 1 & 5 \end{vmatrix} \xrightarrow[=]{-C_1 + C_4} \begin{vmatrix} 2 & 2 & -1 & 1 \\ 1 & 1 & -1 & 0 \\ 1 & -1 & 1 & 0 \\ 4 & 2 & 1 & 1 \end{vmatrix} \xrightarrow[=]{-R_1 + R_4} \begin{vmatrix} 2 & 2 & -1 & 1 \\ 1 & 1 & -1 & 0 \\ 1 & -1 & 1 & 0 \\ 2 & 0 & 2 & 0 \end{vmatrix}$

$= a_{14}C_{14} = 1C_{14} = C_{14}$

$= (-1)^5 \begin{vmatrix} 1 & 1 & -1 \\ 1 & -1 & 1 \\ 2 & 0 & 2 \end{vmatrix} = -\begin{vmatrix} 1 & 1 & -1 \\ 1 & -1 & 1 \\ 2 & 0 & 2 \end{vmatrix} \xrightarrow[=]{-C_1 + C_3} -\begin{vmatrix} 1 & 1 & -2 \\ 1 & -1 & 0 \\ 2 & 0 & 0 \end{vmatrix}$

$= -a_{31}C_{31} = -2C_{31} = -2(-1)^4 \begin{vmatrix} 1 & -2 \\ -1 & 0 \end{vmatrix}$

$= -2(0 - 2)$

$= 4$

33. $\begin{vmatrix} 1 & 1 & 1 \\ x & y & z \\ x^2 & y^2 & z^2 \end{vmatrix} \xrightarrow[=]{-xR_1 + R_2} \begin{vmatrix} 1 & 1 & 1 \\ 0 & y - x & z - x \\ x^2 & y^2 & z^2 \end{vmatrix} \xrightarrow[=]{-x^2 R_1 + R_3} \begin{vmatrix} 1 & 1 & 1 \\ 0 & y - x & z - x \\ 0 & y^2 - x^2 & z^2 - x^2 \end{vmatrix}$

$= a_{11}C_{11} = 1C_{11} = C_{11} = (-1)^2 \begin{vmatrix} y - x & z - x \\ y^2 - x^2 & z^2 - x^2 \end{vmatrix}$

$= (y - x)(z^2 - x^2) - (z - x)(y^2 - x^2)$

$= (y - x)(z - x)(z + x) - (z - x)(y - x)(y + x)$

$= (y - x)(z - x)[(z + x) - (y + x)]$

$= (y - x)(z - x)(z + x - y - x)$

$= (y - x)(z - x)(z - y)$

or $(x - y)(-1)(z - x)(y - z)(-1)$

$= (x - y)(z - x)(y - z)$

35. Let $A = \begin{bmatrix} a_{11} & a_{12} & a_{13} \\ a_{21} & a_{22} & a_{23} \\ a_{31} & a_{32} & a_{33} \end{bmatrix}$

Then $|A| = \begin{vmatrix} a_{11} & a_{12} & a_{13} \\ a_{21} & a_{22} & a_{23} \\ a_{31} & a_{32} & a_{33} \end{vmatrix} = a_{11}C_{11} + a_{21}C_{21} + a_{31}C_{31}$

$= a_{11}(-1)^2 \begin{vmatrix} a_{22} & a_{23} \\ a_{32} & a_{33} \end{vmatrix} + a_{21}(-1)^3 \begin{vmatrix} a_{12} & a_{13} \\ a_{32} & a_{33} \end{vmatrix} + a_{31}(-1)^4 \begin{vmatrix} a_{12} & a_{13} \\ a_{22} & a_{23} \end{vmatrix}$

$= a_{11}(a_{22}a_{33} - a_{23}a_{32}) - a_{21}(a_{12}a_{33} - a_{13}a_{32}) + a_{31}(a_{12}a_{23} - a_{13}a_{22})$ (*)

Now let B be A with rows and columns interchanged. That is,

$B = \begin{bmatrix} a_{11} & a_{21} & a_{31} \\ a_{12} & a_{22} & a_{32} \\ a_{13} & a_{23} & a_{33} \end{bmatrix}$

and $|B| = \begin{vmatrix} a_{11} & a_{21} & a_{31} \\ a_{12} & a_{22} & a_{32} \\ a_{13} & a_{23} & a_{33} \end{vmatrix} = a_{11}C_{11} + a_{21}C_{21} + a_{31}C_{31}$

$$= a_{11}(-1)^2 \begin{vmatrix} a_{22} & a_{32} \\ a_{23} & a_{33} \end{vmatrix} + a_{21}(-1)^3 \begin{vmatrix} a_{12} & a_{32} \\ a_{13} & a_{33} \end{vmatrix} + a_{31}(-1)^4 \begin{vmatrix} a_{12} & a_{22} \\ a_{13} & a_{23} \end{vmatrix}$$

$$= a_{11}(a_{22}a_{33} - a_{32}a_{23}) - a_{21}(a_{12}a_{33} - a_{13}a_{32}) + a_{31}(a_{12}a_{23} - a_{22}a_{13}) \quad (**)$$

Notice that (*) is the same as (**). That is, $|A| = |B|$.

37. Let $A = \begin{bmatrix} a_{11} & a_{12} & a_{13} \\ a_{21} & a_{22} & a_{23} \\ a_{31} & a_{32} & a_{33} \end{bmatrix}$.

Then $|A| = a_{11}a_{22}a_{33} + a_{12}a_{23}a_{31} + a_{13}a_{21}a_{32} - a_{11}a_{23}a_{32} - a_{12}a_{21}a_{33} - a_{13}a_{22}a_{31}$ *

Suppose B is obtained by interchanging the 1st and 2nd rows. Then,

$B = \begin{bmatrix} a_{21} & a_{22} & a_{23} \\ a_{11} & a_{12} & a_{13} \\ a_{31} & a_{32} & a_{33} \end{bmatrix}$.

Then, $|B| = a_{21}a_{12}a_{33} + a_{22}a_{13}a_{31} + a_{23}a_{11}a_{32} - a_{21}a_{13}a_{32} - a_{22}a_{11}a_{33} - a_{23}a_{12}a_{31}$ **

Comparing * with ** we see that one quantity is the negative of the other, and, thus,

$$|B| = -|A| \quad \text{or} \quad |A| = -|B|.$$

39. a) Let $A = \begin{bmatrix} a_1 & a_2 \\ a_3 & a_4 \end{bmatrix}$ and $B = \begin{bmatrix} b_1 & b_2 \\ b_3 & b_4 \end{bmatrix}$

Then $|A| = \begin{vmatrix} a_1 & a_2 \\ a_3 & a_4 \end{vmatrix} = a_1a_4 - a_2a_3$,

$|B| = \begin{vmatrix} b_1 & b_2 \\ b_3 & b_4 \end{vmatrix} = b_1b_4 - b_2b_3$,

and consequently, $|A||B| = (a_1a_4 - a_2a_3)(b_1b_4 - b_2b_3)$
$$= a_1a_4b_1b_4 - a_1a_4b_2b_3 - a_2a_3b_1b_4 + a_2a_3b_2b_3$$

Now $AB = \begin{bmatrix} a_1 & a_2 \\ a_3 & a_4 \end{bmatrix}\begin{bmatrix} b_1 & b_2 \\ b_3 & b_4 \end{bmatrix} = \begin{bmatrix} a_1b_1 + a_2b_3 & a_1b_2 + a_2b_4 \\ a_3b_1 + a_4b_3 & a_3b_2 + a_4b_4 \end{bmatrix}$

and $|AB| = (a_1b_1 + a_2b_3)(a_3b_2 + a_4b_4) - (a_1b_2 + a_2b_4)(a_3b_1 + a_4b_3)$
$$= a_1b_1a_3b_2 + a_1b_1a_4b_4 + a_2b_3a_3b_2 + a_2b_3a_4b_4 - a_1b_2a_3b_1$$
$$- a_2b_4a_3b_1 - a_2b_4a_4b_3$$
$$= a_1a_4b_1b_4 + a_2a_3b_2b_3 - a_1a_4b_2b_3 - a_2a_3b_1b_4$$

observe that both results are identical.

b) Start with $|AB| = |A||B|$ and replace B with A^{-1} to get
$$|AA^{-1}| = |A||A^{-1}|$$
$$|I| = |A||A^{-1}|$$
$$1 = |A||A^{-1}| \quad \text{Now} \div \text{ by } |A|$$
$$\frac{1}{|A|} = |A^{-1}|$$
$$\text{or} \quad |A|^{-1} = |A^{-1}|$$

Exercises 8.7

1. $2x + 9y = 3$
   $3x - 2y = -11$
   $\qquad D = \begin{vmatrix} 2 & 9 \\ 3 & -2 \end{vmatrix} = -4 - 27 = -31$

   $D_x = \begin{vmatrix} 3 & 9 \\ -11 & -2 \end{vmatrix} = -6 + 99 = 93 \qquad D_y = \begin{vmatrix} 2 & 3 \\ 3 & -11 \end{vmatrix} = -22 - 9 = -31$

   Then $x = \dfrac{D_x}{D} = \dfrac{93}{-31} = -3$ and $y = \dfrac{D_y}{D} = \dfrac{-31}{-31} = 1$

3. $7x - y = 9$
   $3x + 5y = -7$
   $\qquad D = \begin{vmatrix} 7 & -1 \\ 3 & 5 \end{vmatrix} = 35 - (-3) = 38$

   $D_x = \begin{vmatrix} 9 & -1 \\ -7 & 5 \end{vmatrix} = 45 - 7 = 38 \qquad D_y = \begin{vmatrix} 7 & 9 \\ 3 & -7 \end{vmatrix} = -49 - 27 = -76$

   Then $x = \dfrac{D_x}{D} = \dfrac{38}{38} = 1$ and $y = \dfrac{D_y}{D} = \dfrac{-76}{38} = -2$

5. $4a + 3b = 1$
   $2a - 5b = -19$
   $\qquad D = \begin{vmatrix} 4 & 3 \\ 2 & -5 \end{vmatrix} = -20 - 6 = -26$

   $D_a = \begin{vmatrix} 1 & 3 \\ -19 & -5 \end{vmatrix} = -5 - (-57) = -5 + 57 = 52 \qquad D_b = \begin{vmatrix} 4 & 1 \\ 2 & -19 \end{vmatrix} = -76 - 2 = -78$

   Then $a = \dfrac{D_a}{D} = \dfrac{52}{-26} = -2$ and $b = \dfrac{D_b}{D} = \dfrac{-78}{-26} = 3$

7. $3x - 12y = 9$
   $-x + 4y = -3$
   $\qquad D = \begin{vmatrix} 3 & -12 \\ -1 & 4 \end{vmatrix} = 12 - (12) = 0$

   $D_x = \begin{vmatrix} 9 & -12 \\ -3 & 4 \end{vmatrix} = 36 - (36) = 0 \qquad D_y = \begin{vmatrix} 3 & 9 \\ -1 & -3 \end{vmatrix} = -9 - (-9) = 0$

   Here, the system is dependent.

9. $7x - y = 1$
   $-14x + 2y = -1$
   $\qquad D = \begin{vmatrix} 7 & -1 \\ -14 & 2 \end{vmatrix} = 14 - (14) = 0$

   $D_x = \begin{vmatrix} 1 & -1 \\ -1 & 2 \end{vmatrix} = 2 - (1) = 1 \qquad D_y = \begin{vmatrix} 7 & 1 \\ -14 & -1 \end{vmatrix} = -7 - (-14) = -7 + 14 = 7$

   Here, the system is inconsistent.

11. $x + y = 3$
    $x + 2y - z = 5$
    $\quad 2y + z = 1$
    $\qquad D = \begin{vmatrix} 1 & 1 & 0 \\ 1 & 2 & -1 \\ 0 & 2 & 1 \end{vmatrix} = 3$

    $D_x = \begin{vmatrix} 3 & 1 & 0 \\ 5 & 2 & -1 \\ 1 & 2 & 1 \end{vmatrix} = 6 \qquad D_y = \begin{vmatrix} 1 & 3 & 0 \\ 1 & 5 & -1 \\ 0 & 1 & 1 \end{vmatrix} = 3$

    $D_z = \begin{vmatrix} 1 & 1 & 3 \\ 1 & 2 & 5 \\ 0 & 2 & 1 \end{vmatrix} = -3 \qquad$ Then $x = \dfrac{D_x}{D} = \dfrac{6}{3} = 2$, $y = \dfrac{D_y}{D} = \dfrac{3}{3} = 1$,

    and $z = \dfrac{D_z}{D} = \dfrac{-3}{3} = -1$

13. $3x + 2y + z = -1 \qquad D = \begin{vmatrix} 3 & 2 & 1 \\ 1 & 0 & -2 \\ 0 & 1 & 2 \end{vmatrix} = 3$

   $x \qquad\;\; - 2z = -3$

   $\qquad\;\; y + 2z = -2$

   $D_x = \begin{vmatrix} -1 & 2 & 1 \\ -3 & 0 & -2 \\ -2 & 1 & 2 \end{vmatrix} = 15 \qquad D_y = \begin{vmatrix} 3 & -1 & 1 \\ 1 & -3 & -2 \\ 0 & -2 & 2 \end{vmatrix} = -30$

   $D_z = \begin{vmatrix} 3 & 2 & -1 \\ 1 & 0 & -3 \\ 0 & 1 & -2 \end{vmatrix} = 12 \qquad$ Then $x = \dfrac{D_x}{D} = \dfrac{15}{3} = 5,\; y = \dfrac{D_y}{D} = \dfrac{-30}{3} = -10,$

   and $z = \dfrac{D_z}{D} = \dfrac{12}{3} = 4$

15. $3x + y + 2z = 3 \qquad D = \begin{vmatrix} 3 & 1 & 2 \\ 1 & -5 & -1 \\ 2 & 3 & 2 \end{vmatrix} = 1$

   $x - 5y - z = 0$

   $2x + 3y + 2z = 0$

   $D_x = \begin{vmatrix} 3 & 1 & 2 \\ 0 & -5 & -1 \\ 0 & 3 & 2 \end{vmatrix} = -21 \qquad D_y = \begin{vmatrix} 3 & 3 & 2 \\ 1 & 0 & -1 \\ 2 & 0 & 2 \end{vmatrix} = -12$

   $D_z = \begin{vmatrix} 3 & 1 & 3 \\ 1 & -5 & 0 \\ 2 & 3 & 0 \end{vmatrix} = 39 \qquad$ Then $x = \dfrac{D_x}{D} = \dfrac{-21}{1} = -21,\; y = \dfrac{D_y}{D} = \dfrac{-12}{1} = -12,$

   and $z = \dfrac{D_z}{D} = \dfrac{39}{1} = 39$

17. $a + 3b - c = 4 \qquad D = \begin{vmatrix} 1 & 3 & -1 \\ 3 & -2 & 4 \\ 2 & 1 & 3 \end{vmatrix} = -20$

   $3a - 2b + 4c = 11$

   $2a + b + 3c = 13$

   $D_a = \begin{vmatrix} 4 & 3 & -1 \\ 11 & -2 & 4 \\ 13 & 1 & 3 \end{vmatrix} = -20 \qquad D_b = \begin{vmatrix} 1 & 4 & -1 \\ 3 & 11 & 4 \\ 2 & 13 & 3 \end{vmatrix} = -40$

   $D_c = \begin{vmatrix} 1 & 3 & 4 \\ 3 & -2 & 11 \\ 2 & 1 & 13 \end{vmatrix} = -60 \qquad$ Then $a = \dfrac{D_a}{D} = \dfrac{-20}{-20} = 1,\; b = \dfrac{D_b}{D} = \dfrac{-40}{-20} = 2,$

   and $c = \dfrac{D_c}{D} = \dfrac{-60}{-20} = 3$

19. $2x - y + 3z = 17 \qquad D = \begin{vmatrix} 2 & -1 & 3 \\ 5 & -2 & 4 \\ 3 & 3 & -1 \end{vmatrix} = 26$

   $5x - 2y + 4z = 28$

   $3x + 3y - z = -1$

   $D_x = \begin{vmatrix} 17 & -1 & 3 \\ 28 & -2 & 4 \\ -1 & 3 & -1 \end{vmatrix} = 52 \qquad D_y = \begin{vmatrix} 2 & 17 & 3 \\ 5 & 28 & 4 \\ 3 & -1 & -1 \end{vmatrix} = -26$

   $D_z = \begin{vmatrix} 2 & -1 & 17 \\ 5 & -2 & 28 \\ 3 & 3 & -1 \end{vmatrix} = 104 \qquad$ Then $x = \dfrac{D_x}{D} = \dfrac{52}{26} = 2,\; y = \dfrac{D_y}{D} = \dfrac{-26}{26} = -1,$

   and $z = \dfrac{D_z}{D} = \dfrac{104}{26} = 4$

21. $2x + y - z = 5$ $\qquad D = \begin{vmatrix} 2 & 1 & -1 \\ 1 & -1 & 2 \\ 0 & -3 & 5 \end{vmatrix} = 0$

$\quad\quad x - y + 2z = -3$

$\quad\quad\quad\quad -3y + 5z = -11$

$D_x = \begin{vmatrix} 5 & 1 & -1 \\ -3 & -1 & 2 \\ -11 & -3 & 5 \end{vmatrix} = 0$ $\qquad D_y = \begin{vmatrix} 2 & 5 & -1 \\ 1 & -3 & 2 \\ 0 & -11 & 5 \end{vmatrix} = 0$

$D_z = \begin{vmatrix} 2 & 1 & 5 \\ 1 & -1 & -3 \\ 0 & -3 & -11 \end{vmatrix} = 0$ $\qquad$ In this case, the system is dependent.

23. $2x_1 - x_2 + 3x_3 = 17$ $\qquad D = \begin{vmatrix} 2 & -1 & 3 \\ 5 & -2 & 4 \\ 3 & 3 & -1 \end{vmatrix} = 26$

$\quad\quad 5x_1 - 2x_2 + 4x_3 = 28$

$\quad\quad 3x_1 + 3x_2 - x_3 = 1$

$D_{x_1} = \begin{vmatrix} 17 & -1 & 3 \\ 28 & -2 & 4 \\ 1 & 3 & -1 \end{vmatrix} = 56$ $\qquad D_{x_2} = \begin{vmatrix} 2 & 17 & 3 \\ 5 & 28 & 4 \\ 3 & 1 & -1 \end{vmatrix} = -12$

$D_{x_3} = \begin{vmatrix} 2 & -1 & 17 \\ 5 & -2 & 28 \\ 3 & 3 & 1 \end{vmatrix} = 106$ $\qquad$ Then $x_1 = \dfrac{D_{x_1}}{D} = \dfrac{56}{26} = \dfrac{28}{13}$, $x_2 = \dfrac{D_{x_2}}{D} = \dfrac{-12}{26} = -\dfrac{6}{13}$,

$\qquad\qquad\qquad\qquad\qquad\qquad\qquad$ and $x_3 = \dfrac{D_{x_3}}{D} = \dfrac{106}{26} = \dfrac{53}{13}$

25. $6x - 4y + 3z = 1$ $\qquad D = \begin{vmatrix} 6 & -4 & 3 \\ 12 & -8 & -9 \\ 12 & -4 & 9 \end{vmatrix} = 360$

$\quad\quad 12x - 8y - 9z = -3$

$\quad\quad 12x - 4y + 9z = 4$

$D_x = \begin{vmatrix} 1 & -4 & 3 \\ -3 & -8 & -9 \\ 4 & -4 & 9 \end{vmatrix} = 60$ $\qquad D_y = \begin{vmatrix} 6 & 1 & 3 \\ 12 & -3 & -9 \\ 12 & 4 & 9 \end{vmatrix} = 90$

$D_z = \begin{vmatrix} 6 & -4 & 1 \\ 12 & -8 & -3 \\ 12 & -4 & 4 \end{vmatrix} = 120$ $\qquad$ Then $x = \dfrac{D_x}{D} = \dfrac{60}{360} = \dfrac{1}{6}$, $y = \dfrac{D_y}{D} = \dfrac{90}{360} = \dfrac{1}{4}$,

$\qquad\qquad\qquad\qquad\qquad\qquad\qquad$ and $z = \dfrac{D_z}{D} = \dfrac{120}{360} = \dfrac{1}{3}$

27. $3x - y + 2z = 0$ $\qquad D = \begin{vmatrix} 3 & -1 & 2 \\ 1 & 2 & 4 \\ 7 & -7 & -2 \end{vmatrix} = 0$

$\quad\quad x + 2y + 4z = -1$

$\quad\quad 7x - 7y - 2z = 1$

$D_x = \begin{vmatrix} 0 & -1 & 2 \\ -1 & 2 & 4 \\ 1 & -7 & -2 \end{vmatrix} = 8$ $\qquad$ Without going any farther, we know that the system is inconsistent.

29. $x - y - 2z = 2$
    $2x + 3y - z = 4$
    $3x - y - 2z = 1$

$D = \begin{vmatrix} 1 & -1 & -2 \\ 2 & 3 & -1 \\ 3 & -1 & -2 \end{vmatrix} = 14$

$D_x = \begin{vmatrix} 2 & -1 & -2 \\ 4 & 3 & -1 \\ 1 & -1 & -2 \end{vmatrix} = -7$ $D_y = \begin{vmatrix} 1 & 2 & -2 \\ 2 & 4 & -1 \\ 3 & 1 & -2 \end{vmatrix} = 15$

$D_z = \begin{vmatrix} 1 & -1 & 2 \\ 2 & 3 & 4 \\ 3 & -1 & 1 \end{vmatrix} = -25$ Then $x = \dfrac{D_x}{D} = \dfrac{-7}{14} = -\dfrac{1}{2}$, $y = \dfrac{D_y}{D} = \dfrac{15}{14}$,

and $z = \dfrac{D_z}{D} = \dfrac{-25}{14} = -\dfrac{25}{14}$

31. $x + 2y + 3z = 0$
    $y + z = 0$
    $x + y + 3z = 1$

$D = \begin{vmatrix} 1 & 2 & 3 \\ 0 & 1 & 1 \\ 1 & 1 & 3 \end{vmatrix} = 1$

$D_x = \begin{vmatrix} 0 & 2 & 3 \\ 0 & 1 & 1 \\ 1 & 1 & 3 \end{vmatrix} = -1$ $D_y = \begin{vmatrix} 1 & 0 & 3 \\ 0 & 0 & 1 \\ 1 & 1 & 3 \end{vmatrix} = -1$

$D_z = \begin{vmatrix} 1 & 2 & 0 \\ 0 & 1 & 0 \\ 1 & 1 & 1 \end{vmatrix} = 1$ Then $x = \dfrac{D_x}{D} = \dfrac{-1}{1} = -1$, $y = \dfrac{D_y}{D} = \dfrac{-1}{1} = -1$,

and $z = \dfrac{D_z}{D} = \dfrac{1}{1} = 1$

33. $3x - y - 2z = -13$
    $5x + 3y - z = 4$
    $2x - 7y + 3z = -36$

$D = \begin{vmatrix} 3 & -1 & -2 \\ 5 & 3 & -1 \\ 2 & -7 & 3 \end{vmatrix} = 105$

$D_x = \begin{vmatrix} -13 & -1 & -2 \\ 4 & 3 & -1 \\ -36 & -7 & 3 \end{vmatrix} = -210$ $D_y = \begin{vmatrix} 3 & -13 & -2 \\ 5 & 4 & -1 \\ 2 & -36 & 3 \end{vmatrix} = 525$

$D_z = \begin{vmatrix} 3 & -1 & -13 \\ 5 & 3 & 4 \\ 2 & -7 & -36 \end{vmatrix} = 105$ Then $x = \dfrac{D_x}{D} = \dfrac{-210}{105} = -2$, $y = \dfrac{D_y}{D} = \dfrac{525}{105} = 5$,

and $z = \dfrac{D_z}{D} = \dfrac{105}{105} = 1$

35. $x - y - z - w = 2$
    $x + y \quad\quad - w = 0$
    $\quad\quad 3y + 2z + w = -2$
    $\quad\quad y \quad\quad + w = 0$

$D = \begin{vmatrix} 1 & -1 & -1 & -1 \\ 1 & 1 & 0 & -1 \\ 0 & 3 & 2 & 1 \\ 0 & 1 & 0 & 1 \end{vmatrix} = 2$

$D_x = \begin{vmatrix} 2 & -1 & -1 & -1 \\ 0 & 1 & 0 & -1 \\ -2 & 3 & 2 & 1 \\ 0 & 1 & 0 & 1 \end{vmatrix} = 4$ $D_y = \begin{vmatrix} 1 & 2 & -1 & -1 \\ 1 & 0 & 0 & -1 \\ 0 & -2 & 2 & 1 \\ 0 & 0 & 0 & 1 \end{vmatrix} = -2$

$$D_z = \begin{vmatrix} 1 & -1 & 2 & -1 \\ 1 & 1 & 0 & -1 \\ 0 & 3 & -2 & 1 \\ 0 & 1 & 0 & 1 \end{vmatrix} = 0 \qquad D_w = \begin{vmatrix} 1 & -1 & -1 & 2 \\ 1 & 1 & 0 & 0 \\ 0 & 3 & 2 & -2 \\ 0 & 1 & 0 & 0 \end{vmatrix} = 2$$

Then $x = \dfrac{D_x}{D} = \dfrac{4}{2} = 2$, $y = \dfrac{D_y}{D} = \dfrac{-2}{2} = -1$, $z = \dfrac{D_z}{D} = \dfrac{0}{2} = 0$, and $w = \dfrac{D_w}{D} = \dfrac{2}{2} = 1$

37.
$$\begin{aligned} x + 3y - 2z + w &= 2 \\ 2x + 4y + 2z - 3w &= -1 \\ x - y - z - 2w &= -1 \\ 2x + 5y - z - w &= 1 \end{aligned} \qquad D = \begin{vmatrix} 1 & 3 & -2 & 1 \\ 2 & 4 & 2 & -3 \\ 1 & -1 & -1 & -2 \\ 2 & 5 & -1 & -1 \end{vmatrix} = -11$$

$$D_x = \begin{vmatrix} 2 & 3 & -2 & 1 \\ -1 & 4 & 2 & -3 \\ -1 & -1 & -1 & -2 \\ 1 & 5 & -1 & -1 \end{vmatrix} = -11 \qquad D_y = \begin{vmatrix} 1 & 2 & -2 & 1 \\ 2 & -1 & 2 & -3 \\ 1 & -1 & -1 & -2 \\ 2 & 1 & -1 & -1 \end{vmatrix} = 0$$

$$D_z = \begin{vmatrix} 1 & 3 & 2 & 1 \\ 2 & 4 & -1 & -3 \\ 1 & -1 & -1 & -2 \\ 2 & 5 & 1 & -1 \end{vmatrix} = 0 \qquad D_w = \begin{vmatrix} 1 & 3 & -2 & 2 \\ 2 & 4 & 2 & -1 \\ 1 & -1 & -1 & -1 \\ 2 & 5 & -1 & 1 \end{vmatrix} = -11$$

Then $x = \dfrac{D_x}{D} = \dfrac{-11}{-11} = 1$, $y = \dfrac{D_y}{D} = \dfrac{0}{-11} = 0$, $z = \dfrac{D_z}{D} = \dfrac{0}{-11} = 0$, and $w = \dfrac{D_w}{D} = \dfrac{-11}{-11} = 1$

39. For $(1, -2)$, let $x = 1$ and $y = -2$:
$$\begin{aligned} y &= mx + b \\ -2 &= m(1) + b \\ -2 &= m + b \end{aligned}$$

For $(3, 2)$, let $x = 3$ and $y = 2$:
$$\begin{aligned} y &= mx + b \\ 2 &= m(3) + b \\ 2 &= 3m + b \end{aligned}$$

The system is $\begin{aligned} m + b &= -2 \\ 3m + b &= 2 \end{aligned}$ $\qquad D = \begin{vmatrix} 1 & 1 \\ 3 & 1 \end{vmatrix} = 1 - 3 = -2$

$$D_m = \begin{vmatrix} -2 & 1 \\ 2 & 1 \end{vmatrix} = -2 - 2 = -4 \qquad D_b = \begin{vmatrix} 1 & -2 \\ 3 & 2 \end{vmatrix} = 2 - (-6) = 8$$

Then, $m = \dfrac{D_m}{D} = \dfrac{-4}{-2} = 2$, $b = \dfrac{D_b}{D} = \dfrac{8}{-2} = -4$

41. For $(2, 10)$, let $x = 2$ and $y = 10$:
$$\begin{aligned} y &= ax^2 + bx + c \\ 10 &= a(2)^2 + b(2) + c \\ 10 &= 4a + 2b + c \end{aligned}$$

For $(-1, -5)$, let $x = -1$ and $y = -5$:
$$\begin{aligned} y &= ax^2 + bx + c \\ -5 &= a(-1)^2 + b(-1) + c \\ -5 &= a - b + c \end{aligned}$$

For $(-3, 5)$, let $x = -3$ and $y = 5$:
$$\begin{aligned} y &= ax^2 + bx + c \\ 5 &= a(-3)^2 + b(-3) + c \\ 5 &= 9a - 3b + c \end{aligned}$$

The system is
$$4a + 2b + c = 10$$
$$a - b + c = -5 \qquad D = \begin{vmatrix} 4 & 2 & 1 \\ 1 & -1 & 1 \\ 9 & -3 & 1 \end{vmatrix} = 30$$
$$9a - 3b + c = 5$$

$$D_a = \begin{vmatrix} 10 & 2 & 1 \\ -5 & -1 & 1 \\ 5 & -3 & 1 \end{vmatrix} = 60 \qquad D_b = \begin{vmatrix} 4 & 10 & 1 \\ 1 & -5 & 1 \\ 9 & 5 & 1 \end{vmatrix} = 90$$

$$D_c = \begin{vmatrix} 4 & 2 & 10 \\ 1 & -1 & -5 \\ 9 & -3 & 5 \end{vmatrix} = -120 \qquad \text{Then,} \quad a = \frac{D_a}{D} = \frac{60}{30} = 2, \quad b = \frac{D_b}{D} = \frac{90}{30} = 3,$$

$$c = \frac{D_c}{D} = \frac{-120}{30} = -4$$

43. In this case, $D_{x_i} = D$ so that $x_i = \dfrac{D_{x_i}}{D} = 1$ and all other $x_j = 0$ for $j \neq i$.

CHAPTER REVIEW

1. $\begin{bmatrix} 2 & -3 \\ -4 & -5 \\ 0 & 2 \end{bmatrix} - 2\begin{bmatrix} -1 & 0 \\ 3 & -11 \\ 1 & 0 \end{bmatrix} = \begin{bmatrix} 2 - 2(-1) & -3 - 2(0) \\ -4 - 2(3) & -5 - 2(-11) \\ 0 - 2(1) & 2 - 2(0) \end{bmatrix} = \begin{bmatrix} 4 & -3 \\ -10 & 17 \\ -2 & 2 \end{bmatrix}$

2. The sum cannot be computed since the dimension of the first matrix is 1×3 and the dimension of the second matrix is 1×2. When the dimensions are different, the sum cannot be determined.

3. $2\begin{bmatrix} -2 & 1 & 3 \\ 1 & 5 & 1 \end{bmatrix} - 3\begin{bmatrix} 1 & 7 & -2 \\ 0 & 1 & -5 \end{bmatrix} = \begin{bmatrix} 2(-2) - 3(1) & 2(1) - 3(7) & 2(3) - 3(-2) \\ 2(1) - 3(0) & 2(5) - 3(1) & 2(1) - 3(-5) \end{bmatrix}$

$$= \begin{bmatrix} -7 & -19 & 12 \\ 2 & 7 & 17 \end{bmatrix}$$

4. The product cannot be computed since the dimension of the first matrix is 2×3 and the dimension of the second matrix is 2×3. The product cannot be determined when the number of columns from the first matrix is different from the number of rows from the second matrix.

5. $\begin{bmatrix} 1 & -2 & 0 \\ 0 & 3 & 2 \\ 5 & 0 & 1 \end{bmatrix}\begin{bmatrix} -1 & 0 \\ 3 & 4 \\ 0 & -1 \end{bmatrix} = \begin{bmatrix} 1(-1) + (-2)(3) + 0(0) & 1(0) + (-2)(4) + 0(-1) \\ 0(-1) + 3(3) + 2(0) & 0(0) + 3(4) + 2(-1) \\ 5(-1) + 0(3) + 1(0) & 5(0) + 0(4) + 1(-1) \end{bmatrix}$

$$= \begin{bmatrix} -7 & -8 \\ 9 & 10 \\ -5 & -1 \end{bmatrix}$$

6. The product cannot be computed since the dimension of the first matrix is 2×3 and the dimension of the second matrix is 2×2. The product cannot be determined when the number of columns from the first matrix is different from the number of rows from the second matrix.

7. $\begin{bmatrix} 2 & 7 \\ 0 & -4 \end{bmatrix}\begin{bmatrix} 2 & 1 & 0 \\ -1 & 3 & 5 \end{bmatrix} = \begin{bmatrix} 2(2) + 7(-1) & 2(1) + 7(3) & 2(0) + 7(5) \\ 0(2) + (-4)(-1) & 0(1) + (-4)(3) & 0(0) + (-4)(5) \end{bmatrix}$

$$= \begin{bmatrix} -3 & 23 & 35 \\ 4 & -12 & -20 \end{bmatrix}$$

8. $\begin{bmatrix} 2 & -1 & \vdots & 2 \\ -1 & 3 & \vdots & 14 \end{bmatrix} \xrightarrow{R_1 \leftrightarrow R_2} \begin{bmatrix} -1 & 3 & \vdots & 14 \\ 2 & -1 & \vdots & 2 \end{bmatrix} \xrightarrow{2R_1 + R_2} \begin{bmatrix} -1 & 3 & \vdots & 14 \\ 0 & 5 & \vdots & 30 \end{bmatrix} \xrightarrow{\frac{1}{5}R_2} \begin{bmatrix} -1 & 3 & \vdots & 14 \\ 0 & 1 & \vdots & 6 \end{bmatrix}$

The second row is $y = 6$.

The first row gives $-x + 3y = 14$
$$-x + 3(6) = 14$$
$$-x + 18 = 14$$
$$-x = -4$$
$$x = 4$$

9. $\begin{bmatrix} 1 & -2 & -4 & \vdots & -1 \\ 3 & 0 & -1 & \vdots & 4 \\ 1 & 4 & 7 & \vdots & 2 \end{bmatrix} \begin{array}{c} \xrightarrow{-3R_1 + R_2} \\ \xrightarrow{-R_1 + R_3} \end{array} \begin{bmatrix} 1 & -2 & -4 & \vdots & -7 \\ 0 & 6 & 11 & \vdots & 7 \\ 0 & 6 & 11 & \vdots & 3 \end{bmatrix} \xrightarrow{-R_2 + R_3} \begin{bmatrix} 1 & -2 & -4 & \vdots & -7 \\ 0 & 6 & 11 & \vdots & 7 \\ 0 & 0 & 0 & \vdots & -4 \end{bmatrix}$

The last row translates into $0x + 0y + 0z = -4$ or $0 = -4$ which is impossible. Thus, there is **no** solution to the system of equations.

10. $\begin{bmatrix} -1 & 2 & 1 & \vdots & -1 \\ 3 & 1 & -1 & \vdots & 7 \\ 0 & 1 & 1 & \vdots & -3 \end{bmatrix} \xrightarrow{3R_1 + R_2} \begin{bmatrix} -1 & 2 & 1 & \vdots & -1 \\ 0 & 7 & 2 & \vdots & 4 \\ 0 & 1 & 1 & \vdots & -3 \end{bmatrix} \xrightarrow{R_2 \leftrightarrow R_3} \begin{bmatrix} -1 & 2 & 1 & \vdots & -1 \\ 0 & 1 & 1 & \vdots & -3 \\ 0 & 7 & 2 & \vdots & 4 \end{bmatrix}$

$\xrightarrow{-7R_2 + R_3} \begin{bmatrix} -1 & 2 & 1 & \vdots & -1 \\ 0 & 1 & 1 & \vdots & -3 \\ 0 & 0 & -5 & \vdots & 25 \end{bmatrix} \xrightarrow{-\frac{1}{5}R_3} \begin{bmatrix} -1 & 2 & 1 & \vdots & -1 \\ 0 & 1 & 1 & \vdots & -3 \\ 0 & 1 & 1 & \vdots & -5 \end{bmatrix}$

The last row is $z = -5$.

The second row gives $y + z = -3$
$$y + (-5) = -3$$
$$y = 2$$

The first row gives $-x + 2y + z = -1$
$$-x + 2(2) + (-5) = -1$$
$$-x - 1 = -1$$
$$-x = 0$$
$$x = 0$$

11. $\begin{bmatrix} 1 & 2 & 1 & \vdots & 1 \\ 2 & 5 & 3 & \vdots & 2 \\ 1 & 0 & -2 & \vdots & 3 \end{bmatrix} \begin{array}{c} \xrightarrow{-2R_1 + R_2} \\ \xrightarrow{-R_1 + R_3} \end{array} \begin{bmatrix} 1 & 2 & 1 & \vdots & 1 \\ 0 & 1 & 1 & \vdots & 0 \\ 0 & -2 & -3 & \vdots & 2 \end{bmatrix} \xrightarrow{2R_2 + R_3} \begin{bmatrix} 1 & 2 & 1 & \vdots & 1 \\ 0 & 1 & 1 & \vdots & 0 \\ 0 & 0 & -1 & \vdots & 2 \end{bmatrix}$

$\xrightarrow{-R_3} \begin{bmatrix} 1 & 2 & 1 & \vdots & 1 \\ 0 & 1 & 1 & \vdots & 0 \\ 0 & 1 & 1 & \vdots & -2 \end{bmatrix}$

The third row is $z = -2$.

The second row gives $y + z = 0$
$$y + (-2) = 0$$
$$y = 2$$

The first row gives $x + 2y + z = 1$
$$x + 2(2) + (-2) = 1$$
$$x + 2 = 1$$
$$x = -1$$

12. $\begin{bmatrix} 1 & 2 & 1 & \vdots & 1 \\ 1 & 1 & -1 & \vdots & 1 \\ 0 & 1 & 3 & \vdots & 1 \end{bmatrix} \xrightarrow{-R_1 + R_2} \begin{bmatrix} 1 & 2 & 1 & \vdots & 1 \\ 0 & -1 & -2 & \vdots & 0 \\ 0 & 1 & 3 & \vdots & 1 \end{bmatrix} \xrightarrow{-R_2} \begin{bmatrix} 1 & 2 & 1 & \vdots & 1 \\ 0 & 1 & 2 & \vdots & 0 \\ 0 & 1 & 3 & \vdots & 1 \end{bmatrix}$

$\xrightarrow{-R_2 + R_3} \begin{bmatrix} 1 & 2 & 1 & \vdots & 1 \\ 0 & 1 & 2 & \vdots & 0 \\ 0 & 0 & 1 & \vdots & 1 \end{bmatrix}$

The last row is $z = 1$.

The second row gives $y + 2z = 0$
$$y + 2(1) = 0$$
$$y + 2 = 0$$
$$y = -2$$

The first row gives $x + 2y + z = 1$
$$x + 2(-2) + 1 = 1$$
$$x - 3 = 1$$
$$x = 4$$

13. $\begin{bmatrix} 1 & 0 & 1 & \vdots & 3 \\ 1 & 1 & 1 & \vdots & 1 \\ 2 & -1 & 3 & \vdots & 12 \end{bmatrix} \begin{array}{c} \xrightarrow{-R_1 + R_2} \\ \xrightarrow{-2R_1 + R_3} \end{array} \begin{bmatrix} 1 & 0 & 1 & \vdots & 3 \\ 0 & 1 & 0 & \vdots & -2 \\ 0 & -1 & 1 & \vdots & 6 \end{bmatrix}$

The second row is $y = -2$.

The third row gives $-y + z = 6$
$$-(-2) + z = 6$$
$$2 + z = 6$$
$$z = 4$$

The first row gives $x + z = 3$
$$x + 4 = 3$$
$$x = -1$$

14. $\delta = (2)(3) - (-1)(-4) = 6 - 4 = 2$

Then, the inverse is $\dfrac{1}{2}\begin{bmatrix} 3 & 4 \\ 1 & 2 \end{bmatrix} = \begin{bmatrix} \frac{3}{2} & 2 \\ \frac{1}{2} & 1 \end{bmatrix}$

15. Let $A = \begin{bmatrix} 8 & 12 \\ 6 & 9 \end{bmatrix}$. Then $\delta(A) = ad - bc = (8)(9) - (12)(6) = 72 - 72 = 0$.

Consequently, the inverse does **not** exist.

16. $\begin{bmatrix} 1 & 0 & 0 & \vdots & 1 & 0 & 0 \\ -1 & 1 & 1 & \vdots & 0 & 1 & 0 \\ 0 & 1 & 2 & \vdots & 0 & 0 & 1 \end{bmatrix} \xrightarrow{R_1 + R_2} \begin{bmatrix} 1 & 0 & 0 & \vdots & 1 & 0 & 0 \\ 0 & 1 & 1 & \vdots & 1 & 1 & 0 \\ 0 & 1 & 2 & \vdots & 0 & 0 & 1 \end{bmatrix}$

$\xrightarrow{-R_2 + R_3} \begin{bmatrix} 1 & 0 & 0 & \vdots & 1 & 0 & 0 \\ 0 & 1 & 1 & \vdots & 1 & 1 & 0 \\ 0 & 0 & 1 & \vdots & -1 & -1 & 1 \end{bmatrix} \xrightarrow{-R_3 + R_2} \begin{bmatrix} 1 & 0 & 0 & \vdots & 1 & 0 & 0 \\ 0 & 1 & 0 & \vdots & 2 & 2 & -1 \\ 0 & 0 & 1 & \vdots & -1 & -1 & 1 \end{bmatrix}$

The inverse is $\begin{bmatrix} 1 & 0 & 0 \\ 2 & 2 & -1 \\ -1 & -1 & 1 \end{bmatrix}$.

17. $\begin{bmatrix} 1 & 0 & 2 & \vdots & 1 & 0 & 0 \\ 2 & 1 & 5 & \vdots & 0 & 1 & 0 \\ 0 & 1 & 2 & \vdots & 0 & 0 & 1 \end{bmatrix} \xrightarrow{-2R_1 + R_2} \begin{bmatrix} 1 & 0 & 2 & \vdots & 1 & 0 & 0 \\ 0 & 1 & 1 & \vdots & -2 & 1 & 0 \\ 0 & 1 & 2 & \vdots & 0 & 0 & 1 \end{bmatrix}$

$\xrightarrow{-R_2 + R_3} \begin{bmatrix} 1 & 0 & 2 & \vdots & 1 & 0 & 0 \\ 0 & 1 & 1 & \vdots & -2 & 1 & 0 \\ 0 & 0 & 1 & \vdots & 2 & -1 & 1 \end{bmatrix} \xrightarrow[-2R_3 + R_1]{-R_3 + R_2} \begin{bmatrix} 1 & 0 & 0 & \vdots & -3 & 2 & -2 \\ 0 & 1 & 0 & \vdots & -4 & 2 & -1 \\ 0 & 0 & 1 & \vdots & 2 & -1 & 1 \end{bmatrix}$

The inverse is $\begin{bmatrix} -3 & 2 & -2 \\ -4 & 2 & -1 \\ 2 & -1 & 1 \end{bmatrix}$.

18. $\begin{bmatrix} 1 & 0 & 2 & \vdots & 1 & 0 & 0 \\ 2 & -2 & 5 & \vdots & 0 & 1 & 0 \\ 0 & 1 & -1 & \vdots & 0 & 0 & 1 \end{bmatrix} \xrightarrow{-2R_1 + R_2} \begin{bmatrix} 1 & 0 & 2 & \vdots & 1 & 0 & 0 \\ 0 & -2 & 1 & \vdots & -2 & 1 & 0 \\ 0 & 1 & -1 & \vdots & 0 & 0 & 1 \end{bmatrix}$

$\xrightarrow{R_2 \leftrightarrow R_3} \begin{bmatrix} 1 & 0 & 2 & \vdots & 1 & 0 & 0 \\ 0 & 1 & -1 & \vdots & 0 & 0 & 1 \\ 0 & -2 & 1 & \vdots & -2 & 1 & 0 \end{bmatrix} \xrightarrow{2R_2 + R_3} \begin{bmatrix} 1 & 0 & 2 & \vdots & 1 & 0 & 0 \\ 0 & 1 & -1 & \vdots & 0 & 0 & 1 \\ 0 & 0 & -1 & \vdots & -2 & 1 & 2 \end{bmatrix}$

$\xrightarrow{-R_3} \begin{bmatrix} 1 & 0 & 2 & \vdots & 1 & 0 & 0 \\ 0 & 1 & -1 & \vdots & 0 & 0 & 1 \\ 0 & 0 & 1 & \vdots & 2 & -1 & -2 \end{bmatrix} \xrightarrow[R_3 + R_2]{-2R_3 + R_1} \begin{bmatrix} 1 & 0 & 0 & \vdots & -3 & 2 & 4 \\ 0 & 1 & 0 & \vdots & 2 & -1 & -1 \\ 0 & 0 & 1 & \vdots & 2 & -1 & -2 \end{bmatrix}$

The inverse is $\begin{bmatrix} -3 & 2 & 4 \\ 2 & -1 & -1 \\ 2 & -1 & -2 \end{bmatrix}$.

19. $\begin{bmatrix} 1 & 1 & -3 & \vdots & 1 & 0 & 0 \\ 1 & 0 & 3 & \vdots & 0 & 1 & 0 \\ -2 & 1 & -12 & \vdots & 0 & 0 & 1 \end{bmatrix} \xrightarrow[2R_1 + R_3]{-R_1 + R_2} \begin{bmatrix} 1 & 1 & -3 & \vdots & 1 & 0 & 0 \\ 0 & -1 & 6 & \vdots & -1 & 1 & 0 \\ 0 & 3 & -18 & \vdots & 2 & 0 & 1 \end{bmatrix}$

$\xrightarrow{-R_2} \begin{bmatrix} 1 & 1 & -3 & \vdots & 1 & 0 & 0 \\ 0 & 1 & -6 & \vdots & 1 & -1 & 0 \\ 0 & 3 & -18 & \vdots & 2 & 0 & 1 \end{bmatrix} \xrightarrow{-3R_2 + R_3} \begin{bmatrix} 1 & 1 & -3 & \vdots & 1 & 0 & 0 \\ 0 & 1 & -6 & \vdots & 1 & -1 & 0 \\ 0 & 0 & 0 & \vdots & -1 & 3 & 1 \end{bmatrix}$

The inverse does **not** exist since three 0's appear on the left side of the dashed line in the third row.

20. $-2r + 3s = 3$
$r + 2s = -19$

Let $A = \begin{bmatrix} -2 & 3 \\ 1 & 2 \end{bmatrix}$, $X = \begin{bmatrix} r \\ s \end{bmatrix}$, and $B = \begin{bmatrix} 3 \\ -19 \end{bmatrix}$.

Now, $\delta(A) = (-2)(2) - (3)(1) = -4 - 3 = -7$

and $A^{-1} = \frac{1}{-7}\begin{bmatrix} 2 & -3 \\ -1 & -2 \end{bmatrix}$.

Finally, $X = A^{-1}B = -\frac{1}{7}\begin{bmatrix} 2 & -3 \\ -1 & -2 \end{bmatrix}\begin{bmatrix} 3 \\ -19 \end{bmatrix} = -\frac{1}{7}\begin{bmatrix} 63 \\ 35 \end{bmatrix} = \begin{bmatrix} -9 \\ -5 \end{bmatrix}$ which means $r = -9$ and $s = -5$.

21. $4x - y = 2$
$7x - 3y = 1$

Let $A = \begin{bmatrix} 4 & -1 \\ 7 & -3 \end{bmatrix}$, $X = \begin{bmatrix} x \\ y \end{bmatrix}$, and $B = \begin{bmatrix} 2 \\ 1 \end{bmatrix}$.

Now, $\delta(A) = (4)(-3) - (-1)(7) = -12 + 7 = -5$ and $A^{-1} = \frac{1}{-5}\begin{bmatrix} -3 & 1 \\ -7 & 4 \end{bmatrix}$.

Finally, $X = A^{-1}B = -\frac{1}{5}\begin{bmatrix} -3 & 1 \\ -7 & 4 \end{bmatrix}\begin{bmatrix} 2 \\ 1 \end{bmatrix} = -\frac{1}{5}\begin{bmatrix} -5 \\ -10 \end{bmatrix} = \begin{bmatrix} 1 \\ 2 \end{bmatrix}$ which means $x = 1$ and $y = 2$.

22. $2x + y - z = 1$ Let $A = \begin{bmatrix} 2 & 1 & -1 \\ 0 & 1 & -2 \\ 1 & 3 & 0 \end{bmatrix}$, $X = \begin{bmatrix} x \\ y \\ z \end{bmatrix}$, and $B = \begin{bmatrix} 1 \\ 9 \\ 1 \end{bmatrix}$
    $y - 2z = 9$
    $x + 3y - = 1$

For the inverse of A, we have

$$\begin{bmatrix} 2 & 1 & -1 & \vdots & 1 & 0 & 0 \\ 0 & 1 & -2 & \vdots & 0 & 1 & 0 \\ 1 & 3 & 0 & \vdots & 0 & 0 & 1 \end{bmatrix} \xrightarrow{R_1 \leftrightarrow R_3} \begin{bmatrix} 1 & 3 & 0 & \vdots & 0 & 0 & 1 \\ 0 & 1 & -2 & \vdots & 0 & 1 & 0 \\ 2 & 1 & -1 & \vdots & 1 & 0 & 0 \end{bmatrix}$$

$$\xrightarrow{-2R_1 + R_3} \begin{bmatrix} 1 & 3 & 0 & \vdots & 0 & 0 & 1 \\ 0 & 1 & -2 & \vdots & 0 & 1 & 0 \\ 0 & -5 & -1 & \vdots & 1 & 0 & -2 \end{bmatrix} \xrightarrow[5R_2 + R_3]{-3R_2 + R_1} \begin{bmatrix} 1 & 0 & 6 & \vdots & 0 & -3 & 1 \\ 0 & 1 & -2 & \vdots & 0 & 1 & 0 \\ 0 & 0 & -11 & \vdots & 1 & 5 & -2 \end{bmatrix}$$

$$\xrightarrow{-\frac{1}{11}R_3} \begin{bmatrix} 1 & 0 & 6 & \vdots & 0 & -3 & 1 \\ 0 & 1 & -2 & \vdots & 0 & 1 & 0 \\ 0 & 0 & 1 & \vdots & -\frac{1}{11} & -\frac{5}{11} & \frac{2}{11} \end{bmatrix} \xrightarrow[2R_3 + R_2]{-6R_3 + R_1} \begin{bmatrix} 1 & 0 & 0 & \vdots & \frac{6}{11} & -\frac{3}{11} & -\frac{1}{11} \\ 0 & 1 & 0 & \vdots & -\frac{2}{11} & \frac{1}{11} & \frac{4}{11} \\ 0 & 0 & 1 & \vdots & -\frac{1}{11} & -\frac{5}{11} & \frac{2}{11} \end{bmatrix}$$

Thus, $A^{-1} = \begin{bmatrix} \frac{6}{11} & -\frac{3}{11} & -\frac{1}{11} \\ -\frac{2}{11} & \frac{1}{11} & \frac{4}{11} \\ -\frac{1}{11} & -\frac{5}{11} & \frac{2}{11} \end{bmatrix}$ and

$$X = A^{-1}B = \begin{bmatrix} \frac{6}{11} & -\frac{3}{11} & -\frac{1}{11} \\ -\frac{2}{11} & \frac{1}{11} & \frac{4}{11} \\ -\frac{1}{11} & -\frac{5}{11} & \frac{2}{11} \end{bmatrix} \begin{bmatrix} 1 \\ 9 \\ 1 \end{bmatrix} = \begin{bmatrix} -2 \\ 1 \\ -4 \end{bmatrix}$$ so that $x = -2$, $y = 1$, $z = -4$.

23. $4x - 5y - 3z = 8$ Let $A = \begin{bmatrix} 4 & -5 & -3 \\ 3 & -3 & -2 \\ 1 & -1 & -1 \end{bmatrix}$, $X = \begin{bmatrix} x \\ y \\ z \end{bmatrix}$, and $B = \begin{bmatrix} 8 \\ 7 \\ 1 \end{bmatrix}$
    $3x - 3y - 2z = 7$
    $x - y - z = 1$

For the inverse of A, we have

$$\begin{bmatrix} 4 & -5 & -3 & \vdots & 1 & 0 & 0 \\ 3 & -3 & -2 & \vdots & 0 & 1 & 0 \\ 1 & -1 & -1 & \vdots & 0 & 0 & 1 \end{bmatrix} \xrightarrow{R_1 \leftrightarrow R_3} \begin{bmatrix} 1 & -1 & -1 & \vdots & 0 & 0 & 1 \\ 3 & -3 & -2 & \vdots & 0 & 1 & 0 \\ 4 & -5 & -3 & \vdots & 1 & 0 & 0 \end{bmatrix}$$

$$\xrightarrow[-4R_1 + R_3]{-3R_1 + R_2} \begin{bmatrix} 1 & -1 & -1 & \vdots & 0 & 0 & 1 \\ 0 & 0 & 1 & \vdots & 0 & 1 & -3 \\ 0 & -1 & 1 & \vdots & 1 & 0 & -4 \end{bmatrix} \xrightarrow{R_2 \leftrightarrow R_3} \begin{bmatrix} 1 & -1 & -1 & \vdots & 0 & 0 & 1 \\ 0 & -1 & 1 & \vdots & 1 & 0 & -4 \\ 0 & 0 & 1 & \vdots & 0 & 1 & -3 \end{bmatrix}$$

$$\xrightarrow{-R_2} \begin{bmatrix} 1 & -1 & -1 & \vdots & 0 & 0 & 1 \\ 0 & 1 & -1 & \vdots & -1 & 0 & 4 \\ 0 & 0 & 1 & \vdots & 0 & 1 & -3 \end{bmatrix} \xrightarrow{R_2 + R_1} \begin{bmatrix} 1 & 0 & -2 & \vdots & -1 & 0 & 5 \\ 0 & 1 & -1 & \vdots & -1 & 0 & 4 \\ 0 & 0 & 1 & \vdots & 0 & 1 & -3 \end{bmatrix}$$

$$\xrightarrow[R_3 + R_2]{2R_3 + R_1} \begin{bmatrix} 1 & 0 & 0 & \vdots & -1 & 2 & -1 \\ 0 & 1 & 0 & \vdots & -1 & 1 & 1 \\ 0 & 0 & 1 & \vdots & 0 & 1 & -3 \end{bmatrix}$$

Then, $A^{-1} = \begin{bmatrix} -1 & 2 & -1 \\ -1 & 1 & 1 \\ 0 & 1 & -3 \end{bmatrix}$.

Now, $X = A^{-1}B = \begin{bmatrix} -1 & 2 & -1 \\ -1 & 1 & 1 \\ 0 & 1 & -3 \end{bmatrix} \begin{bmatrix} 8 \\ 7 \\ 1 \end{bmatrix} = \begin{bmatrix} 5 \\ 0 \\ 4 \end{bmatrix}$ which means that $x = 5$, $y = 0$, $z = 4$.

24. $\begin{vmatrix} 4 & -2 \\ -1 & -3 \end{vmatrix} = (4)(-3) - (-2)(-1) = -12 - 2 = -14$

25. $\begin{vmatrix} -1 & 2 & 3 \\ 4 & -2 & 1 \\ 1 & 5 & 0 \end{vmatrix} = a_{13}C_{13} + a_{23}C_{23} + a_{33}C_{33}$

$= 3C_{13} + 1C_{23} + 0C_{33}$

$= 3C_{13} + C_{23} = 3(-1)^4 \begin{vmatrix} 4 & -2 \\ 1 & 5 \end{vmatrix} + (-1)^5 \begin{vmatrix} -1 & 2 \\ 1 & 5 \end{vmatrix}$

$= 3[20 - (-2)] + (-1)[-5 - 2]$

$= 3(22) - (-7) = 66 + 7 = 73$

26. $\begin{vmatrix} 2 & 1 & -2 \\ 1 & 5 & 0 \\ 1 & -1 & 3 \end{vmatrix} = C_{13}A_{13} + C_{23}A_{23} + C_{33}A_{33}$

$= -2A_{13} + 0A_{23} + 3A_{33}$

$= -2A_{13} + 3A_{33} = -2(-1)^4 \begin{vmatrix} 1 & 5 \\ 1 & -1 \end{vmatrix} + 3(-1)^6 \begin{vmatrix} 2 & 1 \\ 1 & 5 \end{vmatrix}$

$= -2[(1)(-1) - (5)(1)] + 3[(2)(5) - (1)(1)]$

$= -2(-6) + 3(9) = 12 + 27 = 39$

27. $\begin{vmatrix} 3 & -2 & 1 \\ 4 & 0 & 1 \\ 2 & 5 & -2 \end{vmatrix} = a_{21}C_{21} + a_{22}C_{22} + a_{23}C_{23}$

$= 4C_{21} + 0C_{22} + 1C_{23}$

$= 4C_{21} + C_{23} = 4(-1)^3 \begin{vmatrix} -2 & 1 \\ 5 & -2 \end{vmatrix} + (-1)^5 \begin{vmatrix} 3 & -2 \\ 2 & 5 \end{vmatrix}$

$= 4(-1)[(-2)(-2) - (1)(5)] + (-1)[(3)(5) - (-2)(2)]$

$= 4(-1)[4 - 5] + (-1)[15 - (-4)]$

$= 4(-1)(-1) + (-1)(19) = 4 - 19 = -15$

28. $\begin{vmatrix} -2 & x & -2 \\ 3 & 1 & 0 \\ 1 & 2x & 1 \end{vmatrix} = 0$

$C_{13}A_{13} + C_{23}A_{23} + C_{33}A_{33} = 0$

$-2A_{13} + 0A_{23} + 1A_{33} = 0$

$-2A_{13} + A_{33} = 0$

$-2(-1)^4 \begin{vmatrix} 3 & 1 \\ 1 & 2x \end{vmatrix} + (-1)^6 \begin{vmatrix} -2 & x \\ 3 & 1 \end{vmatrix} = 0$

$-2[(3)(2x) - (1)(1)] + [(-2)(1) - (x)(3)] = 0$

$-2[6x - 1] + [-2 - 3x] = 0$

$-12x + 2 - 2 - 3x = 0$

$-15x + 0 = 0$

$x = 0$

29. $\begin{vmatrix} 2 & 1 & 1 \\ -3 & 5 & 3 \\ 4 & 2 & 1 \end{vmatrix} = \begin{vmatrix} 2 & 1 & 1 \\ -3 & 5 & 3 \\ 0 & 0 & x \end{vmatrix}$

$\downarrow$ use $-2R_1 + R_3$

$\begin{vmatrix} 2 & 1 & 1 \\ -3 & 5 & 3 \\ 0 & 0 & -1 \end{vmatrix} = \begin{vmatrix} 2 & 1 & 1 \\ -3 & 5 & 3 \\ 0 & 0 & x \end{vmatrix}$

and by equality, $x = -1$

CHAPTER 8: MATRICES AND DETERMINANTS

30. $\begin{vmatrix} 1 & -2 & -1 \\ 3 & 0 & 2 \\ 2 & 1 & 1 \end{vmatrix} = x \begin{vmatrix} -1 & -2 & 1 \\ 2 & 0 & 3 \\ 1 & 1 & 2 \end{vmatrix}$

↓ interchange 1st and 3rd columns

$-\begin{vmatrix} -1 & -2 & 1 \\ 2 & 0 & 3 \\ 1 & 1 & 2 \end{vmatrix} = x \begin{vmatrix} -1 & -2 & 1 \\ 2 & 0 & 3 \\ 1 & 1 & 2 \end{vmatrix}$

and by equality, $x = -1$

31. $\begin{vmatrix} -2 & 4 & 8 \\ -4 & 2 & -6 \\ 2 & 2 & -4 \end{vmatrix} = x \begin{vmatrix} -1 & 2 & 4 \\ -2 & 1 & -3 \\ 1 & 1 & -2 \end{vmatrix}$

↓ use multiplication property: Factor out 2 from 1st, 2nd, and 3rd rows (or columns)

$2 \cdot 2 \cdot 2 \begin{vmatrix} -1 & 2 & 4 \\ -2 & 1 & -3 \\ 1 & 1 & -2 \end{vmatrix} = x \begin{vmatrix} -1 & 2 & 4 \\ -2 & 1 & -3 \\ 1 & 1 & -2 \end{vmatrix}$

and by equality, $x = 2 \cdot 2 \cdot 2 = 2^3 = 8$

32. $\begin{vmatrix} -1 & 0 & 2 & 0 \\ 2 & 3 & 1 & 2 \\ 1 & 1 & 1 & 0 \\ -1 & 2 & -1 & 3 \end{vmatrix} \xrightarrow[R_3 + R_4]{-R_3 + R_2} \begin{vmatrix} -1 & 0 & 2 & 0 \\ 1 & 2 & 0 & 2 \\ 1 & 1 & 1 & 0 \\ 0 & 3 & 0 & 3 \end{vmatrix} \xrightarrow{R_1 + R_3} \begin{vmatrix} -1 & 0 & 2 & 0 \\ 1 & 2 & 0 & 2 \\ 0 & 1 & 3 & 0 \\ 0 & 3 & 0 & 3 \end{vmatrix}$

$= C_{11}A_{11} + C_{21}A_{21} + C_{31}A_{31} + C_{41}A_{41}$

$= -1A_{11} + 1A_{21} + 0A_{31} + 0A_{41}$

$= -A_{11} + A_{21}$

$= -1(-1)^2 \begin{vmatrix} 2 & 0 & 2 \\ 1 & 3 & 0 \\ 3 & 0 & 3 \end{vmatrix} + (-1)^3 \begin{vmatrix} 0 & 2 & 0 \\ 1 & 3 & 0 \\ 3 & 0 & 3 \end{vmatrix}$

$= -\begin{vmatrix} 2 & 0 & 2 \\ 1 & 3 & 0 \\ 3 & 0 & 3 \end{vmatrix} - \begin{vmatrix} 0 & 2 & 0 \\ 1 & 3 & 0 \\ 3 & 0 & 3 \end{vmatrix}$

$= -(C_{11}A_{11} + C_{12}A_{12} + C_{13}A_{13}) - (C_{11}B_{11} + C_{12}B_{12} + C_{13}B_{13})$

$= -(2A_{11} + 0A_{12} + 2A_{13}) - (0B_{11} + 2B_{12} + 0B_{13})$

$= -2A_{11} - 2A_{13} - 2B_{12}$

$= -2(-1)^2 \begin{vmatrix} 3 & 0 \\ 0 & 3 \end{vmatrix} - 2(-1)^4 \begin{vmatrix} 1 & 3 \\ 3 & 0 \end{vmatrix} - 2(-1)^3 \begin{vmatrix} 1 & 0 \\ 3 & 3 \end{vmatrix}$

$= -2[(3)(3) - (0)(0)] - 2[(1)(0) - (3)(3)] + 2[(1)(3) - (0)(3)]$

$= -2(9) - 2(-9) + 2(3)$

$= -18 + 18 + 6$

$= 6$

33. $3x - y = -2$ $D = \begin{vmatrix} 3 & -1 \\ 6 & -2 \end{vmatrix} = -6 - (-6) = -6 + 6 = 0$
$6x - 2y = -4$

$D_x = \begin{vmatrix} -2 & -1 \\ -4 & -2 \end{vmatrix} = 4 - (4) = 0$ $D_y = \begin{vmatrix} 3 & -2 \\ 6 & -4 \end{vmatrix} = -12 - (-12) = 0$

In this case, the system is dependent.

34. $5x + 2y = 4$
    $-x + 2y = -8$
    $D = \begin{vmatrix} 5 & 2 \\ -1 & 2 \end{vmatrix} = 10 - (-2) = 10 + 2 = 12$

    $D_x = \begin{vmatrix} 4 & 2 \\ -8 & 2 \end{vmatrix} = 8 - (-16) = 8 + 16 = 24$

    $D_y = \begin{vmatrix} 5 & 4 \\ -1 & -8 \end{vmatrix} = -40 - (-4) = -40 + 4 = -36$

Now, $x = \dfrac{D_x}{D} = \dfrac{24}{12} = 2$ and $y = \dfrac{D_y}{D} = \dfrac{-36}{12} = -3$

35. $x + y + 2z = 3$
    $-x + 2y + z = 9$
    $3x - y + 3z = 0$
    $D = \begin{vmatrix} 1 & 1 & 2 \\ -1 & 2 & 1 \\ 3 & 1 & 3 \end{vmatrix} = -3$

    $D_x = \begin{vmatrix} 3 & 1 & 2 \\ 9 & 2 & 1 \\ 0 & 1 & 3 \end{vmatrix} = 6$
    $D_y = \begin{vmatrix} 1 & 3 & 2 \\ -1 & 9 & 1 \\ 3 & 0 & 3 \end{vmatrix} = -9$

    $D_z = \begin{vmatrix} 1 & 1 & 3 \\ -1 & 2 & 9 \\ 3 & 1 & 0 \end{vmatrix} = -3$
    Now, $x = \dfrac{D_x}{D} = \dfrac{6}{-3} = -2$, $y = \dfrac{D_y}{D} = \dfrac{-9}{-3} = 3$,
    and $z = \dfrac{D_z}{D} = \dfrac{-3}{-3} = 1$

36. $2x - y = 3$
    $x + y - z = 1$
    $3x - 3y + z = 5$
    $D = \begin{vmatrix} 2 & -1 & 0 \\ 1 & 1 & -1 \\ 3 & -3 & 1 \end{vmatrix} = 0$

    $D_x = \begin{vmatrix} 3 & -1 & 0 \\ 1 & 1 & -1 \\ 5 & -3 & 1 \end{vmatrix} = 0$
    $D_y = \begin{vmatrix} 2 & 3 & 0 \\ 1 & 1 & -1 \\ 3 & 5 & 1 \end{vmatrix} = 0$
    $D_z = \begin{vmatrix} 2 & -1 & 3 \\ 1 & 1 & 1 \\ 3 & -3 & 5 \end{vmatrix} = 0$

In this case, the system is dependent.

CHAPTER 9   FURTHER TOPICS

Exercises 9.1

1. $a_n = \dfrac{1}{n+1}$ $a_1 = \dfrac{1}{1+1} = \dfrac{1}{2}$

$a_2 = \dfrac{1}{1+2} = \dfrac{1}{3}$

$a_3 = \dfrac{1}{1+3} = \dfrac{1}{4}$

$a_4 = \dfrac{1}{1+4} = \dfrac{1}{5}$

$a_5 = \dfrac{1}{1+5} = \dfrac{1}{6}$

3. $a_n = \dfrac{n-1}{n}$

$a_1 = \dfrac{1-1}{1} = \dfrac{0}{1} = 0$

$a_2 = \dfrac{2-1}{2} = \dfrac{1}{2}$

$a_3 = \dfrac{3-1}{3} = \dfrac{2}{3}$

$a_4 = \dfrac{4-1}{4} = \dfrac{3}{4}$

$a_5 = \dfrac{5-1}{5} = \dfrac{4}{5}$

5. $a_j = (-2)^j$
$a_1 = (-2)^1 = -2$
$a_2 = (-2)^2 = 4$
$a_3 = (-2)^3 = -8$
$a_4 = (-2)^4 = 16$
$a_5 = (-2)^5 = -32$

7. $a_n = 2$
$a_1 = 2$
$a_2 = 2$
$a_3 = 2$
$a_4 = 2$
$a_5 = 2$

9. $a_i = \dfrac{(-1)^i}{i^2}$

$a_1 = \dfrac{(-1)^1}{1^2} = \dfrac{-1}{1} = -1$

$a_2 = \dfrac{(-1)^2}{2^2} = \dfrac{1}{4}$

$a_3 = \dfrac{(-1)^3}{3^2} = -\dfrac{1}{9}$

$a_4 = \dfrac{(-1)^4}{4^2} = \dfrac{1}{16}$

$a_5 = \dfrac{(-1)^5}{5^2} = -\dfrac{1}{25}$

11. $a_j = \dfrac{j}{2j-1}$

$a_1 = \dfrac{1}{2 \cdot 1 - 1} = \dfrac{1}{2-1} = 1$

$a_2 = \dfrac{2}{2 \cdot 2 - 1} = \dfrac{2}{4-1} = \dfrac{2}{3}$

$a_3 = \dfrac{3}{2 \cdot 3 - 1} = \dfrac{3}{6-1} = \dfrac{3}{5}$

$a_4 = \dfrac{4}{2 \cdot 4 - 1} = \dfrac{4}{8-1} = \dfrac{4}{7}$

$a_5 = \dfrac{5}{2 \cdot 5 - 1} = \dfrac{5}{10-1} = \dfrac{5}{9}$

13. $a_i = 3i - 2$
$a_1 = 3 \cdot 1 - 2 = 3 - 2 = 1$
$a_2 = 3 \cdot 2 - 2 = 6 - 2 = 4$
$a_3 = 3 \cdot 3 - 2 = 9 - 2 = 7$
$a_4 = 3 \cdot 4 - 2 = 12 - 2 = 10$
$a_5 = 3 \cdot 5 - 2 = 15 - 2 = 13$

15. $a_n = (x-1)^n$
$a_1 = (x-1)^1 = x - 1$
$a_2 = (x-1)^2$
$a_3 = (x-1)^3$
$a_4 = (x-1)^4$
$a_5 = (x-1)^5$

17. $a_j = (-1)^j x^{2j}$
$a_1 = (-1)^1 x^{2 \cdot 1} = -x^2$
$a_2 = (-1)^2 x^{2 \cdot 2} = x^4$
$a_3 = (-1)^3 x^{2 \cdot 3} = -x^6$
$a_4 = (-1)^4 x^{2 \cdot 4} = x^8$
$a_5 = (-1)^5 x^{2 \cdot 5} = -x^{10}$

19. $a_n = 2n + 1$
$a_1 = 2 \cdot 1 + 1 = 2 + 1 = 3$
$a_2 = 2 \cdot 2 + 1 = 4 + 1 = 5$
$a_3 = 2 \cdot 3 + 1 = 6 + 1 = 7$
$a_4 = 2 \cdot 4 + 1 = 8 + 1 = 9$

The points are $(1, 3)$, $(2, 5)$, $(3, 7)$, $(4, 9)$, and the graph appears as shown at the right.

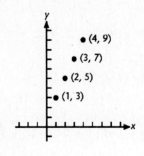

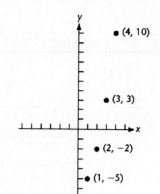

21. $a_n = n^2 - 6$

$a_1 = 1^2 - 6 = 1 - 6 = -5$

$a_2 = 2^2 - 6 = 4 - 6 = -2$

$a_3 = 3^2 - 6 = 9 - 6 = 3$

$a_4 = 4^2 - 6 = 16 - 6 = 10$

The points are $(1, -5)$, $(2, -2)$, $(3, 3)$, $(4, 10)$, and the graph appears as shown at the right.

23. $a_1 = 1$

$a_2 = 2a_1 = 2 \cdot 1 = 2$

$a_3 = 2a_2 = 2 \cdot 2 = 4$

$a_4 = 2a_3 = 2 \cdot 4 = 8$

$a_5 = 2a_4 = 2 \cdot 8 = 16$

$a_6 = 2a_5 = 2 \cdot 16 = 32$

25. $a_1 = -5$

$a_2 = (-1)^1 a_1 = (-1)(-5) = 5$

$a_3 = (-1)^2 a_2 = 1 \cdot 5 = 5$

$a_4 = (-1)^3 a_3 = (-1)5 = -5$

$a_5 = (-1)^4 a_4 = 1(-5) = -5$

$a_6 = (-1)^5 a_5 = (-1)(-5) = 5$

27. $a_1 = 2$

$a_2 = 3$

$a_3 = 2a_2 + a_1 = 2 \cdot 3 + 2 = 6 + 2 = 8$

$a_4 = 2a_3 + a_2 = 2 \cdot 8 + 3 = 16 + 3 = 19$

$a_5 = 2a_4 + a_3 = 2 \cdot 19 + 8 = 38 + 8 = 46$

$a_6 = 2a_5 + a_4 = 2 \cdot 46 + 19 = 92 + 19 = 111$

29. $\displaystyle\sum_{n=1}^{5} 2^n = 2 + 4 + 8 + 16 + 32 = 62$

31. $\displaystyle\sum_{n=3}^{9} \frac{n-2}{n+1} = \frac{1}{4} + \frac{2}{5} + \frac{3}{6} + \frac{4}{7} + \frac{5}{8} + \frac{6}{9} + \frac{7}{10}$

$= \frac{210}{840} + \frac{336}{840} + \frac{420}{840} + \frac{480}{840} + \frac{525}{840} + \frac{560}{840} + \frac{588}{840} = \frac{3119}{840}$

33. $\displaystyle\sum_{j=1}^{7} \frac{j^2 - j}{2} = \frac{0}{2} + \frac{4-2}{2} + \frac{9-3}{2} + \frac{16-4}{2} + \frac{25-5}{2} + \frac{36-6}{2} + \frac{49-7}{2}$

$= 0 + 1 + 3 + 6 + 10 + 15 + 21 = 56$

35. $\displaystyle\sum_{i=1}^{7} (-1)^i = (-1)^1 + (-1)^2 + (-1)^3 + (-1)^4 + (-1)^5 + (-1)^6 + (-1)^7$

$= -1 + 1 - 1 + 1 - 1 + 1 - 1 = -1$

37. $\displaystyle\sum_{n=2}^{7} 2 = 2 + 2 + 2 + 2 + 2 + 2 = 12$

39. $\displaystyle\sum_{j=2}^{7} j^2 (-1)^j = 2^2 (-1)^2 + 3^2 (-1)^3 + 4^2 (-1)^4 + 5^2 (-1)^5 + 6^2 (-1)^6 + 7^2 (-1)^7$

$= 4 - 9 + 16 - 25 + 36 - 49 = -27$

41. $\displaystyle\sum_{i=-1}^{5} i^3 = (-1)^3 + 0^3 + 1^3 + 2^3 + 3^3 + 4^3 + 5^3 = -1 + 0 + 1 + 8 + 27 + 64 + 125 = 224$

43. $\displaystyle\sum_{n=1}^{6}\left(\frac{1}{n}-\frac{1}{n+1}\right)=\left(1-\frac{1}{2}\right)+\left(\frac{1}{2}-\frac{1}{3}\right)+\left(\frac{1}{3}-\frac{1}{4}\right)+\left(\frac{1}{4}-\frac{1}{5}\right)+\left(\frac{1}{5}-\frac{1}{6}\right)+\left(\frac{1}{6}-\frac{1}{7}\right)$

$$=1-\frac{1}{7}=\frac{6}{7}$$

45. $\displaystyle\sum_{n=2}^{5}(-1)^{n}\cdot 2^{-n}=(-1)^{2}\cdot 2^{-2}+(-1)^{3}\cdot 2^{-3}+(-1)^{4}\cdot 2^{-4}+(-1)^{5}\cdot 2^{-5}$

$$=\frac{1}{4}-\frac{1}{8}+\frac{1}{16}-\frac{1}{32}=\frac{8}{32}-\frac{4}{32}+\frac{2}{32}-\frac{1}{32}=\frac{5}{32}$$

47. $\displaystyle\sum_{k=1}^{17}k$ 49. $\displaystyle\sum_{j=1}^{7}2^{j}$ 51. $\displaystyle\sum_{n=1}^{7}(-1)^{n+1}(2n-1)$ 53. $\displaystyle\sum_{i=1}^{5}\frac{i}{i+1}$

55. $\displaystyle\sum_{j=0}^{5}(-1)^{j}\frac{2j+1}{2^{j+1}}$ 57. $\displaystyle\sum_{k=1}^{43}\frac{1}{(2k-1)(2k+1)}$ 59. $\displaystyle\sum_{i=0}^{n}\frac{x^{i}}{i+1}$

61.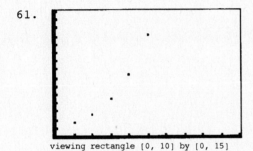

viewing rectangle [0, 10] by [0, 15]

63.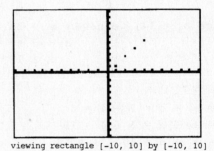

viewing rectangle [-10, 10] by [-10, 10]

65. The two graphs have the same shape. However, the graph of the sequence is a collection of discrete points which are not connected while the graph of the new function is a continuous line.

Exercises 9.2

1. $a_1 = 1$
$a_2 = a_1 + d = 1 + 1 = 2$
$a_3 = a_2 + d = 2 + 1 = 3$
$a_4 = a_3 + d = 3 + 1 = 4$
$a_5 = a_4 + d = 4 + 1 = 5$
$a_6 = a_5 + d = 5 + 1 = 6$

3. $a_4 = 7$ can be expressed as $a_1 + 3d = 7$. But, $d = -2$ which gives:
$a_1 + 3(-2) = 7$
$a_1 - 6 = 7$
$a_1 = 13$
Thus, $a_1 = 13$
$a_2 = a_1 + d = 13 - 2 = 11$
$a_3 = a_2 + d = 11 - 2 = 9$
$a_4 = a_3 + d = 9 - 2 = 7$
$a_5 = a_4 + d = 7 - 2 = 5$
$a_6 = a_5 + d = 5 - 2 = 3$

5. $a_3 = -5$ can be expressed as $a_1 + 2d = -5$
    $a_6 = -17$ $a_1 + 5d = -17$

    Subtract 1st equation from the 2nd: $3d = -12$
                                         $d = -4$

    Also, substitute $d = -4$ into $a_1 + 2d = -5$ to have
    $$a_1 + 2(-4) = -5$$
    $$a_1 - 8 = -5$$
    $$a_1 = 3$$

    Thus, $a_1 = 3$
    $$a_2 = a_1 + d = 3 + (-4) = -1$$
    $$a_3 = a_2 + d = -1 + (-4) = -5$$
    $$a_4 = a_3 + d = -5 + (-4) = -9$$
    $$a_5 = a_4 + d = -9 + (-4) = -13$$
    $$a_6 = a_5 + d = -13 + (-4) = -17$$

7. $a_3 - a_2 = 4$ gives $[a_1 + 2d] - [a_1 + d] = 4$
    $$d = 4$$
    Since $a_4 = 14$, we have $a_1 + 3d = 14$
    $$a_1 + 3(4) = 14$$
    $$a_1 + 12 = 14$$
    $$a_1 = 2$$

    $a_1 = 2$
    $a_2 = a_1 + d = 2 + 4 = 6$
    $a_3 = a_2 + d = 6 + 4 = 10$
    $a_4 = a_3 + d = 10 + 4 = 14$
    $a_5 = a_4 + d = 14 + 4 = 18$
    $a_6 = a_5 + d = 18 + 4 = 22$

9. It is an arithmetic sequence with $a_1 = -17$, $d = -12 - (-17) = -12 + 17 = 5$
    and $a_n = a_1 + (n - 1)d = -17 + (n - 1)(5) = -17 + 5n - 5 = 5n - 22$.

11. It is an arithmetic sequence with $a_1 = -6$, $d = -9 - (-6) = -9 + 6 = -3$ and
    $a_n = a_1 + (n - 1)d = -6 + (n - 1)(-3) = -6 - 3n + 3 = -3n - 3$ or $-3(n + 1)$.

13. Not an arithmetic sequence.

15. It is an arithmetic sequence with $a_1 = 7$, $d = -7 - 7 = -14$, and
    $a_n = a_1 + (n - 1)d = 7 + (n - 1)(-14) = 7 - 14n + 14 = 21 - 14n$.

17. It is an arithmetic sequence with $a_1 = x$, $d = (x + y) - x = y$ and
    $a_n = a_1 + (n - 1)d = x + (n - 1)y = x + ny - y$.

19. Not an arithmetic sequence (unless $x = 0$ or $x = 1$).
    When $x = 0$, then $a_1 = 0$, $d = 0$, and $a_n = 0$.
    When $x = 1$, then $a_1 = 1$, $d = 0$, and $a_n = 1$.

21. It is an arithmetic sequence with $a_1 = x$, $d = 2x - x = x$ and
$a_n = a_1 + (n - 1)d = x + (n - 1)x = x + nx - x = nx$.

23. $a_1 = -5$ and $d = -2$

Now, $a_{11} = a_1 + 10d$　　　$S_{11} = \dfrac{11}{2}(a_1 + a_{11})$

$\qquad = -5 + 10(-2)$　　　　$= \dfrac{11}{2}(-5 - 25)$

$\qquad = -5 - 20$　　　　　　$= \dfrac{11}{2}(-30)$

$\qquad = -25$　　　　　　　　$= 11(-15) = -165$

25. $a_7 = -5$ and $d = 3$

$a_7 = -5$ becomes　　Now $a_{11} = a_1 + 10d$　　$S_{11} = \dfrac{11}{2}(a_1 + a_{11})$

$a_1 + 6d = -5$　　　　　　$= -23 + 10(3)$　　$= \dfrac{11}{2}(-23 + 7)$

$a_1 + 6(3) = -5$　　　　　$= -23 + 30$　　　　$= \dfrac{11}{2}(-16)$

$a_1 + 18 = -5$　　　　　　$= 7$　　　　　　　$= 11(-8)$

$\qquad a_1 = -23$　　　　　　　　　　　　　　$= -88$

27. $a_5 = 5$　becomes　$a_1 + 4d = 5$
$a_9 = -7$　　　　　　$a_1 + 8d = -7$

Subtract 1st equation from the 2nd: $4d = -12$
$\qquad\qquad\qquad\qquad\qquad\qquad\qquad d = -3$

Now, substitute $d = -3$ into $a_1 + 4d = 5$ to get
$\qquad\qquad\qquad a_1 + 4(-3) = 5$
$\qquad\qquad\qquad\quad a_1 - 12 = 5$
$\qquad\qquad\qquad\qquad\quad a_1 = 17$

Thus, $a_{11} = a_1 + 10d$　　　$S_{11} = \dfrac{11}{2}(a_1 + a_{11})$

$\qquad = 17 + 10(-3)$　　　　$= \dfrac{11}{2}(-13 + 17)$

$\qquad = 17 - 30$　　　　　　$= \dfrac{11}{2}(4)$

$\qquad = -13$　　　　　　　　$= 11(2)$
$\qquad\qquad\qquad\qquad\qquad\quad = 22$

29. $a_1 = m$ and $d = -x$

$a_{11} = a_1 + 10d$　　　　$S_{11} = \dfrac{11}{2}(a_1 + a_{11})$

$\qquad = m + 10(-x)$　　　　$= \dfrac{11}{2}(m + m - 10x)$

$\qquad = m - 10x$　　　　　$= \dfrac{11}{2}(2m - 10x)$

$\qquad\qquad\qquad\qquad\qquad\quad = 11(m - 5x)$
$\qquad\qquad\qquad\qquad\qquad\quad = 11m - 55x$

31. $a_5 = x + 2k$ becomes $a_1 + 4d = x + 2k$
     $a_7 = x - 2k$ $a_1 + 6d = x - 2k$

Subtract 1st equation from the 2nd: $2d = -4k$

$d = -2k$

Now, substitute $d = -2k$ into $a_1 + 4d = x + 2k$ to get

$$a_1 + 4(-2k) = x + 2k$$
$$a_1 - 8k = x + 2k$$
$$a_1 = x + 10k$$

Thus, $a_{11} = a_1 + 10d$ $\qquad$ $S_{11} = \dfrac{11}{2}(a_1 + a_{11})$

$\qquad = (x + 10k) + 10(-2k)$ $\qquad = \dfrac{11}{2}[(x + 10k) + (x - 10k)]$

$\qquad = x + 10k - 20k$ $\qquad = \dfrac{11}{2}(2x)$

$\qquad = x - 10k$ $\qquad = 11x$

33. $a_{21} = a_1 + 20d = -7 + 20(3) = -7 + 60 = 53$

35. $a_3 = -7$ can be expressed as $a_1 + 2d = -7$. But, $d = 16$
$$a_1 + 2(16) = -7$$
$$a_1 + 32 = -7$$
$$a_1 = -39$$

Now, $a_4 = a_1 + 3d = -39 + 3(16) = -39 + 48 = 9$

and $S_4 = \dfrac{4}{2}(a_1 + a_4) = 2(-39 + 9) = 2(-30) = -60$

37. Here, $a_1 = -4$ and $d = -1 - (-4) = -1 + 4 = 3$.
Then, $a_{10} = a_1 + 9d = -4 + 9(3) = -4 + 27 = 23$.

39. $a_{17} = -3$ can be expressed as $a_1 + 16d = -3$. But $a_1 = 16$
$$16 + 16d = -3$$
$$16d = -19$$
$$d = -\frac{19}{16}$$

Now, $a_{57} = a_1 + 56d = 16 + 56\left(-\dfrac{19}{16}\right)$

$\qquad = 16 + \dfrac{7(-19)}{2} = \dfrac{32}{2} + \dfrac{7(-19)}{2} = \dfrac{32 + 7(-19)}{2} = -\dfrac{101}{2}$

and $S_{57} = \dfrac{57}{2}(a_1 + a_{57}) = \dfrac{57}{2}\left(16 - \dfrac{101}{2}\right) = \dfrac{57}{2}\left(\dfrac{32}{2} - \dfrac{101}{2}\right)$

$\qquad = \dfrac{57}{2}\left(-\dfrac{69}{2}\right) = -\dfrac{3933}{4}$

41. $\left.\begin{array}{l} S_5 = -30 \\ a_7 = -4 \end{array}\right\}$ can be expressed as $\left\{\begin{array}{l} \dfrac{5}{2}[2a_1 + 4d] = -30 \\ a_1 + 6d = -4 \end{array}\right.$

or $\left.\begin{array}{l} 5a_1 + 10d = -30 \\ a_1 + 6d = -4 \end{array}\right\}$ or $\begin{array}{l} a_1 + 2d = -6 \\ \underline{a_1 + 6d = -4} \end{array}$

subtract: $-4d = -2$

$$d = \dfrac{-2}{-4} = \dfrac{1}{2}$$

Substitute $d = \frac{1}{2}$ into $a_1 + 6d = -4$ to get

$$a_1 + 6\left(\frac{1}{2}\right) = -4$$
$$a_1 + 3 = -4$$
$$a_1 = -7$$

43. $\left.\begin{array}{l} S_7 = 14 \\ a_{10} = -28 \end{array}\right\}$ can be expressed as $\left\{\begin{array}{l} \frac{7}{2}[2a_1 + 6d] = 14 \\ a_1 + 9d = -28 \end{array}\right.$

or $\left\{\begin{array}{l} 7a_1 + 21d = 14 \\ a_1 + 9d = -28 \end{array}\right.$ or $\left\{\begin{array}{l} a_1 + 3d = 2 \\ \underline{a_1 + 9d = -28} \end{array}\right.$

subtract: $-6d = 30$
$$d = \frac{30}{-6} = -5$$

Substitute $d = -5$ into $a_1 + 3d = 2$ to get
$$a_1 + 3(-5) = 2$$
$$a_1 - 15 = 2$$
$$a_1 = 17$$

45. Let $f(i) = 2i - 3$

Then $a_1 = f(5) = 2 \cdot 5 - 3$ $S_7 = \frac{7}{2}(a_1 + a_7)$
$$= 10 - 3$$
$$= 7$$
and $a_7 = f(11) = 2 \cdot 11 - 3$ $= \frac{7}{2}(7 + 19)$
$$= 22 - 3$$
$$= 19$$ $= \frac{7}{2}(26)$
$$= 7(13) = 91$$

47. Let $f(k) = 5 - \frac{k}{3}$

Then $a_1 = f(0) = 5 - \frac{0}{3}$ $S_{11} = \frac{11}{2}(a_1 + a_{11})$
$$= 5$$
and $a_{11} = f(10) = 5 - \frac{10}{3}$ $= \frac{11}{2}\left(5 + \frac{5}{3}\right)$
$$= \frac{15}{3} - \frac{10}{3}$$ $= \frac{11}{2}\left(\frac{15}{3} + \frac{5}{3}\right)$
$$= \frac{5}{3}$$ $= \frac{11}{2}\left(\frac{20}{3}\right)$

$$= 11\left(\frac{10}{3}\right)$$
$$= \frac{110}{3}$$

49. Let $f(n) = 2n$

Then $a_1 = f(1) = 2 \cdot 1$ $S_{50} = \frac{50}{2}(a_1 + a_{50})$
$$= 2$$
and $a_{50} = f(50) = 2 \cdot 50$ $= 25(2 + 100)$
$$= 100$$ $= 25(102)$
$$= 2550$$

51. Let $f(n) = 5 - 2n$

    Then $a_1 = 5 - 2 \cdot 1$
    $= 5 - 2$
    $= 3$
    and $a_{13} = f(13) = 5 - 2 \cdot 13$
    $= 5 - 26 = -21$

    $S_{13} = \dfrac{13}{2}(a_1 + a_{13})$
    $= \dfrac{13}{2}(3 + (-21))$
    $= \dfrac{13}{2}(-18)$
    $= 13(-9) = -117$

53. Let $a_n = an + b$
    Then $a_1 = f(1) = a \cdot 1 + b = a + b$
    $a_2 = f(2) = a \cdot 2 + b = 2a + b$
    $a_3 = f(3) = a \cdot 3 + b = 3a + b$
    $\vdots$
    $a_n = f(n) = a \cdot n + b = na + b$

    Here, $a_1 = a + b$ and $d = a$ thus producing an arithmetic sequence.

55. $a_1 = -15$ and $a_6 = -40$
    Since $a_6 = -40$, we have $a_1 + 5d = -40$
    $-15 + 5d = -40$
    $5d = -25$
    $d = -5$

    Thus, $a_1 = -15$
    $a_2 = a_1 + d = -15 - 5 = -20$
    $a_3 = a_2 + d = -20 - 5 = -25$
    $a_4 = a_3 + d = -25 - 5 = -30$
    $a_5 = a_4 + d = -30 - 5 = -35$
    $a_6 = a_5 + d = -35 - 5 = -40$

57. The first multiple of 5 is 10, the next multiple is 15, etc. The last multiple of 5 is 210.

    Here, $a_1 = 10$, $d = 5$, and $a_n = 210$. But $a_n = 210$ can be written as
    $a_1 + (n - 1)d = 210$
    $10 + (n - 1)5 = 210$
    $10 + 5n - 5 = 210$
    $5n + 5 = 210$
    $5n = 205$
    $n = 41$

    The sum of the 41 terms $(n = 41)$ is $S_{41} = \dfrac{41}{2}(a_1 + a_{41})$
    $= \dfrac{41}{2}(10 + 210)$
    $= \dfrac{41}{2}(220)$
    $= 41(110) = 4510$

59. The first multiple of 5 is 20, the second multiple is 25, etc. The last multiple is 195.

    Here, $a_1 = 20$, $d = 5$, and $a_n = 195$. But $a_n = 195$ can be written as
    $a_1 + (n - 1)d = 195$
    $20 + (n - 1)5 = 195$
    $20 + 5n - 5 = 195$
    $5n + 15 = 195$
    $5n = 180$
    $n = 36$

61. Use $a_1 = 1$, $d = 2$, and $a_{20} = a_1 + 19d$

$$= 1 + 19(2)$$
$$= 1 + 38$$
$$= 39 \text{ pennies}$$

63. Use $a_1 = 1$ and $d = 1$. Let n be the number of rows needed to obtain a sum of 45 cans. The formula $\quad S_n = \frac{n}{2}[2a_1 + (n - 1)d]$

$$\text{becomes} \quad 45 = \frac{n}{2}[2(1) + (n - 1)1]$$

$$45 = \frac{n}{2}[2 + n - 1]$$

$$45 = \frac{n}{2}(n + 1)$$

$$90 = n(n + 1)$$
$$90 = n^2 + n$$
$$0 = n^2 + n - 90$$
$$0 = (n + 10)(n - 9)$$
$$0 = n + 10, \quad 0 = n - 9$$
$$-10 = n, \qquad 9 = n$$

The meaningful value is $n = 9$ (9 rows) and
$$a_9 = a_1 + 8d$$
$$= 1 + 8 \cdot 1$$
$$= 1 + 8 = 9 \text{ cans}$$

65. Use $a_1 = 50$. The top row has 21 logs. We need to find which row this is. That is, we want to find n for which $\quad a_n = 21$

$$a_1 + (n - 1)d = 21$$
$$50 + (n - 1)(-1) = 21$$
$$50 - n + 1 = 21$$
$$-n + 51 = 21$$
$$-n = -30$$
$$n = 30$$

Thus, the 30th row has 21 logs ($a_{30} = 21$) and the sum is
$$S_{30} = \frac{30}{2}(a_1 + a_{30})$$
$$= 15(50 + 21)$$
$$= 15(71)$$
$$= 1065$$

67. Use $a_1 = 34$ and $a_{20} = 110$. The sum is $S_{20} = \frac{20}{2}(a_1 + a_{20})$

$$= 10(34 + 110)$$
$$= 10(144)$$
$$= 1440 \text{ seats}$$

69. You should be able to find several examples of situations where objects are arranged in patterns with numbers that form arithmetic sequences.

Exercises 9.3

1. $a_1 = -1$

$a_2 = a_1 r = (-1)(-2) = 2$

$a_3 = a_1 r^2 = (-1)(-2)^2 = -4$

$a_4 = a_1 r^3 = (-1)(-2)^3 = 8$

$a_5 = a_1 r^4 = (-1)(-2)^4 = -16$

3. $a_1 = 4$

$a_2 = a_1 r = 4\left(\dfrac{1}{4}\right) = 1$

$a_3 = a_1 r^2 = 4\left(\dfrac{1}{4}\right)^2 = 4\left(\dfrac{1}{16}\right) = \dfrac{1}{4}$

$a_4 = a_1 r^3 = 4\left(\dfrac{1}{4}\right)^3 = 4\left(\dfrac{1}{64}\right) = \dfrac{1}{16}$

5. $a_1 = \dfrac{3}{4}$

$a_2 = a_1 r = \left(\dfrac{3}{4}\right)(4) = 3$

$a_3 = a_1 r^2 = \left(\dfrac{3}{4}\right)(4)^2 = \dfrac{3}{4}(16) = 12$

$a_4 = a_1 r^3 = \left(\dfrac{3}{4}\right)(4)^3 = \dfrac{3}{4}(64) = 48$

7. $a_2 = -\dfrac{1}{2}$ and $a_3 = 1$

Here $a_2 = -\dfrac{1}{2}$ $\qquad a_1 r = -\dfrac{1}{2}$

$\quad a_3 = 1 \quad$ becomes $\quad a_1 r^2 = 1$

Divide 2nd equation by 1st: $r = \dfrac{1}{-\dfrac{1}{2}} = -2$

Substitute $r = -2$ into $a_1 r = -\dfrac{1}{2}$ to get

$$a_1(-2) = -\dfrac{1}{2}$$

$$a_1 = \dfrac{1}{4}$$

Thus, $a_1 = \dfrac{1}{4}$

$a_2 = a_1 r = \dfrac{1}{4}(-2) = -\dfrac{2}{4} = -\dfrac{1}{2}$

$a_3 = a_1 r^2 = \dfrac{1}{4}(-2)^2 = \dfrac{1}{4}(4) = 1$

$a_4 = a_1 r^3 = \dfrac{1}{4}(-2)^3 = \dfrac{1}{4}(-8) = -2$

$a_5 = a_1 r^4 = \dfrac{1}{4}(-2)^4 = \dfrac{1}{4}(16) = 4$

9. $a_1 = -3$

Since $a_3 = -12$, this can be written as

$a_1 r^2 = -12$

$-3(r^2) = -12$

$r^2 = 4$

$r = \pm\sqrt{4} = \pm 2$

Two sequences are possible:

For $r = 2$, $a_1 = -3$

$a_2 = a_1 r = -3(2) = -6$

$a_3 = a_1 r^2 = -3(2)^2$

$= -3(4) = -12$

For $r = -2$, $a_1 = -3$

$a_2 = a_1 r = (-3)(-2) = 6$

$a_3 = a_1 r^2 = (-3)(-2)^2$

$= (-3)(4) = -12$

11. $a_1 = \dfrac{1}{4}$ and $r = 2$

    Now, $a_5 = a_1 r^4 = \left(\dfrac{1}{4}\right)(2)^4 = \dfrac{1}{4}(16) = 4$

$$a_n = a_1 r^{n-1} = \left(\dfrac{1}{4}\right)(2)^{n-1} = \dfrac{2^{n-1}}{4} = \dfrac{2^n \cdot 2^{-1}}{4} = \dfrac{2^n}{4 \cdot 2} = \dfrac{2^n}{8}$$

$$S_5 = a_1\left(\dfrac{1-r^5}{1-r}\right) = \dfrac{1}{4}\left(\dfrac{1-2^5}{1-2}\right) = \dfrac{1}{4}\left(\dfrac{1-32}{-1}\right) = \dfrac{1}{4}\left(\dfrac{-31}{-1}\right) = \dfrac{31}{4}$$

13. $a_1 = \dfrac{1}{3}$ and $r = -3$

$$a_5 = a_1 r^4 = \dfrac{1}{3}(-3)^4 = \dfrac{1}{3}(81) = 27$$

$$a_n = a_1 r^{n-1} = \dfrac{1}{3}(-3)^{n-1} = \dfrac{(-3)^{n-1}}{3} = \dfrac{(-3)^n (-3)^{-1}}{3} = \dfrac{(-3)^n}{3(-3)} = -\dfrac{(-3)^n}{9}$$

$$S_5 = a_1\left(\dfrac{1-r^5}{1-r}\right) = \dfrac{1}{3}\left[\dfrac{1-(-3)^5}{1-(-3)}\right] = \dfrac{1}{3}\left[\dfrac{1-(-243)}{4}\right] = \dfrac{1}{3}\left(\dfrac{244}{4}\right) = \dfrac{1}{3}(61) = \dfrac{61}{3}$$

15. $a_3 = -2$ becomes $a_1 r^2 = -2$
    $a_4 = 4$ $a_1 r^3 = 4$

    Divide the 2nd equation by the 1st: $r = \dfrac{4}{-2} = -2$

    Substitute $r = -2$ into $a_1 r^2 = -2$ to get

$$a_1(-2)^2 = -2$$
$$a_1(4) = -2$$
$$a_1 = \dfrac{-2}{4} = -\dfrac{1}{2}$$

    Now, $a_1 = -\dfrac{1}{2}$

$$a_5 = a_1 r^4 = \left(-\dfrac{1}{2}\right)(-2)^4 = \left(-\dfrac{1}{2}\right)(16) = -8$$

$$a_n = a_1 r^{n-1} = \left(-\dfrac{1}{2}\right)(-2)^{n-1} = \dfrac{(-2)^{n-1}}{-2} = \dfrac{(-2)^n(-2)^{-1}}{-2} = \dfrac{(-2)^n}{(-2)(-2)} = \dfrac{(-2)^n}{4}$$

$$S_5 = a_1\left(\dfrac{1-r^5}{1-r}\right) = -\dfrac{1}{2}\left[\dfrac{1-(-2)^5}{1-(-2)}\right] = -\dfrac{1}{2}\left[\dfrac{1-(-32)}{3}\right] = -\dfrac{1}{2}\left(\dfrac{33}{3}\right) = -\dfrac{1}{2}(11) = -\dfrac{11}{2}$$

17. It is a geometric sequence with $a_1 = 7$ and $r = \dfrac{a_2}{a_1} = \dfrac{7/2}{7} = \dfrac{1}{2}$.

    Also, $a_n = a_1 r^{n-1} = 7\left(\dfrac{1}{2}\right)^{n-1} = \dfrac{7}{2^{n-1}} = \dfrac{7}{2^n 2^{-1}} = \dfrac{7 \cdot 2}{2^n} = \dfrac{14}{2^n}$

19. It is a geometric sequence with $a_1 = 4$ and $r = \dfrac{a_2}{a_1} = \dfrac{2\sqrt{2}}{4} = \dfrac{\sqrt{2}}{2}$. Also,

$$a_n = a_1 r^{n-1} = 4\left(\dfrac{\sqrt{2}}{2}\right)^{n-1} = 4\left(\dfrac{1}{\sqrt{2}}\right)^{n-1} = \dfrac{4}{(\sqrt{2})^{n-1}} = \dfrac{4}{(\sqrt{2})^{n-5}(\sqrt{2})^4} = \dfrac{\overset{1}{\cancel{4}}}{(\sqrt{2})^{n-5} \cdot \underset{1}{\cancel{4}}}$$

$$= \dfrac{1}{(\sqrt{2})^{n-5}} = (\sqrt{2})^{5-n}$$

21. Not a geometric sequence.   **23.** Not a geometric sequence.

25. It is a geometric sequence with $a_1 = -4$ and $r = \dfrac{a_2}{a_1} = \dfrac{2}{-4} = -\dfrac{1}{2}$.

Also, $a_n = a_1 r^{n-1} = (-4)\left(-\dfrac{1}{2}\right)^{n-1}$.

27. It is geometric with $a_1 = 343$ and $r = \dfrac{a_2}{a_1} = \dfrac{-49}{343} = -\dfrac{1}{7}$.

Also, $a_n = a_1 r^{n-1} = 343\left(-\dfrac{1}{7}\right)^{n-1} = \dfrac{(-1)^{n-1}\,343}{7^{n-1}} = \dfrac{(-1)^{n-1}\cdot 7^3}{7^{n-1}}$

$$= \dfrac{(-1)^{n-1}}{7^{n-4}}$$

29. $\displaystyle\sum_{i=1}^{5} 2^{i-1} = 2^0 + 2^1 + 2^2 + 2^3 + 2^4$. Here, $a_1 = 2^0 = 1$, $r = 2$, and $n = 5$.

The sum is $S_5 = a_1\left(\dfrac{1-r^5}{1-r}\right) = 1\left(\dfrac{1-2^5}{1-2}\right) = \dfrac{1-32}{-1} = 31$.

31. $\displaystyle\sum_{i=1}^{5}\left(\dfrac{3}{5}\right)^i = \dfrac{3}{5} + \left(\dfrac{3}{5}\right)^2 + \left(\dfrac{3}{5}\right)^3 + \left(\dfrac{3}{5}\right)^4 + \left(\dfrac{3}{5}\right)^5$. Here, $a_1 = \dfrac{3}{5}$, $r = \dfrac{3}{5}$, and $n = 5$.

The sum is $S_5 = a_1\left(\dfrac{1-r^5}{1-r}\right) = \dfrac{3}{5}\left[\dfrac{1-\left(\dfrac{3}{5}\right)^5}{1-\dfrac{3}{5}}\right] = \dfrac{3}{5}\left[\dfrac{1-\dfrac{243}{3125}}{\dfrac{2}{5}}\right] = \dfrac{3}{5}\left[\dfrac{\dfrac{2882}{3125}}{\dfrac{2}{5}}\right]$

$$= \dfrac{3}{5}\left(\dfrac{\overset{1441}{\cancel{2882}}}{3125}\right)\left(\dfrac{\cancel{5}}{\cancel{2}}\right) = \dfrac{4323}{3125}$$

33. $\displaystyle\sum_{n=0}^{4} 5\left(-\dfrac{2}{3}\right)^n = 5\left(-\dfrac{2}{3}\right)^0 + 5\left(-\dfrac{2}{3}\right)^1 + 5\left(-\dfrac{2}{3}\right)^2 + 5\left(-\dfrac{2}{3}\right)^3 + 5\left(-\dfrac{2}{3}\right)^4$. Here, $a_1 = 5\left(-\dfrac{2}{3}\right)^0 = 5\cdot 1 = 5$,

$r = -\dfrac{2}{3}$, and $n = 5$. The sum is $S_5 = a_1\left(\dfrac{1-r^5}{1-r}\right) = 5\left[\dfrac{1-\left(-\dfrac{2}{3}\right)^5}{1-\left(-\dfrac{2}{3}\right)}\right] = 5\left[\dfrac{1-\left(-\dfrac{32}{243}\right)}{1+\dfrac{2}{3}}\right]$

$$= 5\left[\dfrac{1+\dfrac{32}{243}}{\dfrac{5}{3}}\right] = 5\left[\dfrac{\dfrac{275}{243}}{\dfrac{5}{3}}\right] = \cancel{5}\left(\dfrac{275}{\underset{81}{\cancel{243}}}\right)\left(\dfrac{\cancel{3}}{\cancel{5}}\right) = \dfrac{275}{81}$$

35. $\displaystyle\sum_{j=0}^{4} 64\left(\dfrac{5}{4}\right)^j = 64\left(\dfrac{5}{4}\right)^0 + 64\left(\dfrac{5}{4}\right)^1 + 64\left(\dfrac{5}{4}\right)^2 + 64\left(\dfrac{5}{4}\right)^3 + 64\left(\dfrac{5}{4}\right)^4$.

Here, $a_1 = 64\left(\dfrac{5}{4}\right)^0 = 64\cdot 1 = 64$, $r = \dfrac{5}{4}$, and $n = 5$. The sum is

$$S_5 = a_1\left(\frac{1 - r^n}{1 - r}\right) = 64\left[\frac{1 - \left(\frac{5}{4}\right)^5}{1 - \frac{5}{4}}\right] = 64\left[\frac{1 - \frac{3125}{1024}}{-\frac{1}{4}}\right] = 64\left[\frac{\frac{1024}{1024} - \frac{3125}{1024}}{-\frac{1}{4}}\right] = 64\left[\frac{-\frac{2101}{1024}}{-\frac{1}{4}}\right]$$

$$= 64\left(-\frac{2101}{1024}\right)\left(-\frac{4}{1}\right) = \frac{2101}{4}$$

37. $\sum\limits_{j=1}^{6} 8 \cdot 2^{1-j} = 8(2^0) + 8(2^{-1}) + 8(2^{-2}) + 8(2^{-3}) + 8(2^{-4}) + 8(2^{-5})$

$$= 8 + 4 + 2 + 1 + \frac{1}{2} + \frac{1}{4}$$

Here, $a_1 = 8$, $r = \frac{1}{2}$, and $n = 6$. The sum is

$$S_5 = a_1\left(\frac{1 - r^n}{1 - r}\right) = 8\left[\frac{1 - \left(\frac{1}{2}\right)^6}{1 - \frac{1}{2}}\right] = 8\left[\frac{1 - \frac{1}{64}}{\frac{1}{2}}\right] = 8\left[\frac{\frac{64}{64} - \frac{1}{64}}{\frac{1}{2}}\right]$$

$$= 8\left[\frac{\frac{63}{64}}{\frac{1}{2}}\right] = 8\left(\frac{63}{64}\right)\left(\frac{2}{1}\right) = \frac{63}{4}$$

39. $a_7 = -9$ is $a_1 r^6 = -9$

$a_{11} = -81$ is $a_1 r^{10} = -81$

Now, $\dfrac{a_{11}}{a_7}$ produces $\dfrac{a_1 r^{10}}{a_1 r^6} = \dfrac{-81}{-9}$ or $r^4 = 9$ or $r = \pm\sqrt[4]{9} = \pm\sqrt{3}$.

Finally, $a_1 r^6 = -9$ becomes $a_1(\pm\sqrt{3})^6 = -9$

$$a_1(27) = -9$$

$$a_1 = -\frac{9}{27} = -\frac{1}{3}$$

41. $a_6 = 4(1.01)^4$ is $a_1 r^5 = 4(1.01)^4$

$a_8 = 4(1.01)^6$ is $a_1 r^7 = 4(1.01)^6$

Now, $\dfrac{a_8}{a_6}$ produces $\dfrac{a_1 r^7}{a_1 r^5} = \dfrac{4(1.01)^6}{4(1.01)^4}$ or $r^2 = (1.01)^2$ or $r = \pm1.01$.

If $r = 1.01$, then $a_1 r^5 = 4(1.01)^4$ becomes $a_1(1.01)^5 = 4(1.01)^4$

$$\text{or} \qquad a_1 = \frac{4(1.01)^4}{(1.01)^5} = \frac{4}{1.01}$$

If $r = -1.01$, then $a_1 r^5 = 4(1.01)^4$ becomes $a_1(-1.01)^5 = 4(1.01)^4$

$$\text{or} \qquad a_1 = \frac{4(1.01)^4}{-(1.01)^5} = -\frac{4}{1.01}$$

43. The sum is $\$1 + \$2 + \$4 + \$8 + \$16 + \$32 + \$64 + \$128 + \$256$

           ↑   ↑

   1st day   9th day

This is a geometric series with $a_1 = 1$, $r = 2$, and $n = 9$. The sum is

$$S_9 = a_1\left(\frac{1 - r^9}{1 - r}\right) = 1\left(\frac{1 - 2^9}{1 - 2}\right) = \frac{1 - 512}{-1} = \frac{-511}{-1} = \$511.$$

45. Earnings $= \$42,000(2)^6 = \$42,000(2^6) = \$42,000(64) = \$2,688,000$

47. Another real-world situation involving a geometric sequence is to deposit P dollars into a savings account earning interest at a rate of 6% compounded annually. The amount in the account at the end of the first year is

$$P + 0.06P = P(1.06).$$

The amount in the account at the end of the second year is

$$P(1.06) + 0.06[P(1.06)] = P(1.06)(1.06) = P(1.06)^2.$$

Continue in this manner to discover that the amount in the account at the end of the third year is $P(1.06)^3$ and the amount in the account at the end of the nth year is

$$P(1.06)^n.$$

In this case, $r = 1.06$.

Exercises 9.4

1. Here, $a_1 = -3$ and $r = \dfrac{a_2}{a_1} = \dfrac{3/2}{-3} = -\dfrac{1}{2}$ so that the sum exists ($|r| < 1$) and is

$$S = \frac{a_1}{1 - r} = \frac{-3}{1 - \left(-\dfrac{1}{2}\right)} = \frac{-3}{1 + \dfrac{1}{2}} = \frac{-3}{\dfrac{3}{2}} = -3\left(\frac{2}{3}\right) = -2.$$

3. Here, $a_1 = \dfrac{9}{2}$ and $r = \dfrac{a_2}{a_1} = \dfrac{-3/2}{9/2} = -\dfrac{3}{2} \cdot \dfrac{2}{9} = -\dfrac{1}{3}$ so that the sum exists

($|r| < 1$) and $S = \dfrac{a_1}{1 - r} = \dfrac{9/2}{1 - \left(-\dfrac{1}{3}\right)} = \dfrac{9/2}{1 + \dfrac{1}{3}} = \dfrac{9/2}{4/3} = \dfrac{9}{2} \cdot \dfrac{3}{4} = \dfrac{27}{8}.$

5. Here, $a_1 = -\dfrac{4}{3}$ and $r = \dfrac{a_2}{a_1} = \dfrac{4}{-4/3} = 4\left(-\dfrac{3}{4}\right) = -3$ so that the sum does **not** exist since $|r| > 1$.

7. Here, $a_1 = 1$ and $r = \dfrac{a_2}{a_1} = \dfrac{1.01}{1} = 1.01$ so that the sum does **not** exist since $|r| > 1$.

9. Here, $a_1 = 2$ and $r = \dfrac{a_2}{a_1} = \dfrac{\sqrt{2}}{2}$ so that the sum exists ($|r| < 1$) and is

$$S = \frac{a_1}{1 - r} = \frac{2}{1 - \dfrac{\sqrt{2}}{2}} = \frac{2}{\dfrac{2 - \sqrt{2}}{2}} = 2 \cdot \frac{2}{2 - \sqrt{2}} = \frac{4}{2 - \sqrt{2}} \cdot \frac{2 + \sqrt{2}}{2 + \sqrt{2}}$$

$$= \frac{4(2 + \sqrt{2})}{4 - 2} = \frac{4(2 + \sqrt{2})}{2} = 2(2 + \sqrt{2})$$

11. Here, $a_1 = 5$ and $r = \dfrac{a_2}{a_1} = \dfrac{0.5}{5} = 0.1$ so that the sum exists ($|r| < 1$) and is

$$S = \frac{a_1}{1 - r} = \frac{5}{1 - 0.1} = \frac{5}{0.9} = \frac{50}{9}.$$

13. Here, $a_1 = \dfrac{5}{3}$ and $r = \dfrac{a_2}{a_1} = \dfrac{10/9}{5/3} = \dfrac{10}{9} \cdot \dfrac{3}{5} = \dfrac{2}{3}$ so that the sum exists ($|r| < 1$)

and is $S = \dfrac{a_1}{1 - r} = \dfrac{5/3}{1 - \dfrac{2}{3}} = \dfrac{5/3}{1/3} = \dfrac{5}{3} \cdot \dfrac{3}{1} = 5.$

15. Here, $a_1 = 15$ and $r = \dfrac{a_2}{a_1} = \dfrac{15/2}{15} = \dfrac{1}{2}$ and the sum is $S = \dfrac{a_1}{1 - r} = \dfrac{15}{1 - \dfrac{1}{2}} = \dfrac{15}{1/2} = 30$.

17. Here, $a_1 = 4$ and $r = \dfrac{a_2}{a_1} = \dfrac{-4/3}{4} = -\dfrac{1}{3}$ and the sum is

$$S = \dfrac{a_1}{1 - r} = \dfrac{4}{1 - \left(-\dfrac{1}{3}\right)} = \dfrac{4}{4/3} = 4 \cdot \dfrac{3}{4} = 3.$$

19. Here, $a_1 = 1$ and $r = \dfrac{a_2}{a_1} = \dfrac{-1.02}{1} = -1.02$ and the sum does **not** exist since this is a geometric series with $|r| > 1$. That is $r = -1.02$ and $|r| = |-1.02| = 1.02 > 1$.

21. Here, $a_1 = 3$ and $r = \dfrac{a_2}{a_1} = \dfrac{\dfrac{-3}{0.99}}{3} = -\dfrac{3}{0.99} \cdot \dfrac{1}{3} = -\dfrac{1}{0.99} = -1.\overline{01}$ and the sum does **not** exist since this is a geometric series with $|r| > 1$. That is, $r = -1.\overline{01}$ and $|r| = |-1.\overline{01}| = 1.\overline{01} > 1$.

23. Here, $a_1 = 1$ and $r = \dfrac{a_2}{a_1} = \dfrac{64}{512} = \dfrac{1}{8}$ so that the sum exists and is

$$S = \dfrac{a_1}{1 - r} = \dfrac{512}{1 - \dfrac{1}{8}} = \dfrac{512}{7/8} = 512 \cdot \dfrac{8}{7} = \dfrac{4096}{7}.$$

25. Here, $a_1 = \dfrac{1}{36}$ and $r = \dfrac{a_2}{a_1} = \dfrac{1/6}{1/36} = \dfrac{1}{6} \cdot \dfrac{36}{1} = 6$ and the sum does **not** exist since this is a geometric series with $|r| > 1$. That is, $r = 6$ and $|r| = |6| = 6 > 1$.

27. $\displaystyle\sum_{i=1}^{\infty} 5\left(\dfrac{1}{2}\right)^i = 5\left(\dfrac{1}{2}\right)^0 + 5\left(\dfrac{1}{2}\right)^1 + 5\left(\dfrac{1}{2}\right)^2 + \ldots$.

Here, $a_1 = 5\left(\dfrac{1}{2}\right)^0 = 5 \cdot 1 = 5$ and $r = \dfrac{a_2}{a_1} = \dfrac{5(1/2)}{5} = \dfrac{1}{2}$ and the sum is

$$S = \dfrac{a_1}{1 - r} = \dfrac{5}{1 - \dfrac{1}{2}} = \dfrac{5}{1/2} = 10.$$

29. $\displaystyle\sum_{i=3}^{\infty} 17(-3)^i = 17(-3)^3 + 17(-3)^4 + 17(-3)^5 + \ldots$.

Here, $a_1 = 17(-3)^3$ and $r = \dfrac{a_2}{a_1} = \dfrac{17(-3)^4}{17(-3)^3} = -3$ and the sum does **not** exist since this is a geometric series with $|r| > 1$. That is $r = -3$ and $|r| = |-3| = 3 > 1$.

31. $\sum\limits_{i=1}^{\infty} \dfrac{13}{4^i} = \dfrac{13}{4} + \dfrac{13}{4^2} + \dfrac{13}{4^3} + \dots$.

Here, $a_1 = \dfrac{13}{4}$, $r = \dfrac{a_2}{a_1} = \dfrac{13/4^2}{13/4} = \dfrac{13}{4^2} \cdot \dfrac{4}{13} = \dfrac{1}{4}$ and the sum is

$S = \dfrac{a_1}{1-r} = \dfrac{13/4}{1-\frac{1}{4}} = \dfrac{13/4}{3/4} = \dfrac{13}{4} \cdot \dfrac{4}{3} = \dfrac{13}{3}$.

33. $\sum\limits_{n=5}^{\infty} 8\left(-\dfrac{1}{5}\right)^{n-4} = 8\left(-\dfrac{1}{5}\right)^1 + 8\left(-\dfrac{1}{5}\right)^2 + 8\left(-\dfrac{1}{5}\right)^3 + \dots$.

Here, $a_1 = 8\left(-\dfrac{1}{5}\right)^1 = -\dfrac{8}{5}$ and $r = \dfrac{a_2}{a_1} = \dfrac{8\left(-\frac{1}{5}\right)^2}{8\left(-\frac{1}{5}\right)^1} = -\dfrac{1}{5}$ and the sum is

$S = \dfrac{a_1}{1-r} = \dfrac{-8/5}{1-\left(-\frac{1}{5}\right)} = \dfrac{-8/5}{1+\frac{1}{5}} = \dfrac{-8/5}{6/5} = -\dfrac{8}{5} \cdot \dfrac{5}{6} = -\dfrac{8}{6} = -\dfrac{4}{3}$.

35. $\sum\limits_{n=2}^{\infty} 6\left(\dfrac{1}{3}\right)^n = 6\left(\dfrac{1}{3}\right)^2 + 6\left(\dfrac{1}{3}\right)^3 + 6\left(\dfrac{1}{3}\right)^4 + \dots$.

Here, $a_1 = 6\left(\dfrac{1}{3}\right)^2 = 6\left(\dfrac{1}{9}\right) = \dfrac{2}{3}$ and $r = \dfrac{a_2}{a_1} = \dfrac{6\left(\frac{1}{3}\right)^3}{6\left(\frac{1}{3}\right)^2} = \dfrac{1}{3}$ and the sum is

$S = \dfrac{a_1}{1-r} = \dfrac{6\left(\frac{1}{3}\right)^2}{1-\frac{1}{3}} = \dfrac{2/3}{2/3} = 1$.

37. $0.999\dots = 0.9 + 0.09 + 0.009 + \dots$
It is a geometric series with $a_1 = 0.9$ and $r = 0.1$. The sum is
$S = \dfrac{a_1}{1-r} = \dfrac{0.9}{1-0.1} = \dfrac{0.9}{0.9} = 1$.

39. $0.010101\dots = 0.01 + 0.0001 + 0.000001 + \dots$
It is a geometric series with $a_1 = 0.01$ and $r = 0.01$. The sum is
$S = \dfrac{a_1}{1-r} = \dfrac{0.01}{1-0.01} = \dfrac{0.01}{0.99} = \dfrac{1}{99}$.

41. $0.013101310131\dots = 0.0131 + 0.00000131 + 0.000000000131 + \dots$
It is a geometric series with $a_1 = 0.0131$ and $r = 0.0001$. The sum is
$S = \dfrac{a_1}{1-r} = \dfrac{0.0131}{1-0.0001} = \dfrac{0.0131}{0.9999} = \dfrac{131}{9999}$.

43. $3.1111\dots = 3 + 0.1 + 0.01 + 0.001 + \dots$
The part after 3 is a geometric series with $a_1 = 0.1$ and $r = 0.1$.

Thus, $0.1 + 0.01 + 0.001 + \dots = \dfrac{a_1}{1-r} = \dfrac{0.1}{1-0.1} = \dfrac{0.1}{0.9} = \dfrac{1}{9}$.

Thus, $3.11111\dots = 3 + \dfrac{1}{9} = 3\dfrac{1}{9} = \dfrac{28}{9}$.

45. $-2.2917917\ldots = -2.2 - 0.0917 - 0.0000917 - 0.0000000917 - \ldots$
The part after -2.2 is a geometric series with $a_1 = -0.0917$ and $r = 0.001$.

Thus, $-0.0917 - 0.0000917 - \ldots = \dfrac{a_1}{1-r} = \dfrac{-0.0917}{1-0.001} = \dfrac{-0.0917}{.999} = -\dfrac{917}{9990}$.

Thus, $-2.2917917\ldots = -2.2 - \dfrac{917}{9990}$

$$= -2\frac{1}{5} - \frac{917}{9990}$$

$$= -\frac{11}{5} - \frac{917}{9990}$$

$$= -\frac{21978}{9990} - \frac{917}{9990}$$

$$= -\frac{22895}{9990} = -\frac{4579}{1998}$$

47. $10.1343434\ldots = 10.1 + 0.034 + 0.00034 + 0.0000034 + \ldots$
The part after 10.1 is a geometric series with $a_1 = 0.034$ and $r = 0.01$.

Thus, $0.034 + 0.00034 + 0.0000034 + \ldots = \dfrac{a_1}{1-r} = \dfrac{0.034}{1-0.01} = \dfrac{0.034}{0.99} = \dfrac{34}{990}$.

Now, $10.1343434\ldots = 10.1 + \dfrac{34}{990}$

$$= \frac{10.1}{1} \cdot \frac{990}{990} + \frac{34}{990}$$

$$= \frac{9999}{990} + \frac{34}{990} = \frac{10033}{990}$$

49. Distance is $4 + 2\left(4 \cdot \dfrac{3}{7}\right) + 2\left[4 \cdot \left(\dfrac{3}{7}\right)^2\right] + 2\left[4 \cdot \left(\dfrac{3}{7}\right)^3\right] + \ldots$

$\uparrow$
drop   1st bounce   2nd bounce   3rd bounce
        up/down   up/down   up/down

Distance $= 4 + 8\left(\dfrac{3}{7}\right) + 8\left(\dfrac{3}{7}\right)^2 + 8\left(\dfrac{3}{7}\right)^3 + \ldots$

The part after 4 is a geometric series with $a_1 = 8\left(\dfrac{3}{7}\right) = \dfrac{24}{7}$ and $r = \dfrac{3}{7}$. Thus,

$8\left(\dfrac{3}{7}\right) + 8\left(\dfrac{3}{7}\right)^2 + 8\left(\dfrac{3}{7}\right)^3 + \ldots = \dfrac{a_1}{1-r} = \dfrac{24/7}{1 - \dfrac{3}{7}} = \dfrac{24/7}{4/7} = \dfrac{24}{4} = 6$.

Now, Distance $= 4 + 6 = 10$ feet.

51. Number of bacteria $= 4000(2)^6 = 4000(64) = 256{,}000$.

53. When $r = 1$, then $S_n = \underbrace{a_1 + a_1 + a_1 + \ldots + a_1}_{n} = n(a_1)$ and $\lim\limits_{n \to \infty} S_n$ does **not** exist.

When $r = -1$, then $S_n = \underbrace{a_1 - a_1 + a_1 - a_1 + \ldots}_{n}$ which is either a_1 or 0 depending upon n being even or odd and, thus, $\lim\limits_{n \to \infty} S_n$ does **not** exist.

Exercises 9.5

1. $(x + y)^7 = x^7 + \binom{7}{1}x^6y^1 + \binom{7}{2}x^5y^2 + \binom{7}{3}x^4y^3 + \binom{7}{4}x^3y^4 + \binom{7}{5}x^2y^5 + \binom{7}{6}x^1y^6 + y^7$

$\qquad = x^7 + 7x^6y + 21x^5y^2 + 35x^4y^3 + 35x^3y^4 + 21x^2y^5 + 7xy^6 + y^7$

3. $(2x + y^4) = (2x)^4 + \binom{4}{1}(2x)^3y + \binom{4}{2}(2x)^2y^2 + \binom{4}{3}(2x)y^3 + y^4$

$\qquad = 16x^4 + 4(8x^3)y + 6(4x^2)y^2 + 4(2x)y^3 + y^4$

$\qquad = 16x^4 + 32x^3y + 24x^2y^2 + 8xy^3 + y^4$

5. $(x^2 + 2y)^5 = (x^2)^5 + \binom{5}{1}(x^2)^4(2y)^1 + \binom{5}{2}(x^2)^3(2y)^2 + \binom{5}{3}(x^2)^2(2y)^3$

$\qquad\qquad\qquad\qquad\qquad + \binom{5}{4}(x^2)^1(2y)^4 + (2y)^5$

$\qquad = x^{10} + 5(x^8)(2y) + 10(x^6)(4y^2) + 10(x^4)(8y^3) + 5(x^2)(16y^4) + 32y^5$

$\qquad = x^{10} + 10x^8y + 40x^6y^2 + 80x^4y^3 + 80x^2y^4 + 32y^5$

7. $(x^2 - y^2)^4 = [x^2 + (-y^2)^4] = (x^2)^4 + \binom{4}{1}(x^2)^3(-y^2)^1 + \binom{4}{2}(x^2)^2(-y^2)^2$

$\qquad\qquad\qquad\qquad\qquad + \binom{4}{3}(x^2)^1(-y^2)^3 + (-y^2)^4$

$\qquad = x^8 + 4(x^6)(-y^2) + 6(x^4)(y^4) + 4(x^2)(-y^6) + y^8$

$\qquad = x^8 - 4x^6y^2 + 6x^4y^4 - 4x^2y^6 + y^8$

9. $\left(2x - \dfrac{1}{y}\right)^4 = \left[2x + \left(-\dfrac{1}{y}\right)\right]^4 = (2x)^4 + \binom{4}{1}(2x)^3\left(-\dfrac{1}{y}\right)^1 + \binom{4}{2}(2x)^2\left(-\dfrac{1}{y}\right)^2$

$\qquad\qquad\qquad\qquad\qquad + \binom{4}{3}(2x)^1\left(-\dfrac{1}{y}\right)^3 + \left(-\dfrac{1}{y}\right)^4$

$\qquad = 16x^4 + 4(8x^3)\left(-\dfrac{1}{y}\right) + 6(4x^2)\left(\dfrac{1}{y^2}\right) + 4(2x)\left(-\dfrac{1}{y^3}\right) + \dfrac{1}{y^4}$

$\qquad = 16x^4 - \dfrac{32x^3}{y} + \dfrac{24x^2}{y^2} - \dfrac{8x}{y^3} + \dfrac{1}{y^4}$

11. $\left(3x + \dfrac{y}{3}\right)^4 = (3x)^4 + \binom{4}{1}(3x)^3\left(\dfrac{y}{3}\right)^1 + \binom{4}{2}(3x)^2\left(\dfrac{y}{3}\right)^2 + \binom{4}{3}(3x)^1\left(\dfrac{y}{3}\right)^3 + \left(\dfrac{y}{3}\right)^4$

$\qquad = 81x^4 + 4(27x^3)\left(\dfrac{y}{3}\right) + 6(9x^2)\left(\dfrac{y^2}{9}\right) + 4(3x)\left(\dfrac{y^3}{27}\right) + \dfrac{y^4}{81}$

$\qquad = 81x^4 + 36x^3y + 6x^2y^2 + \dfrac{4xy^3}{9} + \dfrac{y^4}{81}$

13. $\left(x^3 - \dfrac{x^2}{2}\right)^6 = \left[x^3 + \left(-\dfrac{x^2}{2}\right)\right]^6 = (x^3)^6 + \binom{6}{1}(x^3)^5\left(-\dfrac{x^2}{2}\right)^1 + \binom{6}{2}(x^3)^4\left(-\dfrac{x^2}{2}\right)^2 + \binom{6}{3}(x^3)^3\left(-\dfrac{x^2}{2}\right)^3$

$\qquad\qquad\qquad\qquad\qquad + \binom{6}{4}(x^3)^2\left(-\dfrac{x^2}{2}\right)^4 + \binom{6}{5}(x^3)^1\left(-\dfrac{x^2}{2}\right)^5 + \left(-\dfrac{x^2}{2}\right)^6$

$$= x^{18} + 6(x^{15})\left(-\frac{x^2}{2}\right) + 15(x^{12})\left(\frac{x^4}{4}\right) + 20(x^9)\left(-\frac{x^6}{8}\right)$$

$$+ 15(x^6)\left(\frac{x^8}{16}\right) + 6(x^3)\left(-\frac{x^{10}}{32}\right) + \frac{x^{12}}{64}$$

$$= x^{18} - 3x^{17} + \frac{15}{4}x^{16} - \frac{5}{2}x^{15} + \frac{15}{16}x^{14} - \frac{3}{16}x^{13} + \frac{1}{64}x^{12}$$

15. $\left(\frac{1}{2}a - 3b^2\right)^5 = \left[\frac{1}{2}a + (-3b^2)\right]^5$

$$= \left(\frac{1}{2}a\right)^5 + \binom{5}{1}\left(\frac{1}{2}a\right)^4(-3b^2) + \binom{5}{2}\left(\frac{1}{2}a\right)^3(-3b^2)^2$$

$$+ \binom{5}{3}\left(\frac{1}{2}a\right)^2(-3b^2)^3 + \binom{5}{4}\left(\frac{1}{2}a\right)^1(-3b^2)^4 + (-3b^2)^5$$

$$= \frac{a^5}{32} + 5\left(\frac{a^4}{16}\right)(-3b^2) + 10\left(\frac{a^3}{8}\right)(9b^4) + 10\left(\frac{a^2}{4}\right)(-27b^6) + 5\left(\frac{1}{2}a\right)(81b^8) - 243b^{10}$$

$$= \frac{1}{32}a^5 - \frac{15}{16}a^4b^2 + \frac{45}{4}a^3b^4 - \frac{135}{2}a^2b^6 + \frac{405}{2}ab^8 - 243b^{10}$$

17. coefficient is $\binom{10}{2} = \frac{10!}{2!8!} = \frac{10 \cdot 9 \cdot 8!}{2 \cdot 1 \cdot 8!} = \frac{\overset{5}{\cancel{10}} \cdot 9}{\cancel{2} \cdot 1} = 45$

19. coefficient is $\binom{5}{3}(2)^3 = \frac{5!}{3!2!}2^3 = \frac{5 \cdot 4 \cdot 3!}{3! \cdot 2 \cdot 1}2^3 = \frac{5 \cdot \overset{2}{\cancel{4}}}{\cancel{2} \cdot 1} \cdot 8 = 10 \cdot 8 = 80$

21. 10th term: $\binom{12}{9}(2x)^3(y)^9 = \frac{12!}{9!3!}(8x^3)(y^9) = \frac{\overset{4}{\cancel{12}} \cdot 11 \cdot \overset{5}{\cancel{10}} \cdot 9!}{9! \cdot 3 \cdot 2 \cdot 1}(8x^3y^9)$

$$= 220(8x^3y^9) = 1760x^3y^9$$

23. 4th term: $\binom{15}{3}(x)^{12}(-4y^2)^3 = \frac{15!}{3!12!}(x^{12})(-64y^6) = \frac{\overset{5}{\cancel{15}} \cdot \overset{7}{\cancel{14}} \cdot 13 \cdot 12!}{3 \cdot 2 \cdot 1 \cdot 12!}(-64x^{12}y^6)$

$$= 455(-64x^{12}y^6) = -29,120x^{12}y^6$$

25. 9th term: $\binom{14}{8}(p^2)^6(q^3)^8 = \frac{14!}{8!6!}(p^{12})(q^{24}) = \frac{\overset{7}{\cancel{14}} \cdot 13 \cdot \cancel{12} \cdot 11 \cdot \cancel{10} \cdot \overset{3}{\cancel{9}} \cdot 8!}{8! \cdot \cancel{6} \cdot \cancel{5} \cdot \cancel{4} \cdot \cancel{3} \cdot \cancel{2} \cdot 1}(p^{12}q^{24})$

$$= 3003p^{12}q^{24}$$

27. 14th term (or last term): $(2y)^{13} = 2^{13}y^{13} = 8192y^{13}$

29. $(1 + x)^{1/4} = 1 + \frac{1}{4}x + \frac{\left(\frac{1}{4}\right)\left(\frac{1}{4} - 1\right)}{2!}x^2 + \frac{\left(\frac{1}{4}\right)\left(\frac{1}{4} - 1\right)\left(\frac{1}{4} - 2\right)}{3!}x^3$

$$+ \frac{\left(\frac{1}{4}\right)\left(\frac{1}{4} - 1\right)\left(\frac{1}{4} - 2\right)\left(\frac{1}{4} - 3\right)}{4!}x^4 + \ldots$$

$$= 1 + \frac{1}{4}x + \frac{\frac{1}{4}\left(-\frac{3}{4}\right)}{2}x^2 + \frac{\frac{1}{4}\left(-\frac{3}{4}\right)\left(-\frac{7}{4}\right)}{3!}x^3 + \frac{\frac{1}{4}\left(-\frac{3}{4}\right)\left(-\frac{7}{4}\right)\left(-\frac{11}{4}\right)}{4!}x^4 + \dots$$

$$= 1 + \frac{1}{4}x - \frac{3}{32}x^2 + \frac{7}{128}x^3 - \frac{77}{2048}x^4 + \dots \text{ for } |x| < 1 \text{ or } -1 < x < 1$$

31. $(1 - y)^{2/3} = [1 + (-y)]^{2/3} = 1 + \frac{2}{3}(-y) + \frac{\frac{2}{3}\left(\frac{2}{3} - 1\right)}{2!}(-y)^2 + \frac{\frac{2}{3}\left(\frac{2}{3} - 1\right)\left(\frac{2}{3} - 2\right)}{3!}(-y)^3$

$$+ \frac{\frac{2}{3}\left(\frac{2}{3} - 1\right)\left(\frac{2}{3} - 2\right)\left(\frac{2}{3} - 3\right)}{4!}(-y)^4 + \dots$$

$$= 1 - \frac{2}{3}y + \frac{\frac{2}{3}\left(-\frac{1}{3}\right)}{2}y^2 + \frac{\frac{2}{3}\left(-\frac{1}{3}\right)\left(-\frac{4}{3}\right)}{6}(-y^3) + \frac{\frac{2}{3}\left(-\frac{1}{3}\right)\left(-\frac{4}{3}\right)\left(-\frac{7}{3}\right)}{24}(y^4) + \dots$$

$$= 1 - \frac{2}{3}y - \frac{1}{9}y^2 - \frac{4}{81}y^3 - \frac{7}{243}y^4 + \dots \text{ for } |y| < 1 \text{ or } -1 < y < 1$$

33. $(1 + 2x)^{-1/2} = 1 - \frac{1}{2}(2x) + \frac{\left(-\frac{1}{2}\right)\left(-\frac{1}{2} - 1\right)}{2!}(2x)^2 + \frac{\left(-\frac{1}{2}\right)\left(-\frac{1}{2} - 1\right)\left(-\frac{1}{2} - 2\right)}{3!}(2x)^3$

$$+ \frac{\left(-\frac{1}{2}\right)\left(-\frac{1}{2} - 1\right)\left(-\frac{1}{2} - 2\right)\left(-\frac{1}{2} - 3\right)}{4!}(2x)^4 + \dots$$

$$= 1 - x + \frac{\left(-\frac{1}{2}\right)\left(-\frac{3}{2}\right)}{2}(4x^2) + \frac{\left(-\frac{1}{2}\right)\left(-\frac{3}{2}\right)\left(-\frac{5}{2}\right)}{6}(8x^3)$$

$$+ \frac{\left(-\frac{1}{2}\right)\left(-\frac{3}{2}\right)\left(-\frac{5}{2}\right)\left(-\frac{7}{2}\right)}{24}(16x^4) + \dots$$

$$= 1 - x + \frac{3}{2}x^2 - \frac{5}{2}x^3 + \frac{35}{8}x^4 + \dots \text{ for } |2x| < 1 \text{ or } -\frac{1}{2} < x < \frac{1}{2}$$

35. $(1 + x^2)^{-2} = 1 - 2(x^2) + \frac{(-2)(-3)}{2!}(x^2)^2 + \frac{(-2)(-3)(-4)}{3!}(x^2)^3$

$$+ \frac{(-2)(-3)(-4)(-5)}{4!}(x^2)^4 + \dots$$

$$= 1 - 2x^2 + \frac{6}{2}(x^4) + \frac{(-24)}{6}(x^6) + \frac{120}{24}(x^8) + \dots$$

$$= 1 - 2x^2 + 3x^4 - 4x^6 + 5x^8 + \dots \text{ for } |x^2| < 1 \text{ or } -1 < x < 1$$

37. $\left(1 - \frac{a}{2}\right)^{-2} = 1 - 2\left(-\frac{a}{2}\right) + \frac{(-2)(-3)}{2!}\left(-\frac{a}{2}\right)^2 + \frac{(-2)(-3)(-4)}{3!}\left(-\frac{a}{2}\right)^3$

$$+ \frac{(-2)(-3)(-4)(-5)}{4!}\left(-\frac{a}{2}\right)^4 + \dots$$

$$= 1 + a + \frac{6}{2}\left(\frac{a}{4}\right) + \frac{(-24)}{6}\left(-\frac{a^3}{8}\right) + \frac{120}{24}\left(\frac{a^4}{16}\right) + \dots$$

$$= 1 + a + \frac{3}{4}a^2 + \frac{1}{2}a^3 + \frac{5}{16}a^4 + \dots \text{ for } \left|\frac{a}{2}\right| < 1 \text{ or } -2 < a < 2$$

39. $(x + y + 1)^4 = (x + y)^4 + \binom{4}{1}(x + y)^3(1)^1 + \binom{4}{2}(x + y)^2(1)^2 + \binom{4}{3}(x + y)^1(1)^3 + 1^4$

$= x^4 + 4x^3y + 6x^2y^2 + 4xy^3 + y^4 + 4x^3 + 12x^2y + 12xy^2 + 4y^3 + 6x^2$
$\qquad + 12xy + 6y^2 + 4x + 4y + 1$

41. $(x - y + z - w)^3 = [(x - y) + (z - w)]^3$

$= (x - y)^3 + \binom{3}{1}(x - y)^2(z - w) + \binom{3}{2}(x - y)(z - w)^2 + (z - w)^3$

$= x^3 - 3x^2y + 3xy^2 - y^3 + 3(x^2 - 2xy + y^2)(z - w)$
$\qquad + 3(x - y)(z^2 - 2zw + w^2) + z^3 - 3z^2w + 3zw^2 - w^3$

$= x^3 - 3x^2y + 3xy^2 - y^3 + 3x^2z - 3x^2w - 6xyz + 6xyw + 3y^2z - 3y^2w$
$\qquad + 3xz^2 - 6xzw + 3xw^2 - 3yz^2 + 6yzw - 3yw^2 + z^3 - 3z^2w$
$\qquad + 3zw^2 - w^3$

43. $(1.01)^6 = (1 + 0.01)^6 \approx 1 + \binom{6}{1}(1)^6(0.01)^1 + \binom{6}{2}(1)^5(0.01)^2 + \binom{6}{3}(1)^4(0.01)^3$

$= 1 + 6(0.01) + 15(0.0001) + 20(0.000001)$
$= 1 + 0.06 + 0.0015 + 0.000020$
$= 1.061520$

45. $(0.99)^8 = [1 + (-0.01)]^8 \approx 1 + \binom{8}{1}(1)^7(-0.01)^1 + \binom{8}{2}(1)^6(-0.01)^2 + \binom{8}{3}(1)^5(-0.01)^3$

$= 1 + 8(-0.01) + 28(0.0001) + 56(-0.000001)$
$= 1 - 0.08 + 0.0028 - 0.000056 = 0.922744$

47. $(3.01)^3 = (3 + 0.01)^3 = 3^3 + \binom{3}{1}(3)^2(0.01)^1 + \binom{3}{2}(3)^1(0.01)^2 + (0.01)^3$

$= 27 + 3(9)(0.01) + 3(3)(0.0001) + (0.000001)$
$= 27 + 0.27 + 0.0009 + 0.000001$
$= 27.270901$

49. $\sqrt{1.02} = (1.02)^{1/2} = (1 + 0.02)^{1/2}$

$\approx 1 + \frac{1}{2}(0.02) + \frac{\left(\frac{1}{2}\right)\left(\frac{1}{2} - 1\right)}{2!}(0.02)^2 + \frac{\left(\frac{1}{2}\right)\left(\frac{1}{2} - 1\right)\left(\frac{1}{2} - 2\right)}{3!}(0.02)^3$

$= 1 + 0.01 + \frac{\frac{1}{2}\left(-\frac{1}{2}\right)}{2}(0.0004) + \frac{\left(\frac{1}{2}\right)\left(-\frac{1}{2}\right)\left(-\frac{3}{2}\right)}{6}(0.000008)$

$= 1 + 0.01 - 0.00005 + 0.0000005$
$= 1.0099505$

51. Using the graphics calculator, $\binom{42}{11} = 4,280,561,376$.

53. Using the graphics calculator, $\binom{105}{99} = 1,609,344,100$.

55. One such case is $n = 1$ and $m \geq 1$. Here, $\frac{m}{n} = \frac{m}{1} = m$ and $\left(\frac{m}{n}\right)! = m!$.

Also, $\frac{m!}{n!} = \frac{m!}{1!} = m!$ and we see that

$$\left(\frac{m}{n}\right)! = \frac{m!}{n!}$$

Observe that the case $n = 0$ must be excluded since $\frac{m}{n}$ is **not** defined when $m \geq 1$ and $n = 0$.

Exercises 9.6

1. $1 + 3 + 5 + \ldots + (2n - 1) = n^2$

    For $n = 1$, the left side is 1 and the right side is $1^2 = 1$.

    Assume true for $n = k$: $1 + 3 + 5 + \ldots + (2k - 1) = k^2$

    Prove true for $n = k + 1$: $1 + 3 + 5 + \ldots + (2k - 1) + [2(k + 1) - 1] = (k + 1)^2$

    Proof: $\underbrace{1 + 3 + 5 + \ldots + (2k - 1)}_{\text{By assumption}} + [2(k + 1) - 1]$

    $$= k^2 + [2(k + 1) - 1]$$
    $$= k^2 + 2k + 2 - 1$$
    $$= k^2 + 2k + 1$$
    $$= (k + 1)^2 \text{ which is the right side}$$

3. $1^2 + 2^2 + 3^2 + \ldots + n^2 = \dfrac{n(n + 1)(2n + 1)}{6}$

    For $n = 1$, the left side is $1^2 = 1$ and the right side is
    $$\frac{1(1 + 1)(2 \cdot 1 + 1)}{6} = \frac{1(2)(3)}{6} = \frac{6}{6} = 1.$$

    Assume true for $n = k$: $1^2 + 2^2 + 3^2 + \ldots + k^2 = \dfrac{k(k + 1)(2k + 1)}{6}$

    Prove true for $n = k + 1$: $1^2 + 2^2 + 3^2 + \ldots + k^2 + (k + 1)^2$
    $$= \frac{(k + 1)(k + 2)(2k + 3)}{6}$$

    Proof: $\underbrace{1^2 + 2^2 + 3^2 + \ldots + k^2}_{\text{By assumption}} + (k + 1)^2$

    $$= \frac{k(k + 1)(2k + 1)}{6} + (k + 1)^2 = \frac{k(k + 1)(2k + 1)}{6} + \frac{6(k + 1)^2}{6}$$
    $$= \frac{k(k + 1)(2k + 1) + 6(k + 1)^2}{6} \quad (k + 1) \text{ is common factor:}$$
    $$= \frac{(k + 1)[k(2k + 1) + 6(k + 1)]}{6}$$
    $$= \frac{(k + 1)[2k^2 + 7k + 6]}{6}$$
    $$= \frac{(k + 1)(k + 2)(2k + 3)}{6} \text{ which is the right side}$$

5. $1 \cdot 2 + 2 \cdot 2^2 + 3 \cdot 2^3 + \ldots + n \cdot 2^n = (n - 1)2^{n+1} + 2$

For $n = 1$, the left side is $1 \cdot 2 = 2$ and the right side is
$$(1 - 1)2^2 + 2 = 0 + 2 = 2.$$
Assume true for $n = k$: $1 \cdot 2 + 2 \cdot 2^2 + 3 \cdot 2^3 + \ldots + k \cdot 2^k = (k - 1)2^{k+1} + 2$

Prove true for $n = k + 1$: $1 \cdot 2 + 2 \cdot 2^2 + 3 \cdot 2^3 + \ldots + k \cdot 2^k + (k + 1)2^{k+1}$
$$= k \cdot 2^{k+2} + 2$$

Proof: $\underbrace{1 \cdot 2 + 2 \cdot 2^2 + 3 \cdot 2^3 + \ldots + k \cdot 2^k}_{\text{By assumption}} + (k + 1)2^{k+1}$

$= (k - 1)2^{k+1} + 2 + (k + 1)2^{k+1}$

$= k \cdot 2^{k+1} - 2^{k+1} + 2 + k \cdot 2^{k+1} + 2^{k+1}$

$= k \cdot 2^{k+1} + 2 + k \cdot 2^{k+1} = 2 \cdot k \cdot 2^{k+1} + 2$

$= k \cdot 2^{k+2} + 2$ or $k2^{k+2} + 2$ which is the right side

7. $4 + 4^2 + 4^3 + \ldots + 4^n = \dfrac{4(4^n - 1)}{3}$

For $n = 1$, the left side is 4 and the right side is $\dfrac{4(4^1 - 1)}{3} = \dfrac{4(3)}{3} = 4.$

Assume true for $n = k$: $4 + 4^2 + 4^3 + \ldots + 4^k = \dfrac{4(4^k - 1)}{3}$

Prove true for $n = k + 1$: $4 + 4^2 + 4^3 + \ldots + 4^k + 4^{k+1} = \dfrac{4(4^{k+1} - 1)}{3}$

Proof: $\underbrace{4 + 4^2 + 4^3 + \ldots + 4^k}_{\text{By assumption}} + 4^{k+1}$

$= \dfrac{4(4^k - 1)}{3} + 4^{k+1}$

$= \dfrac{4(4^k - 1)}{3} + \dfrac{3 \cdot 4^{k+1}}{3} = \dfrac{4(4^k - 1) + 3 \cdot 4^{k+1}}{3}$

$= \dfrac{4^{k+1} - 4 + 3 \cdot 4^{k+1}}{3}$

$= \dfrac{4 \cdot 4^{k+1} - 4}{3} = \dfrac{4(4^{k+1} - 1)}{3}$ which is the right side

9. $3 + 6 + 9 + \ldots + 3n = \dfrac{3n(n + 1)}{2}$

For $n = 1$, the left side is 3 and the right side is $\dfrac{3(1)(1 + 1)}{2} = \dfrac{3(1)(2)}{2} = 3.$

Assume true for $n = k$: $3 + 6 + 9 + \ldots + 3k = \dfrac{3k(k + 1)}{2}$

Prove for $n = k + 1$: $3 + 6 + 9 + 3k + 3(k + 1) = \dfrac{3(k + 1)(k + 2)}{2}$

Proof: $\underbrace{3 + 6 + 9 + \ldots + 3k}_{\text{By assumption}} + 3(k + 1)$

$= \dfrac{3k(k + 1)}{2} + 3(k + 1)$

$$= \frac{3k(k+1)}{2} + \frac{2 \cdot 3(k+1)}{2} = \frac{3k(k+1) + 6(k+1)}{2}$$

$$= \frac{3(k+1)(k+2)}{2} \text{ which is the right side}$$

11. $\dfrac{1}{1 \cdot 2 \cdot 3} + \dfrac{1}{2 \cdot 3 \cdot 4} + \dfrac{1}{3 \cdot 4 \cdot 5} + \dots + \dfrac{1}{n(n+1)(n+2)} = \dfrac{n(n+3)}{4(n+1)(n+2)}$

For $n = 1$, the left side is $\dfrac{1}{1 \cdot 2 \cdot 3} = \dfrac{1}{6}$ and the right side is

$$\frac{1(1+3)}{4(1+1)(1+2)} = \frac{1(4)}{4(2)(3)} = \frac{1}{6}.$$

Assume true for $n = k$: $\dfrac{1}{1 \cdot 2 \cdot 3} + \dfrac{1}{2 \cdot 3 \cdot 4} + \dots + \dfrac{1}{k(k+1)(k+2)}$

$$= \frac{k(k+3)}{4(k+1)(k+2)}$$

Prove true for $n = k + 1$: $\dfrac{1}{1 \cdot 2 \cdot 3} + \dfrac{1}{2 \cdot 3 \cdot 4} + \dots + \dfrac{1}{k(k+1)(k+2)}$

$$+ \frac{1}{(k+1)(k+2)(k+3)} = \frac{(k+1)(k+4)}{4(k+2)(k+3)}$$

Proof: $\dfrac{1}{1 \cdot 2 \cdot 3} + \dfrac{1}{2 \cdot 3 \cdot 4} + \dots + \underbrace{\dfrac{1}{k(k+1)(k+2)}}_{\text{By assumption}} + \dfrac{1}{(k+1)(k+2)(k+3)}$

$$= \frac{k(k+3)}{4(k+1)(k+2)} + \frac{1}{(k+1)(k+2)(k+3)}$$

$$= \frac{k(k+3)(k+3)}{4(k+1)(k+2)(k+3)} + \frac{4}{4(k+1)(k+2)(k+3)}$$

$$= \frac{k^3 + 6k^2 + 9k + 4}{4(k+1)(k+2)(k+3)} = \frac{\cancel{(k+1)}(k+1)(k+4)}{4\cancel{(k+1)}(k+2)(k+3)} \quad \leftarrow \text{Factor numerator}$$

$$= \frac{(k+1)(k+4)}{4(k+2)(k+3)} \text{ which is the right side}$$

13. $\dfrac{1}{3} + \dfrac{1}{3^2} + \dfrac{1}{3^3} + \dots + \dfrac{1}{3^n} = \dfrac{1}{2}\left[1 - \left(\dfrac{1}{3}\right)^n\right].$

For $n = 1$, the left side is $\dfrac{1}{3}$ and the right side is $\dfrac{1}{2}\left[1 - \dfrac{1}{3}\right] = \dfrac{1}{2}\left[\dfrac{2}{3}\right] = \dfrac{1}{3}.$

Assume true for $n = k$: $\dfrac{1}{3} + \dfrac{1}{3^2} + \dfrac{1}{3^3} + \dots + \dfrac{1}{3^k} = \dfrac{1}{2}\left[1 - \left(\dfrac{1}{3}\right)^k\right]$

Prove true for $n = k + 1$: $\dfrac{1}{3} + \dfrac{1}{3^2} + \dfrac{1}{3^3} + \dots + \dfrac{1}{3^k} + \dfrac{1}{3^{k+1}} = \dfrac{1}{2}\left[1 - \left(\dfrac{1}{3}\right)^{k+1}\right]$

Proof: $\underbrace{\dfrac{1}{3} + \dfrac{1}{3^2} + \dfrac{1}{3^3} + \dots + \dfrac{1}{3^k}}_{\text{By assumption}} + \dfrac{1}{3^{k+1}}$

$$= \frac{1}{2}\left[1 - \left(\frac{1}{3}\right)^k\right] + \frac{1}{3^{k+1}}$$

$$= \frac{1}{2} - \frac{1}{2}\left(\frac{1}{3}\right)^k + \left(\frac{1}{3}\right)^{k+1}$$

$$= \frac{1}{2} - \frac{1}{2} \cdot 3 \cdot \frac{1}{3}\left(\frac{1}{3}\right)^k + \left(\frac{1}{3}\right)^{k+1}$$

$$= \frac{1}{2} - \frac{3}{2}\left(\frac{1}{3}\right)^{k+1} + \left(\frac{1}{3}\right)^{k+1} = \frac{1}{2} - \frac{1}{2}\left(\frac{1}{3}\right)^{k+1} = \frac{1}{2}\left[1 - \left(\frac{1}{3}\right)^{k+1}\right]$$ which is the right side

15. $a + ar + ar^2 + \ldots + ar^{n-1} = a\left(\frac{1 - r^n}{1 - r}\right)$

For $n = 1$, the left side is a and the right side is $a\left(\frac{1 - r}{1 - r}\right) = a$.

Assume true for $n = k$: $a + ar + ar^2 + \ldots + ar^{k-1} = a\left(\frac{1 - r^k}{1 - r}\right)$

Prove true for $n = k + 1$: $a + ar + ar^2 + \ldots + ar^{k-1} + ar^k = a\left(\frac{1 - r^{k+1}}{1 - r}\right)$

Proof: $\underbrace{a + ar + ar^2 + \ldots + ar^{k-1}}_{\text{By assumption}} + ar^k$

$$= a\left(\frac{1 - r^k}{1 - r}\right) + ar^k$$

$$= a\left(\frac{1 - r^k}{1 - r}\right) + \frac{ar^k(1 - r)}{1 - r}$$

$$= \frac{a(1 - r^k) + ar^k(1 - r)}{1 - r} = \frac{a - ar^k + ar^k - ar^{k+1}}{1 - r}$$

$$= \frac{a - ar^{k+1}}{1 - r} = \frac{a(1 - r^{k+1})}{1 - r}$$ which is the right side

17. 3 is a factor of $n^3 + 2n$.
   For $n = 1$, $n^3 + 2n$ becomes $1^3 + 2(1) = 1 + 2 = 3$ which is $3 \cdot 1$ where 3 is a factor. Assume true for $n = k$: 3 is a factor of $k^3 + 2k$
   Prove true for $n = k + 1$: 3 is a factor of $(k + 1)^3 + 2(k + 1)$
   Proof: $(k + 1)^3 + 2(k + 1)$
   $$= k^3 + 3k^2 + 3k + 1 + 2k + 2$$
   $$= k^3 + 2k + 3k^2 + 3k + 3$$
   $$= \underbrace{(k^3 + 2k)}_{\substack{\text{By assumption} \\ \text{3 is a factor}}} + 3(k^2 + k + 1)$$
   Here, 3 is a factor.

Thus, $(k + 1)^3 + 2(k + 1)$ has a factor of 3.

19. $a - b$ is a factor of $a^n - b^n$.

    For $n = 1$, then $a^n - b^n$ becomes $a - b$ and it is a factor of $a - b$.

    Assume true for $n = k$: $a - b$ is a factor of $a^k - b^k$

    Prove true for $n = k + 1$: $a - b$ is a factor of $a^{k+1} - b^{k+1}$

    Proof: $a^{k+1} - b^{k+1} = a^k(a - b) + (a^k - b^k)b$

    $$\underset{\substack{a - b \text{ is} \\ \text{a factor}}}{\downarrow} \qquad \underset{\substack{\text{By assumption, } a - b \\ \text{is a factor}}}{\downarrow}$$

    Thus, $a^{k+1} - b^{k+1}$ has a factor of $a - b$.

21. $x^{2n} > 0$ if $x \neq 0$.

    For $n = 1$, then x^{2n} becomes x^2 which is greater than zero.

    Assume true for $n = k$: $x^{2k} > 0$

    Prove true for $n = k + 1$: $x^{2(k+1)} > 0$

    Proof: $x^{2(k+1)} = x^{2k} x^2 > 0$ since $x^{2k} > 0$ by assumption and $x^2 > 0$ and the product is greater than zero.

23. $1 + 2n \leq 3^n$.

    For $n = 1$, the left side is $1 + 2 \cdot 1 = 1 + 2 = 3$ and the right side is $3^1 = 3$.

    Assume true for $n = k$: $1 + 2k \leq 3^k$

    Prove true for $n = k + 1$: $1 + 2(k + 1) \leq 3^{k+1}$

    Proof: $1 + 2(k + 1) = 1 + 2k + 2 \leq 3^k + 2 \leq 3^k + 3 = 3^{k+1}$

    $$\qquad\qquad \underbrace{\qquad\qquad}_{} \uparrow\uparrow \qquad \underbrace{\qquad\qquad}_{} \uparrow$$
    $$\qquad\qquad \text{By assumption} \qquad \text{Here, } 2 \leq 3$$

    Thus, $1 + 2(k + 1) \leq 3^{k+1}$

25. $|a_1 + a_2 + \ldots + a_n| \leq |a_1| + |a_2| + \ldots + |a_n|$

    For $n = 1$, the left side is $|a_1|$ and the right side is $|a_1|$.

    Assume true for $n = k$: $|a_1 + a_2 + \ldots + a_k| \leq |a_1| + |a_2| + \ldots + |a_k|$

    Also, assume $|a + b| \leq |a| + |b|$.

    Prove true for $n = k + 1$: $|a_1 + a_2 + \ldots + a_k + a_{k+1}| \leq |a_1| + |a_2| + \ldots + |a_{k+1}|$

    Proof:

    $|a_1 + a_2 + \ldots + a_k + a_{k+1}| \leq |a_1 + a_2 + \ldots + a_k| + |a_{k+1}| \leq |a_1| + |a_2| + \ldots + |a_k| + |a_{k+1}|$

    $$\underbrace{\qquad\qquad}_{\text{By one assumption}} \uparrow \qquad \underbrace{\qquad\qquad}_{} \qquad\qquad \uparrow \qquad \uparrow$$
    $$\qquad\qquad\qquad \text{By other assumption}$$

27. $\displaystyle\sum_{i=1}^{n} ca_i = c \sum_{i=1}^{n} a_i$

    For $n = 1$, the left side is ca_1 and the right side is ca_1.

    Assume true for $n = k$: $\displaystyle\sum_{i=1}^{k} ca_i = c \sum_{i=1}^{k} a_i$

    Prove true for $n = k + 1$: $\displaystyle\sum_{i=1}^{k+1} ca_i = c \sum_{i=1}^{k+1} a_i$

    Proof: $\displaystyle\sum_{i=1}^{k+1} ca_i = \sum_{i=1}^{k} ca_i + ca_{k+1} = c\sum_{i=1}^{k} a_i + c(a_{k+1}) = c\left[\sum_{i=1}^{k} a_i + a_{k+1}\right] = c\sum_{i=1}^{k+1} a_i$

    $$\qquad\qquad\qquad \underbrace{\qquad\qquad}_{} \uparrow \qquad\qquad \uparrow$$
    $$\qquad\qquad\qquad\qquad \text{By assumption}$$

29. $\sum_{i=1}^{n} (a_i - b_i) = \sum_{i=1}^{n} a_i - \sum_{i=1}^{n} b_i$

For $n = 1$, the left side is $a_1 - b_1$ and the right side is $a_1 - b_1$.

Assume true for $n = k$: $\sum_{i=1}^{k} (a_i - b_i) = \sum_{i=1}^{k} a_i - \sum_{i=1}^{k} b_i$

Prove true for $n = k + 1$: $\sum_{i=1}^{k+1} (a_i - b_i) = \sum_{i=1}^{k+1} a_i - \sum_{i=1}^{k+1} b_i$

Proof: $\sum_{i=1}^{k+1} (a_i - b_i) = \sum_{i=1}^{k} (a_i - b_i) + a_{k+1} - b_{k+1} = \underbrace{\sum_{i=1}^{k} a_i - \sum_{i=1}^{k} b_i}_{\text{By assumption}} + a_{k+1} - b_{k+1}$

$= \sum_{i=1}^{k} a_i + a_{k+1} - \sum_{i=1}^{k} b_i - b_{k+1}$

$= \sum_{i=1}^{k+1} a_i - \sum_{i=1}^{k+1} b_i$ which is the right side

31. $n^2 - n + 5$. For $n = 1$: $1^2 - 1 + 5 = 1 - 1 + 5 = 5$, a prime number

$n = 2$: $2^2 - 2 + 5 = 4 - 2 + 5 = 7$, a prime number

$n = 3$: $3^2 - 3 + 5 = 9 - 3 + 5 = 11$, a prime number

$n = 4$: $4^2 - 4 + 5 = 16 - 4 + 5 = 17$, a prime number

$n = 5$: $5^2 - 5 + 5 = 25$ which is **not** a prime number

Exercises 9.7

1. 12 different T-shirts are available

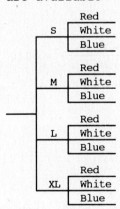

3. 0
1
0, 1
1, 0

5.
1, 2	2, 1	3, 1	4, 1	5, 1
1, 3	2, 3	3, 2	4, 2	5, 2
1, 4	2, 4	3, 4	4, 3	5, 3
1, 5	2, 5	3, 5	4, 5	5, 4

7. $P(4, 4) = 4! = 4 \cdot 3 \cdot 2 \cdot 1 = 24$

9. $P(15, 2) = \dfrac{15!}{13!} = \dfrac{15 \cdot 14 \cdot 13!}{13!} = 15 \cdot 14 = 210$

11. $P(15, 0) = \dfrac{15!}{15!} = 1$

13. $P(n, 2) = \dfrac{n!}{(n-2)!} = \dfrac{n(n-1)(n-2)!}{(n-2)!} = n(n-1)$

15. $P(n, n-2) = \dfrac{n!}{2!} = \dfrac{n!}{2}$

17. $\dfrac{n!}{(n-2)!} = \dfrac{n(n-1)(n-2)!}{(n-2)!} = n(n-1)$

19. $\dfrac{n!}{(n+2)!} = \dfrac{n!}{(n+2)(n+1)n!} = \dfrac{1}{(n+2)(n+1)}$

21. $4 \cdot 3 = 12$

CHAPTER 9: FURTHER TOPICS

23. $4 \cdot 6 \cdot 5 = 120$

25. $2 \cdot 2 \cdot 2 \cdot 2 \cdot 2 \cdot 2 \cdot 2 \cdot 2 \cdot 2 \cdot 2 = 1024$

27. $2 \cdot 2 \cdot 2 \cdot 2 = 16$

29. $6 \cdot 5 \cdot 4 = 120$

31. $10 \cdot 10 \cdot 10 \cdot 10 \cdot 10 \cdot 10 \cdot 10 \cdot 10 \cdot 10 = 10^9 = 1,000,000,000$

33. The numbers greater than or equal to 2000 would be $8 \cdot 10 \cdot 10 \cdot 10 = 8000$. Now, exclude the possibility of 2000 to get $8000 - 1 = 7999$.

35. $6 \cdot 6 \cdot 3 = 108$

37. $P(6, 6) = 6! = 6 \cdot 5 \cdot 4 \cdot 3 \cdot 2 \cdot 1 = 720$
$P(7, 6) = \dfrac{7!}{1!} = 7! = 7 \cdot 6 \cdot 5 \cdot 4 \cdot 3 \cdot 2 = 5040$

39. $P(11, 11) = 11! = 11 \cdot 10 \cdot 9 \cdot 8 \cdot 7 \cdot 6 \cdot 5 \cdot 4 \cdot 3 \cdot 2 \cdot 1 = 39,916,800$

41. Consider the three math books to be together. This group of 3 books along with the other three books will give 4 books to be arranged.
$P(4, 4) = 4! = 4 \cdot 3 \cdot 2 \cdot 1 = 24$. But, also, the three math books can be arranged as $P(3, 3) = 3! = 3 \cdot 2 \cdot 1 = 6$. The total number of possibilities is $P(4, 4)P(3, 3) = 24 \cdot 6 = 144$.

43. If two students are to be together, then the arrangements are
$P(6, 6) = 6! = 6 \cdot 5 \cdot 4 \cdot 3 \cdot 2 \cdot 1 = 720$. The two students can be arranged at $P(2, 2) = 2! = 2 \cdot 1 = 2$. The total number of possibilities is
$P(6, 6)P(2, 2) = 720 \cdot 2 = 1440$.

45. $P(5, 5) = 5! = 5 \cdot 4 \cdot 3 \cdot 2 \cdot 1 = 120$

47. a) $\dfrac{7!}{2!} = \dfrac{7 \cdot 6 \cdot 5 \cdot 4 \cdot 3 \cdot 2}{2} = 2520$ b) $\dfrac{8!}{2!2!2!} = \dfrac{8 \cdot 7 \cdot 6 \cdot 5 \cdot 4 \cdot 3 \cdot 2 \cdot 1}{2 \cdot 1 \cdot 2 \cdot 1 \cdot 2 \cdot 1} = 5040$
c) $\dfrac{3!}{2!} = \dfrac{3 \cdot 2 \cdot 1}{2 \cdot 1} = 3$ d) $\dfrac{6!}{2!2!} = \dfrac{6 \cdot 5 \cdot 4 \cdot 3 \cdot 2 \cdot 1}{2 \cdot 1 \cdot 2 \cdot 1} = 180$

49. $\dfrac{15!}{6!4!5!} = \dfrac{15 \cdot 14 \cdot 13 \cdot 12 \cdot 11 \cdot 10 \cdot 9 \cdot 8 \cdot 7 \cdot 6!}{6! \cdot 4 \cdot 3 \cdot 2 \cdot 1 \cdot 5 \cdot 4 \cdot 3 \cdot 2 \cdot 1} = 630,630$

51. $\dfrac{10!}{2!4!4!} = \dfrac{10 \cdot 9 \cdot 8 \cdot 7 \cdot 6 \cdot 5 \cdot 4!}{2 \cdot 1 \cdot 4 \cdot 3 \cdot 2 \cdot 1 \cdot 4!} = 3150$

53. If boys and girls must alternate and there are 4 girls and 3 boys, then we must start with a girl and end with a girl. The pattern is $G \cdot B \cdot G \cdot B \cdot G \cdot B \cdot G$ and the number of possibilities is $4 \cdot 3 \cdot 3 \cdot 2 \cdot 2 \cdot 1 \cdot 1 = 144$.

55. Number of circular possibilities is $(n - 1)! = 4! = 24$.

57. Alternate: $4 \cdot 3 \cdot 3 \cdot 2 \cdot 2 \cdot 1 \cdot 1 = 144$
or use: $\dfrac{P(4, 4)P(4, 4)}{4} = \dfrac{4!4!}{4} = \dfrac{4 \cdot 3 \cdot 2 \cdot 1 \cdot 4 \cdot 3 \cdot 2 \cdot 1}{4} = 3 \cdot 2 \cdot 1 \cdot 4 \cdot 3 \cdot 2 \cdot 1 = 144$
seated together $P(4, 4)P(4, 4) = 4!4! = 4 \cdot 3 \cdot 2 \cdot 1 \cdot 4 \cdot 3 \cdot 2 \cdot 1 = 576$.

59. $P(29, 7) = 7.8663312 \times 10^9 = 7,866,331,200$

330

61. $P(42, 11) = 1.708663123 \times 10^{17} = 170,866,312,300,000,000$

63. Number of possibilities is $(n - 1)!$

Exercises 9.8

1. $\{0\}, \{1\}, \{0, 1\}$

3. $\{1, 2\}, \{1, 3\}, \{1, 4\}, \{1, 5\}, \{2, 3\}, \{2, 4\}, \{2, 5\}, \{3, 4\}, \{3, 5\}, \{4, 5\}$

5. $C(6, 5) = \dfrac{6!}{1!5!} = \dfrac{6!}{5!} = \dfrac{6 \cdot 5!}{5!} = 6$ 7. $C(10, 10) = \dfrac{10!}{0!10!} = \dfrac{10!}{10!} = 1,$ where $0! = 1$

9. $\dbinom{5}{0} = \dfrac{5!}{5!0!} = \dfrac{5!}{5!} = 1$, where $0! = 1$ 11. $\dbinom{6}{4} = \dfrac{6!}{2!4!} = \dfrac{6 \cdot 5 \cdot 4!}{2 \cdot 1 \cdot 4!} = \dfrac{6 \cdot 5}{2 \cdot 1} = 15$

13. $\dbinom{n}{n-1} = \dfrac{n!}{1!(n-1)!} = \dfrac{n(n-1)!}{1 \cdot (n-1)!} = n$

15. $C(n, n) = \dfrac{n!}{0!n!} = \dfrac{n!}{n!} = 1$, where $0! = 1$

17. $\dbinom{12}{4} = \dfrac{12!}{8!4!} = \dfrac{12 \cdot 11 \cdot 10 \cdot 9 \cdot 8!}{8! \, 4 \cdot 3 \cdot 2 \cdot 1} = 495$

19. $\dbinom{15}{6} = \dfrac{15!}{9!6!} = \dfrac{15 \cdot 14 \cdot 13 \cdot 12 \cdot 11 \cdot 10 \cdot 9!}{9! \, 6 \cdot 5 \cdot 4 \cdot 3 \cdot 2 \cdot 1} = 5005$

$\dbinom{15}{10} = \dfrac{15!}{5!10!} = \dfrac{15 \cdot 14 \cdot 13 \cdot 12 \cdot 11 \cdot 10!}{5 \cdot 4 \cdot 3 \cdot 2 \cdot 1 \cdot 10!} = 3003$

21. $\dbinom{8}{4} = \dfrac{8!}{4!4!} = \dfrac{8 \cdot 7 \cdot 6 \cdot 5 \cdot 4!}{4 \cdot 3 \cdot 2 \cdot 1 \cdot 4!} = 70$

Since one of the four positions must be the chairperson, this produces 4 possibilities. The number of committees is $70 \cdot 4 = 280$.

23. $\underset{\substack{\uparrow \quad \uparrow \\ \text{men women}}}{\dbinom{6}{2}\dbinom{12}{3}} = \dfrac{6!}{2!4!} \cdot \dfrac{12!}{9!3!}$

$= \dfrac{6 \cdot 5 \cdot 4!}{2 \cdot 1 \cdot 4!} \cdot \dfrac{12 \cdot 11 \cdot 10 \cdot 9!}{9! \cdot 3 \cdot 2 \cdot 1} = 15 \cdot 220 = 3300$

25. $\underset{\substack{\uparrow \quad \uparrow \quad \uparrow \\ \text{math} \\ \text{novels} \\ \text{psychology}}}{\dbinom{6}{2}\dbinom{7}{4}\dbinom{9}{3}} = \dfrac{6!}{4!2!} \cdot \dfrac{7!}{3!4!} \cdot \dfrac{9!}{6!3!} = \dfrac{6 \cdot 5 \cdot 4!}{4! \, 2 \cdot 1} \cdot \dfrac{7 \cdot 6 \cdot 5 \cdot 4!}{3 \cdot 2 \cdot 1 \cdot 4!} \cdot \dfrac{9 \cdot 8 \cdot 7 \cdot 6!}{6! \cdot 3 \cdot 2 \cdot 1}$

$= \dfrac{6 \cdot 5}{2 \cdot 1} \cdot \dfrac{7 \cdot 6 \cdot 5}{3 \cdot 2 \cdot 1} \cdot \dfrac{9 \cdot 8 \cdot 7}{3 \cdot 2 \cdot 1} = 15 \cdot 35 \cdot 84 = 44,100$

27. $\displaystyle \binom{10}{3}\binom{8}{2}\binom{15}{5} = \frac{10!}{7!3!} \cdot \frac{8!}{6!2!} \cdot \frac{15!}{10!5!}$

$\displaystyle \qquad = \frac{10 \cdot 9 \cdot 8 \cdot 7!}{7! \cdot 3 \cdot 2 \cdot 1} \cdot \frac{8 \cdot 7 \cdot 6!}{6! \cdot 2 \cdot 1} \cdot \frac{15 \cdot 14 \cdot 13 \cdot 12 \cdot 11 \cdot 10!}{10! \cdot 5 \cdot 4 \cdot 3 \cdot 2 \cdot 1}$

$\displaystyle \qquad = \frac{10 \cdot 9 \cdot 8}{3 \cdot 2 \cdot 1} \cdot \frac{8 \cdot 7}{2 \cdot 1} \cdot \frac{15 \cdot 14 \cdot 13 \cdot 12 \cdot 11}{5 \cdot 4 \cdot 3 \cdot 2 \cdot 1}$

$\displaystyle \qquad = 120 \cdot 28 \cdot 3003$

$\displaystyle \qquad = 10,090,080$

29. a) $\displaystyle \binom{7}{1}\binom{5}{2}\binom{2}{0} = \frac{7!}{6!1!} \cdot \frac{5!}{3!2!} \cdot \frac{2!}{2!0!} = 7 \cdot 10 \cdot 1 = 70$

b) $\displaystyle \binom{7}{0}\binom{5}{3}\binom{2}{0} = \frac{7!}{7!0!} \cdot \frac{5!}{2!3!} \cdot \frac{2!}{2!0!} = 1 \cdot 10 \cdot 1 = 10$

c) $\displaystyle \binom{7}{1}\binom{5}{1}\binom{2}{1} = \frac{7!}{6!1!} \cdot \frac{5!}{4!1!} \cdot \frac{2!}{1!1!} = 7 \cdot 5 \cdot 2 = 70$

d) Not possible since there are only two purple balls. It is **not** possible to select three purple balls. Number of possibilities is 0.

31. $\displaystyle \binom{9}{2} = \frac{9!}{7!2!} = \frac{9 \cdot 8 \cdot 7!}{7! \cdot 2 \cdot 1} = \frac{9 \cdot 8}{2 \cdot 1} = 36$

33. a) $\displaystyle \binom{13}{5}\binom{13}{3}\binom{13}{2}\binom{13}{3} = \frac{13!}{8!5!} \cdot \frac{13!}{10!3!} \cdot \frac{13!}{11!2!} \cdot \frac{13!}{10!3!}$

$\displaystyle \qquad = \frac{13 \cdot 12 \cdot 11 \cdot 10 \cdot 9 \cdot 8!}{8! \cdot 5 \cdot 4 \cdot 3 \cdot 2 \cdot 1} \cdot \frac{13 \cdot 12 \cdot 11 \cdot 10!}{10! \cdot 3 \cdot 2 \cdot 1} \cdot \frac{13 \cdot 12 \cdot 11!}{11! \cdot 2 \cdot 1} \cdot \frac{13 \cdot 12 \cdot 11 \cdot 10!}{10! \cdot 3 \cdot 2 \cdot 1}$

$\displaystyle \qquad = \frac{13 \cdot 12 \cdot 11 \cdot 10 \cdot 9}{5 \cdot 4 \cdot 3 \cdot 2 \cdot 1} \cdot \frac{13 \cdot 12 \cdot 11}{3 \cdot 2 \cdot 1} \cdot \frac{13 \cdot 12}{2 \cdot 1} \cdot \frac{13 \cdot 12 \cdot 11}{3 \cdot 2 \cdot 1}$

$\displaystyle \qquad = 1287 \cdot 286 \cdot 78 \cdot 286$

$\displaystyle \qquad = 8,211,173,256$

b) $\displaystyle \binom{13}{12}\binom{13}{1}\binom{13}{0}\binom{13}{0} = \frac{13!}{1!12!} \cdot \frac{13!}{12!1!} \cdot \frac{13!}{13!0!} \cdot \frac{13!}{13!0!}$

$\displaystyle \qquad = 13 \cdot 13 \cdot 1 \cdot 1 = 169$

c) $\displaystyle \binom{13}{7}\binom{13}{6}\binom{13}{0}\binom{13}{0} = \frac{13!}{6!7!} \cdot \frac{13!}{7!6!} \cdot \frac{13!}{13!0!} \cdot \frac{13!}{13!0!}$

$\displaystyle \qquad = \left[\frac{13!}{6!7!}\right]^2 \cdot 1 \cdot 1 = \left[\frac{13!}{6!7!}\right]^2$

$\displaystyle \qquad = \left[\frac{13 \cdot 12 \cdot 11 \cdot 10 \cdot 9 \cdot 8 \cdot 7!}{6 \cdot 5 \cdot 4 \cdot 3 \cdot 2 \cdot 1 \cdot 7!}\right]^2 = (1716)^2 = 2,944,656$

35. $\displaystyle C(n, r) = \frac{n}{(n-r)!r!}$ and $\displaystyle C(n, n-r) = \frac{n!}{[n-(n-r)]!(n-r)!} = \frac{n!}{r!(n-r)!}$ and both outcomes are the same

37. $\dbinom{100}{5} = \dfrac{100!}{95!5!} = \dfrac{100 \cdot 99 \cdot 98 \cdot 97 \cdot 96 \cdot 95!}{5 \cdot 4 \cdot 3 \cdot 2 \cdot 1 \cdot 95!} = \dfrac{100 \cdot 99 \cdot 98 \cdot 97 \cdot 96}{5 \cdot 4 \cdot 3 \cdot 2 \cdot 1} = 75{,}287{,}520$

39. $\dbinom{52}{13} = 635{,}013{,}559{,}600$ 41. Each one is a permutation.

Exercises 9.9

1. a) $\{(H, 1), (H, 2), (H, 3), (H, 4), (H, 5), (H, 6), (T, 1), (T, 2), (T, 3),$ $(T, 4), (T, 5), (T, 6)\}$

   b) $\{(H, 2), (H, 4), (H, 6)\}$

3. The sample space has $5 \cdot 5 = 25$ outcomes

   a) Let A be the event that both tags are even. Then $N(A) = 4$ since
   $A = \{2 - 2, \ 2 - 4, \ 4 - 2, \ 4 - 4\}$. Now, $P(A) = \dfrac{N(A)}{N(S)} = \dfrac{4}{25}$.

   b) Let B be the event that the sum is 6. Then $N(B) = 5$ since
   $B = \{1 - 5, \ 2 - 4, \ 3 - 3, \ 4 - 2, \ 5 - 1\}$. Now, $P(B) = \dfrac{N(B)}{N(S)} = \dfrac{5}{25} = \dfrac{1}{5}$.

   c) Let C be the event that the sum is less than 6. Then $N(C) = 10$ since
   $C = \{1 - 1, \ 1 - 2, \ 1 - 3, \ 1 - 4, \ 2 - 1, \ 2 - 2, \ 2 - 3, \ 3 - 1, \ 3 - 2, \ 4 - 1\}$.
   Now, $P(C) = \dfrac{N(C)}{N(S)} = \dfrac{10}{25} = \dfrac{2}{5}$.

5. The sample space is $\{HH, \ HT, \ TH, \ TT\}$ and $N(S) = 4$.

   a) Let A be the event that 2 heads are obtained. Then $N(A) = 1$ since
   $A = \{HH\}$ and $P(A) = \dfrac{N(A)}{N(S)} = \dfrac{1}{4}$.

   b) Let B be the event that no heads is obtained. Then $N(B) = 1$ since
   $B = \{TT\}$ and $P(B) = \dfrac{N(B)}{N(S)} = \dfrac{1}{4}$.

   c) Let C be the event that at least one tail is obtained.
   Then $N(C) = 3$ since $C = \{HT, \ TH, \ TT\}$ and $P(C) = \dfrac{N(C)}{N(S)} = \dfrac{3}{4}$.

7.

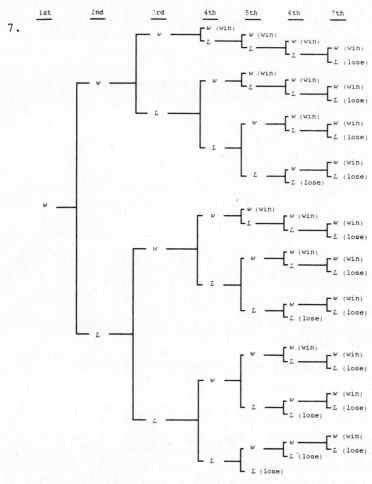

| 1st | 2nd | 3rd | 4th | 5th | 6th | 7th |

There are 35 outcomes with 20 wins and 15 losses. Now $P(\text{win}) = \dfrac{20}{35} = \dfrac{4}{7}$.

9. $P(\text{Tail}) = 1 - P(\text{Head}) = 1 - \dfrac{3}{4} = \dfrac{1}{4}$

11. Let S be the sample space consisting of a deck of cards. Then $N(S) = 52$.

   a) Let A be the event that a card with 2 is selected. Then $N(A) = 4$ and
   $$P(A) = \frac{N(A)}{N(S)} = \frac{4}{52} = \frac{1}{13}.$$

   b) Let B be the event that a black face card is selected. Then $N(B) = 6$ and
   $$P(B) = \frac{N(B)}{N(S)} = \frac{6}{52} = \frac{3}{26}.$$

   c) Let C be the event that a space or a club is selected. Then $N(C) = 26$ and
   $$P(C) = \frac{N(C)}{N(S)} = \frac{26}{52} = \frac{1}{2}.$$

   d) Let D be the event that a face card or a club is selected. Then $N(D) = 22$
   and $P(D) = \dfrac{N(D)}{N(S)} = \dfrac{22}{52} = \dfrac{11}{26}.$

334

13. Let S be the sample space consisting of the bag with 13 marbles. Then

$$N(S) = \binom{13}{3} = \frac{13 \cdot 12 \cdot 11 \cdot 10!}{10! \cdot 3 \cdot 2 \cdot 1} = \frac{13 \cdot 12 \cdot 11}{3 \cdot 2 \cdot 1} = 286$$

a) Let A be the event that 3 black marbles are selected.

Then $N(A) = \binom{2}{0}\binom{5}{3}\binom{6}{0} = 1 \cdot \binom{5}{3} \cdot 1 = \binom{5}{3} = 10$ and $P(A) = \frac{N(A)}{N(S)} = \frac{10}{286} = \frac{5}{143}$.

b) Let B be the event that no red marbles are selected.

Then $N(B) = \binom{2}{0}\binom{11}{3} = 1\binom{11}{3} = \frac{11 \cdot 10 \cdot 9 \cdot 8!}{8! \cdot 3 \cdot 2 \cdot 1} = \frac{11 \cdot 10 \cdot 9}{3 \cdot 2 \cdot 1} = 165$

and $P(B) = \frac{N(B)}{N(S)} = \frac{165}{286} = \frac{15}{26}$.

c) Let C be the event that 1 red, 1 black, and 1 blue are selected.

Then $N(C) = \binom{2}{1}\binom{5}{1}\binom{6}{1} = 2 \cdot 5 \cdot 6 = 60$ and $P(C) = \frac{N(C)}{N(S)} = \frac{60}{286} = \frac{30}{143}$.

d) Let D be the event that at least 1 red is selected.

Then $P(D) = 1 - P(\text{no red}) = 1 - \frac{15}{26} = \frac{26}{26} - \frac{15}{26} = \frac{11}{26}$

part b) above

e) Let E be the event that 3 red marbles are selected. Since there are only 2 red marbles in the bag, it is impossible to select 3 red marbles. Consequently, $P(E) = 0$.

15. Let S be the sample space of selecting two cards (without replacement) from a deck of cards. Then $N(S) = 52 \cdot 51$.

a) Let A be the event that both cards are clubs. Then $N(A) = 13 \cdot 12$ and $P(A) = \frac{13 \cdot 12}{52 \cdot 51} = \frac{1}{17}$.

b) Let B be the event that both cards are kings. Then $N(B) = 4 \cdot 3$ and $P(B) = \frac{N(B)}{N(S)} = \frac{4 \cdot 3}{52 \cdot 51} = \frac{1}{13 \cdot 17} = \frac{1}{221}$.

c) Let C be the event that both cards are face cards. Then $N(C) = 12 \cdot 11$ and $P(C) = \frac{N(C)}{N(S)} = \frac{12 \cdot 11}{52 \cdot 51} = \frac{11}{221}$.

17. Let S be the sample space of selecting a poker hand of 5 cards from a deck of cards. Then $N(S) = \binom{52}{5} = \frac{52 \cdot 51 \cdot 50 \cdot 49 \cdot 48 \cdot 47!}{47! \cdot 5 \cdot 4 \cdot 3 \cdot 2 \cdot 1}$

$$= \frac{52 \cdot 51 \cdot 50 \cdot 49 \cdot 48}{5 \cdot 4 \cdot 3 \cdot 2 \cdot 1} = 2,598,960$$

a) Let A be the event that 5 cards are selected from the same suit.

Then $N(A) = 4\binom{13}{5} = 4\frac{13 \cdot 12 \cdot 11 \cdot 10 \cdot 9 \cdot 8!}{8! \cdot 5 \cdot 4 \cdot 3 \cdot 2 \cdot 1} = 4\frac{13 \cdot 12 \cdot 11 \cdot 10 \cdot 9}{5 \cdot 4 \cdot 3 \cdot 2 \cdot 1}$

$$= 4(1287) = 5148$$

and $P(A) = \frac{N(A)}{N(S)} = \frac{5148}{2,598,960} = \frac{33}{16,660}$

b) Let B be the event that 3 aces and 2 kings are selected.

Then $N(B) = \binom{4}{3}\binom{4}{2} = 4 \cdot 6 = 24$ and $P(B) = \dfrac{N(B)}{N(S)} = \dfrac{24}{2,598,960} = \dfrac{1}{108,290}$.

19. Let S be the sample space consisting of selecting 4 students from a group of 11 students. Then $N(S) = \binom{11}{4} = \dfrac{11 \cdot 10 \cdot 9 \cdot 8 \cdot 7!}{7! \cdot 4 \cdot 3 \cdot 2 \cdot 1} = 330$

a) Let A be the event that all 4 people selected are juniors.

Then $N(A) = \binom{6}{4}\binom{5}{0} = \binom{6}{4} \cdot 1 = \binom{6}{4} = 15$ and $P(A) = \dfrac{N(A)}{N(S)} = \dfrac{15}{330} = \dfrac{1}{22}$.

b) Let B be the event that 2 juniors and 2 seniors are selected.

Then $N(B) = \binom{6}{2}\binom{5}{2} = 15 \cdot 10 = 150$ and $P(B) = \dfrac{N(B)}{N(S)} = \dfrac{150}{330} = \dfrac{5}{11}$.

c) Let C be the event that 1 junior and 3 seniors are selected.

Then $N(C) = \binom{6}{1}\binom{5}{3} = 6 \cdot 10 = 60$ and $P(C) = \dfrac{N(C)}{N(S)} = \dfrac{60}{330} = \dfrac{2}{11}$.

21. The sample space is $\{1, 2\}$, $\{1, 3\}$, $\{1, 4\}$, $\{1, 5\}$, $\{1, 6\}$, $\{1, 7\}$, $\{1, 8\}$, $\{1, 9\}$, $\{1, 10\}$, $\{2, 3\}$, $\{2, 4\}$, $\{2, 5\}$, $\{2, 6\}$, $\{2, 7\}$, $\{2, 8\}$, $\{2, 9\}$, $\{2, 10\}$, $\{3, 4\}$, $\{3, 5\}$, $\{3, 6\}$, $\{3, 7\}$, $\{3, 8\}$, $\{3, 9\}$, $\{3, 10\}$, $\{4, 5\}$, $\{4, 6\}$, $\{4, 7\}$, $\{4, 8\}$, $\{4, 9\}$, $\{4, 10\}$, $\{5, 6\}$, $\{5, 7\}$, $\{5, 8\}$, $\{5, 9\}$, $\{5, 10\}$, $\{6, 7\}$, $\{6, 8\}$, $\{6, 9\}$, $\{6, 10\}$, $\{7, 8\}$, $\{7, 9\}$, $\{7, 10\}$, $\{8, 9\}$, $\{8, 10\}$, $\{9, 10\}$

Here, $N(S) = 45$.

a) Let A be the event that the sum is even. Now A consists of $\{1, 3\}$, $\{1, 5\}$, $\{1, 7\}$, $\{1, 9\}$, $\{2, 4\}$, $\{2, 6\}$, $\{2, 8\}$, $\{2, 10\}$, $\{3, 5\}$, $\{3, 7\}$, $\{3, 9\}$, $\{4, 6\}$, $\{4, 8\}$, $\{4, 10\}$, $\{5, 7\}$, $\{5, 9\}$, $\{6, 8\}$, $\{6, 10\}$, $\{7, 9\}$, $\{8, 10\}$.

Then $N(A) = 20$ and $P(A) = \dfrac{N(A)}{N(S)} = \dfrac{20}{45} = \dfrac{4}{9}$.

b) Let B be the event that the product is even. Now B consists of $\{1, 2\}$, $\{1, 4\}$, $\{1, 6\}$, $\{1, 8\}$, $\{1, 10\}$, $\{2, 3\}$, $\{2, 4\}$, $\{2, 5\}$, $\{2, 6\}$, $\{2, 7\}$, $\{2, 8\}$, $\{2, 9\}$, $\{2, 10\}$, $\{3, 4\}$, $\{3, 6\}$, $\{3, 8\}$, $\{3, 10\}$, $\{4, 5\}$, $\{4, 6\}$, $\{4, 7\}$, $\{4, 8\}$, $\{4, 9\}$, $\{4, 10\}$, $\{5, 6\}$, $\{5, 8\}$, $\{5, 10\}$, $\{6, 7\}$, $\{6, 8\}$, $\{6, 9\}$, $\{6, 10\}$, $\{7, 8\}$, $\{7, 10\}$, $\{8, 9\}$, $\{8, 10\}$, $\{9, 10\}$.

Then $N(B) = 35$ and $P(B) = \dfrac{N(B)}{N(S)} = \dfrac{35}{45} = \dfrac{7}{9}$.

23. If $S = A \cup B$, then $P(S) = P(A \cup B) = 1$.

Now, $P(A \cap B) = P(A) + P(B) - P(A \cup B)$

$= 0.4 + 0.9 - 1.0$

$= 0.3$

25. Be sure you use the right resources to obtain the appropriate permutations and/or combinations for computing the probabilities for having winning entries.

CHAPTER REVIEW

1. $a_1 = \dfrac{1-1}{1+1} = \dfrac{0}{2} = 0$

 $a_2 = \dfrac{2-1}{2+1} = \dfrac{1}{3}$

 $a_3 = \dfrac{3-1}{3+1} = \dfrac{2}{4} = \dfrac{1}{2}$

 $a_4 = \dfrac{4-1}{4+1} = \dfrac{3}{5}$

 $a_5 = \dfrac{5-1}{5+1} = \dfrac{4}{6} = \dfrac{2}{3}$

2. $a_1 = \dfrac{(-2)^1}{1!} = \dfrac{-2}{1} = -2$

 $a_2 = \dfrac{(-2)^2}{2!} = \dfrac{4}{2 \cdot 1} = 2$

 $a_3 = \dfrac{(-2)^3}{3!} = \dfrac{-8}{3 \cdot 2 \cdot 1} = -\dfrac{8}{6} = -\dfrac{4}{3}$

 $a_4 = \dfrac{(-2)^4}{4!} = \dfrac{16}{4 \cdot 3 \cdot 2 \cdot 1} = \dfrac{16}{24} = \dfrac{2}{3}$

 $a_5 = \dfrac{(-2)^5}{5!} = \dfrac{-32}{5 \cdot 4 \cdot 3 \cdot 2 \cdot 1} = -\dfrac{32}{120} = -\dfrac{4}{15}$

3. $a_1 = 1$

 $a_2 = a_1 + 2 = 1 + 2 = 3$

 $a_3 = a_2 + 2 = 3 + 2 = 5$

 $a_4 = a_3 + 2 = 5 + 2 = 7$

 $a_5 = a_4 + 2 = 7 + 2 = 9$

 $a_6 = a_5 + 2 = 9 + 2 = 11$

4. $a_1 = 2$

 $a_2 = 2a_1 - 3 = 2(2) - 3 = 4 - 3 = 1$

 $a_3 = 2a_2 - 3 = 2(1) - 3 = 2 - 3 = -1$

 $a_4 = 2a_3 - 3 = 2(-1) - 3 = -2 - 3 = -5$

 $a_5 = 2a_4 - 3 = 2(-5) - 3 = -10 - 3 = -13$

 $a_6 = 2a_5 - 3 = 2(-13) - 3 = -26 - 3 = -29$

5. $\displaystyle\sum_{n=1}^{5} (-2)^n = (-2)^1 + (-2)^2 + (-2)^3 + (-2)^4 + (-2)^5 = -2 + 4 - 8 + 16 - 32 = -22$

6. $\displaystyle\sum_{n=1}^{6} \dfrac{1}{2^{n-1}} = \dfrac{1}{2^0} + \dfrac{1}{2^1} + \dfrac{1}{2^2} + \dfrac{1}{2^3} + \dfrac{1}{2^4} + \dfrac{1}{2^5}$

 $= 1 + \dfrac{1}{2} + \dfrac{1}{4} + \dfrac{1}{8} + \dfrac{1}{16} + \dfrac{1}{32}$

 $= \dfrac{32 + 16 + 8 + 4 + 2 + 1}{32} = \dfrac{63}{32}$

7. $\displaystyle\sum_{n=1}^{7} 5 = 5 + 5 + 5 + 5 + 5 + 5 + 5$

 $= 7(5) = 35$

8. $\displaystyle\sum_{n=1}^{4} \dfrac{(2n)!}{2^n} = \dfrac{2!}{2^1} + \dfrac{4!}{2^2} + \dfrac{6!}{2^3} + \dfrac{8!}{2^4}$

 $= \dfrac{2 \cdot 1}{2} + \dfrac{4 \cdot 3 \cdot 2 \cdot 1}{4} + \dfrac{6 \cdot 5 \cdot 4 \cdot 3 \cdot 2 \cdot 1}{8} + \dfrac{8 \cdot 7 \cdot 6 \cdot 5 \cdot 4 \cdot 3 \cdot 2 \cdot 1}{16}$

 $= 1 + 6 + 90 + 2520$

 $= 2617$

9. $\displaystyle\sum_{n=1}^{5} \dfrac{n!}{(-1)^n} = \dfrac{1!}{(-1)^1} + \dfrac{2!}{(-1)^2} + \dfrac{3!}{(-1)^3} + \dfrac{4!}{(-1)^4} + \dfrac{5!}{(-1)^5} = -1! + 2! - 3! + 4! - 5!$

 $= -1 + 2 - 6 + 24 - 120 = -101$

10. $\displaystyle\sum_{k=1}^{7} 2k$ 11. $\displaystyle\sum_{j=1}^{5} 2^{j-1}(2j)$ 12. $\displaystyle\sum_{n=1}^{6} (-1)^{n+1}(2n-1)$ 13. $\displaystyle\sum_{i=1}^{7} \dfrac{i^2}{i+1}$ 14. $\displaystyle\sum_{k=1}^{6} \dfrac{k!}{2^{k-1}}$

15. $a_4 = 6$ becomes $a_1 + 3d = 6$
    $a_9 = -4$ $a_1 + 8d = -4$

    Subtract the first equation from the second: $5d = -10$
    $$d = -2$$

    Substitute $d = -2$ into $a_1 + 3d = 6$ to get
    $$a_1 + 3(-2) = 6$$
    $$a_1 - 6 = 6$$
    $$a_1 = 12$$
    Now, $a_{11} = a_1 + 10d = 12 + 10(-2) = 12 - 20 = -8$
    and $S_{11} = \frac{11}{2}(a_1 + a_{11}) = \frac{11}{2}[12 + (-8)] = \frac{11}{2}(4) = 22$.

16. $a_6 = 8$ can be expressed as $a_1 + 5d = 8$
    $S_8 = 40$ $4[2a_1 + 7d] = 40$

    or $a_1 + 5d = 8$ $\xrightarrow{\text{Mult by -2}}$ $-2a_1 - 10d = -16$
     $2a_1 + 7d = 10$ $\underline{2a_1 + 7d = 10}$
     Add: $-3d = -6$
    $$d = \frac{-6}{-3} = 2$$

    Now, substitute $d = 2$ into $a_1 + 5d = 8$ to get
    $$a_1 + 5(2) = 8$$
    $$a_1 + 10 = 8$$
    $$a_1 = -2$$

17. $a_1 = 5$, $a_2 = a_1 + d = 5 + (-2) = 3$
    $a_3 = a_1 + 2d = 5 + 2(-2) = 5 - 4 = 1$
    $a_4 = a_1 + 3d = 5 + 3(-2) = 5 - 6 = -1$
    $a_5 = a_1 + 4d = 5 + 4(-2) = 5 - 8 = -3$

18. Here, $a_1 = 3$ and $d = 1 - 3 = -2$.
    Now, $a_{19} = a_1 + 18d$ $a_n = a_1 + (n - 1)d$
    $ = 3 + 18(-2)$ $ = 3 + (n - 1)(-2)$
    $ = 3 - 36$ $ = 3 - 2n + 2$
    $ = -33$ $ = 5 - 2n$

19. a) Geometric sequence with $a_1 = 24$, $r = \frac{-36}{24} = -\frac{3}{2}$,

    and $a_n = a_1(r)^{n-1} = 24\left(-\frac{3}{2}\right)^{n-1}$

    $= 24\left(-\frac{3}{2}\right)^n \cdot \left(-\frac{3}{2}\right)^{-1}$

    $= 24\left(-\frac{3}{2}\right)^n \left(-\frac{2}{3}\right) = -16\left(-\frac{3}{2}\right)^n$

    b) Not a geometric sequence.

20. $a_3 = 81$ can be expressed as $a_1 r^2 = 81$
    $a_6 = -24$ $a_1 r^5 = -24$

Now, $\dfrac{a_6}{a_3} = \dfrac{-24}{81}$ gives $\dfrac{a_1 r^5}{a_1 r^2} = -\dfrac{24}{81}$

or $r^3 = -\dfrac{8}{27}$

so that $r = \sqrt[3]{-\dfrac{8}{27}} = -\dfrac{2}{3}$

Finally, substitute $r = -\dfrac{2}{3}$ into $a_1 r^2 = 81$ to obtain

$$a_1\left(-\frac{2}{3}\right)^2 = 81$$

$$a_1\left(\frac{4}{9}\right) = 81$$

$$a_1 = 81\left(\frac{9}{4}\right) = \frac{729}{4}$$

21. $a_6 = a_1 r^5 = 4\left(-\dfrac{1}{2}\right)^5 = 4\left(-\dfrac{1}{32}\right) = -\dfrac{1}{8}$

$a_n = a_1 r^{n-1} = 4\left(-\dfrac{1}{2}\right)^{n-1} = 2^2\dfrac{(-1)^{n-1}}{2^{n-1}} = \dfrac{(-1)^{n-1}}{2^{n-3}}$

22. a) $r = \dfrac{a_2}{a_1} = \dfrac{3/4}{9/16} = \dfrac{3}{4}\cdot\dfrac{16}{9} = \dfrac{4}{3}$ and the sum does **not** exist since $|r| > 1$.

b) $r = \dfrac{a_2}{a_1} = -\dfrac{6}{9} = -\dfrac{2}{3}$. Now, $S = \dfrac{a_1}{1 - r} = \dfrac{9}{1 - \left(-\dfrac{2}{3}\right)} = \dfrac{9}{1 + \dfrac{2}{3}} = \dfrac{9}{5/3} = 9\cdot\dfrac{3}{5} = \dfrac{27}{5}$

23. $S = \dfrac{a_1}{1 - r} = \dfrac{25\left(\dfrac{2}{3}\right)}{1 - \left(\dfrac{2}{3}\right)} = \dfrac{25\left(\dfrac{2}{3}\right)}{1/3} = 25\left(\dfrac{2}{3}\right)\left(\dfrac{3}{1}\right) = 25\cdot2 = 50$

24. $3.212121\ldots = 3 + 0.21 + 0.0021 + 0.000021 + \ldots$
    The sum starting with 0.21 is a geometric series with $a_1 = 0.21$ and $r = 0.01$.
    Now, $0.21 + 0.0021 + 0.000021 + \ldots = \dfrac{0.21}{1 - 0.01} = \dfrac{0.21}{0.99} = \dfrac{21}{99} = \dfrac{7}{33}$.
    Finally, $3.212121\ldots = 3 + \dfrac{7}{33} = 3\dfrac{7}{33}$ or $\dfrac{106}{33}$.

25. $a_1 = 1$, $r = 2$ so that $a_{15} = a_1 r^{14} = 1(2)^{14} = 16384$¢ or \$163.84.
    $S_{15} = \dfrac{a_1(1 - r^{15})}{1 - r} = \dfrac{1(1 - 2^{15})}{1 - 2} = \dfrac{1 - 2^{15}}{-1} = \dfrac{1 - 32768}{-1} = 32767$¢ or \$327.67

26. Distance $= 2 + 2\left[2\left(\dfrac{2}{5}\right)\right] + 2\left[2\left(\dfrac{2}{5}\right)^2\right] + 2\left[2\left(\dfrac{2}{5}\right)^3\right] + \ldots$

$= 2 + 4\left(\dfrac{2}{5}\right) + 4\left(\dfrac{2}{5}\right)^2 + 4\left(\dfrac{2}{5}\right)^3 + \ldots$

The sum starting with $4\left(\dfrac{2}{5}\right)$ is a geometric series with $a_1 = \dfrac{8}{5}$ and $r = \dfrac{2}{5}$.

Now, $4\left(\dfrac{2}{5}\right) + 4\left(\dfrac{2}{5}\right)^2 + 4\left(\dfrac{2}{5}\right)^3 + \ldots = \dfrac{8/5}{1 - \left(\dfrac{2}{5}\right)} = \dfrac{8/5}{3/5} = \dfrac{8}{5} \cdot \dfrac{5}{3} = \dfrac{8}{3} = 2\dfrac{2}{3}$ meters.

Finally, distance $= 2 + \underbrace{4\left(\dfrac{2}{5}\right) + 4\left(\dfrac{2}{5}\right)^2 + 4\left(\dfrac{2}{5}\right)^3 + \ldots}$

$= 2 + 2\dfrac{2}{3} = 4\dfrac{2}{3}$ meters

27. $(x - 2y)^6 = x^6 + \binom{6}{1}x^5(-2y)^1 + \binom{6}{2}x^4(-2y)^2 + \binom{6}{3}x^3(-2y)^3 + \binom{6}{4}x^2(-2y)^4$

$+ \binom{6}{5}x(-2y)^5 + (-2y)^6$

$= x^6 + 6x^5(-2y) + 15x^4(4y^2) + 20x^3(-8y^3) + 15x^2(16y^4)$
$+ 6x(-32y^5) + 64y^6$

$= x^6 - 12x^5y + 60x^4y^2 - 160x^3y^3 + 240x^2y^4 - 192xy^5 + 64y^6$

28. Sixth term: $\binom{9}{5}(2x)^4(-y^2)^5 = \dfrac{9!}{5!4!}16x^4(-y^{10})$

$= -\dfrac{9 \cdot \overset{}{8} \cdot 7 \cdot \overset{2}{6} \cdot 5!}{5! \cdot 4 \cdot 3 \cdot 2 \cdot 1}16x^4y^{10}$

$= -(126)16x^4y^{10}$

$= -2016x^4y^{10}$

29. $(1.02)^7 \approx 1^7 + \binom{7}{1}(1)^6(0.02) + \binom{7}{2}(1)^5(0.02)^2 + \binom{7}{3}(1)^4(0.02)^3$

$= 1 + 7(1)(0.02) + 21(1)(0.0004) + 35(1)(0.000008)$

$= 1 + 0.14 + 0.0084 + 0.000280$

$= 1.148680$ or 1.14868

30. For $n = 1$: the left side is $\dfrac{1}{1 \cdot 3} = \dfrac{1}{3}$ and the right side is

$\dfrac{1}{2(1) + 1} = \dfrac{1}{2 + 1} = \dfrac{1}{3}$.

Assume true for $n = k$: $\dfrac{1}{1 \cdot 3} + \dfrac{1}{3 \cdot 5} + \ldots + \dfrac{1}{(2k - 1)(2k + 1)} = \dfrac{k}{2k + 1}$

Prove it true for $n = k + 1$: $\dfrac{1}{1 \cdot 3} + \dfrac{1}{3 \cdot 5} + \ldots + \dfrac{1}{(2k - 1)(2k + 1)}$

$+ \dfrac{1}{(2k + 1)(2k + 3)} = \dfrac{k + 1}{2k + 3}$

Proof: $\dfrac{1}{1 \cdot 3} + \dfrac{1}{3 \cdot 5} + \dots + \underbrace{\dfrac{1}{(2k - 1)(2k + 1)}}_{\text{By assumption}} + \dfrac{1}{(2k + 1)(2k + 3)}$

$$= \dfrac{k}{2k + 1} + \dfrac{1}{(2k + 1)(2k + 3)}$$

$$= \dfrac{k}{2k + 1} \cdot \dfrac{2k + 3}{2k + 3} + \dfrac{1}{(2k + 1)(2k + 3)} = \dfrac{k(2k + 3) + 1}{(2k + 1)(2k + 3)}$$

$$= \dfrac{2k^2 + 3k + 1}{(2k + 1)(2k + 3)} = \dfrac{(2k + 1)(k + 1)}{(2k + 1)(2k + 3)} = \dfrac{k + 1}{2k + 3}$$

31. When $n = 1$, the left side is 3 and the right side is $\dfrac{1(2 + 3 + 13)}{6} = \dfrac{18}{6} = 3$.

Assume true for $n = k$: $3 + 6 + 11 + \dots + (k^2 + 2) = \dfrac{k(2k^2 + 3k + 13)}{6}$

Prove it true for $n = k + 1$: $3 + 6 + 11 + \dots + (k^2 + 2) + [(k + 1)^2 + 2]$

$$= \dfrac{(k + 1)[2(k + 1)^2 + 3(k + 1) + 13]}{6}$$

$$= \dfrac{(k + 1)(2k^2 + 7k + 18)}{6}$$

$$= \dfrac{2k^3 + 9k^2 + 25k + 18}{6}$$

Proof: $\underbrace{3 + 6 + 11 + \dots + (k^2 + 2)}_{\text{By assumption}} + [(k + 1)^2 + 2]$

$$= \dfrac{k(2k^2 + 3k + 13)}{6} + [(k + 1)^2 + 2]$$

$$= \dfrac{k(2k^2 + 3k + 13)}{6} + \dfrac{6[(k + 1)^2 + 2]}{6}$$

$$= \dfrac{2k^3 + 3k^2 + 13k}{6} + \dfrac{6k^2 + 12k + 6 + 12}{6}$$

$$= \dfrac{2k^3 + 9k^2 + 25k + 18}{6}$$ which is the desired result for the right side.

32. For $n = 4$: the left side is $4! = 4 \cdot 3 \cdot 2 \cdot 1 = 24$ and the right side is $4^2 = 16$.
Assume true for $n = k$: $k! > k^2$ for $k \geq 4$
Prove true for $n = k + 1$: $(k + 1)! > (k + 1)^2$
Proof: $(k + 1)! = (k + 1)k! > \underbrace{(k + 1)k^2}_{\text{By assumption}} = k^3 + k^2 > \underbrace{2k + 1}_{\text{True for } k \geq 4} + k^2 = (k + 1)^2$

33. When $n = 1$, then $a^{2n} - b^{2n} = a^2 - b^2 = (a + b)(a - b)$ and, clearly, $a - b$ is a factor. Assume true for $n = k$: Assume $a - b$ is a factor of $a^{2k} - b^{2k}$
Prove it true for $n = k + 1$: Prove that $a - b$ is a factor of $a^{2k+2} - b^{2k+2}$
Proof: $a^{2k+2} - b^{2k+2} = a^{2k}(a^2 - b^2) + (a^{2k} - b^{2k})b^2$
Since $a - b$ is a factor of $a^2 - b^2 = (a - b)(a + b)$ and is a factor of $a^{2k} - b^{2k}$ by assumption, it is a factor of $a^{2k+2} - b^{2k+2}$.
This completes the proof.

34. Possibilities: $10 \cdot 10 \cdot 10 \cdot 9 \cdot 10 \cdot 10 \cdot 10 = 9(10)^6$
$$= 9,000,000$$

35. a) $P(26, 3) = \dfrac{26!}{(26 - 3)!} = \dfrac{26!}{23!} = \dfrac{26 \cdot 25 \cdot 24 \cdot 23!}{23!} = 26 \cdot 25 \cdot 24 = 15,600$
   b) $14 \cdot 25 \cdot 24 = 8,400$

36. a) Possibilities: $26 \cdot 10 \cdot 9 \cdot 8 = 18,720$
   b) Possibilities: $26 \cdot 10 \cdot 10 \cdot 5 = 13,000$

37. $P(8, 3) = \dfrac{8!}{(8 - 3)!} = \dfrac{8!}{5!} = \dfrac{8 \cdot 7 \cdot 6 \cdot 5!}{5!} = 8 \cdot 7 \cdot 6 = 336$

38. $P(5, 5) = \dfrac{5!}{0!} = \dfrac{5!}{1} = 5! = 5 \cdot 4 \cdot 3 \cdot 2 \cdot 1 = 120$

39. $P(8, 6) = \dfrac{8!}{(8 - 6)!} = \dfrac{8!}{2!} = \dfrac{8 \cdot 7 \cdot 6 \cdot 5 \cdot 4 \cdot 3 \cdot 2!}{2!} = 8 \cdot 7 \cdot 6 \cdot 5 \cdot 4 \cdot 3 = 20,160$

40. Possibilities: $7 \cdot 7 \cdot 3 = 147$

41. a) $0! = 1$, by definition $\qquad$ b) $\dfrac{6!}{2!} = 6 \cdot 5 \cdot 4 \cdot 3 = 360$

   c) $\dfrac{(n + 1)!}{n!} = \dfrac{(n + 1)n!}{n!} = n + 1$ $\qquad$ d) $\dfrac{(n + 1)!}{(n - 1)!} = \dfrac{(n + 1)(n)(n - 1)!}{(n - 1)!}$
   $$= (n + 1)n \text{ or } n(n + 1)$$

42. a) $P(4, 2) = \dfrac{4!}{(4 - 2)!} = \dfrac{4!}{2!} = \dfrac{4 \cdot 3 \cdot 2!}{2!} = 4 \cdot 3 = 12$

   b) $P(5, 3) = \dfrac{5!}{(5 - 3)!} = \dfrac{5!}{2!} = \dfrac{5 \cdot 4 \cdot 3 \cdot 2!}{2!} = 5 \cdot 4 \cdot 3 = 60$

   c) $C(8, 3) = \dfrac{8!}{5! 3!} = \dfrac{8 \cdot 7 \cdot 6 \cdot 5!}{5! \cdot 3 \cdot 2 \cdot 1} = \dfrac{8 \cdot 7 \cdot 6}{3 \cdot 2 \cdot 1} = 8 \cdot 7 = 56$

   d) $\dbinom{7}{3} = \dfrac{7!}{4! 3!} = \dfrac{7 \cdot 6 \cdot 5 \cdot 4!}{4! \cdot 3 \cdot 2 \cdot 1} = \dfrac{7 \cdot 6 \cdot 5}{3 \cdot 2 \cdot 1} = 7 \cdot 5 = 35$

   e) $C(n, n - 1) = \dfrac{n!}{1!(n - 1)!} = \dfrac{n(n - 1)!}{(n - 1)!} = n$

   f) $\dbinom{n + 1}{n - 1} = \dfrac{(n + 1)!}{2!(n - 1)!} = \dfrac{(n + 1)(n)(n - 1)!}{2(n - 1)!} = \dfrac{(n + 1)(n)}{2}$ or $\dfrac{n(n + 1)}{2}$

43. $\dfrac{7!}{2! 2!} = \dfrac{7 \cdot 6 \cdot 5 \cdot \overset{1}{\cancel{4}} \cdot 3 \cdot 2 \cdot 1}{\underset{1}{\cancel{2 \cdot 1 \cdot 2 \cdot 1}}} = 7 \cdot 6 \cdot 5 \cdot 3 \cdot 2 \cdot 1 = 1260$

44. Possibilities: $\dfrac{8!}{3!} = \dfrac{8 \cdot 7 \cdot 6 \cdot 5 \cdot 4 \cdot 3!}{3!} = 8 \cdot 7 \cdot 6 \cdot 5 \cdot 4 = 6720$

45. $\dfrac{8!}{3! 2!} = \dfrac{8 \cdot 7 \cdot 6 \cdot 5 \cdot 4 \cdot 3 \cdot 2 \cdot 1}{3 \cdot 2 \cdot 1 \cdot 2 \cdot 1} = 3360$

46. Possibilities: $\dfrac{8!}{2! 2!} = \dfrac{8 \cdot 7 \cdot 6 \cdot 5 \cdot \overset{1}{\cancel{4}} \cdot 3 \cdot 2 \cdot 1}{\underset{1}{\cancel{2 \cdot 1 \cdot 2 \cdot 1}}} = 8 \cdot 7 \cdot 6 \cdot 5 \cdot 3 \cdot 2 \cdot 1 = 10,080$

47. $\dbinom{12}{4} = \dfrac{12!}{8!\,4!} = \dfrac{12 \cdot 11 \cdot 10 \cdot 9 \cdot 8!}{4 \cdot 3 \cdot 2 \cdot 1 \cdot 8!} = \dfrac{12 \cdot 11 \cdot 10 \cdot 9}{4 \cdot 3 \cdot 2 \cdot 1} = 495$

48. Possibilities: $\dbinom{12}{1}\dbinom{8}{3} = 12 \cdot 56 = 672$

49. $\dbinom{12}{5} = \dfrac{12!}{7!\,5!} = \dfrac{12 \cdot 11 \cdot 10 \cdot 9 \cdot 8 \cdot 7!}{5 \cdot 4 \cdot 3 \cdot 2 \cdot 1 \cdot 7!} = \dfrac{12 \cdot 11 \cdot 10 \cdot 9 \cdot 8}{5 \cdot 4 \cdot 3 \cdot 2 \cdot 1} = 792$

50. $P(A \cap B) = P(A) + P(B) - P(A \cup B) = 0.8 + 0.6 - 1 = 0.4$

51. Let S be the sample space for selecting 3 balls.

    Then $N(S) = \dbinom{8}{3} = \dfrac{8!}{5!\,3!} = \dfrac{8 \cdot 7 \cdot 6}{3 \cdot 2 \cdot 1} = 56$.

    a) Let A be the event that 2 blue balls and 1 red ball are selected.

    Then $N(A) = \dbinom{5}{2}\dbinom{3}{1} = \dfrac{5!}{3!\,2!} \cdot \dfrac{3!}{2!\,1!} = 10 \cdot 3 = 30$ and $P(A) = \dfrac{N(A)}{N(S)} = \dfrac{30}{56} = \dfrac{15}{28}$.

    b) Let B be the event that all 3 blue balls are selected.

    Then $N(B) = \dbinom{5}{3}\dbinom{3}{0} = \dfrac{5!}{2!\,3!} \cdot \dfrac{3!}{3!\,0!} = 10 \cdot 1 = 10$ and $P(B) = \dfrac{N(B)}{N(S)} = \dfrac{10}{56} = \dfrac{5}{28}$.

52. Let S be the sample space of selecting 3 people.

    Then $N(S) = \dbinom{11}{3} = \dfrac{11!}{8!\,3!} = \dfrac{11 \cdot 10 \cdot 9}{3 \cdot 2 \cdot 1} = 165$

    a) Let A be the event that one girl and 2 boys are selected.

    Then $N(A) = \dbinom{6}{1}\dbinom{5}{2} = \dfrac{6!}{5!\,1!} \cdot \dfrac{5!}{3!\,2!} = 6 \cdot 10 = 60$ and $P(A) = \dfrac{N(A)}{N(S)} = \dfrac{60}{165} = \dfrac{4}{11}$.

    b) Let B be the event that 3 girls (no boys) are selected.

    Then $N(B) = \dbinom{6}{3}\dbinom{5}{0} = \dfrac{6!}{3!\,3!} \cdot \dfrac{5!}{5!\,0!} = 20 \cdot 1 = 20$ and $P(B) = \dfrac{N(B)}{N(S)} = \dfrac{20}{165} = \dfrac{4}{33}$.

APPENDIX   TABLE EVALUATION OF LOGARITHMS

Appendix A.1

1. For $\log[4.16 \times 10^{-9}]$, the mantissa is 0.6191 and the characteristic is -9 so that the final value is
$$\log[4.16 \times 10^{-9}] = -9 + 0.6191$$
$$= 0.6191 - 9$$
$$\text{or} \quad 1.6191 - 10$$

3. Express 30.7 as 3.07×10^1. For $\log 30.7 = \log[3.07 \times 10^1]$, the characteristic is 1 and the mantissa is 0.4871 so that the final value is
$$\log 30.7 = 1 + 0.4871 = 1.4871$$

5. Direct use of the tables gives $\log 4.51 = 0.6542$.

7. Express 10.0 as 1.0×10^1. For $\log 10.0 = \log[1.0 \times 10^1]$, the characteristic is 1 and the mantissa is 0.0000 so that $\log 10.0 = 1 + 0.0000 = 1.0000$ or 1. *Note:* If you are familiar with properties of common logs, then you know that $\log 10.0 = \log 10 = 1$.

9. Express 1,070,000 as 1.07×10^6. For $\log 1{,}070{,}000 = \log[1.07 \times 10^6]$, the characteristic is 6 and the mantissa is 0.0294 so that the final value is
$$\log 1{,}070{,}000 = 6 + 0.0294 = 6.0294$$

11. Express 0.00107 as 1.07×10^{-3}. For $\log 0.00107 = \log[1.07 \times 10^{-3}]$, the characteristic is -3 and the mantissa is 0.0294 so that the final value is
$$\log 0.00107 = -3 + 0.0294$$
$$= 0.0294 - 3$$
$$\text{or } 7.0294 - 10$$

13. $\log N = 0.8561$. Here, the mantissa is 0.8561 and the characteristic is 0. Using the table, the final value is
$$N = 7.18 \times 10^0 = 7.18 \times 1 = 7.18$$

15. $\log N = 8.4518$. The mantissa is 0.4518 and the characteristic is 8. Using the table, the final value is
$$N = 2.83 \times 10^8 = 283{,}000{,}000$$
Notice that the decimal point was moved 8 places to the right.

17. $\log N = 7.6776 - 10$. The mantissa is 0.6776 and the characteristic is $7 - 10 = -3$. Using the table, the final value is
$$N = 4.76 \times 10^{-3} = 0.00476$$
Notice that the decimal point was moved 3 places to the left.

19. $\log N = -3$. The mantissa is 0.0000 and the characteristic is -3. Using the table, the final value is $1.00 \times 10^{-3} = 0.001$. Again, the decimal point was moved 3 places to the left.

21. To find $\log 10.11$, observe that 10.11 falls between 10.10 and 10.20 where the logs are $\log 10.10 = 1.0043$ and $\log 10.20 = 1.0086$. Construct the following table:
$$10\left[1\left[\begin{array}{l} \log 10.10 = 1.0043 \\ \log 10.11 = \\ \log 10.20 = 1.0086 \end{array} \right]d \right]0.0043$$

To find d, use the following proportion:

$$\frac{1}{10} = \frac{d}{0.0043} \qquad \text{or} \qquad 10d = 1 \times 0.0043$$

$$10d = 0.0043$$

$$d = \frac{0.0043}{10} = 0.00043 \text{ which rounds to } 0.0004$$

Thus, $\log 10.11 = 1.0043 + 0.0004$

$$= 1.0047$$

23. To find $\log 4.171$, observe that 4.171 falls between 4.170 and 4.180 where the logs are $\log 4.170 = 0.6201$ and $\log 4.180 = 0.6212$. Construct the following table:

$$10\left[1\left[\begin{array}{l}\log 4.170 = 0.6201 \\ \log 4.171 = \\ \log 4.180 = 0.6212\end{array}\right]d\right]0.0011$$

To find d, use the following proportion:

$$\frac{1}{10} = \frac{d}{0.0011} \qquad \text{or} \qquad 10d = 1 \times 0.0011$$

$$10d = 0.0011$$

$$d = \frac{0.0011}{10} = 0.00011 \text{ which rounds to } 0.0001$$

Thus, $\log 4.171 = 0.6201 + 0.0001$

$$= 0.6202$$

25. To find $\log 417{,}800$, observe that 417,800 falls between 417,000 and 418,000 where the logs are $\log 417{,}000 = 5.6201$ and $\log 418{,}000 = 5.6212$. Construct the following table:

$$10\left[8\left[\begin{array}{l}\log 417{,}000 = 5.6201 \\ \log 417{,}800 = \\ \log 418{,}000 = 5.6212\end{array}\right]d\right]0.0011$$

To find d, use the following proportion:

$$\frac{8}{10} = \frac{d}{0.0011} \qquad \text{or} \qquad 10d = 8 \times 0.0011$$

$$10d = 0.0088$$

$$d = \frac{0.0088}{10} = 0.00088 \text{ which rounds to } 0.0009$$

Thus, $\log 417{,}800 = 5.6201 + 0.0009$

$$= 5.6210$$

27. To find $\log 0.007717$, observe that 0.007717 falls between 0.00771 and 0.00772 where the logs are $\log 0.007710 = 7.8871 - 10$ and $\log 0.007720 = 7.8876 - 10$. Construct the following table:

$$10\left[7\left[\begin{array}{l}\log 0.007710 = 7.8871 - 10 \\ \log 0.007717 = \\ \log 0.007720 = 7.8876 - 10\end{array}\right]d\right]0.0005$$

To find d, use the following proportion:

$$\frac{7}{10} = \frac{d}{0.0005} \quad \text{or} \quad 10d = 7 \times 0.0005$$

$$10d = 0.0035$$

$$d = \frac{0.0035}{10} = 0.00035 \text{ which rounds to } 0.0004$$

Thus, $\log 0.00717 = 7.8871 - 10 + 0.0004$
$$= 7.8875 - 10$$

29. $\log N = 0.1113$. The mantissa 0.1113 is located between 0.1106 and 0.1139 and these correspond to the values 1.29 and 1.30. The characteristic is 0. Construct the following table:

$$10\left[x\left[\begin{array}{l} \log 1.290 = 0.1106 \\ \log N \quad\;\; = 0.1113 \\ \log 1.300 = 0.1139 \end{array} \right] 0.0007 \right] 0.0033$$

To find x, use the following proportion:

$$\frac{x}{10} = \frac{0.0007}{0.0033} \quad \text{or} \quad x(0.0033) = 10(0.0007)$$

$$x = \frac{10(0.0007)}{0.0033} \approx 2.12 \text{ which rounds to } 2$$

Thus, $N = 1.290 + 0.002$
$$= 1.292$$

(*Notice that the 2 was placed in the thousandths position.*)

31. $\log N = 4.9433$. The mantissa 0.9433 is located between 0.9430 and 0.9435 and these correspond to the values 8.77 and 8.78. The characteristic is 4. Construct the following table:

$$10\left[x\left[\begin{array}{l} \log 87,700 = 4.9430 \\ \log N \quad\quad\;\; = 4.9433 \\ \log 87,800 = 4.9435 \end{array} \right] 0.0003 \right] 0.0005$$

To find x, use the following proportion:

$$\frac{x}{10} = \frac{0.0003}{0.0005} \quad \text{or} \quad x(0.0005) = 10(0.0003)$$

$$x = \frac{10(0.0003)}{0.0005} = \frac{30}{5} = 6$$

Thus, $N = 87,700 + 60$
$$= 87,760$$

(*Notice that the 6 was placed in the tens position.*)

33. $\log N = 7.6950 - 10$. The mantissa 0.6950 falls between 0.6946 and 0.6955 and these correspond to the values 4.95 and 4.96. The characteristic is $7 - 10 = -3$. Construct the following table:

$$10\left[x\left[\begin{array}{l} \log 0.004950 = 7.6946 - 10 \\ \log N \quad\quad\quad = 7.6950 - 10 \\ \log 0.004960 = 7.6955 - 10 \end{array} \right] 0.0004 \right] 0.0009$$

To find x, use the following proportion:

$$\frac{x}{10} = \frac{0.0004}{0.0009} \quad \text{or} \quad x(0.0009) = 10(0.0004)$$

$$x = \frac{10(0.0004)}{0.0009} = \frac{40}{9} \approx 4.44 \text{ which rounds to 4}$$

Thus, N = 0.004950 + 0.000004
     = 0.004954

35. log N = 9.3881 - 10. The mantissa 0.3881 falls between 0.3874 and 0.3892
     and these correspond to the values 2.44 and 2.45. The characteristic is
     9 - 10 = -1. Construct the following table:

$$10 \left[x \left[\begin{matrix} \log 0.2440 = 9.3874 - 10 \\ \log N \quad\;\; = 9.3381 - 10 \\ \log 0.2450 - 9.3892 - 10 \end{matrix} \right] 0.0007 \right] 0.0018$$

To find x, use the following proportion:

$$\frac{x}{10} = \frac{0.0007}{0.0018} \quad \text{or} \quad x(0.0018) = 10(0.0007)$$

$$x = \frac{10(0.0007)}{0.0018} = \frac{70}{18} \approx 3.89 \text{ which rounds to 4}$$

Thus, N = 0.2440 + 0.0004
     = 0.2444